Tensor Geometry

The Geometric Viewpoint and its Uses

SURVEYS AND REFERENCE WORKS IN MATHEMATICS

Tensor Geometry

The Geometric Viewpoint and its Uses

C T J Dodson
University of Lancaster

T Poston
Battelle, Geneva

Pitman
LONDON · SAN FRANCISCO · MELBOURNE

PITMAN PUBLISHING LIMITED
39 Parker Street, London WC2B 5PB

FEARON—PITMAN PUBLISHERS INC.
6 Davis Drive, Belmont, California 94002, USA

Associated Companies
Copp Clark Ltd, Toronto
Pitman Publishing Co. SA (Pty) Ltd, Johannesburg
Pitman Publishing New Zealand Ltd, Wellington
Pitman Publishing Pty Ltd, Melbourne

First published 1977

AMS Subject Classifications: (main) 53—XX
(subsidiary) 15—XX, 83—53

Library of Congress Cataloging in Publication Data

Dodson, C T J
Tensor geometry.

(Surveys and reference works in mathematics)
Bibliography: p.
Includes index.
1. Geometry, Differential. 2. Calculus of tensors.
I. Poston, T., joint author. II. Title. III. Series.
QA649.D6 516'.36 77—6370
ISBN 0—273—00317—8

Reproduced and printed by photolithography
in Great Britain at Biddles of Guildford

Contents

Introduction

The title of this book is misleading.

Any possible title would mislead somebody. "Tensor Analysis" suggests to a mathematician an ungeometric, manipulative debauch of indices, with tensors illdefined as "quantities that transform according to" unspeakable formulae. "Differential Geometry" would leave many a physicist unaware that the book is about matters with which he is very much concerned. We hope that "Tensor Geometry" will at least lure both groups to look more closely.

Most modern "differential geometry" texts use a coordinate-free notation almost throughout. This is excellent for a coherent understanding, but leaves the physics student quite unequipped for the physical literature, or for the specific physical computations in which coordinates are unavoidable. Even when the relation to classical notation is explained, as in the magnificent [Spivak], <u>pseudo</u>-Riemannian geometry is barely touched on. This is crippling to the physicist, for whom spacetime is the most important example, and perverse even for the geometer. Indefinite metrics arise as easily within pure mathematics (for instance in Lie group theory) as in applications, and the mathematician should know the differences between such geometries and the positive definite type. In this book therefore we treat both cases equally, and describe both relativity theory and (in Ch.IX, §6) an important "abstract" pseudo Riemannian space, SL(2;R).

The argument is largely carried in modern, intrinsic notation which lends itself to an intensely geometric (even pictorial) presentation, but a running translation into indexed notation explains and derives the manipulation rules so beloved of, and necessary to, the physical community. Our basic notations are summarised in Ch. 0, along with some basic physics.

Einstein's system of 1905 deduced everything from the Principle of Relativity:

that no experiment whatever can define for an observer his "absolute speed". Minkowski published in 1907 a geometric synthesis of this work, replacing the once separately absolute space and time of physics by an absolute four dimensional spacetime. Einstein initially resisted this shift away from argument by comparison of observers, but was driven to a more "spacetime geometric" view in his effort to account for gravitation, which culminated in 1915 with General Relativity. For a brilliant account of the power of the Principle of Relativity used directly, see [Feynman]; particularly the deduction (vol. 2, p.13-16) of magnetic effects from the laws of electrostatics. It is harder to maintain this approach when dealing with the General theory. The Equivalence Principle (the most physical assumption used) is hard even to state precisely without the geometric language of covariant differentiation, while Einstein's Equation involves sophisticated geometric objects. Before any detailed physics, therefore, we develop the geometrical setting: Chapters I - X are a geometry text, whose material is chosen with an eye to physical usefulness. The motivation is largely geometric also, for accessibility to mathematics students, but since physical thinking occasionally offers the most direct insight into the geometry we cover in Ch. 0, §3 those elementary facts about special relativity that we refer to before Ch. XI. British students of either mathematics or physics should usually know this much before reaching university, but variations in educational systems - and students - are immense.

The book's prerequisites are some mathematical or physical sophistication, the elementary functions (log, exp, cos, cosh, etc), plus the elements of vector algebra and differential calculus, taught in any style at all. Chapter I will be a recapitulation and compendium of known facts, geometrically expressed, for the student who has learnt "Linear Algebra". The student who knows the same material as "Matrix Theory" will need to read it more carefully, as the style of argument will be less familiar. (S)he will be well advised to do a proportion of the exercises, to consolidate understanding on matters like "how matrices multiply" which we assume familiar from <u>some</u> point of view. The next three chapters develop affine and linear geometry, with material new to most students and so more slowly taken. Chapter V sets up the algebra of tensors, handling both ends and the middle of the communication gap that made 874 U.S. "active research physicists" [Miller] rank "tensor analysis" ninth among all Math courses needed for physics Ph.D. students, more than 80% considering it necessary, while "multilinear algebra" is not among the first 25, less than 20% in each specialisation recommending it. "Multilinear algebra" is just the algebra of the manipulations, differentiation excepted, that make up "tensor analysis".

Chapter VI covers those facts about continuity, compactness and so on needed for precise argument later; we resisted the temptation to write a topology text. Chapter VII treats differential calculus "in several variables", namely between affine spaces. The affine setting makes the "local linear approximation" character of the derivative much more perspicuous than does a use of vector spaces only, which permit much more ambiguity as to "where vectors are". This advantage is increased when we go on to construct manifolds; modelling them on affine spaces gives an unusually neat and geometric construction of the tangent bundle and its own manifold structure. These once set up, we treat the key facts about vector fields, previously met as "first order differential equations" by many readers. To keep the book self-contained we show the existence and smoothness of flows for vector fields (solutions to equations) in an Appendix, by a recent, simple and attractively geometric proof due to Sotomayor. The mathematical sophistication called for is greater than for the body of the book, but so is that which makes a student want a proof of this result.

Chapter VIII begins differential geometry proper with the theory of connections, and their several interrelated geometric interpretations. The "rolling tangent planes without slipping" picture allows us to "see" the connection between tangent spaces along a curve in an ordinary embedded surface, while the intrinsic geometry of the tangent bundle formulation gives a tool both mathematically simpler in the end, and more appropriate to physics.

Chapter IX discusses geodesics both locally and variationally, and examines some special features of indefinite metric geometry (such as geodesics <u>never</u> "the shortest distance between two points"). Geodesics provide the key to analysis of a wealth of illuminating examples.

In Chapter X the Riemann curvature tensor is introduced as a measure of the failure of a manifold-with-connection to have locally the flat geometry of an affine space. We explore its geometry, and that of the related objects (scalar curvature, Ricci tensor, etc.) important in mathematics and physics.

Chapter XI is concerned chiefly with a geometric treatment of how matter and its motion must be described, once the Newtonian separation of space and time dissolves into one absolute spacetime. It concludes with an explanation of the geometric incompatibility of gravitation with any simple flat view of spacetime, so leading on to general relativity.

Chapter XII uses all of the geometry (and many of the examples) previously set up, to make the interaction of matter and spacetime something like a visual experience. After introducing the equivalence principle and Einstein's equation, and discussing their cosmic implications, we derive the Schwarzschild solution and consider planetary motion. By this point we are equipped both to <u>compute</u> physical quantities like orbital periods and the famous advance of the perihelion of Mercury, and to <u>see</u> that the paths of the planets (which to the flat or Riemannian intuition have little in common with straight lines) correspond indeed to geodesics.

Space did not permit the coherent inclusion of differential forms and integration. Their use in geometry involves connection and curvature forms with values not in the real numbers but in the Lie algebra of the appropriate Lie group. A second volume will treat these topics and develop the clear exposition of the tensor geometric tools of solid state physics, which has suffered worse than most subjects from index debauchery.

The only feature in which this book is richer than in pictures (to strengthen geometric insight) is exercises (to strengthen detailed comprehension). Many of the longer and more intricate proofs have been broken down into carefully programmed exercises. To work through a proof in this way teaches the mind, while a displayed page of calculation merely blunts the eye.

Thus, the exercises are an integral part of the text. The reader need not do them all, perhaps not even many, but should read them at least as carefully as the main text, and think hard about any that seem difficult. If the "really hard" proportion seems to grow, reread the recent parts of the text - doing more exercises.

We are grateful to various sources of support during the writing of this book: Poston to the Instituto de Matemática Pura e Aplicada in Rio de Janeiro, Montroll's "Institute for Fundamental Studies" in Rochester, N.Y., the University of Oporto, and at Battelle Geneva to the Fonds National Suisse de la Recherche Scientifique (Grant no.2.461-0.75) and to Battelle Institute, Ohio (Grant no. 333-207); Dodson to the University of Lancaster and (1976-77) the International Centre for Theoretical Physics for hospitality during a European Science Exchange Programme Fellowship sabbatical year. We learned from conversation with too many people to begin to list. Each author, as usual, is convinced that any remaining errors are the responsibility of the other, but errors in the diagrams are due to the draughtsman, Poston, alone.

Finally, admiration, gratitude and sympathy are due Sylvia Brennan for the vast job well done of preparing camera ready copy in Lancaster with the authors in two other countries.

Kit Dodson
ITCP, Trieste

Tim Poston
Battelle, Geneva.

0 Fundamental Not(at)ions

"Therefore is the name of it called Babel;
because the Lord did there confound the language
of all the earth"

Genesis 11, 9.

Please at least skim through this chapter; if a mathematician, your habits are probably different somewhere (maybe f^{-1} not $\overleftarrow{f}$) and if a physicist, perhaps almost everywhere.

1. SETS

A <u>set</u>, or <u>class</u>, or <u>family</u> is a collection of things, called <u>members</u>, <u>elements</u>, or <u>points</u> of it. Brackets like { } will always denote a set, with the elements either listed between them (as, {1,3,1,2}, the set whose elements are the number 1, 2 and 3 - repetition, and order, make no difference) or specified by a rule, in the form $\{x|x$ is an integer, $x^2 = 1\}$ or $\{\text{Integer } x|x^2 = 1\}$, which are abbreviations of "the set of all those things x such that x is an integer and $x^2 = 1$" which is exactly the set {1,-1}. Read the vertical line | as "such that" when it appears in a specification of a set by a rule.

Sets can be collections of numbers (as above), of people ({Peter Kropotkin, Henry Crun, Balthazar Vorster}), of sets ({{1,-1}, {Major Bludnok, Oberon}, {this book}}), or of things with little in common beyond their declared membership of the set ({passive resistance, the set of all wigs, 3, Isaac Newton }) though this is uncommon in everyday mathematics.

We abbreviate "x is a member of the set S" to "x is <u>in</u> S" or $x \in S$, and "x is not in S" to $x \notin S$. (Thus for instance if $S = \{1,3,1,2,2\}$ then $x \in S$ means that x is the number 1, or 2, or 3.) If x, y and z are all members of S, we write briefly $x, y, z \in S$. A singleton set contains just one element.

If every $x \in S$ is also in another set T, we write $S \subseteq T$, and say S is a subset of T. This includes the possibility that S = T; that is when $T \subseteq S$ as well as $S \subseteq T$.

Some sets have special standard symbols. The set of all natural, or "counting", numbers like 1,2,3,...,666,... etc. is always N (not vice versa, but when N means anything else this should be clear by context. Life is short, and the alphabet shorter.) There is no consensus whether to include 0 in N; on the grounds of its invention several millenia after the other counting numbers, and certain points of convenience, we choose not to. The set of all real (as opposed to complex) numbers like 1, $\sqrt{\tfrac{1}{2}}$, $-\pi$, 8.2736 etc. is called R. The empty set $\emptyset$ by definition has no members; thus if $S = \{x \in N | x^2 = -1\}$ then $S = \emptyset$. Note that $\emptyset \subseteq N \subseteq R$. ($\emptyset$ is a subset of any other set: for "$\emptyset \not\subseteq N$" would mean "there is an $x \in \emptyset$ which is not a natural number". This is false, as there is no $x \in \emptyset$ which is, or is not, anything: hence $\emptyset \subseteq N$.) Various other subsets of R have special symbols. We agree as usual that among real numbers

$a < b$ means "a is strictly less than b" or "b-a is not zero or negative"

$a \leq b$ means "a is less than or equal to b" or "b-a is not negative"

(note that for any $a \in R$, $a \leq a$). Then we define the intervals

		a b	
$[a,b]$	$= \{x \in R \mid a \leq x \leq b\}$	——+————+——	including ends
$]a,b[$	$= \{x \in R \mid a < x < b\}$	——+————+——	not including ends
$[a,b[$	$= \{x \in R \mid a \leq x < b\}$	——+————+——	including one end.
$]a,b]$	$= \{x \in R \mid a < x \leq b\}$	——+————+——	

When $b < a$, the definitions imply that all of these sets equal $\emptyset$: if $a = b$, then $[a,b] = \{a\} = \{b\}$ and the rest are empty. By convention the <u>half-unbounded</u> intervals are written similarly : if $a,b \in R$ then

$$]-\infty,b]=\{x|x\leq b\},\ ]-\infty,b[=\{x|x<b\},\ [a,\infty[=\{x|x\geq a\},\ ]a,\infty[=\{x|x>a\}$$

by definition, without thereby allowing $-\infty$ or ∞ as "numbers". We also call R itself an interval. (We may define the term <u>interval</u> itself either by gathering together the above definitions of all particular cases or - anticipating Chapter III - as a convex subset of R.)

By $a > b$, $a \geq b$ we mean $b < a$, $b \leq a$ respectively.

A <u>finite</u> subset $S = \{a_1,a_2,\ldots,a_n\} \subseteq R$ must have a least member, min S, and a greatest, max S. An infinite set may, but need not have extreme members. For example, min $[0,1] = 0$, max $[0,1] = 1$, but neither min $]0,1[$ nor max $]0,1[$ exists. For any $t \in]0,1[$, $\frac{1}{2}t < t < \frac{1}{2}(t+1)$ which gives elements of $]0,1[$ strictly less and greater than t. So t can be neither a minimum nor a maximum.

We shall be thinking of R far more as a <u>geometric</u> object, with its points as <u>positions</u>, than as algebraic with its elements as numbers. (These different viewpoints are represented by different names for it, as the <u>real line</u> or the <u>real number system</u> or <u>field</u>.) Its geometry, which we partly explore in VII. §4, has more subtlety than high school treatments lead one to realise.

If S and T are any two sets their <u>intersection</u> is the set (Fig. 1.1a)

$$S \cap T = \{x \in S \mid x \in T\}$$

and their <u>union</u> is (Fig. 1.1b)

$$S \cup T = \{x|x \in S, \text{ or } x \in T, \text{ or both}\}.$$

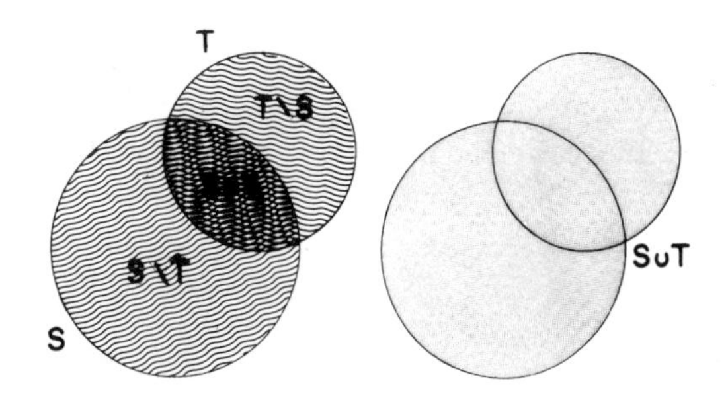

Fig. 1.1

By S <u>less</u> T we mean the set (Fig. 1.1a)

$$S \setminus T = \{x \in S \mid x \in T\}.$$

If we have an <u>indexing set</u> K such as {1,2,3,4} or {3, Fred, Jam} labelling sets S_3, S_{Fred}, S_{Jam} (one for each $k \in K$) we denote the resulting set of sets $\{S_3, S_{Fred}, S_{Jam}\}$ by $\{S_k\}_{k \in K}$. K may well be infinite (for instance K = N or K = R). The <u>union</u> of all of the S_k is

$$\bigcup_{k \in K} S_k = \{x \mid x \in S_k \text{ for } \underline{\text{some}}\ k \in K\}$$

and their <u>intersection</u> is

$$\bigcap_{k \in K} S_k = \{x \mid x \in S_k \text{ for } \underline{\text{all}}\ k \in K\},$$

which obviously reduce to the previous definitions when k has exactly two members.

To abbreviate expressions like those above, we sometimes write "for all" as $\forall$, "there exists" as $\exists$, and abbreviate "such that" to "s.t.". Then

$$\bigcap_{k \in K} S_k = \{x \mid x \in S_k\ \forall k \in K\}, \qquad \bigcup_{k \in K} S_k = \{x \mid \exists k \in K \text{ s.t. } x \in S_k\}.$$

If $S \cap T = \emptyset$, S and T are <u>disjoint</u>; $\{S_k\}_{k \in K}$ is disjoint if $S_k \cap S_\ell = \emptyset, \forall k \neq \ell \in K$.

When $K = \{1,\ldots,n\}$ we write $\bigcup_{k \in K} S_k$ as $\bigcup_{i=1}^{n} S_i$ or $S_1 \cup S_2 \cup \ldots \cup S_n$, by analogy with the expression $\sum_{i=1}^{n} x_i = x_1 + x_2 + \ldots + x_n$ where the x_i are things that can be added, such as members of N, of R, or (cf Ch. I) of a vector space; similarly for $\bigcap_{i=1}^{n} S_i = S_1 \cap S_2 \cap \ldots \cap S_n$.

We shorten "implies" to =>, "is implied by" to <=, and "=> and <=" to <=>. Thus, for example,

$$x \in N \Rightarrow x^2 \in N, \qquad x \in R \Leftarrow x \in N,$$

I was married to John $\iff$ John was married to me, or in compound use

$$[x \in S \Rightarrow x \in T] \iff [x \in T \Leftarrow x \in S].$$

The <u>product</u> of two sets X and Y is the set of <u>ordered pairs</u>

$$X \times Y = \{(x,y) \mid x \in X,\ y \in Y\}.$$

The commonest example is the description of the Euclidean plane by Cartesian coordinates $(x,y) \in R \times R$. Note the importance of the ordering: though $\{1,0\}$ and $\{0,1\}$ are the same <u>subset</u> of R, $(1,0)$ and $(0,1)$ are different <u>elements</u> of $R \times R$ (one "on the x-axis" the other "on the y-axis"). $R \times R$ is often written R^2. We generally identify $(R \times R) \times R$ and $R \times (R \times R)$, whose elements are strictly of the forms $((x,y),z)$ and $(x,(y,z))$, with the set R^3 of ordered triples labelled (x,y,z), or (x^1,x^2,x^3) according to taste and convenience. Here the $^1,^2,^3$ on the x's are position labels for numbers and not powers. Similarly for the set

$$R^n = R \times R \times \ldots \times R = \{(x^1,x^2,\ldots,x^n) \mid x^1,\ldots,x^n \in R\}$$

of <u>ordered n-tuples</u>. (Note that the set R^1 of one-tuples is just R.)

A less "flat" illustration arises from the unit circle

$$S^1 = \{(x,y) \in R^2 \mid x^2+y^2 = 1\}.$$

The product $S^1 \times [1,2]$ is a subset of $R^2 \times R$, since $S^1 \subseteq R^2$, $[1,2] \subseteq R$. (Fig. 1.2, with some sample points labelled.)

S^1 is one of the n-<u>spheres</u>:

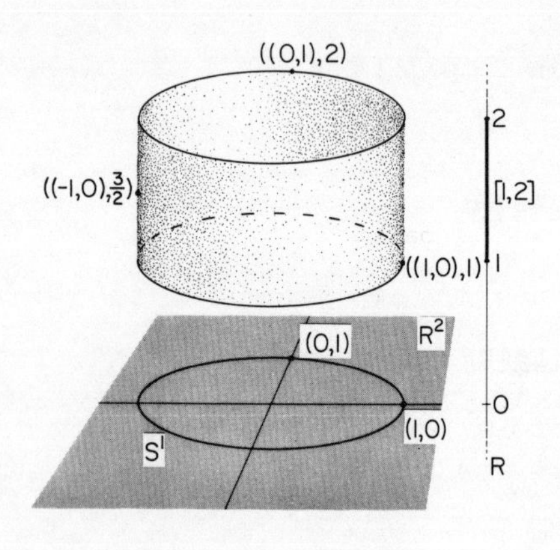

Fig 1.2

$$S^n = \{(x^1,\ldots,x^{n+1}) \mid (x^1)^2+\ldots+(x^{n+1})^2=1\} \subseteq R^{n+1}.$$

S^2 is the usual "unit sphere" of 3-dimensional Cartesian geometry, and S^0 is simply $\{-1,1\} \subseteq R^1$. The higher spheres are logically no different, but take a little practice to "visualise".

A relation ρ on a set X is a subset of $X \times X$. We generally abbreviate $(x,y) \in \rho$ to $x\rho y$. Typical cases are

$$\{(x,y) \in R^2 \mid y-x \text{ is not negative}\} \subseteq R \times R,$$

this is the relation $\leq$ used above, and

$$\{(x,y) \mid x,y \text{ are people, } x \text{ is married to } y\},$$

on the set of people. Various kinds of relation have special names; for instance, $\leq$ is an example of an order relation. We need only define one kind in detail here:

An equivalence relation $\sim$ on X is a relation such that

(i) $x \in X \Rightarrow x \sim x$

(ii) $x \sim y \Rightarrow y \sim x$

(iii) $x \sim y$ and $y \sim z \Rightarrow x \sim z$.

For example, $\{(x,y) \mid x^2 = y^2\}$ is an equivalence relation on a set of numbers, and $\rho = \{(x,y) \mid x \text{ has the same birthday as } y\}$ is an equivalence relation on the set of mammals. On the other hand, $\sigma = \{(x,y) \mid x \text{ is married to the husband of } y\}$ is an equivalence relation on the set of wives in many cultures, but not on the set of women, by the failure of (i).

The important feature of an equivalence relation is that it partitions X into equivalence classes. These are the subsets $[x] = \{y \in X \mid y \sim x\}$, with the properties

(i) $x \in [x]$, we say x is a representative of $[x]$,

(ii) either $[x] \cap [y] = \emptyset$, or $[x] = [y]$,

(iii) the union of all the classes is X.

This device is used endlessly in mathematics, from the construction of the integers on up. For, very often, the set of equivalence classes possesses a nicer structure than X itself. We construct some vector spaces with it (in II §3, VIII §1). The example on mammals produces classes that interest astrologers, and σ partitions wives into ... ?

2. FUNCTIONS

A function, mapping or map $f : X \to Y$ between the sets X and Y may be thought of as a rule, however specified, giving for each $x \in X$ exactly one $y \in Y$. Technically, it is best described as a subset $f \subseteq X \times Y$ such that

Fi) $x \in X \Rightarrow \exists\, y \in Y$ s.t. $(x,y) \in f$

Fii) $(x,y),(x,y') \in f \Rightarrow y = y'$.

These rules say that for each $x \in X$, (i) there is a (ii) unique $y \in Y$ that we may label $f(x)$ or fx. A map may be specified simply by a list, such as f: {Peter Kropotkin, Henry Crun, Balthazar Vorster} → {x|x is a possible place}

Peter Kropotkin ⤳ Switzerland

Henry Crun ⤳ Balham Gas Works

Balthazar Vorster ⤳ Robben Island

An example of a function specified by a rule allowing for several possibilities is

(part of) RxR
(1,g(1))
(-2,0)
(1,0)
(-2,g(-2))
(a)

$$g: R \to R: x \rightsquigarrow \begin{cases} 1 \text{ if } x \in N \text{ and } x \geq 0 \\ -1 \text{ if } x \in N \text{ and } x < 0 \\ \frac{1}{2} \text{ if } x \notin N \end{cases}$$

(Fig. 2.1a uses artistic license in representing the "zero width" gaps in the graph of g - which, as a subset of R × R, technically is g.) Often we shall specify a map by one or more formulae, for example (Fig. 2.1b,c)

RxR
(1,h(1))
(1,0)
(b)

$$h: R \to R: x \rightsquigarrow \begin{cases} x^2 \text{ if } \geq 0 \\ \\ 0 \text{ if } \leq 0 \end{cases},$$

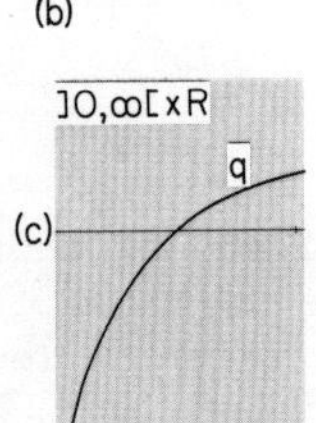

(c)

q:]0,∞[→ R:x ⤳ log x.

All these satisfy Fi) and Fii). Notice the way we have used → to specify the sets a function is between and ⤳ to specify its "rule".

Fig. 2.1

(Technically, q: x ⤳ log x is short for q = {(x,y)|y = log x} ⊆]0,∞[× R.) This distinction between → and ⤳ will be consistent throughout the book. We also read "f: X → Y" as the statement "f is a function from X to Y".

If f: X → Y, we call X its domain, Y its range, and the subset

$\{y|y = f(x)$ for some $x \in X\}$ of Y its image, denoted by $f(X)$ or fX. We generalise this last notation: if S is any subset of X, set

$$fS = f(S) = \{y|y = f(x) \text{ for some } x \in S\}$$

the image of S by f. Note that $f(\{x\}) = \{f(x)\}$ for any $x \in X$, as sets.

We are committing a slight "abuse of language" in using f to denote both a map $X \to Y$ and the function

$$\{S|S\subseteq X\} \to \{T|T\subseteq Y\} : S \rightsquigarrow \{y|\exists x \in S \text{ s.t. } f(x) = y\}$$

that it defines between sets of subsets: generally we shall insist firmly that the domain and range are parts of the function's identity, just as much as the rule giving it, and $S \rightsquigarrow fS$ is different in all these ways from $x \rightsquigarrow fx$. This precision about domain and range becomes crucial when we define the composite of two maps $g:X \to Y$ and $f:Y \to Z$ by

$$f\circ g : X \to Z : x \rightsquigarrow f(g(x)); \quad \text{so} \quad (f\circ g)(x) \text{ is "f of g of x".}$$

If, say, we wish to compose q and h above, we have

$$h\circ q:]0,\infty[\to R: x \rightsquigarrow h(qx) = \begin{cases} (\log x)^2 \text{ if } x \geq 1 \\ 0 \quad \text{if } 1 \geq x > 0 \end{cases}$$

but $q\circ h$ cannot satisfy Fi; how can we define $q\circ h(-1)$, since log 0 does not exist? Or consider $s : R \to R : x \rightsquigarrow \sin x$:-

$$\left[x\in R\right] \Rightarrow \left[sx \leq 1\right] \Rightarrow \left[\log(sx) \leq 0 \text{ \underline{when} \underline{defined}}\right] \Rightarrow \left[\log(\log(sx)) \text{ \underline{never} defined}\right]$$

so we cannot define $q\circ q\circ s$ anywhere. (Note that formally "differentiating" $x \rightsquigarrow \log\log\sin x$ by the rules of school calculus gives a formula that does define something for some values of x. What, if anything, does the rate of change with x of a nowhere-defined function mean? What is the sound of one hand clapping?) So insisting that X and Y are "part of" $f: X \to Y$ is a vital safety measure, not pedantry.

So we should not write down $f\circ g$ unless (range of g) = (domain of f). We may so far abuse language as to write $f\circ g$ for $x \rightsquigarrow f(gx)$ when (image of g) = (domain of f) or when (range of g) $\subseteq$ (domain of f); this latter is really the triple composite $f\circ i\circ g$ with the inclusion map

$$i:(\text{range of } g) \hookrightarrow (\text{domain of } f) : x \rightsquigarrow x$$

quietly suppressed. Note also the amalgam of $\subset$ and $\to$ for inclusions.

We sometimes want to change a function by reducing its domain; if $f:X \to Y$ and $S \subseteq X$ we defined f restricted to S or the restriction of f to S as

$$f|S : S \to Y : x \rightsquigarrow f(x)$$

or equivalently $f|S = f\circ i$, where i is the inclusion $S \hookrightarrow X$.

Notice that $f|S$ may have a simpler expression than f: for $h:R \to R$ as above, $h|[0,\infty[$ is given simply by $x \rightsquigarrow x^2$. It thus coincides with $k|[0,\infty[$ where $k:R \to R : x \rightsquigarrow x^2$, though $h(x)$ is not the same as $k(x)$, (we write $h(x) \neq k(x)$ for short) if $x < 0$. This is another reason for considering the domain as "part of" the function : if a change in domain can make different functions the same, the change is not trivial. (To regard f and $f|S$ as the same function and allow them the same name would lead to "$h : R \to R$ is the same as $h|[0,\infty[$ is the same as

$k|[0,\infty[$ is the same as k" which is ridiculous.) When we have this situation of two functions $f,g : X \to Y$, $S \subseteq X$, and $f|S = g|S$, we say f and g agree on S.

A function $f : X \to Y$ defines, besides $S \rightsquigarrow fS$ from subsets of X to subsets of Y, a map in the other direction between subsets. It is defined for all $T \subseteq Y$ by

$$\overleftarrow{f}(T) = \{x | f(x) \in T\} \subseteq X,$$

the inverse image of T by f. If $fX \cap T = \emptyset$, then $\overleftarrow{f}(T) = \emptyset$; the inverse image of a set outside the image of f is empty. (Likewise if $fS \cap T = \emptyset$, $\overleftarrow{(f|S)}(T) = \emptyset$.) Some images and inverse images are illustrated in Fig. 2.2. There f is represented as taking any $x \in X$ to the point directly below it - a pictorial device we shall use constantly.

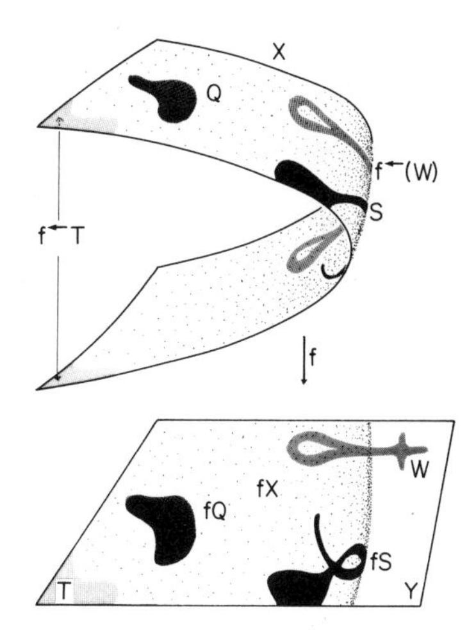

Fig. 2.2

In general $\overleftarrow{f}$, a map taking subsets of Y to subsets of X, does not come from a map $Y \to X$ in the way that $S \rightsquigarrow fS$ does come from $f: X \to Y$. If for every $y \in Y$ we had $\overleftarrow{f}(\{y\})$ a set containing exactly one point, as we have for y on the line C in Fig. 2.2 (rather than none, as to the right of C, or more than one, as to the left) then we can define $\overleftarrow{f} : Y \to X$ by the condition $\overleftarrow{f}(y)$ = the unique member of $\overleftarrow{f}(\{y\})$; otherwise not. We can break this necessary condition "every $\overleftarrow{f}(\{x\})$ contains exactly one point" into two, that are often useful separately:

$f:X \to Y$ is injective or into or an injection if for any $y \in Y$, $\overleftarrow{f}(\{y\})$ contains at most one point.

Equivalently, if $f(x) = f(x') \in Y \Rightarrow x = x'$.

$f:X \to Y$ is <u>onto</u> or <u>surjective</u> or a <u>surjection</u> (dog latin for "throwing onto") if for any $y \in Y$ $\overleftarrow{f}(\{y\}) \neq \emptyset$. This means it contains <u>at least</u> one point. Equivalently, if $fX=Y$ (not just $fX \subseteq Y$, which is true by definition).

$f:X \to Y$ is <u>bijective</u> or a <u>bijection</u> if it is both injective and surjective.

There exists a function $\overleftarrow{f} : Y \to X$ such that $\{\overleftarrow{f}(y)\} = \overleftarrow{f}(\{y\})$ $\forall$ $y \in Y$, if and only if f is bijective. For if there is such an $\overleftarrow{f}$, each $\overleftarrow{f}(\{y\}) = \{\overleftarrow{f}(y)\} \neq \emptyset$ since $\overleftarrow{f}$ satisfies Fi, and

$$f(x) = f(x') = y, \text{ say} \Rightarrow x,x' \in \overleftarrow{f}(\{y\}) = \{\overleftarrow{f}(y)\}$$
$$\Rightarrow x = x' \text{ since } \overleftarrow{f} \text{ satisfies Fii}$$

so f is bijective. Conversely if f is bijective the subset $g = \{(y,x) \mid (x,y) \in f\} \subseteq Y \times X$ satisfies Fi, Fii for a function $Y \to X$ and $\{g(y)\} = \overleftarrow{f}(\{y\})$ $\forall$ $y \in Y$, so we can put $\overleftarrow{f} = g$. Notice that $\overleftarrow{f}$, when it exists, is also a bijection.

(It is common to write f^{-1} for $\overleftarrow{f}$, but this habit leads to all sorts of confusion between $\overleftarrow{f}(x)$ and $(f(x))^{-1} = 1/f(x)$, and should be stamped out.)

We can state these ideas in terms of functions alone, not mentioning members of sets, if we define for any set X the <u>identity</u> <u>map</u> $I_X: X \to X: x \rightsquigarrow x$. Now the following two statements should be obvious, otherwise the reader should prove them as a worthwhile exercise:

A function $f: X \to Y$ is injective if and only if $\exists$ $g: Y \to X$ s.t.

$g \circ f = I_X : X \to X$.

A function $f: X \to Y$ is surjective if and only if $\exists$ $g: Y \to X$ s.t.

$f \circ g = I_Y : Y \to Y$.

Neither case need involve a <u>unique</u> g. If $X = \{0,1\}$, $Y = [0,1]$ then $i : X \hookrightarrow Y : x \rightsquigarrow x$ (Fig. 2.3a) is injective with infinitely many candidates for g such that $g \circ f = I_X$. (For instance take all of $[0,\frac{3}{4}]$ to 0 and all of $]\frac{3}{4},1]$ to 1.) Similarly the unique (why?) map $[0,1] \rightarrow \{0\}$ is surjective, and <u>any</u> $g : \{0\} \rightarrow [0,1]$ (say, $0 \rightsquigarrow \frac{1}{2}$) has $f \circ g = I_{\{0\}}$ (Fig. 2.3b). But if f is bijective by the existence of $g : Y \rightarrow X$ such that $g \circ f = I_X$ and $g' : Y \rightarrow X$ such that $f \circ g' = I_Y$ then we have

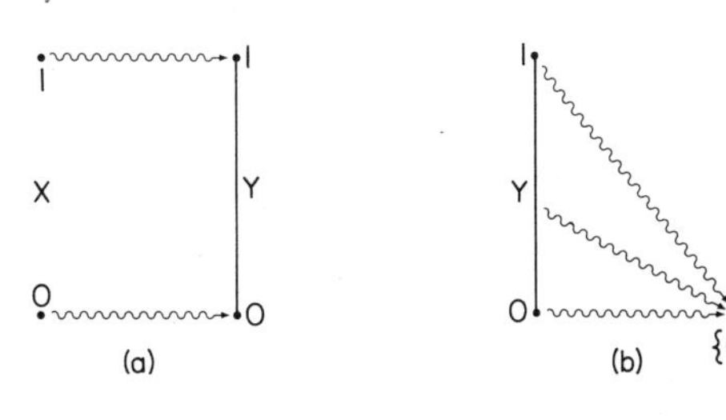

Fig. 2.3

$$g = g \circ I_Y = g \circ (f \circ g') = (g \circ f) \circ g' = I_X \circ g' = g'.$$

By the same argument, if $h : Y \rightarrow X$ is any other map with $h \circ f = I_X$ then it must equal g' and hence g, or with $f \circ h = I_Y$ it must equal g. Then we have a unique <u>inverse map</u> that we may call $\overleftarrow{f}$ as above, with $\overleftarrow{f} \circ f = I_X$, $f \circ \overleftarrow{f} = I_Y$. We may omit the subscript when the domain of the identity is plain from the context.

When maps with various ranges and domains are around, we shall sometimes gather them into a composite diagram such as

$$X \xrightarrow{f} W \xrightarrow{g} Z \xrightarrow{h} Y \xrightarrow{q} T \ , \qquad \text{or} \qquad \begin{array}{ccc} X & \xrightarrow{f} & W \\ {\scriptstyle F}\downarrow & & \downarrow{\scriptstyle g} \\ M & \xrightarrow[G]{} & T \end{array}$$

where the domain and range of each map are given by the beginning and end, respectively, of its arrow.

This helps keep track of which compositions are legitimate. For instance, if

$f : X \to Y$ and $g : Y \to Z$ are both bijections, then we have two diagrams

$$X \xrightarrow{f} Y \xrightarrow{g} Z \quad \text{and} \quad X \xleftarrow{\overleftarrow{f}} Y \xleftarrow{\overleftarrow{g}} Z$$

which make clear that we can form the composites $g \circ f$ and $\overleftarrow{f} \circ \overleftarrow{g}$, but not $f \circ g$ (since $g(y) \in Z$, and $f(z)$ is not defined for $z \in Z$) or $\overleftarrow{g} \circ \overleftarrow{f}$. The composite $g \circ f$ is again a bijection, with inverse $\overleftarrow{(g \circ f)} = \overleftarrow{f} \circ \overleftarrow{g}$, since

$$(\overleftarrow{f} \circ \overleftarrow{g}) \circ (g \circ f) = \overleftarrow{f} \circ (\overleftarrow{g} \circ g) \circ f = \overleftarrow{f} \circ I_Y \circ f = \overleftarrow{f} \circ f = I_X$$

$$(g \circ f) \circ (\overleftarrow{f} \circ \overleftarrow{g}) = g \circ (f \circ \overleftarrow{f}) \circ \overleftarrow{g} = g \circ I_Y \circ \overleftarrow{g} = g \circ \overleftarrow{g} = I_Z.$$

We assume the existence and familiar properties of certain common functions : notably

$$+ : R \times R \to R : (x,y) \rightsquigarrow x+y , \qquad \times : R \times R \to R : (x,y) \rightsquigarrow xy,$$

$$- : R \to R : x \rightsquigarrow -x, \qquad \text{modulus} : R \to R : x \rightsquigarrow |x| = \begin{cases} x \text{ if } x \geq 0 \\ -x \text{ if } x < 0, \end{cases}$$

whose precise definitions involve that of R itself, and the corresponding division, subtraction and polynomial functions (such as $x \rightsquigarrow x^3+x$) that can be defined from them. When constructing examples we shall often use (as already above) the functions

$$\exp : R \to]0,\infty[: x \rightsquigarrow \exp(x) = e^x, \text{ (its series is mentioned only in IX.6.2)},$$

its inverse the natural logarithm

$$\log :]0,\infty[\to R : x \rightsquigarrow (\text{that } y \text{ s.t. } e^y = x)$$

the trigonometrical functions

$$\sin : R \to R, \quad \cos : R \to R,$$

and (in IX §6 only) the hyperbolic functions

$$\sinh : R \to R, \quad \cosh : R \to R,$$

taking as given their standard properties (various identities are stated in Ex **IX.6.2**). Among these properties we include their differentials

$$\frac{d}{dt}(\exp)(x) = \exp(x), \quad \frac{d}{dt}(\log)(x) = \frac{1}{x}, \quad \frac{d}{dt}(\sin)(x) = \cos x,$$

$$\frac{d}{dt}(\cos)(x) = -\sin x, \quad \frac{d}{dt}(\sinh)(x) = \cosh x, \quad \frac{d}{dt}(\cosh)(x) = \sinh x,$$

since to prove these would involve adding to the precise treatment of differentiation in Ch. VIII the material on infinite sums necessary to define exp, log, sin and cos rigorously. This seems unnecessary - when the functions are already familiar - for the purposes of this book. (The physics student, who may not have seen them precisely defined, should, if assailed by Doubt, refer to any elementary Analysis text, such as [Moss and Roberts].)

Finally we define the map named after Kronecker,

$$\delta : N \times N \to \{0,1\} : (i,j) \rightsquigarrow \begin{cases} 0 \text{ if } i \neq j \\ 1 \text{ if } i = j \end{cases}$$

and the standard abbreviations δ^i_j, δ_{ij} and δ^{ij} (according to varying convenience) for the real number $\delta(i,j)$. Thus, for instance,

$$\delta^1_1 = \delta_{22} = \delta^{55} = 1, \quad \delta^2_3 = \delta^1_8 = \delta_{23} = \delta^{34} = 0.$$

3. PHYSICAL BACKGROUND

In 1887, Albert Abraham Michelson and Edward Williams Morley tried to measure the absolute velocity of the Earth through space, as follows.

Light was believed to consist of movements, analogous to water or sound waves, of a _luminiferous_ (= light-carrying) _ether_. (The name is descended deviously from the theories of Aristotle, in which heavenly bodies - only - are made of a _luminous_ element, "ether" or "aether", instead of terrestrial earth, air, fire and water. Such an element is rather unlike the 19th Century omnipresent something, whose only discernible property was carrying light by its oscillations.) Any attempt to allow currents or eddies in the ether led to the prediction of unobserved effects. Therefore it seemed reasonable to allow the ether to enjoy absolute rest, apart from its light-carrying oscillations. Hence an absolute velocity could be assigned to the Earth, as its velocity relative to the ether. Thus the crucial experiment is equivalent to measuring the flow of ether through the Earth. Since the ether was detectable only by its luminiferosity, any such measurements had to be of light waves.

The problem is analogous to that of measuring the speed of a river by timing swimmers who move at constant speed, relative to the water, as light waves were believed to, relative to the ether. This constancy followed from the wave theory of light; Newton's "light corpuscles" had no more reason for constant speed than bullets have. (In what follows, remember that "speed" is a number, while "velocity" is speed in a particular direction: the man who said he had been fined for a "velocity offence" had been driving below the speed limit, but down a wrong-way street.) The wave characteristics of light were also used essentially in the experiment; the times involved were too short to measure directly, but could be compared through wave interference effects. For the optical details we refer the reader to [Feynman], and limit ourselves here to the way the time comparisons were used.

Suppose (Fig. 3.1) that we have three rigidly linked rafts moored in water flowing at uniform speed v, all in the same direction. The raft separations AB, AC are at right angles, and each of length L. If a swimmer's speed, relative to the water, is always c, her time from A to C and back will be

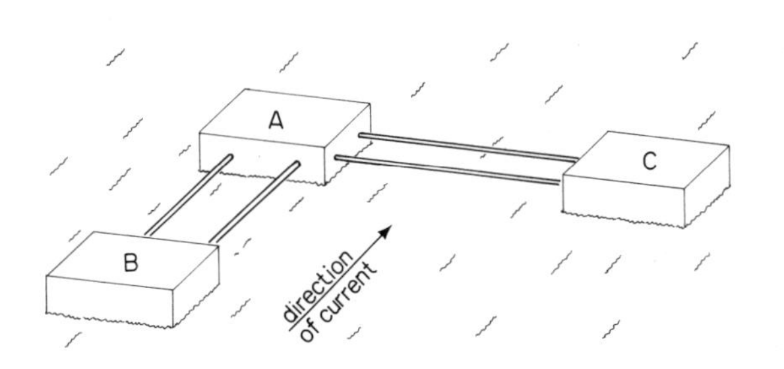

Fig. 3.1

$$T_1 = \frac{L}{\text{upstream speed}} + \frac{L}{\text{downstream speed}} = \frac{L}{c-v} + \frac{L}{c+v} = \frac{2\ v\ L}{c^2-v^2}$$

where the speeds c-v, c+v are relative to the rafts. For swims from A to B, the velocities add more awkwardly (Fig. 3.2). Then she achieves a cross current speed of $(c^2-v^2)^{\frac{1}{2}}$ giving a time from A to B and back of

$$T_2 = \frac{2L}{\sqrt{c^2-v^2}} .$$

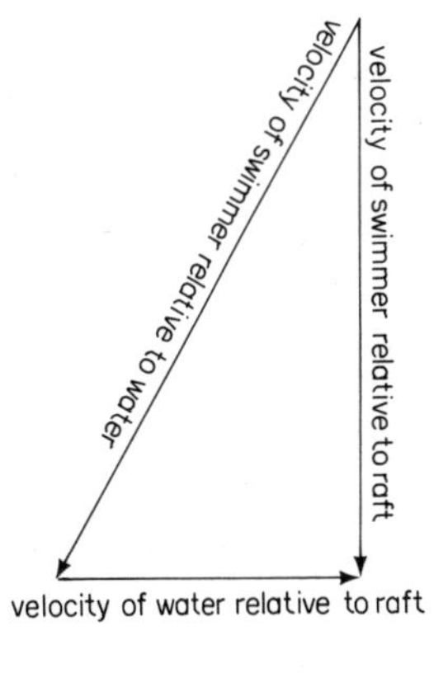

Fig. 3.2

Simple algebra then gives

$$v = c\sqrt{1 - \frac{T_2^2}{T_1^2}},$$

so that measurement of the ratio T_1/T_2 gives v as a multiple of the "measuring standard" velocity c. Minor elaborations involving turning the apparatus take care of not knowing the current direction in advance, and the possibility that $AB \neq AC$.

The analogous experiment with c as the enormous speed of light (which Michelson was brilliant at measuring) and v as the relative velocity of Earth and ether, required great skill. Repeated attempts, ever more refined, gave $v = 0$, even when the margin of error was held well below Earth's orbital speed and the experiment repeated six months later with the Earth, halfway round the sun, going the other way. Thus two different velocities, v and -v, both appeared to be zero relative to the unmoving ether!

In retrospect, this experiment is seen as changing physics utterly (though it did not strike Michelson that way). More and more ad hoc hypotheses had to be added to conventional physics to cope with it. The Irish physicist Fitzgerald proposed that velocity v in any direction shrunk an object's length in that direction by $(1 - v^2/c^2)^{\frac{1}{2}}$. Hendrik Antoon Lorentz suggested the same (the effect is now known as the Lorentz-Fitzgerald contraction). He also saw that to save Newton's law "force = mass times acceleration" the mass of a moving object had also to change, this time <u>increased</u> by the same factor.

Every effort to get round these effects and find an absolute velocity hit a new contradiction or a similar "fudge factor", as though there were a conspiracy to conceal the answer. Henri Poincaré pointed out that "a successful conspiracy is itself a law of nature", and in 1905 Albert Einstein proposed the theory now called Special Relativity. He assumed that it is <u>completely</u> impossible, by any means whatever, to discover for oneself an absolute velocity. Any velocity at all may be treated as "rest". From this "Principle of Relativity" he deduced all the previously ad hoc fudge factors in a coherent way. Moreover, he accounted effectively for a wide range of experimental facts - both those then known, and many learnt since. His theory is now firmly established, in the sense that any future theories must at least include it as a special case. For no experiment has contradicted these consequences of the theory that have been elaborated to date.

One such consequence caused great surprise at the time, and leads to a "spacetime geometry" which - even before gravity is considered - is different from the "space geometries" studied up to that time. It even differs from the generalised (non-Euclidean and n-dimensional) ones investigated in the 19th Century. By the Relativity Principle, every observer measuring the speed of light in vacuum must find the same answer. (Or Michelson and Morley would have got the results they expected.) Consider a flash of light travelling at uniform speed c, straight from a point X_1 to a point X_2. Then any observer will find the equation

$$\bigstar \qquad c = \frac{\text{distance from } X_1 \text{ to } X_2}{\text{time taken by light flash}}$$

exactly satisfied. But another observer may easily measure the distance differently, even on Newtonian assumptions, since "arrival" is later than "departure". (A minister in a Concorde drops his champagne glass and it hits the floor after travelling - to him - just three feet, downwards. But he drops it as he booms over one taxpayer, and it breaks over another, more than 500 ft away.) Then the Principle requires that the time taken <u>also</u> be measured differently, to keep the same ratio c (using, we must obviously insist, the same units for length, and time - otherwise <u>one</u> observer can change c however he chooses). This created controversy, above all because it implied that two identical systems (clocks or twins for instance) could leave the same point at the same moment, travel differently, and meet later after the passage of more time (measured by ticks or biological growth) for one than for the other. This contradicted previous opinion so strongly as to be miscalled the Clocks, or Twins, Paradox; cf. IX. 4.05. (Strictly a paradox must be *self*-contradictory, like the logical difficulties that were troubling mathematics at the time. Physics has had its share of paradoxes, such as finite quantities proved infinite, but this is not one of them. There is nothing *logically* wrong about contradicting authority, whether Church, State or Received Opinion, though it may be found *morally* objectionable.) With the techniques for producing very high speeds developed since 1905 - near lightspeed in particle accelerators - and atomic clocks of extreme accuracy, this dependence of elapsed time on the measurer has been confirmed to many decimal places in innumerable experiments of very various kinds. We consider its geometrical aspects in Ch. IX (since it is a failure of geometric rather than physical insight that gives the feeling of "paradox") and its more physical, quantitative aspects in Ch. XI. The following remarks explain some terminology chosen in Ch. IV.

Choose, for our first observer of the above movement of a light flash, coordinates (x,y,z) for space and t for time, with $t = x = y = z = 0$ labelling "departure". (We choose rectangular coordinates (x,y,z) if we can, though this is usually only locally and approximately possible in the General theory. The discussion below then leads to the structure we attribute to spacetime "in the limit of smallness" where the approximations disappear, so it remains satisfactory for motivation.) In these coordinates, "arrival" is labelled by the four numbers (t,x,y,z). Then equation ★ becomes, using Pythagoras's theorem,

$$c = \frac{\sqrt{x^2 + y^2 + z^2}}{t}$$

or equivalently

$$c^2t^2 - x^2 - y^2 - z^2 = 0.$$

The Principle requires this to be equally true for an observer using different coordinates with the same origin, "departure" labelled by (0,0,0,0), but giving a new label (t',x',y',z') to "arrival". As remarked above, t' will in general be different from t. But we must still have

$$c^2(t')^2 - (x')^2 - (y')^2 - (z')^2 = 0$$

with the same value of c. It follows fairly easily (the more mathematical reader should prove it) that there is a positive number S such that for <u>any</u> "time and position" labelled (T,X,Y,Z) by one system and (T',X',Y',Z') by the other, not just the possible "arrival" points of light flashes with "departure" (0,0,0,0), we have

$$c^2(T')^2 - (X')^2 - (Y')^2 - (Z')^2 = S(c^2T^2 - X^2 - Y^2 - Z^2).$$

Now the Principle requires that both systems use the same units; in particular they must give lengths

$$\sqrt{(X')^2 + (Y')^2 + (Z')^2} = \sqrt{X^2 + Y^2 + Z^2}$$

to points for which they agree are at time zero, so that $T' = T = 0$. There always will be such points (a proof needs some machinery from Chapters I - IV), so S can only be 1. Up to choice of unit, then,

$$c^2T^2 - X^2 - Y^2 - Z^2$$

is a quantity which, unlike T,X,Y,Z individually, does <u>not</u> depend on the labelling system. This is in close analogy to the familiar fact of three-dimensional analytic geometry, used above, that

$$x^2 + y^2 + z^2 = (\text{distance from origin})^2$$

does not depend on the rectangular axes chosen. It is common nowadays to strengthen the analogy by choosing units to make $c = 1$. For instance, measuring time in years and distance in light years, the speed of light becomes exactly 1 light year per year by definition. Or, as in [Misner, Thorne and Wheeler], time may be measured in centimeters - in multiples of the time in which light travels 1 cm in vacuum. (Such mingling of space and time is ancient in English, though "a length of time" is untranslatable into some languages, but is only fully consummated in Relativity.) This practice gives the above quantity the standard form

$$T^2 - X^2 - Y^2 - Z^2$$

independently even of units (though its value at a point will depend on whether your scale derives from the year or from the Pyramid Inch.) We examine the geometry of spaces with label-independent quantities like this one and like Euclidean length, from Ch.IV onwards.

Two ironies : Michelson lived to 1933 without ever accepting Relativity. Modern astronomers, who accept it almost completely, expect in the next decade or so to measure something very like an "absolute velocity" for the Earth. This derives from the Doppler shift (cf. XI. 2.09) of the amazingly isotropic universal background of cosmic black body radiation.

I Real Vector Spaces

"To banish reality is to sink deeper into the real;
allegiance to the void implies denial of its voidness."
Seng-ts'an.

1. SPACES

1.01 Definition

A <u>real vector space</u> is a non-empty set X of things we call <u>vectors</u> and two functions

$$\text{"vector addition"} \ : \ X \times X \to X : (\underset{\sim}{x},\underset{\sim}{y}) \rightsquigarrow \underset{\sim}{x} + \underset{\sim}{y}$$

$$\text{"scalar multiplication"} \ : \ X \times R \to X : (\underset{\sim}{x},a) \rightsquigarrow \underset{\sim}{x}a$$

such that for $\underset{\sim}{x},\underset{\sim}{y},\underset{\sim}{z} \in Z$ and $a, b \in R$ we have

(i) $\underset{\sim}{x} + \underset{\sim}{y} = \underset{\sim}{y} + \underset{\sim}{x}$, (commutativity of +).

(ii) $(\underset{\sim}{x} + \underset{\sim}{y}) + \underset{\sim}{z} = \underset{\sim}{x} + (\underset{\sim}{y} + \underset{\sim}{z})$, (associativity of +).

(iii) There exists a unique element $\underset{\sim}{0} \in X$, the <u>zero vector</u>, such that for any $\underset{\sim}{x} \in X$ we have $\underset{\sim}{x} + \underset{\sim}{0} = \underset{\sim}{x}$, (+ has an identity).

(iv) For any $\underset{\sim}{x} \in X$, there exists $(-\underset{\sim}{x}) \in X$ such that $\underset{\sim}{x} + (-\underset{\sim}{x}) = \underset{\sim}{0}$, (+ admits inverses).

(v) For any $\underset{\sim}{x} \in X$, $\underset{\sim}{x}1 = \underset{\sim}{x}$.

(vi) $(\underset{\sim}{x} + \underset{\sim}{y})a = \underset{\sim}{x}a + \underset{\sim}{y}b$, ⎫

⎬ (distributivity).

(vii) $\underset{\sim}{x}(a + b) = \underset{\sim}{x}a + \underset{\sim}{x}b$, ⎭

(viii) $(\underset{\sim}{x}a)b = \underset{\sim}{x}(ab)$.

This long list of axioms does not mean that a vector space is immensely complicated. Each one of them, properly considered, is a rule that something difficult should not happen. English breaks (i), since

killer rat ≠ rat killer

and similarly (ii), since

killer of young rats = (young rat)killer ≠ young(rat killer) = young killer of rats.

In consequence the objects that obey all of them are beautifully simple, and the theory of them is the most perfect and complete in all of mathematics (particularly for "finite-dimensional" ones, which we come to in a moment in 1.09). The theory of objects obeying only some of these rules is very much harder. English, which obeys none of them, is only beginning to acquire a formal theory.

If a vector space is "finite-dimensional" it may be thought of simply and effectively as a geometrical rather than an algebraic object; the vectors are "directed distances" from a point O called the <u>origin</u>, vector addition is defined by the parallelogram rule and scalar multiplication by $\underset{\sim}{x}a$ = "(length of $\underset{\sim}{x}$)×a in the direction of $\underset{\sim}{x}$". All of linear algebra (alias, sometimes, "matrix theory") is just a way of getting a grip with the aid of numbers on this geometrical object. We shall thus talk of geometrical vectors as line segments: they all have one end at O, and we shall always draw them with an arrowhead on the other. To forget the geometry and stop drawing pictures is <u>voluntarily</u> to create enormous problems for yourself - equal and opposite to the

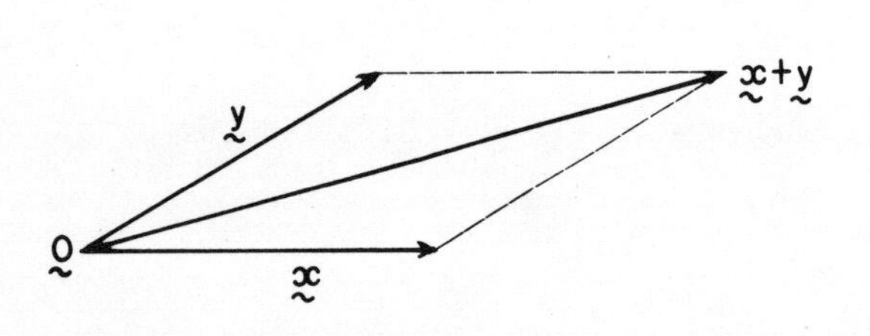

Fig. 1.1

difficulties the Greeks had in working with raw geometry alone, with no use of coordinates at all. (Often other pictures than arrows will be appropriate, as with vectors in the dual spaces discussed in Chapter III. But reasoning motivated by the arrow pictures, within any particular vector space, remains useful.)

In this context, the real numbers used are called scalars. The only reason for not calling them just "numbers", which would adequately distinguish them from vectors, is that for historical reasons nobody else does, and in mathematics as in other languages the idea is to be understood.

The term real vector space refers to our use of R as the source of scalars. We shall use no others (and so henceforth we banish the "real" from the name), but other number systems can replace it: for instance, in quantum mechanics vector spaces with complex scalars are important. We recall that R is algebraically a field (cf.Ex. 10).

Notice that properties (ii), (iii) and (iv) are sufficient axioms for a vector space to be a group under addition; property (i) implies that this group is commutative. (cf. Ex. 10).

1.02 Definition

The standard real n-space R^n is the vector space consisting of ordered n-tuples $(x^1,\ldots,x^n)$ of real numbers as on p.5, with its operations defined by

$$(x^1,\ldots,x^n) + (y^1,\ldots,y^n) = (x^1 + y^1,\ldots,x^n + y^n)$$
$$(x^1,\ldots,x^n)a = (ax^1,\ldots,ax^n)$$

(cf. Ex. 1).

1.03 Definition

A subspace of a vector space X is a non-empty subset $S \subseteq X$ such that

$$x, y \in S \Rightarrow (x + y) \in S$$

$$x \in S,\ a \in R \Rightarrow xa \in S.$$

For instance, in a three-dimensional geometrical picture, the only subspaces are the following.

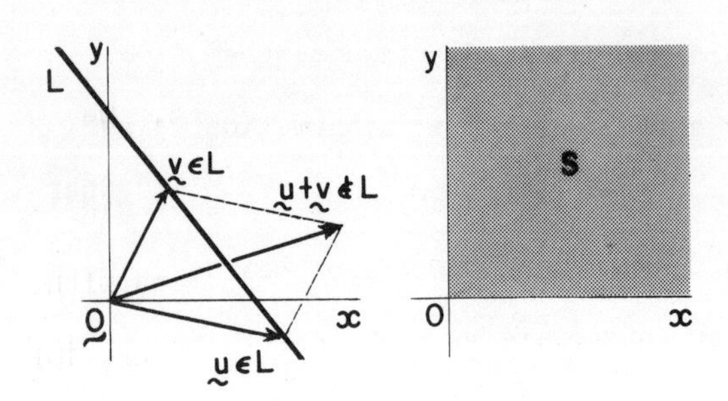

Fig. 1.2

(1) The directed distances from origin 0 to points in a line through 0, (a line subspace).

(2) The directed distances from origin 0 to points in a plane through 0, (a plane subspace).

(3) The trivial subspaces: the whole space itself, and the zero subspace (consisting of the zero vector 0 alone). By Ex. 2a this is contained in every other subspace.

Sets of directed distances to lines and planes not through 0 are examples of subsets which are not subspaces, (cf. Ex.2). Nor are sets like S (cf. Fig. 1.2).

1.04 Definition

The linear hull of any set $S \subseteq X$ is the intersection of all the subspaces containing S. It is always a subspace of X (cf. Ex.3a). We shall also say that it is the subspace spanned by the vectors in S.

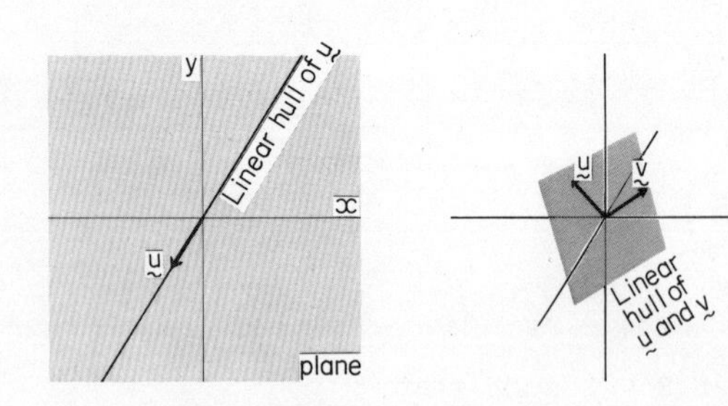

Fig. 1.3

Thus, for instance, the linear hull of a single vector is the intersection of all line subspaces and plane subspaces etc. that contain it, which is clearly just the line subspace in the direction of the vector. Similarly the linear hull of two non-zero vectors is the plane subspace they define as two line segments (Fig. 1.3) unless they are in the same or precisely opposite direction, in which case it is the line subspace in that direction. The linear hull of three vectors may be three-dimensional, a plane subspace, a line subspace or (if they are all zero) the zero subspace.

1.05 Definition

A <u>linear combination</u> of vectors in a set $S \subseteq X$ is a finite sum $\underset{\sim}{x}_1 a^1 + \underset{\sim}{x}_2 a^2 + \ldots + \underset{\sim}{x}_n a^n$, where $\underset{\sim}{x}_1, \ldots, \underset{\sim}{x}_n \in S$ and $a^1, \ldots, a^n \in R$.
(cf. Ex.3b)

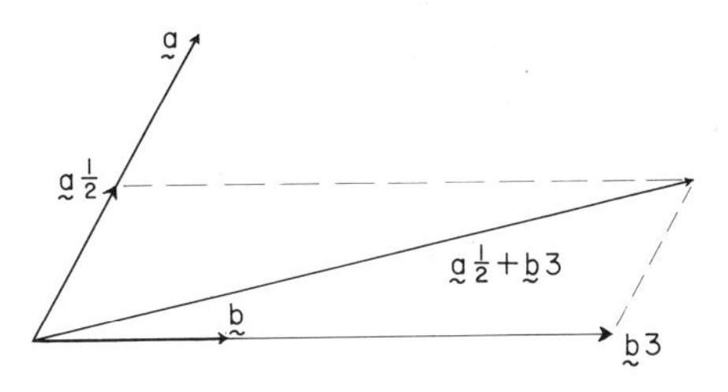

Fig. 1.4

1.06 Notation

The <u>summation convention</u> (invented by Einstein) represents $\underset{\sim}{x}_1 a^1 + \ldots + \underset{\sim}{x}_n a^1$ by $\underset{\sim}{x}_i a^i$, and in this book $\underset{\sim}{x}_i a^i$ will <u>always</u> represent such a sum. (Be warned: this is mainly a physicist's habit. Mathematicians mostly use $\sum_{i=1}^{n} \underset{\sim}{x}_i a^i$ for sums, and by $\underset{\sim}{x}_i a^i$ would mean $\underset{\sim}{x}_1 a^1$, <u>or</u> $x_2 a^2$, <u>or</u> $\underset{\sim}{x}_n a^n$. There are good arguments for either, and if you go further you will meet both. We shall always favour physicist's notation in this book, unless it is hopelessly destructive of clarity.) Evidently $\underset{\sim}{x}_j a^j$ or $\underset{\sim}{x}_\alpha a^\alpha$ represent the same sum equally well, as long as we know what $\underset{\sim}{x}_1, \ldots, \underset{\sim}{x}_n$ and $a^1, \ldots, a^n$ are, and so i, j, α etc are often called <u>dummy indices</u>, to emphasise that while $\underset{\sim}{x}_i$ need not be the same vector as $\underset{\sim}{x}_j$, $\underset{\sim}{x}_i a^i$ is always the same as $\underset{\sim}{x}_j a^j$. It is often convenient to "change dummy index" in the middle of a computation - this makes use without explicit mention of the identity $\underset{\sim}{x}_i a^i = \underset{\sim}{x}_j a^j$.

The convention does not apply only to writing down linear combinations. For example, if we have real-valued functions $f_1, \ldots, f_n$ and $g^1, \ldots, g^n$, then $f_i g^i$ is short for $f_1 g + \ldots + f_n g^n$. (This expression will emerge in later chapters as the value of a covariant vector field applied to a contravariant one - we have not forsaken geometry.) Invariably, even if there are a lot of other indices around, $a^{ijk}_{pq} b^{rs}_j$ for instance will mean $a^{i1k}_{pq} b^{rs} + \ldots + a^{ink}_{pq} b^{rs}_n$, where n is (hopefully) clear from the context. The Kronecker function (cf. Ch. I.2) often crops up with the summation convention, as $\underset{\sim}{x}_i \delta^i_j = \underset{\sim}{x}_j$ in change of index for example; beware however, $\delta^i_i = n$.

Notice that the convention applies only if we have one upper and lower index (though $a_i b^i$ means the same as $b^i a_i$ - order does not signify); if we want to abbreviate $x^1 y^1 + x^2 y^2 + \ldots + x^n y^n$ we must use $\sum\limits_{i=n}^{n} x^i y^i$. This is not as daft as it seems. The positions of the indices usually have geometrical significance, and the two kinds of sum then represent quite different geometrical ideas: Chapter III is about one, Chapter IV about the other.

1.07 Definition

A subset $S \subseteq X$ is <u>linearly dependent</u> if some vector in S is a linear combination of other vectors in S. (Notice that $\underset{\sim}{0}$ is always a linear combination of any other vectors, for $\underset{\sim}{0} = \underset{\sim}{x}0 + \underset{\sim}{y}0$, so any set containing 0 is linearly dependent if we say for tidiness that $\{\underset{\sim}{0}\}$ is linearly dependent too.) Equivalently (Ex. 4), S is linearly dependent if and only if there is a linear combination $\underset{\sim}{x}_i a^i = \underset{\sim}{0}$ of vectors $\underset{\sim}{x}_i \in S$, with not all the $a^i = 0$. If S is not linearly dependent, it is <u>linearly independent</u>.

Geometrical example: a set of three vectors in the same plane through the origin is always linearly dependent. To have three independent directions (only the directions of the vectors in S matter for dependence, not their lengths; why?)

we need more room. This leads us to

1.08 Definition

A subset $\beta \subseteq X$ is a basis for X if the linear hull of β is all of X, and β is linearly independent.

Intuitively, it is clear that a basis for a line subspace must have exactly one member, whereas a plane subspace requires vectors in two directions, and so forth; the number of independent vectors you can get, and the number you need to span the space, will correspond to the "dimension" of the space. Now our concept of dimension does not rely on linear algebra. It is much older and more fundamental. What we must check, then, is not so much that our ideas of dimension are right as that linear algebra models nicely our geometrical intuition. The algebraic proof of the geometrically visible statement that if X has a basis consisting of a set of n vectors, any other basis also contains n vectors, is indicated in Exercises 5-7. (The same sort of proof goes through for infinite dimensions, but we shall stick to finite ones.) Hence we can define dimension algebraically, which is a great deal easier than making precise within geometry the "concept of dimension" we have just been so free with. But remember that this is an algebraic convenience for handling a geometrical idea.

1.09 Definition

If X has a basis with a finite number n of vectors, then X is finite-dimensional and in particular n-dimensional. Thus R^3 is 3-dimensional, by Ex. 9. The number n is the dimension of X. We shall assume all vector spaces we mention to be finite-dimensional unless we specifically indicate that they are not. If a subspace of X has dimension (dimX - 1) it is called a hyperplane of X by analogy with plane subspaces of R^3. (What are the hyperplanes of R^2?)

1.10 Definition

The standard basis $\mathcal{E}$ for R^n (cf. Defn. 1.02) is the set of n vectors $e_1, e_2, \ldots, e_n$ where $e_i = (0, \ldots, 0,1,0, \ldots, 0)$ with the 1 in the i-th place. (cf. Ex. 9).

Exercises II.1

1) The standard real n-space is indeed a vector space.

2a) Any subspace of a vector space must include the zero 0.

b) The set $\{0\} \subseteq X$ is always a subspace of X.

3a) The linear hull of $S \subseteq X$ is a subspace of X.

b) The linear hull of $S \subseteq X$ is exactly the set of all linear combinations of vectors in S.

c) A subset S is a subspace of X if and only if it coincides with its linear hull.

4) Prove the equivalence of the alternative definitions given in 1.07.

5) A subset of a linearly independent set of vectors is also independent.

6) If β is a basis for X then no subset of β (other than β itself) is also a basis for X.

7a) If $\beta = \{x_1, \ldots, x_n\}$ and $\beta' = \{y_1, \ldots, y_m\}$ are bases for X then so is $\{y_1, x_1, \ldots, x_{j-1}, x_{j+1}, \ldots, x_n\}$ for some omitted x_j. Notice that the new basis, like β, has n members.

b) Prove that if $k < n$, then a set consisting of $y_1, \ldots, y_k$ and some suitable set of (n-k) of the x_i's is a basis for X. Deduce that $m \leq n$.

c) Prove that $m = n$.

8) If β is a basis for X then any vector in X is, in a unique way, a linear combination of vectors in β. If therefore,

$$x_i a^i = y = x'_j b^j, \text{ where the } x_i, x'_j \in \beta$$

then the non-zero a^i's are equal to non-zero b^j's and multiply the same vectors. They are called the components of y with respect to β.

9) Prove that $\{\underset{\sim}{e}_1, \ldots, \underset{\sim}{e}_n\}$ is a basis for R^n.

10) A <u>group</u> $(X,*)$ is a non-empty set X and a map $* : X \times X \to X$ such that $*$ is associative, has an identity, and admits inverses. Thus $(R,+)$ and $(R\backslash\{0\},\times)$ are groups and in fact this double group structure makes the real numbers a <u>field</u> because + and × interact in a distributive way.

2. MAPS

In almost all mathematical theories we have two basic tools: sets with a particular kind of structure, and functions between them that respect their structure. We shall meet several examples of this in the course of the book. For sets with a vector space structure, the functions we want are as follows.

2.01 Definition

A function $\underset{\sim}{A}:X \to Y$ is <u>linear</u> if for all $\underset{\sim}{x},\underset{\sim}{y}, \in X$ and $a \in R$ we have

$$\underset{\sim}{A}(\underset{\sim}{x} + \underset{\sim}{y}) = \underset{\sim}{A}\underset{\sim}{x} + \underset{\sim}{A}\underset{\sim}{y}$$
$$\underset{\sim}{A}(\underset{\sim}{x}a) = (\underset{\sim}{A}\underset{\sim}{x})a \ .$$

The terms linear map or mapping, linear transformation and linear operator for such functions are frequent, though the latter is generally reserved for maps $\underset{\sim}{A}:\underset{\sim}{X} \to \underset{\sim}{X}$, which "operate" on X. (It is also the favourite term in books which discuss, for example, quantum mechanics in terms of operators without ever saying what they operate on. This is perhaps intended to make things easier.) We shall use "linear map" for a general linear function $X \to Y$, "linear <u>operator</u>" in the case $X \to X$, omitting "linear" like "real" where no confusion is created.

The set $L(X;Y)$ of <u>all</u> linear maps $X \to Y$ itself forms a vector space under the addition and scalar multiplication

$$(\underset{\sim}{A}:X \to Y) + (\underset{\sim}{B}:X \to Y) = \underset{\sim}{A} + \underset{\sim}{B} : X \to Y : \underset{\sim}{x} \rightsquigarrow \underset{\sim}{Ax} + \underset{\sim}{Bx}$$

$$(\underset{\sim}{A}:X \to Y)a = \underset{\sim}{A}a \quad : X \to Y : \underset{\sim}{x} \rightsquigarrow (\underset{\sim}{Ax})a$$

as is easily checked. So is the fact that the composite $\underset{\sim}{BA}$ of linear maps $\underset{\sim}{A}:X \to Y$, $\underset{\sim}{B}:Y \to Z$ is again linear. We show in 2.07 that dim L(X;Y) = dim X dim Y.

2.02 Definition

The identity operator $\underset{\sim}{I}_X$ on X is defined by $\underset{\sim}{I}_X(\underset{\sim}{x}) = \underset{\sim}{x}$, for all $\underset{\sim}{x}$. We shall denote it by just $\underset{\sim}{I}$ when it is clear which space is involved. A scalar operator is defined for every $a \in R$ by $(\underset{\sim}{I}a)\underset{\sim}{x} = \underset{\sim}{x}a$ for all $\underset{\sim}{x} \in X$. Such an operator is abbreviated to a, so that $\underset{\sim}{x}a = a\underset{\sim}{x}$.

The zero map $\underset{\sim}{0}:X \to Y$ is defined by $\underset{\sim}{0x} = \underset{\sim}{0}$.

A linear map $\underset{\sim}{A}:X \to Y$ is an isomorphism if there is a map $\underset{\sim}{B}:Y \to X$ such that both $\underset{\sim}{AB} = \underset{\sim}{I}_Y$ and $\underset{\sim}{BA} = \underset{\sim}{I}_X$. (Notice that it is possible to have one but not the other: if $\underset{\sim}{A}:R^2 \to R^3 : (x,y) \rightsquigarrow (x,y,0)$ and $\underset{\sim}{B}:R^3 \to R^2 : (x,y,z) \rightsquigarrow (x,y)$, then $\underset{\sim}{BA} = \underset{\sim}{I}_{R^2}$ but $\underset{\sim}{AB} \neq \underset{\sim}{I}_{R^3}$.) We read $\underset{\sim}{A}:X \cong Y$ as "$\underset{\sim}{A}$ is an isomorphism from X to Y".

Such a $\underset{\sim}{B}$ is the inverse of $\underset{\sim}{A}$ and we write $\underset{\sim}{B} = \overset{\leftarrow}{\underset{\sim}{A}}$. $\underset{\sim}{A}$ is then invertible.

$\underset{\sim}{A}:\underset{\sim}{X} \to \underset{\sim}{Y}$ is non-singular if $\underset{\sim}{Ax} = \underset{\sim}{0}$ implies $\underset{\sim}{x} = \underset{\sim}{0}$, otherwise singular. (cf. Ex.1) Evidently an invertible map is non-singular.

If $\underset{\sim}{x} \neq \underset{\sim}{0}$, $\underset{\sim}{Ax} = \underset{\sim}{0}$ then $\underset{\sim}{x}$ is a singular vector of $\underset{\sim}{A}$.

2.03 Lemma

A function $\underset{\sim}{A}:X \to Y$ is an isomorphism if and only if it is a linear bijection.

Proof

If $\underset{\sim}{A}$ is a bijection, there exists $\underset{\sim}{B}:Y \to X$ (not necessarily linear) such that $\underset{\sim}{AB} = \underset{\sim}{I}_Y$, $\underset{\sim}{BA} = \underset{\sim}{I}_X$. If $\underset{\sim}{A}$ is also linear, consider $\underset{\sim}{y},\underset{\sim}{y}' \in Y$. For some $\underset{\sim}{x},\underset{\sim}{x}' \in X$ we have $\underset{\sim}{y} = \underset{\sim}{Ax}$, $\underset{\sim}{y}' = \underset{\sim}{Ax}'$, since $\underset{\sim}{A}$ is surjective, and $\underset{\sim}{y} + \underset{\sim}{y}' = \underset{\sim}{Ax} + \underset{\sim}{Ax}' = \underset{\sim}{A}(\underset{\sim}{x} + \underset{\sim}{x}')$,

($\underset{\sim}{A}$ linear). So $\underset{\sim}{B}(\underset{\sim}{y} + \underset{\sim}{y}') = \underset{\sim}{BA}(\underset{\sim}{x} + \underset{\sim}{x}') = \underset{\sim}{I}(\underset{\sim}{x} + \underset{\sim}{x}') = \underset{\sim}{x} + \underset{\sim}{x}' = (\underset{\sim}{BA})\underset{\sim}{x} + (\underset{\sim}{BA})\underset{\sim}{x}'$ $= \underset{\sim}{B}(\underset{\sim}{Ax}) + \underset{\sim}{B}(\underset{\sim}{Ax}') = \underset{\sim}{By} + \underset{\sim}{By}'$. Similarly $\underset{\sim}{B}(\underset{\sim}{y}a) = (\underset{\sim}{By})a$, and hence $\underset{\sim}{B}$ is linear. Conversely, an isomorphism is linear by definition and a bijection by the existence of its inverse. ■

2.04 Corollary

If $\underset{\sim}{A}$ is non-singular and surjective, it is an isomorphism.

Proof

Non-singularity implies that $\underset{\sim}{A}$ is injective by Ex. 1, and hence bijective. The result follows. ■

2.05 Lemma

If β is a basis for X, then (i) any linear map $\underset{\sim}{A}:X \to Y$ is completely specified by its values on β, and (ii) any function $A:\beta \to Y$ extends uniquely to a linear map $\underset{\sim}{A}:X \to Y$.

Proof

Let $\underset{\sim}{x} \in X$ be the linear combination $\underset{\sim}{b}_i a^i$ of elements of β. By linearity of $\underset{\sim}{A}$, $\underset{\sim}{Ax} = \underset{\sim}{A}(\underset{\sim}{b}_i a^i) = (\underset{\sim}{Ab}_i)a^i$, which depends only on $\underset{\sim}{x}$ (via the scalars a^i) and the vectors $\underset{\sim}{Ab}_i$. Thus $\underset{\sim}{A}$ is fixed if we know its values on β, and since $\underset{\sim}{x} = \underset{\sim}{b}_i a^i$ in a unique way (Ex. 1.8), we can without ambiguity define $\underset{\sim}{A}$ by $\underset{\sim}{A}x = (A\underset{\sim}{b}_i)a^i$, and check that $\underset{\sim}{A}$ so defined is linear. ■

Geometrically: think of a parallelogram or parallelepiped linkage attached to the origin

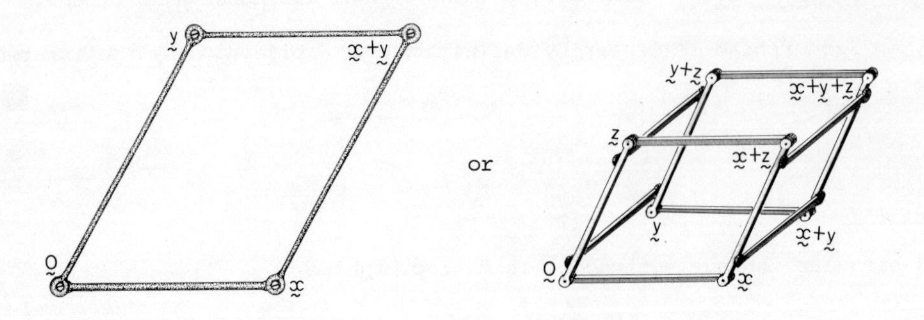

Fig. 2.1

Move the basis vectors $\underset{\sim}{x}, \underset{\sim}{y}, \underset{\sim}{z}$ around, and their sums (given by the parallelogram law) are forced to more in a corresponding way. If not only the parallelogram law but also scalar multiplication is to be preserved, it is clear that defining an operation on basis vectors is enough to determine it everywhere.

2.06 Corollary

If X is an n-dimensional vector space, there is an isomorphism $\underset{\sim}{A}: X \to R^n$.

Proof

Pick any basis $\{\underset{\sim}{b}_1, \ldots, \underset{\sim}{b}_n\}$ for X, then define

$$\{\underset{\sim}{b}_1, \ldots, \underset{\sim}{b}_n\} \underset{B}{\overset{A}{\rightleftarrows}} \{\underset{\sim}{e}_1, \ldots, \underset{\sim}{e}_n\} : \underset{\sim}{b}_i \rightleftarrows \underset{\sim}{e}_i .$$

The functions A and B extend to $\underset{\sim}{A}$ and $\underset{\sim}{B}$ between X and R^n, and if $\underset{\sim}{x} \in X$, $\underset{\sim}{y} \in R^n$ we have

$$\begin{aligned}\underset{\approx}{BAx} &= \underset{\approx}{BA}(\underset{\sim}{b}_i a^i) \\ &= \underset{\approx}{B}((\underset{\approx}{A}\underset{\sim}{b}_i)a^i) \\ &= \underset{\approx}{B}((\underset{\sim}{e}_i)a^i) \\ &= (\underset{\approx}{B}\underset{\sim}{e}_i)a^i \\ &= \underset{\sim}{b}_i a^i \\ &= \underset{\sim}{x}\end{aligned}$$

so that $\underset{\approx}{BA} = \underset{\sim}{I}_X$, and similarly $\underset{\approx}{AB} = \underset{\sim}{I}_{R^n}$. ■

2.07 Matrices

By the last lemma any finite-dimensional space is a copy of R^n - so why not just use R^n, instead of all this stuff about vector spaces? The reason is that to get the isomorphism $\underset{\sim}{A}$ you had to choose a basis, and an ordering for it. Once you have done that, you have "chosen coordinates" on X, because you can label a vector by its image $\underset{\approx}{Ax} = (a^1, \ldots, a^n)$. (In the presence of a basis we shall use such labels quite often, sometimes abbreviating them to a single representative a^i.) But there may be no particular reason for choosing any one ordered basis (as in interplanetary space, for instance) or - worse - good reasons for several different ones. Moreover, it is often easier to see what is going on if a basis is not brought in. However for specific computations a basis is usually essential, so the best approach is to work with a general vector space and bring in or change a basis as and when convenient.

A basis enables us to write down vectors in an n-dimensional vector space X conveniently as n-tuples of numbers, and to specify a map $\underset{\sim}{A}$ to an n-dimensional space Y by what it does to just the set of n basis vectors $\{\underset{\sim}{b}_1, \ldots, \underset{\sim}{b}_n\}$. This involves giving an ordered list of the n vectors

$\underset{\sim}{A}(\underset{\sim}{b}_j) = \underset{\sim}{c}_i a^i_j = \underset{\sim}{c}_1 a^1_j + \ldots + \underset{\sim}{c}_m a^m_j = (a^1_j, \ldots, a^m_j)$ "in coordinates" according to an ordered basis $\underset{\sim}{c}_1, \ldots, \underset{\sim}{c}_m$ for Y. Given this, we know that for a general $\underset{\sim}{x} = \underset{\sim}{b}_j a^j = (x^1, \ldots, x^n)$ "in coordinates" we have

$$\underset{\approx}{A}\underset{\sim}{x} = \underset{\sim}{A}(\underset{\sim}{b}_j x^j) = (\underset{\approx}{A}\underset{\sim}{b}_j)x^j = (a^1_j, \ldots, a^m_j)x^j = (a^1_j x^j, \ldots, a^m_j x^j).$$

Thus $\underset{\sim}{A}$ is specified in this choice of coordinates by the mn numbers a^i_j. It is convenient to lay these out in the m by n rectangle, or <u>matrix</u>

$$\begin{bmatrix} a^1_1 & \cdots\cdots\cdots & a^1_n \\ a^2_1 & \cdots\cdots\cdots\cdots & \\ \cdots\cdots & \cdots\cdots\cdots & \cdots \\ \cdots\cdots & \cdots\cdots\cdots & \cdots \\ a^m_1 & \cdots\cdots\cdots & a^m_n \end{bmatrix}; \quad [a^i_j], \quad [\underset{\sim}{A}], \text{ or } A \text{ for short.}$$

If we have not already labelled the entries of, say, $[\underset{\sim}{B}]$, and want to refer to its entry in the i-th row and j-th column we shall call it $[\underset{\sim}{B}]^i_j$. Notice that the columns here are just the vectors $\underset{\sim}{A}(\underset{\sim}{b}_j)$, written in "$\underset{\sim}{c}_1, \ldots, \underset{\sim}{c}_m$ coordinates". If in a similar way the vector $\underset{\sim}{x} = (x^1, \ldots, x^n)$ in "$\underset{\sim}{b}_1, \ldots, \underset{\sim}{b}_n$ coordinates" is written as a column matrix

$\begin{pmatrix} x^1 \\ \vdots \\ x^n \end{pmatrix}$, then the rule for finding $\underset{\approx}{A}\underset{\sim}{x}$ in coordinates is exactly the rule for "matrix multiplication" (cf. also Ex.2). By 2.04, once we have chosen bases every map $\underset{\sim}{A}$ has a matrix A and every matrix defines a map.

If we define, in terms of these bases for X and Y, the mn linear maps $\underset{\sim}{L}^j_i$ such that

$$\underset{\sim}{L}_i^j(x^1\underset{\sim}{b}_1 + \ldots + x^n\underset{\sim}{b}_n) = x^i\underset{\sim}{c}_j$$

with the matrix for $\underset{\sim}{L}_i^j$ being

$$\begin{bmatrix} 0 & \cdots & 0 & \cdots & 0 \\ \vdots & & \vdots & & \\ & & 0 & & \\ 0 & \cdots\ 0 & 1 & 0\ \cdots & 0 \\ & & 0 & & \\ \vdots & & \vdots & & \vdots \\ 0 & \cdots & 0 & \cdots & 0 \end{bmatrix} \leftarrow \text{j-th row}$$

$$\uparrow$$

i-th column

we get a basis for $L(X;Y)$ since

$$\underset{\sim}{A} = a_j^i\,\underset{\sim}{L}_i^j$$

using the usual addition for maps (2.01; cf. also Ex.3).

Thus the a_j^i are just the components of $\underset{\sim}{A}$, considered as a vector in the mn-dimensional space $L(X;Y)$, with respect to the basis induced by those chosen for X and Y. Notice that we have proved, for general finite-dimensional X and Y, that

$$\dim(L(X;Y)) = \dim X \dim Y.$$

The identity on X, regarded as a map from "X with basis β" to "X with basis β" must always have the matrix

$$\begin{bmatrix} 1 & 0 & \dots\dots\dots & 0 \\ 0 & 1 & \dots\dots\dots & \\ 0 & & 1 \quad \dots\dots & \\ \dots\dots\dots & & 1 & \dots \\ 0 & \dots\dots\dots & & 1 \end{bmatrix} = [\delta^i_j], \text{ whatever } \beta \text{ is.}$$

All that is involved is the use of the same basis at each "end" of the identity map. This matrix is therefore called the <u>identity matrix</u>.

Notice that the matrix representing a map from X to Y depends on the particular basis chosen for each. If several bases have got involved, it is sometimes useful to label a matrix representation according to the particular bases we are using. Thus we write the matrix for $\underset{\sim}{A}$, via bases β,β' for X,Y respectively, as $[a^i_j]^{\beta'}_{\beta}$. Then, if we have the matrix $[b^r_s]^{\beta''}_{\beta'}$ similarly representing $\underset{\sim}{B}:Y \to Z$, the representations fit nicely and we have $\underset{\sim}{BA}$ represented by $[b^r_s a^s_j]^{\beta''}_{\beta} = [b^r_s]^{\beta''}_{\beta'}[a^i_j]^{\beta'}_{\beta}$, with the basis β' "summed over" and vanishing in the final result like the number s or i it is indexed by. If two different bases for Y are involved in defining the two matrices we can still algebraically "multiply" them but we cannot expect it to mean very much. For this and other purposes, we need to be able to change basis.

2.08 Change of basis

If we have bases β,β' for X, changing from β to β' involves simply looking at the identity map $\underset{\sim}{I}:X \to X$ as a map from "X with basis β" (call it (X,β) for short) to (X,β'). This we can represent, just as in the last section, by $[\underset{\sim}{I}]^{\beta'}_{\beta}$. That is a matrix whose columns are the vector $\underset{\sim}{I}(\underset{\sim}{b}_i)$, for $\underset{\sim}{b}_i \in \beta$, written in β' - coordinates. But as $\underset{\sim}{I}(\underset{\sim}{b}_i) = \underset{\sim}{b}_i$, this just means the coordinates of the vectors in β in terms of the basis β'. Multiplying the column matrix $[\underset{\sim}{x}]^{\beta}$, representing $\underset{\sim}{x}$ according to β, by the $n \times n$ matrix $[\underset{\sim}{I}]^{\beta'}_{\beta}$ gives the column matrix representing $\underset{\sim}{x}$ according to β':

$$[\underset{\sim}{I}]^{\beta'}_{\beta}[\underset{\sim}{x}]^{\beta} = [\underset{\sim}{x}]^{\beta'} .$$

There is a sneaky point here: most often when changing bases you are given the new basis vectors, β', in terms of the old basis β, rather than the other way about. Putting these n-tuples of numbers straight in as columns of a matrix gives you not the matrix $[\underset{\sim}{I}]^{\beta'}_{\beta}$ of the change you want, but the matrix $[\underset{\sim}{I}]^{\beta}_{\beta'}$ for changing back. To get $[\underset{\sim}{I}]^{\beta'}_{\beta}$ you need to find the <u>inverse</u> of $[\underset{\sim}{I}]^{\beta}_{\beta'}$, since clearly

$$[\underset{\sim}{I}]^{\beta}_{\beta'}[\underset{\sim}{I}]^{\beta'}_{\beta} = [\underset{\sim}{I}]^{\beta}_{\beta} = [\delta^i_j] = [\underset{\sim}{I}]^{\beta'}_{\beta'} = [\underset{\sim}{I}]^{\beta'}_{\beta}[\underset{\sim}{I}]^{\beta}_{\beta'} .$$

(Fortunately this inversion is one computation we shall not need to do explicitly; we shall just denote the inverse of any $[a^i_j]$, if it has one, by $[\tilde{a}^j_i]$, "defining the $\tilde{a}^j_i$'s as the solutions of the n equations $\tilde{a}^j_i a^i_k = \delta^j_k$". This is a physicist's habit, except for the ~'s we have put in. In many physics books and articles you have to remember that "a^i_j when j = 2, i = 3" is not the same as "a^j_i when j = 3 and i = 2". If you find that peculiar, put the ~'s in yourself.) This need for the inverse is somewhat unexpected when first met and you should get as clear as you can where it comes from. The nineteenth century workers were really bogged down in it, in the absence of the right pictures. The worst pieces of language we are stuck with in tensor analysis started right there. We discuss this further in III.1.07 and VII.4.04.

It is important to be conscious that although matrix multiplication generalises the ordinary kind, each entry $\tilde{a}^j_i$ of $[a^i_j]^{\leftarrow}$ depends on the <u>whole</u> matrix $[a^i_j]$. It is not just the multiplicative inverse $(a^i_j)^{-1}$ (as is emphasised by Ex.7). This point does not seem deep when we are discussing only linear algebra, but in the differential calculus of several variables it has sometimes caused real confusion (see VII. 4.04(2)).

So, $[I]_{\beta}^{\beta'}$ changes the representation of a vector. To change that of an operator, so as to apply it to vectors given in terms of β', just change the vectors to the old coordinates, operate, and change back:

$$[\underset{\sim}{A}]_{\beta'}^{\beta'} = [\underset{\sim}{I}]_{\beta}^{\beta'}[\underset{\sim}{A}]_{\beta}^{\beta}[\underset{\sim}{I}]_{\beta'}^{\beta}$$

or equivalently $a_j^i = b_k^i a_l^k b_j^l$, where $b_k^i \delta_l^k b_j^l = \delta_j^i$.

When two matrices are related by an equation of this kind, $P = RQ\overset{\leftarrow}{R}$ for some invertible matrix R, they are called <u>similar</u>. Thus we have shown that the matrices representing a map according to different bases are similar. Conversely, any pair of similar matrices can be obtained as representations of the same map (Ex. 6), so the two concepts correspond precisely.

2.09 Definition

The <u>kernel</u> $\ker\underset{\sim}{A}$ of $\underset{\sim}{A}: X \to Y$ is the subspace $\{\underset{\sim}{x} \in X \mid \underset{\sim}{A}\underset{\sim}{x} = \underset{\sim}{0}\}$, of singular vectors of $\underset{\sim}{A}$. Note that by Ex. 1 (an easy but very important exercise), $\underset{\sim}{A}$ is injective if and only if $\ker\underset{\sim}{A} = \{\underset{\sim}{0}\}$.

The <u>image</u> $\underset{\sim}{A}X$ of $\underset{\sim}{A}$ is the subspace $\{\underset{\sim}{y} \in Y \mid \underset{\sim}{y} = \underset{\sim}{A}\underset{\sim}{x}$ for some $\underset{\sim}{x} \in X\}$. (cf. Ex.4)

The <u>nullity</u> $n(\underset{\sim}{A})$ of $\underset{\sim}{A}$ is $\dim(\ker\underset{\sim}{A})$, the dimension of the kernel.

The <u>rank</u> $r(\underset{\sim}{A})$ of $\underset{\sim}{A}$ is $\dim(\underset{\sim}{A}X)$, the dimension of the image.

Geometrically, in the case $\underset{\sim}{A}: R^2 \to R^2$: $(x,y) \rightsquigarrow (2(x-y),(x-y))$, for example, see Fig. 2.2.

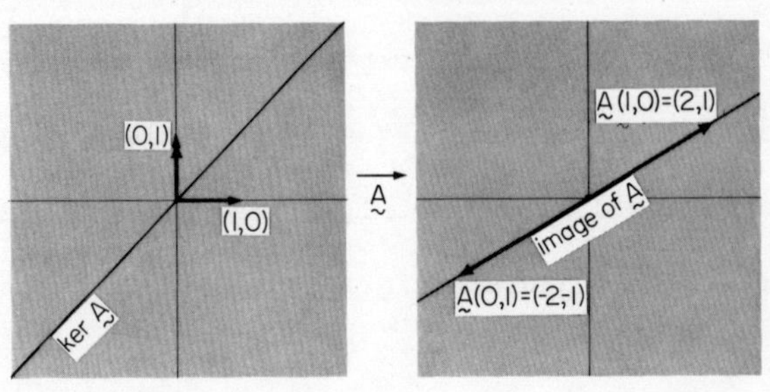

Fig. 2.2

The image and the kernel are as shown, and the rank and the nullity each 1. This illustrates a general proposition; the number of directions you squash flat, plus the number of directions you are left pointing in, is the number of directions you started with. More formally:

2.10 Theorem

For any finite-dimensional vector space X and linear map $\underset{\sim}{A}:X \to Y$, we have

$$n(\underset{\sim}{A}) + r(\underset{\sim}{A}) = \dim X.$$

Proof

An exercise in shuffling bases, and left as such. (Ex.5) ∎

2.11 Corollary

A linear map $\underset{\sim}{A}:X \to Y$ is non-singular if and only if $r(\underset{\sim}{A}) = \dim X$. ∎

2.12 Corollary

An operator $\underset{\sim}{A}:X \to X$ is non-singular if and only if $\underset{\sim}{A}$ is an isomorphism. ∎

2.13 Corollary

Suppose $\dim X = \dim Y$, and $\underset{\sim}{A}:X \to Y$ is linear. Then

$\underset{\sim}{A}$ is an isomorphism if and only if it is injective

AND

$\underset{\sim}{A}$ is an isomorphism if and only if it is surjective.

∎

Exercises II.2.

1) A linear map $\underset{\sim}{A}:X \to Y$ is non-singular if and only if $\underset{\sim}{A}$ is injective (if $\underset{\sim}{A}\underset{\sim}{x} = \underset{\sim}{A}\underset{\sim}{x}'$ what is $\underset{\sim}{A}(\underset{\sim}{x} - \underset{\sim}{x}')$?).

2) If, with bases chosen for X,Y,Z we have maps $\underset{\sim}{A}:X \to Y$ and $\underset{\sim}{B}:Y \to Z$ represented by matrices $[a^i_j]$, $[b^r_s]$, then their composite $\underset{\sim}{B}\underset{\sim}{A}:X \to Z$ is represented by the matrix $[b^r_s a^s_i]$.

3) If, with bases chosen for X and Y the maps $\underset{\sim}{A},\underset{\sim}{B}$ from X to Y have matrices $[a^i_j]$, $[b^i_j]$ respectively, then the matrix of $\underset{\sim}{A} + \underset{\sim}{B}:X \to Y : \underset{\sim}{x} \rightsquigarrow \underset{\sim}{A}\underset{\sim}{x} + \underset{\sim}{B}\underset{\sim}{x}$ is $[a^i_j + b^i_j]$.

4a) The kernel of a linear map $X \to Y$ is a subspace (not just a subset) of X.

b) The image of a linear map $X \to Y$ is a subspace of Y.

5) If $\beta = \{\underset{\sim}{b}_1,\ldots,\underset{\sim}{b}_n\}$ is a basis for X, $\omega = \{\underset{\sim}{d}_1,\ldots,\underset{\sim}{d}_k\}$ is a basis for $X' \subseteq X$ and $\underset{\sim}{A}:X \to Y$ is a linear map, then the following hold.

a) There is a basis $\{\underset{\sim}{d}_1,\ldots,\underset{\sim}{d}_n\}$ for X including all the vectors in ω (cf. Ex. 1.7);

b) the vectors $\underset{\sim}{A}\underset{\sim}{d}_1,\ldots,\underset{\sim}{A}\underset{\sim}{d}_n$ span the image of $\underset{\sim}{A}$;

c) if $\underset{\sim}{y}$ is in the image of $\underset{\sim}{A}$, as the image of both the vectors $\underset{\sim}{x}$ and $\underset{\sim}{x}'$ in X (so $\underset{\sim}{y} = \underset{\sim}{A}\underset{\sim}{x} = \underset{\sim}{A}\underset{\sim}{x}'$), then $\underset{\sim}{x}' = \underset{\sim}{x} + \underset{\sim}{z}$, where $\underset{\sim}{z}$ is in the kernel of $\underset{\sim}{A}$.

d) If $\ker\underset{\sim}{A} = X'$, deduce from (b) that the vectors $\underset{\sim}{A}\underset{\sim}{d}_{k+1},\ldots,\underset{\sim}{A}\underset{\sim}{d}_n$ span the image of $\underset{\sim}{A}$, and thence and from (c) that they are a basis for it.

e) Deduce Theorem 2.10.

6a) Deduce from 5(d) that if $\beta = \{\underset{\sim}{b}_1,\ldots,\underset{\sim}{b}_n\}$ is a basis for X, and $\underset{\sim}{A}:X \to Y$ is an isomorphism, then $\underset{\sim}{A}\beta = \{\underset{\sim}{A}\underset{\sim}{b}_1,\ldots,\underset{\sim}{A}\underset{\sim}{b}_n\}$ is a basis for Y.

b) If β is a basis for X and $\underset{\sim}{A}:X \to X$ an isomorphism, the change of basis matrix $[\underset{\sim}{I}]^{\underset{\sim}{A}\beta}_{\beta}$ is exactly the matrix $([\underset{\sim}{A}]^{\beta}_{\beta})^{\leftarrow}$.

c) Hence, if matrices P,Q are similar by $P = AQ\overset{\leftarrow}{A}$, and $\underset{\sim}{P},\underset{\sim}{Q},\underset{\sim}{A}$ are the maps $X \to X$ defined by P,Q,A via the basis β, then Q represents $\underset{\sim}{P}$ in the basis $\underset{\sim}{A}\beta$.

7) Defining $A = \begin{bmatrix} 1 & 2 \\ 1 & 1 \end{bmatrix}$, $B = \begin{bmatrix} 1 & \frac{1}{2} \\ 1 & 1 \end{bmatrix}$, $C = \begin{bmatrix} -1 & 2 \\ 1 & -1 \end{bmatrix}$, show that

$$AC = CA = \begin{bmatrix} 1 & 0 \\ 0 & 1 \end{bmatrix}, \quad AB = \begin{bmatrix} 3 & 2\frac{1}{2} \\ 2 & 1\frac{1}{2} \end{bmatrix}, \quad BA = \begin{bmatrix} 1\frac{1}{2} & 2\frac{1}{2} \\ 2 & 3 \end{bmatrix}.$$

3. OPERATORS

Operators (linear maps from a vector space to itself) have a very special role. Among the definitions involving only this special class of maps are

3.01 Definition

An operator on X which is an isomorphism is called an <u>automorphism</u>. The set GL(X) of all automorphisms of X form a (Lie) group, the <u>general linear group</u> of X, under composition (cf. Ex. 1.10). (<u>Not</u> under addition; $\underset{\sim}{I} + (-\underset{\sim}{I}) = \underset{\sim}{0}$, which is not an automorphism.)

3.02 Definition

An operator $\underset{\sim}{A}: X \to X$ is <u>idempotent</u> if $\underset{\sim}{A}\underset{\sim}{A} = \underset{\sim}{A}$. Essentially, this means that $\underset{\sim}{A}$ is projecting X onto a subspace, as in the figure (where the ⇝'s indicate the movement under $\underset{\sim}{A}$ of a sample of vectors), and a vector having arrived in the subspace L is then left alone by further application of $\underset{\sim}{A}$. Hence we shall often call $\underset{\sim}{A}$ a <u>projection</u> onto $\underset{\sim}{A}(X)$. An important class of such operators will concern us in Chapter IV.

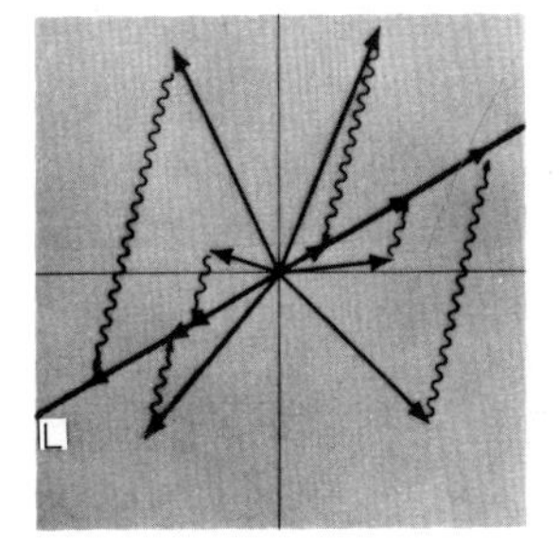

Fig. 3.1

3.03 Definition

A vector $\underset{\sim}{x} \neq \underset{\sim}{0}$ is an eigenvector of $\underset{\sim}{A}: X \to X$ if $\underset{\sim}{A}\underset{\sim}{x} = \underset{\sim}{x}\lambda$ for some scalar λ. Then λ is an eigenvalue of $\underset{\sim}{A}$, and $\underset{\sim}{x}$ is an eigenvector belonging to λ. The set of eigenvectors belonging to λ, together with $\underset{\sim}{0}$, is a subspace of X (easily checked), the eigenspace belonging to λ.

(Eigenvectors are sometimes called characteristic vectors, and correspondingly eigenvalues are called characteristic roots or values. This conveys the feel of the German "eigen-" but is more cumbersome and less sonorous. However, ...values are almost always denoted by λ, just as unknowns are by x and beautiful Russian spies by Olga.)

We have already met one example; $\ker\underset{\sim}{A}$ is the eigenspace belonging to 0. Another is familiar; a rotation in three dimensions must leave some direction - the axis of rotation - fixed, and so we have eigenvectors in that direction belonging to the eigenvalue 1. If $\underset{\sim}{A}$ is the identity, then the whole of X belongs to the eigenvalue 1. Reflection in the line x = y is

$$\underset{\sim}{A} : R^2 \to R^2 : (x,y) \rightsquigarrow (y,x)$$

in this case we have eigenvalues ±1.

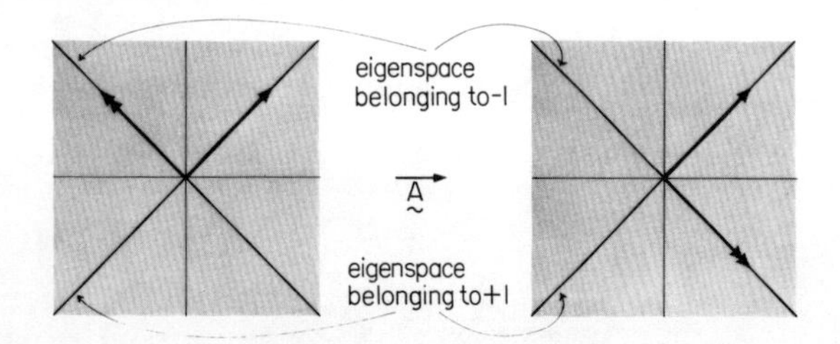

Fig. 3.2

3.04 Definition

For L(X;X) we have not only addition and scalar multiplication as for L(X;Y) but a "multiplication" defined by composition. For any operators $\underset{\sim}{A},\underset{\sim}{B}$ their composites $\underset{\sim}{A}\underset{\sim}{B}$ and $\underset{\sim}{B}\underset{\sim}{A}$ are again operators on X. The operator algebra of X is the set L(X;X) with these three operations. This is an "algebra with identity" : for all

$\underset{\sim}{A},\underset{\sim}{B},\underset{\sim}{C} \subseteq L(X;X)$, $a \in R$ we have

$$\underset{\sim}{A}(\underset{\sim}{B}\underset{\sim}{C}) = (\underset{\sim}{A}\underset{\sim}{B})\underset{\sim}{C} \qquad \text{(associativity of composition)}$$

$$\left.\begin{array}{l}\underset{\sim}{A}(\underset{\sim}{B}+\underset{\sim}{C}) = \underset{\sim}{A}\underset{\sim}{B} + \underset{\sim}{A}\underset{\sim}{C} \\ (\underset{\sim}{A}+B)\underset{\sim}{C} = \underset{\sim}{A}\underset{\sim}{C} + \underset{\sim}{B}\underset{\sim}{C}\end{array}\right\} \text{(distributivity)}$$

$$a(\underset{\sim}{A}\underset{\sim}{B}) = (a\underset{\sim}{A})\underset{\sim}{B} = \underset{\sim}{A}(a\underset{\sim}{B})$$

$$\underset{\sim}{A}\underset{\sim}{I} = \underset{\sim}{A} = \underset{\sim}{I}\underset{\sim}{A} \qquad \text{(composition has an identity)}$$

as for numbers. Unlike that of numbers, this multiplication is <u>not</u> commutative ($\underset{\sim}{A}\underset{\sim}{B} \neq \underset{\sim}{B}\underset{\sim}{A}$ in general). Either multiply $\begin{bmatrix} 0 & 0 \\ 0 & 1 \end{bmatrix}$ and $\begin{bmatrix} 0 & 0 \\ 1 & 0 \end{bmatrix}$ both ways round, if you are best convinced by algebra, or wave your hands in the air: If $\underset{\sim}{A}$ is "rotation through 90° about a vertical axis", and $\underset{\sim}{B}$ is "rotation through 90° about a northward axis", both clockwise, experiments with your elbow as origin will show that $\underset{\sim}{A}\underset{\sim}{B} \neq \underset{\sim}{B}\underset{\sim}{A}$.

There are two important functions from $L(X;X)$ to R, one preserving multiplication and the other addition; the determinant and the trace.

3.05 Determinants

The determinant function may be regarded in several ways. Algebraically, one may start with either matrices or linear maps. We shall give here a geometric account of it, with the matrix proof of its properties (the least instructive but most direct) indicated in the exercises. (Manipulations of this kind are unilluminating to see, but essential practice.) In Ex V. 1.11 it emerges from some rather more sophisticated algebra, which corresponds more closely to the geometry below and amounts to a rigorous version of the same ideas.

Consider the map $\underset{\sim}{A}:R^2 \to R^2$ with matrix $\begin{bmatrix} a & c \\ b & d \end{bmatrix}$ in the standard basis, and examine its effect on the unit square. (Fig. 3.3) The area of the unit square is 1; the area of the parallelogram to which it is taken may be found, for instance, by adding and subtracting rectangles and right-angled triangles.

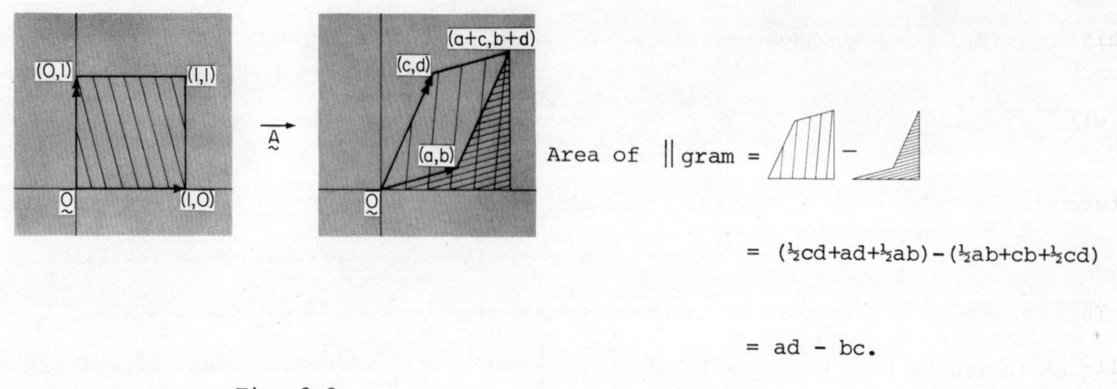

$$= (\tfrac{1}{2}cd+ad+\tfrac{1}{2}ab)-(\tfrac{1}{2}ab+cb+\tfrac{1}{2}cd)$$

$$= ad - bc.$$

Fig. 3.3

Now any other shape may be approximated by squares. The area of these squares is evidently changed by $\underset{\sim}{A}$ in the same proportion as the unit square, so taking a high-handed Ancient Greek attitude to limits it is clear that the area of <u>any</u> figure is multiplied by the same quantity (ad-bc), which we shall call $\det\underset{\sim}{A}$.

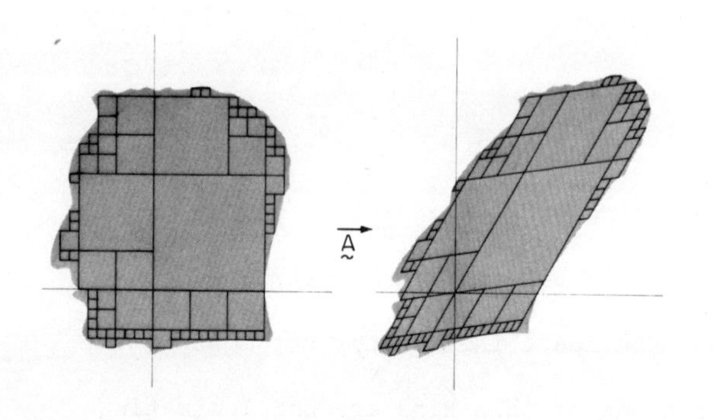

Fig. 3.4

Thus "what $\underset{\sim}{A}$ does to area" is to multiply it by $\det\underset{\sim}{A}$, which quantity therefore, although given as (ad-bc) in terms of the entries for a matrix for $\underset{\sim}{A}$, does not depend as those entries do on the particular basis chosen for R^2. Conveniently, not only the number $\det\underset{\sim}{A}$ but the formula for it are independent of the basis. For any basis at all,

$$\text{if } [\underset{\sim}{A}] = \begin{bmatrix} a^1_1 & a^1_2 \\ a^2_1 & a^2_2 \end{bmatrix} \text{ then}$$

$$\det\underset{\sim}{A} = a^1_1 a^2_2 - a^1_2 a^2_1. \qquad \text{(Equation D2)}$$

(This is a manipulative algebraic fact, and as such left to the exercises.)

It will often be useful to write the determinant of a matrix $A = \begin{bmatrix} a^1_1 & a^1_2 \\ a^2_1 & a^2_2 \end{bmatrix}$ (or the determinant of any map A represents) as $\begin{vmatrix} a^1_1 & a^1_2 \\ a^2_1 & a^2_2 \end{vmatrix}$.

The alert reader may have noticed that we have sneakily assumed that "area" is well-defined, which for R^2 is true, but how about an arbitrary two dimensional space? In fact, more than one measure of "area" _is_ possible, but the "multilinear" ones appropriate to a vector space are all scalar multiples one of another (Ex.V. 1.11), so "what $\underset{\sim}{A}$ does to area" is independent of which measure we pick - they are all multiplied in the same proportion.

In a similar fashion, if X is 3-dimensional "what $\underset{\sim}{A}$ does to volume" is naturally independent of basis, and is represented by the equally invariant formula

$$\det\underset{\sim}{A} = a^1_1 \begin{vmatrix} a^2_2 & a^2_3 \\ a^3_2 & a^3_3 \end{vmatrix} - a^1_2 \begin{vmatrix} a^2_1 & a^2_3 \\ a^3_1 & a^3_3 \end{vmatrix} + a^1_3 \begin{vmatrix} a^2_1 & a^2_2 \\ a^3_1 & a^3_2 \end{vmatrix}, \text{ where } [\underset{\sim}{A}] = \begin{vmatrix} a^1_1 & a^1_2 & a^1_3 \\ a^2_1 & a^2_2 & a^2_3 \\ a^3_1 & a^3_2 & a^3_3 \end{vmatrix}.$$

Or, expanded in detail,

$$\det\underset{\sim}{A} = a^1_1 a^2_2 a^3_3 - a^1_1 a^2_3 a^3_2 - a^1_2 a^2_1 a^3_3 + a^1_2 a^2_3 a^3_1 + a^1_3 a^2_1 a^3_2 - a^1_3 a^2_2 a^3_1 .$$

(Equation D3)

This can be checked by "Euclidean geometry" calculations of the volume of the parallelepiped to which $\underset{\sim}{A}$ takes the unit cube (Fig. 3.5).

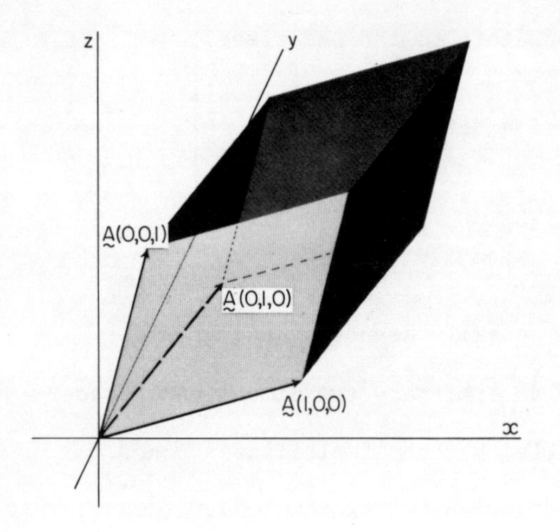

Fig. 3.5

In four dimensions, starting with the obvious definition for "hypervolume" of a "hyperbrick" by multiplying all four edge lengths together, the same approach leads to a determinant for $\underset{\sim}{A}$. In any basis we have

$$\det\underset{\sim}{A} = \begin{vmatrix} a^1_1 & a^1_2 & a^1_3 & a^1_4 \\ a^2_1 & a^2_2 & a^2_3 & a^2_4 \\ a^3_1 & a^3_2 & a^3_3 & a^3_4 \\ a^4_1 & a^4_2 & a^4_3 & a^4_4 \end{vmatrix}$$

$$= a^1_1 \begin{vmatrix} a^2_2 & a^2_3 & a^2_4 \\ a^3_2 & a^3_3 & a^3_4 \\ a^4_2 & a^4_3 & a^4_4 \end{vmatrix} - a^1_2 \begin{vmatrix} a^2_1 & a^2_3 & a^2_4 \\ a^3_1 & a^3_3 & a^3_4 \\ a^4_1 & a^4_3 & a^4_4 \end{vmatrix} + a^1_3 \begin{vmatrix} a^2_1 & a^2_2 & a^2_4 \\ a^3_1 & a^3_2 & a^3_4 \\ a^4_1 & a^4_2 & a^4_4 \end{vmatrix} - a^1_4 \begin{vmatrix} a^2_1 & a^2_2 & a^2_3 \\ a^3_1 & a^3_2 & a^3_3 \\ a^4_1 & a^4_2 & a^4_3 \end{vmatrix}, \quad \text{(Equation D4)}$$

and so forth for higher dimensions. If you are ready to believe that $\det\underset{\sim}{A}$ is only and exactly "what $\underset{\sim}{A}$ does to volume" you can ignore the next section, as being preparation for proving the obvious. The important thing about determinants is that they exist and have nice properties, not the algebra which justifies the properties.

3.06 Formulae

The general way to find the determinant of any operator $\underset{\sim}{A}$ from an n × n matrix representing it should now be clear: go along the top row taking alternately + and - each entry times the determinant of the (n-1) × (n-1) matrix got from $\underset{\sim}{A}$ by leaving out the top row and the column that the entry is in (Fig. 3.6). (Notice that the number of multiplications needed altogether is n! which increases rather fast with n; for example 5! = 120, 7! = 5,040. This is why finding the determinant of matrices bigger than 4 × 4 occurs as an exercise only in computer textbooks. Reducing the number of multiplications is an art in itself.) This describes well how to compute it, though rather uneconomically, but does not lead straight to a formula convenient for proving general properties of n × n determinants. To get such a formula, it is best to back off and approach matters a little more symmetrically.

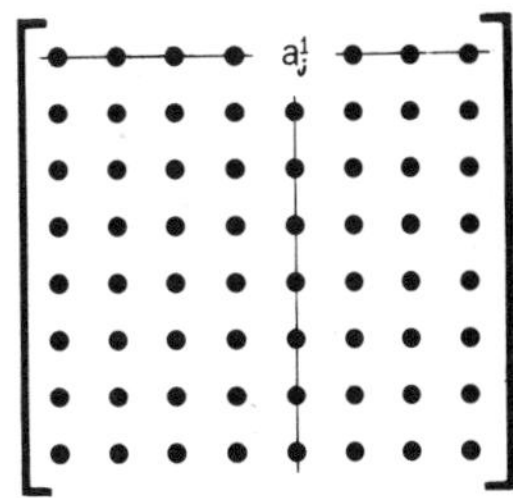

Fig. 3.6

Firstly, notice that in any one of the actual multiples that are involved (in for instance D3 above) no two of the entries multiplied are in the same row or the same column. Typically, they appear arranged like the asterisks in Fig. 3.7 - exactly one entry in each row and column. Moreover, all such arrangements of n entries do get multiplied up and added, with either a + or a - sign, to get the determinant. If they all had + signs, we'd be home, but we must find a systematic way to indicate which multiple has which sign. Now, since each such set of entries, M say, has exactly one member in each column, we

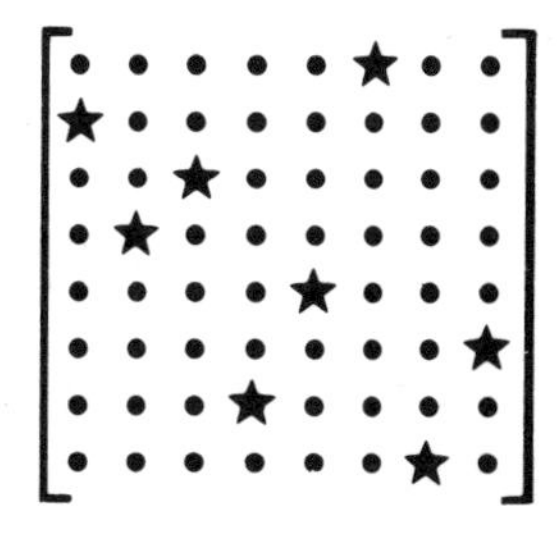

Fig. 3.7

can list M in the order of the columns containing its members:

$$M = \left\{a_1^{m_1}, \ldots, a_n^{m_n}\right\}$$

say, where m_i means "the number of the row in which the element of M in column i sits". Clearly, M is completely specified by $m_1, \ldots, m_n$, or to put it a little differently, by the function

$$m: \{1, \ldots, n\} \to \{1, \ldots, n\} : i \rightsquigarrow m_i.$$

Since the elements of M are all in different rows, m is a bijection from the finite set $\{1, \ldots, n\}$ to itself - that is, a <u>permutation</u> of the numbers 1, ..., n. (This and its properties could also be related to the geometry, at the expense of greater space. At the moment we want the quickest possible algebraic back-up for the geometry that will follow this section.) Now (Ex. 1a), any permutation can be built up by successively switching neighbouring pairs (1,2,3,4,5 goes to 1,2,4,3,5 for example), and this can generally be done in several different ways. Moreover (Ex. 1b), the number of such switches required in any such building up is for a given permutation either always even or always odd. This lets us define the <u>sign</u> of m:

$$\operatorname{sgn}(m) = \begin{Bmatrix} 1 = (-1)^{\text{even}} \\ -1 = (-1)^{\text{odd}} \end{Bmatrix} \text{ according as } m \text{ is an } \begin{Bmatrix} \text{even} \\ \text{odd} \end{Bmatrix} \text{ combination of switches.}$$

Finally, it turns out that the sign of m is exactly the sign we want for our multiple. (Ex. 1c,d) So if we denote the set of all permutations of 1, ..., n by S_n (it is in fact a group (cf. Ex. 1.10) - the <u>symmetric group</u> on 1, ..., n) we can at last write down a nice closed formula

$$\det[a^i_j] = \sum_{m \in S_n} \operatorname{sgn}(m) a_1^{m_1} a_2^{m_2} \cdots a_n^{m_n}$$

for the determinant of a matrix.

With this we can prove algebraically the important properties of determinants that are geometrically obvious, but harder to prove rigorously. (Exercises 2-5). Returning to the geometrical viewpoint, we can see these properties directly.

3.07 Lemma

$\det \underset{\sim}{I} = 1$.

Proof

Either calculate from the matrix $[\delta^i_j]$, or observe that $\underset{\sim}{I}$ leaves volume, along with everything else, unchanged. ■

3.08 Theorem (The Product Rule)

For any two operators $\underset{\sim}{A}, \underset{\sim}{B}$ on X,

$$\det(\underset{\sim}{A}\underset{\sim}{B}) = \det\underset{\sim}{A}\,\det\underset{\sim}{B}.$$

Proof

$\det\underset{\sim}{A}$ is what applying $\underset{\sim}{A}$ multiplies volumes by.

$\det\underset{\sim}{B}$ is what applying $\underset{\sim}{B}$ multiplies volumes by.

$\det(\underset{\sim}{A}\underset{\sim}{B})$ is what the operation of (applying $\underset{\sim}{B}$ and then applying $\underset{\sim}{A}$) multiplies volumes by.

End of proof. (Compare Ex. 2). ■

3.09 Lemma

If $\underset{\sim}{A}:X \to X$ and $\dim X = n$, then $\det(a\underset{\sim}{A}) = a^n \det\underset{\sim}{A}$.

Proof

$\det(a\underset{\sim}{A}) = \det(a(\underset{\sim}{I}\underset{\sim}{A})) = \det((a\underset{\sim}{I})\underset{\sim}{A}) = \det(a\underset{\sim}{I})\det\underset{\sim}{A}$.

Evidently $a\underset{\sim}{I}$, which multiplies the length of each side of the n-cube by a, multiplies its volume by a^n.

3.10 Theorem

An operator $\underset{\sim}{A}$ on X is an automorphism if and only if $\det\underset{\sim}{A} \neq 0$.

Proof

If $\underset{\sim}{A}$ is an automorphism, then there exists $\underset{\sim}{A}^{\leftarrow}$ such that $\underset{\sim}{A}\underset{\sim}{A}^{\leftarrow} = \underset{\sim}{I}$.

Hence

$$\det\underset{\sim}{A}\det(\underset{\sim}{A}^{\leftarrow}) = \det(\underset{\sim}{A}\underset{\sim}{A}^{\leftarrow}) = \det\underset{\sim}{I} = 1 \quad \text{and thus } \det\underset{\sim}{A} = \frac{1}{\det(\underset{\sim}{A}^{\leftarrow})} \neq 0.$$

Conversely, if $\underset{\sim}{A}$ is singular, the unit cube is squashed flat by $\underset{\sim}{A}$ in the direction of some singular vector. Thus its image has zero volume, so $\det\underset{\sim}{A} = 0$. (This argument is made rigorous, via algebra, in Ex. 4.) Thus if $\det\underset{\sim}{A} \neq 0$, $\underset{\sim}{A}$ is non-singular and hence by 2.12 $\underset{\sim}{A}$ is an automorphism.

3.11 Orientation

By 3.10 all automorphisms have non-zero determinant. Hence they fall naturally into two classes - those with positive and those with negative determinant. Now if X is 1-dimensional, $\underset{\sim}{A}:X \to X$ reduces to multiplication by some scalar a. The determinant $\det\underset{\sim}{A}$ is just a, and is positive

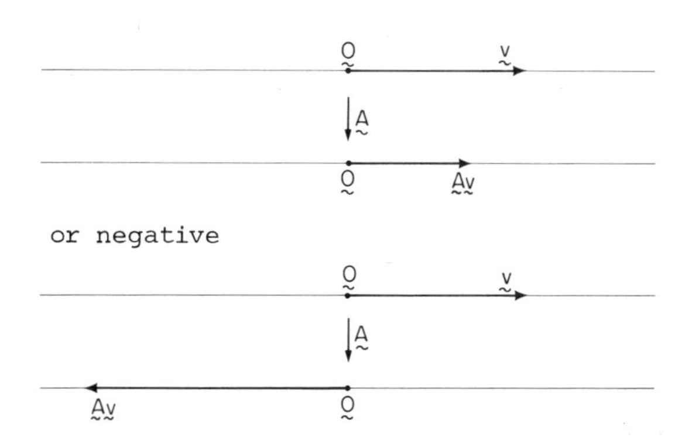

Fig. 3.8

according as $\mathbf{A}$ preserves or reverses direction. In two dimensions $\det\mathbf{A}$ is positive according as $\mathbf{A}$ merely distorts the unit square into a parallelogram or turns it over as well. In three dimensions, $\det\mathbf{A}$ is positive or negative according to whether $\mathbf{A}$ preserves or exchanges left and right handedness, apart from warping hands. We are led to a general definition:

An automorphism $\mathbf{A}: X \to X$ is <u>orientation preserving</u> or <u>reversing</u> according as $\det\mathbf{A}$ is positive or negative. (A precise definition of "orientation" is given in Ex. XI 3.1)

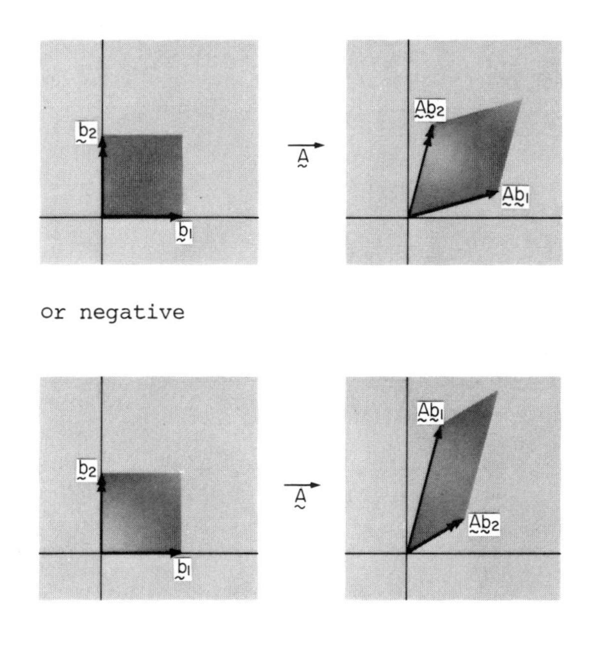

Fig. 3.9

<u>3.12 Remark</u>

If $\mathbf{A}: X \to Y$ is a linear map from X to a <u>different</u> space Y, even with $\dim X = \dim Y$, then $\det\mathbf{A}$ is not defined; we could change "what $\mathbf{A}$ does to volume" by altering our measure of volume at one end but not at the other. However, if we pick ordered bases β, β' for X and Y they define an isomorphism $\mathbf{B}: Y \to X$ (cf. proof of 2.06), and hence a quantity $\det(\underset{\sim}{B}\underset{\sim}{A})$ since $\underset{\sim}{B}\underset{\sim}{A}: X \to X$. This is exactly the result of computing the determinant of the <u>matrix</u> $[\underset{\sim}{A}]_{\beta}^{\beta'}$. Now since $\underset{\sim}{B}$ is an isomorphism, $\underset{\sim}{A}$ is an isomorphism if and only if $\underset{\sim}{B}\underset{\sim}{A}$ is an automorphism ($\underset{\sim}{A}\underset{\sim}{x} = \underset{\sim}{0} \iff \underset{\sim}{B}\underset{\sim}{A}\underset{\sim}{x} = \underset{\sim}{0}$, since $\underset{\sim}{B}$ is non-singular: then apply 2.12), that is if and only if $\det(\underset{\sim}{B}\underset{\sim}{A}) \neq 0$. Thus the determinant of any matrix representing $\underset{\sim}{A}$ remains a valid test for singularity.

If we have a measure of volume already chosen at each end, with a little care $\det\underset{\sim}{A}$ can be reinstated in its full glory as "what $\underset{\sim}{A}$ does to volume". (cf. Ex. V 1.12)

3.13 Characteristic equation

One of det's many uses is concerned with eigenvalues:

$$\begin{aligned}
\lambda \text{ is an eigenvalue of } \underset{\sim}{A} &\iff \underset{\sim}{A}\underset{\sim}{x} = \lambda\underset{\sim}{x} \text{ for some non-zero } \underset{\sim}{x} \\
&\iff \underset{\sim}{A}\underset{\sim}{x} - \lambda\underset{\sim}{x} = \underset{\sim}{0} \text{ for some non-zero } \underset{\sim}{x} \\
&\iff (\underset{\sim}{A} - \lambda\underset{\sim}{I})\underset{\sim}{x} = \underset{\sim}{0} \text{ for some non-zero } \underset{\sim}{x} \\
&\iff \det(\underset{\sim}{A} - \lambda\underset{\sim}{I}) = 0. \quad \text{(by 3.10)}
\end{aligned}$$

Now for any choice of basis, giving a matrix $[a_j^i]$ for $\underset{\sim}{A}$, $\det(\underset{\sim}{A} - \lambda\underset{\sim}{I})$ is a polynomial in λ. Its coefficients are various terms built up from the a_j^i's. Hence λ is an eigenvalue of $\underset{\sim}{A}$ if and only if λ is a real root of the n-th order polynomial equation $\det(\underset{\sim}{A} - \lambda\underset{\sim}{I}) = 0$, which is therefore called the <u>characteristic equation</u> of $\underset{\sim}{A}$. (With real vector spaces, complex roots are irrelevant. For :

$$\begin{bmatrix} 0 & -1 \\ 1 & 0 \end{bmatrix} : \mathbb{R}^2 \to \mathbb{R}^2, \text{ rotation through}$$

90°, has characteristic equation

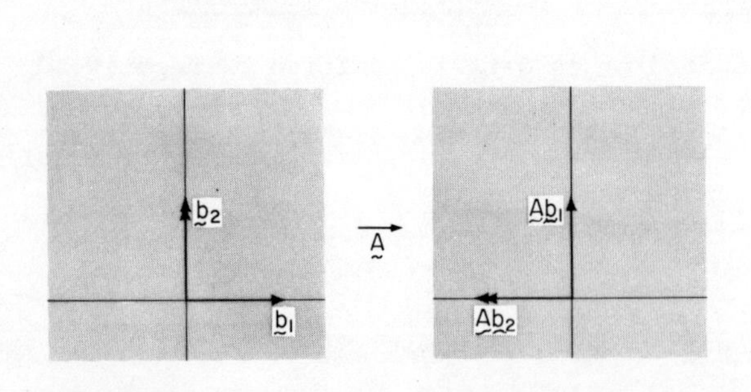

Fig. 3.10

$$\begin{aligned}
0 &= \det\left(\begin{bmatrix} 0 & -1 \\ 1 & 0 \end{bmatrix} - \begin{bmatrix} \lambda & 0 \\ 0 & \lambda \end{bmatrix}\right) \\
&= \det\begin{bmatrix} -\lambda & -1 \\ 1 & -\lambda \end{bmatrix} \\
&= \lambda^2 + 1
\end{aligned}$$

with no real roots. Clearly from the picture there are no eigenvectors or corresponding eigenvalues.)

3.14 Trace

The meaning of the trace of an operator is less clear geometrically than that of the determinant. Algebraically, it is very simply defined: if $[\underset{\sim}{A}] = [a^i_j]$,

$$\underline{\text{trace}}\ \underset{\sim}{A} = \text{tr}\underset{\sim}{A} = a^1_1 + a^2_2 + \ldots + a^n_n$$

$$= a^i_i \text{ in the summation convention.}$$

It is obvious that $\text{tr}(\underset{\sim}{A}+\underset{\sim}{B}) = \text{tr}\underset{\sim}{A} + \text{tr}\underset{\sim}{B}$, and a simple check (Ex. 6) shows that this formula, like that for determinant, gives the same answer regardless of the basis in terms of which $\underset{\sim}{A}$ is expressed.

Trace can partly be thought of by its role in an important special case, where it measures how "close to the identity" an operator is. Each diagonal entry such as a^3_3 is "the 3rd component of the image $\underset{\sim}{A}\underset{\sim}{b}_3$ of the basis vector $\underset{\sim}{b}_3$" in $\underset{\sim}{b}_1,\ldots,\underset{\sim}{b}_n$ coordinates. If $\underset{\sim}{A}$ is a rotation, so keeping all vectors the same length (and while we're assuming we are in a situation with length defined we might as well take the basis vectors to be of unit length), this comes to exactly $\cos\alpha_3$ where α_3 is the angle $\underset{\sim}{b}_3$ has been turned through. (If we have lengths defined, then we have angles, by $c^2 = a^2 + b^2 - 2ab\cos\alpha$ for a triangle.)

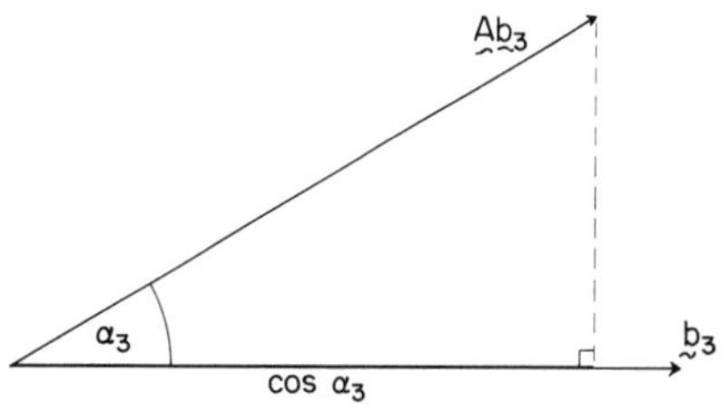

Fig. 3.11

The trace is the sum of these cosines, and thus for a rotation varies from $n = \text{tr}\underset{\sim}{I} = \dim X$ to $-n$. For the rotation in 3.13, all vectors turn through 90°, and the trace is 0 + 0 = 0.

This description is complicated for general operators by the fact that, trivially, $\mathrm{tr}(a\underset{\sim}{A}) = a(\mathrm{tr}\underset{\sim}{A})$, so that the "size" of an operator comes into play; the trace function is the only major one in linear algebra that seems to be genuinely more algebraic than geometric. Like the determinant, it can be defined without reference to a basis (cf. V. 1.12) but this takes more theory than the coordinate approach and in this instance is no more intuitive. (An example in which trace is intimately involved is discussed at length in Chapter IX.6.) If $\underset{\sim}{A}$ is a projection, $\mathrm{tr}\underset{\sim}{A}$ is just the dimension of its image, as is obvious by a convenient choice of basis.

Exercises II.3

1a) Any permutation m can be produced by successively switching neighbouring pairs. (Hint: get the number $\overleftarrow{m}(1)$ into 1st place and proceed inductively.)

b) If $t_1t_2\ldots t_h$ is a composite of neighbour-switches t_i, then $(t_1t_2\ldots t_h)^{\leftarrow} = t_ht_{h-1}\ldots t_1$. Show that if such a composite ends up with everything where it started, h must be even. (Hint: show that any given switch must be used an even number of times.) Deduce that if

$$s_1s_2\ldots s_k = m = t_1t_2\ldots t_h$$

then k+h is even and hence $(-1)^k = (-1)^h$.

c) Check that the signs in equations D2, D3 and D4 coincide with the signs obtained by considering permutations.

d) Prove by induction that this holds in general.

2) Let $[a_j^i],[b_j^i]$ be square matrices with determinants detA, detB, respectively.

a) Show that if $m \in S_n$ then $\mathrm{sgn}(m) = \mathrm{sgn}(m^{-1})$ and deduce that

$$\det A = \sum_{m\in S_n} \mathrm{sgn}(m)a^1_{m_1}a^2_{m_2}\ldots a^n_{m_n}.$$

That is, rows and columns may be interchanged without altering the determinant.

b) Prove that the determinant function on matrices defines a linear function on the space of possible i-th columns for any fixed choice of the other columns, for any $i = 1, \ldots, n$.

c) Consider $[a^i_j]$ as the ordered set of columns $(\underset{\sim}{a}^i_1, \underset{\sim}{a}^i_2, \ldots, \underset{\sim}{a}^i_n)$ where for instance $\underset{\sim}{a}^i_3 = \begin{bmatrix} a^1_3 \\ a^2_3 \\ \vdots \\ a^n_3 \end{bmatrix}$.

For any $m \in S_n$, prove that $\text{sgn}(m) \det A = \det(\underset{\sim}{a}^i_{m_1}, \underset{\sim}{a}^i_{m_2}, \ldots, \underset{\sim}{a}^i_{m_n})$.

d) Prove that if $[c^i_j] = [a^i_k b^k_j]$ then, from part (b)

$$\det C = \det[c^i_j] = \sum_{m \in S_n} b_1^{m_1} b_2^{m_2} \ldots b_n^{m_n} \det(\underset{\sim}{a}^i_{m_1}, \underset{\sim}{a}^i_{m_2}, \ldots, \underset{\sim}{a}^i_{m_n}).$$

e) Deduce Theorem 3.08 : $\det\underset{\sim}{C} = \det\underset{\sim}{A} \det\underset{\sim}{B}$.

3a) Deduce from 3.07 and 3.08 that if $\underset{\sim}{A}$ is invertible then $\det\underset{\sim}{A}^{\leftarrow} = (\det\underset{\sim}{A})^{-1}$.

b) Deduce from (a) and 3.08 that for matrices P,Q,R with R invertible, if $P = RQR^{\leftarrow}$ then $\det P = \det Q$. Hence deduce that if P represents $\underset{\sim}{P}$ according to the basis β, for any other basis β' we have $\det([\underset{\sim}{P}]^{\beta'}_{\beta'}) = \det P$.

4) If $\underset{\sim}{A}: X \to X$ is singular then X has an ordered basis β whose first member is a singular vector for $\underset{\sim}{A}$. Deduce by considering $[A]^{\beta}_{\beta}$ that $\det\underset{\sim}{A} = 0$.

5) Prove, from the general formula, that for any $n \times n$ matrix $A = [a^i_j]$:

a) $\det(bA) = b^n \det A$, where $bA = [ba^i_j]$.

b) If B is obtained by multiplying some row or column of A by b, $\det B = b\det A$.

c) If B is obtained by adding some multiple of one row [or column] of A to another row [or column] of A, $\det B = \det A$.

(Results (b) and (c) can save a great deal of work in computing large determinants by hand. But who does, nowadays?)

6) If $b^i_j = c^i_k a^{k\sim\ell}_\ell c_j$, where $c^i_k \delta^{k\sim\ell}_\ell c_j = \delta^i_j$ (so $[\overset{\sim\ell}{c}_j] = [c^i_k]^{\leftarrow}$), then $b^i_i = a^k_k$.

7) Let $\underset{\sim}{x}_1, \ldots, \underset{\sim}{x}_k$ be eigenvectors belonging to eigenvalues $\lambda_1, \ldots, \lambda_k$ of an operator $\underset{\sim}{A}$.

If $\underset{\sim}{x}_1 = \sum_{i=2}^{k} \underset{\sim}{x}_i b_i$, for some $b_2, \ldots, b_k \in R$, then $\underset{\sim}{0} = \sum_{i=2}^{k} \underset{\sim}{x}_i b_i (\lambda_i - \lambda_1)$,

so that $\underset{\sim}{x}_2 = \sum_{i=3}^{k} \underset{\sim}{x}_i \dfrac{b_i(\lambda_i - \lambda_1)}{b(\lambda_1 - \lambda_2)}$ if $\lambda_2 \neq \lambda_1$, $a_2 \neq 0$.

Deduce inductively that if $\lambda_1, \ldots, \lambda_k$ are distinct, $\underset{\sim}{x}_1, \ldots, \underset{\sim}{x}_k$ are independent.

8) For any projection $\underset{\sim}{P}: X \to X$, the only choices of $\underset{\sim}{y} \in \underset{\sim}{P}(X)$, $\underset{\sim}{z} \in \ker\underset{\sim}{P}$ such that $\underset{\sim}{x} = \underset{\sim}{y} + \underset{\sim}{z}$ are $\underset{\sim}{y} = \underset{\sim}{P}(\underset{\sim}{x})$, $\underset{\sim}{z} = \underset{\sim}{x} - \underset{\sim}{y}$. (cf. VII. Ex.3.1d)

II Affine Spaces

"Let the thought of the dharmas as all one bring you
to the So in Itself: thus their origin is forgotten
and nothing is left to make us pit one against another."

Seng-ts'an

1. SPACES

Our geometrical idea (I.1.01) of a vector space depended on a choice of some point $\underset{\sim}{0}$ as origin. However, just as for bases, there may be more than one plausible choice of origin. Similarly, it may be useful to avoid committing oneself on the question (a fact discovered by Galileo). For this purpose, and for the sake of some language useful even when we have an origin, we shall consider affine spaces.

The basic idea is a return to the school notion of a vector, as going from one point A to another point B in space. Points are just points, without direction, but their separations have direction and length. Thus we define:-

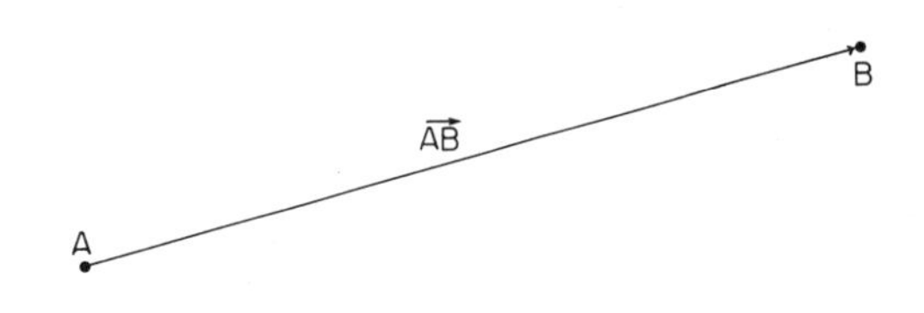

Fig. 1.1

1.01 Definition

An affine space with vector space T is a non-empty set X of points and a map

$$\underset{\sim}{d} : X \times X \to T,$$

called a difference function, such that for any x, y, z $\in$ X :-

Ai) $\underset{\sim}{d}(x,y) + \underset{\sim}{d}(y,z) = \underset{\sim}{d}(x,z)$

Aii) The restricted map $\underset{\sim}{d}_x : \{x\} \times X \to T : (x,y) \rightsquigarrow \underset{\sim}{d}(x,y)$ is bijective.

Condition Ai) says that "going from x to y, then y to z" is a change by the same directed distance as going directly to z. It has two important immediate consequences:

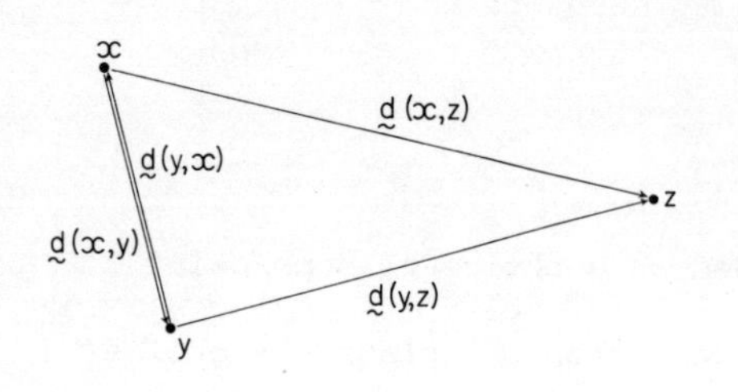

Fig. 1.2

(a) Put y = z = x, then

$2\underset{\sim}{d}(x,x) = \underset{\sim}{d}(x,x) + \underset{\sim}{d}(x,x) = \underset{\sim}{d}(x,x)$.

So $\underset{\sim}{d}(x,x) = \underset{\sim}{0}$ for all $x \in X$, hence

(b) putting z = x,

$\underset{\sim}{d}(x,y) + \underset{\sim}{d}(y,x) = \underset{\sim}{d}(x,x) = \underset{\sim}{0}$

so $\underset{\sim}{d}(x,y) = -\underset{\sim}{d}(y,x)$ for all $x,y \in X$.

Condition Aii) just says that given $x \in X$ and $\underset{\sim}{t} \in T$, there is a unique point to be reached by "going the directed distance $\underset{\sim}{t}$, starting from x". We denote this point by $x+\underset{\sim}{t}$: if $\underset{\sim}{t} = \underset{\sim}{0}$, $x+\underset{\sim}{t} = x$ by (b) above. Similarly, if V is a subset of T we denote $\{x+\underset{\sim}{t} \mid \underset{\sim}{t} \in V\}$ by x+V.

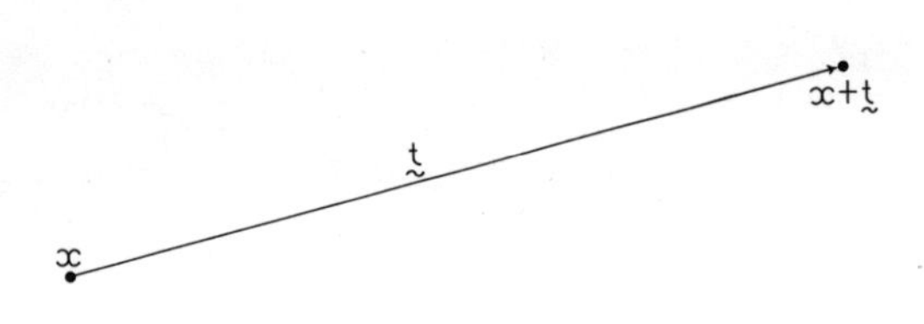

Fig. 1.3

1.02 Tangent spaces

We can use the bijection $\underset{\sim}{d}_x$ given by Aii) to define a vector space structure for $\{x\} \times X$ from that of T, by

$$(x,y) + (x,z) = \overset{\leftarrow}{\underset{\sim}{d}_x}(\underset{\sim}{d}_x(x,y) + \underset{\sim}{d}_x(x,z))$$
$$(x,y)a = \overset{\leftarrow}{\underset{\sim}{d}_x}((\underset{\sim}{d}(x,y))a)$$

(cf. Ex.1). The set $\{x\} \times X$ with this structure will be called the <u>tangent space</u> to X at x, and denoted by T_xX.

For a one-dimensional T we have a picture, but two dimensions of T require four for the analogous diagram. If $\underset{\sim}{v} \in T_x X$ we denote $x + \underset{\sim}{d}_x(\underset{\sim}{v})$ also by $x + \underset{\sim}{v}$.

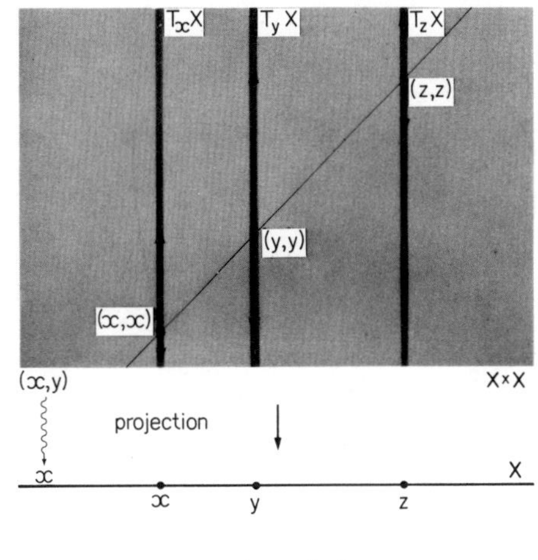

Fig. 1.4

The vectors in $T_x X$ are called <u>tangent</u> or <u>bound</u> vectors <u>at</u> x; the vectors in T are called <u>free</u>. The reason for the word "tangent" will become apparent when we start bending pieces of affine spaces around and sticking them together to make manifolds. (Even if an affine space X has a vector space with some other symbol, S say, we shall still call $\{x\} \times X$ with this vector space structure $T_x X$, to keep the association with "tangent".) We call $\underset{\sim}{d}_x$ a <u>freeing</u> map, its inverse a <u>binding</u> map.

1.03 Subspaces

The requirements for a subspace of a vector space X were essentially that it should be again a vector space (I.1.03.) Since we already knew that $\underset{\sim}{x} + \underset{\sim}{y}$ and $\underset{\sim}{y} + \underset{\sim}{x}$, etc, were equal, it was only needful to require that <u>within</u> the subspace they should still be well defined.

In the same way, $X' \subseteq X$ is an <u>affine subspace</u> or <u>flat</u> of X if

i) $\underset{\sim}{d}(X' \times X')$ is a vector subspace of the vector space T for X, and

ii) X' is an affine space, with vector space $\underset{\sim}{d}(X' \times X')$ and difference function

$$\underset{\sim}{d}: X' \times X' \to \underset{\sim}{d}(X \times X) : (x,y) \rightsquigarrow \underset{\sim}{d}(x,y).$$

If $\underset{\sim}{d}(X' \times X')$ is a hyperplane of T, then X' is an <u>affine hyperplane</u> of X.

Evidently if V is a vector subspace of T and $x \in X$, then $x + V$ is an affine subspace of X.

Notice that any vector space X has a natural affine structure, with vector space X itself and the difference function

$$X \times X \to X : (\underset{\sim}{x}, \underset{\sim}{y}) \rightsquigarrow \underset{\sim}{y} - \underset{\sim}{x}.$$

Hence we may talk of an affine subspace, or hyperplane, of a vector space X, which need not be a vector subspace of X. (cf. 1.06 below)

The affine subspace generated by a set S, or affine hull H(S) of S (compare I.1.04), is the smallest affine subspace containing S. (It is easy to show that the intersection of any set of subspaces is again a subspace, so

$$\cap\{X' \mid X' \text{ an affine subspace of } X\}$$

is a subspace. Evidently it contains S and it is contained in every other such; so it is the smallest, (cf. Ex. I.1.3a).) Pairs of points generate lines, non-collinear triples generate planes, etc. We could define "affine independence" analogously to Def^n I.1.07 (for instance three points in a straight line are dependent) together with "affine rank" etc: we shall not develop this beyond a consistency check in Ex. 2.3.

1.04 Definition

The translate $X' + \underset{\sim}{t}$ of an affine subspace X' of X by a vector $\underset{\sim}{t} \in T$ is defined as the affine subspace $\{x + \underset{\sim}{t} \mid x \in X'\}$. (cf. Ex. 2b, and Defn 3.03.)

Two affine subspaces X', X'' of X are parallel if $\underset{\sim}{d}(X' \times X') = \underset{\sim}{d}(X'' \times X'')$.

1.05 Lemma

Two affine subspaces X', X'' of X are parallel if and only if $X'' = X' + \underset{\sim}{t}$ for some $\underset{\sim}{t} \in T$ (not necessarily unique).

Proof

1) If X', X'' are parallel, choose $x' \in X'$, $x'' \in X''$ and set $\underset{\sim}{t} = \underset{\sim}{d}(x',x'')$.

Then $y'' \in X'' \iff \underset{\sim}{d}(x'',y'') \in \underset{\sim}{d}(X'' \times X'')$

$\iff \underset{\sim}{d}(x'',y'') \in \underset{\sim}{d}(X' \times X')$

$\iff \underset{\sim}{d}(x',y'') = \underset{\sim}{d}(x',x'') + d(x'',y'')$ by A i)

$= \underset{\sim}{s} + \underset{\sim}{t}$, where $\underset{\sim}{s} \in \underset{\sim}{d}(X' \times X')$

$\iff y'' \in X' + \underset{\sim}{t}$. (cf. Ex. 2b)

Hence $X'' = X' + \underset{\sim}{t}$.

2) If $X'' = X' + \underset{\sim}{t}$, and $x'', y'' \in X''$ then

$\underset{\sim}{d}(x'',y'') = \underset{\sim}{d}(\overset{\leftarrow}{\underset{\sim}{d}}_{x'}(\underset{\sim}{t}), \overset{\leftarrow}{\underset{\sim}{d}}_{y'}(\underset{\sim}{t}))$ for some $x', y' \in X'$ (def^n of $X' + \underset{\sim}{t}$)

$= \underset{\sim}{d}(\overset{\leftarrow}{\underset{\sim}{d}}_{x'}(\underset{\sim}{t}), \overset{\leftarrow}{\underset{\sim}{d}}_{x'}(\underset{\sim}{t} - \underset{\sim}{d}(x',y')))$ (cf. Ex. 2d)

$= \underset{\sim}{t} - (\underset{\sim}{t} - \underset{\sim}{d}(x',y'))$

$= \underset{\sim}{d}(x',y')$

$\in \underset{\sim}{d}(X' \times X')$.

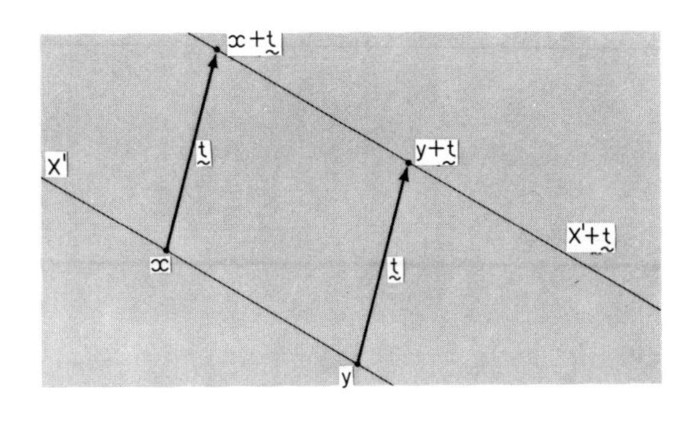

Fig. 1.5

Hence, $\underset{\sim}{d}(X'' \times X'') \subseteq \underset{\sim}{d}(X' \times X')$

Similarly

$\underset{\sim}{d}(X' \times X') \subseteq \underset{\sim}{d}(X'' \times X'')$.

Hence, $\underset{\sim}{d}(X' \times X') = \underset{\sim}{d}(X'' \times X'')$. ■

1.06 Lemma

For X a vector space, $X' \subseteq X$ is an affine subspace of X if and only if X' is a translate of some vector subspace of X.

Proof

If X' is an affine subspace of X, set $X'' = \{\underset{\sim}{x} - \underset{\sim}{y} \mid \underset{\sim}{x}, \underset{\sim}{y} \in X'\} = \underset{\sim}{d}(X' \times X')$.

Then X" is a vector subspace of X by def^n (1.03), and

$$\underset{\sim}{d}(X'' \times X'') = X'' \quad \text{since } \underset{\sim}{0} \in X''$$
$$= \underset{\sim}{d}(X' \times X')$$

and X' is a translate of X" by Lemma 1.05.

If X' is a translate of a vector subspace then it is an affine subspace by Ex. 2c.

1.07 Definition

The dimension, dim X, of an affine space X is the dimension of its space of free vectors. (cf. also Ex. 2.3)

1.08 Coordinates

If we choose an ordered basis for T, we have an isomorphism $\underset{\sim}{A}:T \to R^n$, by I.2.06.
If we then "choose as origin" some point $a \in X$, the composite bijection

$$C_a : X \to T_a X \xrightarrow{\underset{\sim}{d}_a} T \xrightarrow{\underset{\sim}{A}} R^n$$
$$x \leadsto (a,x) \leadsto \underset{\sim}{d}(a,x)$$

defines a "choice of coordinates" for X, or chart on X. (A chart of an ocean assigns, as labels to points of the ocean, pairs of numbers - 23°N, 15°W etc. - and thus is essentially a function : Ocean $\to R^2$. Unlike charting a plane, however, we cannot choose coordinates nicely all over the Earth; longitude, for instance, is not defined at the poles. This leads us to the notion of <u>local</u> chart that we use for manifolds.) Notice that we are <u>not</u> using C_a to make X a vector space, in contrast to the way we make $T_a = \{a\} \times X$ a vector space by $\underset{\sim}{d}_a$; we are just using it for labelling points. In the same way one does not add the coordinates of Greenwich to those of Montreal and get anything of any significance. Fixing an origin for X and a basis for T, we label points $x \in X$ by their images in R^n; this is illustrated in Fig. 1.6 for two such choices for the plane (if you don't bend it) of this page.

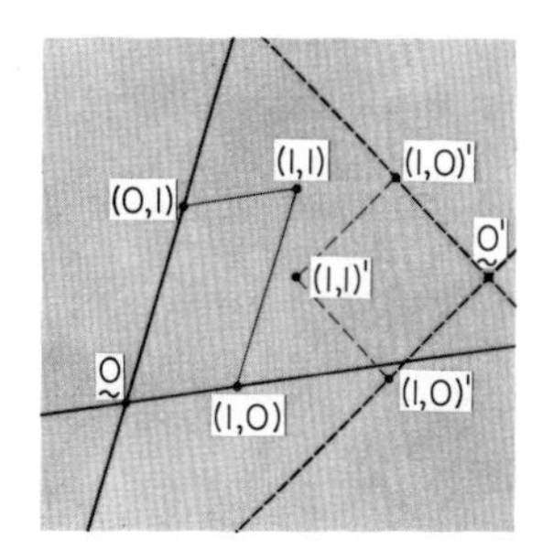

Fig. 1.6

The basis $\beta = (\underset{\sim}{b}_1, \ldots, \underset{\sim}{b}_n)$ for T defines a basis $(\overset{\leftarrow}{\underset{\sim}{d}}_x(\underset{\sim}{b}_1), \ldots, \overset{\leftarrow}{\underset{\sim}{d}}_x(\underset{\sim}{b}_n))$ for each T_xX. We denote this basis by β_x, and its members also by $\underset{\sim}{b}_{1x}, \ldots, \underset{\sim}{b}_{nx}$.

If we have two different choices of origin and basis, a,β and a',β' say, to change from the first system of coordinates to the second we must apply

$$R^n \to R^n : (x^1, \ldots, x^n) \rightsquigarrow [I]^{\beta'}_{\beta} \begin{bmatrix} x_1 \\ \vdots \\ x_n \end{bmatrix} + \underset{\sim}{A}(\underset{\sim}{d}(a',a))$$

where $\underset{\sim}{A}$ is the map $T \to R^n$ given by the basis β'. In the formula for individual coordinates, this becomes

$$x^{i'} = b^i_j x^j + a^i$$

where b^i_j is the i-th coordinate in the β' system of the j-th vector in β, and a^i is the i-th coordinate of the vector $\underset{\sim}{d}(a',a)$ from a' to a in the β' system. We shall not often need this particular operation.

Exercises II.1

1) If T_x has the vector space structure defined in 1.02, then $\underset{\sim}{d}_x : T_xX \to T$ is an isomorphism.

2a) Find a subset $S \subseteq R$ such that any vector $\underset{\sim}{v} \in R$ (treating R as a real vector space) occurs as $\underset{\sim}{u} - \underset{\sim}{w}$ for some $\underset{\sim}{u}$ and $\underset{\sim}{w} \in S$, so that S satisfies 1.03i but S is not a flat of R.

Show that if $X' \subseteq X$ satisfies 1.03i, and also $X' = x + \overleftarrow{\underset{\sim}{d}_x}(X' \times X')$ for some $x \in X'$, X' is a flat of X.

b) For any $x \in X'$, $X' + \underset{\sim}{t} = \{(x+\underset{\sim}{t}) + \underset{\sim}{s} \mid \underset{\sim}{s} \in \underset{\sim}{d}(X' \times X')\}$.

c) Prove that $X' + \underset{\sim}{t}$ is an affine subspace of X.

d) Prove that if $x,x' \in X$, $\underset{\sim}{t} \in T$, then $x' + \underset{\sim}{t} = x + \underset{\sim}{t} + \underset{\sim}{d}(x,x') = x + \underset{\sim}{t} - d(x,x')$.

e) Prove that if $\underset{\sim}{t}_1,\underset{\sim}{t}_2 \in T$, $(x+\underset{\sim}{t}_1) + \underset{\sim}{t}_2 = x + (\underset{\sim}{t}_1+\underset{\sim}{t}_2)$.

2. COMBINATIONS OF POINTS

We cannot add two points x,y in an affine space X, any more than we can add the positions of London and Glasgow. But we can talk about the point "midway between them" : namely, the point $\underset{\sim}{x} + \frac{1}{2}\underset{\sim}{d}(x,y)$ reached from x by going halfway to y (translating by half the difference vector). But $x + \frac{1}{2}\underset{\sim}{d}(x,y)$ is a rather asymmetrical name for a symmetrical notion; so is $y + \frac{1}{2}\underset{\sim}{d}(y,x)$, which ought to be the same point. Indeed,

$$y = x + \underset{\sim}{d}(x,y),$$

$$\text{so } y + \tfrac{1}{2}\underset{\sim}{d}(y,x) = x + \underset{\sim}{d}(x,y) + \tfrac{1}{2}(-\underset{\sim}{d}(x,y)), \quad \text{by Ai) in 1.01,}$$

$$\text{hence} \qquad y + \tfrac{1}{2}\underset{\sim}{d}(y,x) = x + \tfrac{1}{2}\underset{\sim}{d}(x,y).$$

So we give it the symmetrical name

$$\tfrac{1}{2}x + \tfrac{1}{2}y$$

without asserting that $\frac{1}{2}x$, $\frac{1}{2}y$, or + here mean anything: we are just abbreviating $x + \frac{1}{2}\underset{\sim}{d}(x,y)$ and $y + \frac{1}{2}\underset{\sim}{d}(y,x)$ symmetrically. (But when they do have separate meanings, because X is a vector space, no ambiguity arises. Giving X its natural affine structure (1.03),

$$\underset{\sim}{x} + \tfrac{1}{2}\underset{\sim}{d}(x,y) = \underset{\sim}{x} + \tfrac{1}{2}(\underset{\sim}{y}-\underset{\sim}{x}) = \tfrac{1}{2}\underset{\sim}{x} + \tfrac{1}{2}\underset{\sim}{y}$$

anyway.)

Of course, $\frac{1}{2}x + \frac{1}{2}y$ lies on the line through x and y. So in fact does any point $x + \lambda\underset{\sim}{d}(x,y)$: this is a special case of Ex. 2b, which we need not prove yet. Again, $x + \lambda\underset{\sim}{d}(x,y)$ is asymmetrical in starting from x, and we have

$$y + (1-\lambda)\underset{\sim}{d}(y,x) = x + \underset{\sim}{d}(x,y) - (1-\lambda)\underset{\sim}{d}(x,y) = x + \lambda\underset{\sim}{d}(x,y)$$

and we would prefer a symmetric notation.

2.01 Definition

The affine combination

$$\mu x + \lambda y, \quad \text{where} \quad \mu + \lambda = 1$$

of $x, y \in X$ is the point defined equivalently by

$$x + \lambda\underset{\sim}{d}(x,y) \quad \text{or} \quad y + \mu\underset{\sim}{d}(y,x).$$

(Notice that $\mu x + \lambda y$ is not defined if $\mu + \lambda \neq 1$.)

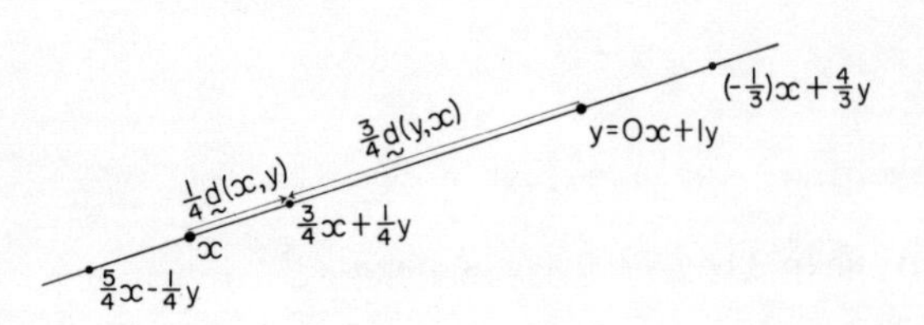

Fig. 2.1

We shall further abbreviate $\mu x + (-\lambda)y$ to $\mu x - \lambda y$, as in Fig. 2.1. Notice that $\mu x + \lambda y$ is <u>between</u> x and y exactly when λ, μ are both positive.

What about repeating this "combination" process? For example, what is "the point midway between z and $\frac{1}{2}x + \frac{1}{2}y$ "? Very conveniently,

$$\tfrac{1}{2}(\tfrac{1}{2}x + \tfrac{1}{2}y) + \tfrac{1}{2}z$$

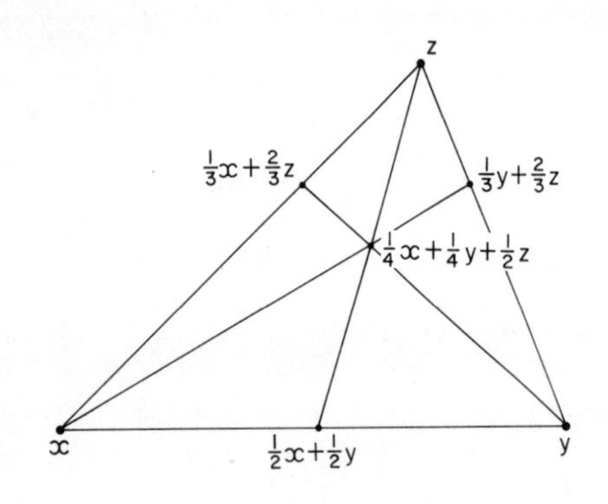

Fig. 2.2

coincides with

$$\frac{1}{4}x + \frac{3}{4}(\frac{1}{3}y + \frac{2}{3}z)$$

and with $\frac{1}{4}y + \frac{3}{4}(\frac{1}{3}x + \frac{2}{3}z)$

(Fig. 2.2, Ex. 1a). We can unambiguously call it

$$\frac{1}{4}x + \frac{1}{4}y + \frac{1}{2}z,$$

multiplying out the brackets. In general, for $\lambda_1, \ldots, \lambda_4 \in R$ we can take a "repeated combination" of $x_1, \ldots, x_5 \in X$

$$\lambda_1 x_1 + (1-\lambda_1)\left(\frac{\lambda_2}{1-\lambda_1}x_2 + \left(1 - \frac{\lambda_2}{(1-\lambda_1)}\right)\left(\frac{\lambda_3}{1-\lambda_1-\lambda_2}x_3 + \left(1 - \frac{\lambda_3}{1-\lambda_1-\lambda_2}\right)\left(\frac{\lambda_4}{1-\lambda_1-\lambda_1-\lambda_3}x_4 + \left(1 - \frac{\lambda_4}{1-\lambda_1-\lambda_2-\lambda_3}\right)x_5\right)\right)\right)$$

which multiplies out to

$$\lambda_1 x_1 + \lambda_2 x_2 + \lambda_3 x + \lambda_4 x_4 + (1-\lambda_1-\lambda_2-\lambda_3-\lambda_4) x_5 .$$

2.02 Definition

Given $x_1, \ldots, x_k \in X$, and $\lambda_1, \ldots, \lambda_k \in R$ such that $\sum_{i=1}^{k} \lambda_i = 1$, the affine combination

$$\lambda_1 x_1 + \ldots + \lambda_n x_n$$

is defined in terms of 2.01 as

$$\lambda_1 x_1 + (1-\lambda_1)\left(\frac{\lambda_2}{1-\lambda_1} x_2 + \ldots + \left(1 - \frac{n-1}{1-\lambda_1-\ldots-\lambda_{n-2}}\right) x_n\right) \ldots\Bigg)$$

(where, by Ex. 1b, the order in which we take the terms $\lambda_i x_i$ does not affect the point defined.)

The requirement that $\sum_{i=1}^{k} \lambda_i = 1$ imposes a little extra care in manipulation. For instance, the statements

$$x = y + w - z, \quad \tfrac{1}{2}x + \tfrac{1}{2}y = \tfrac{1}{2}w + \tfrac{1}{2}z ,$$

are meaningful, since they have names of points on each side. But the superficially equivalent

$$x - w = y - z \qquad \tfrac{1}{2}x - \tfrac{1}{2}w = \tfrac{1}{2}z - \tfrac{1}{2}y$$

$$x + z = y + w \qquad x + y = w + z$$

equate expressions we have not defined. (We *could* define them, but expressions like x - y would have to refer to "points at infinity" - can you see why? - and would take us into projective, not affine, geometry.)

This gives us an "internal" expression (Ex. 2a) for the affine hull (1.03) of $S \subseteq X$, as the set of affine combinations of points in S. This is precisely analogous to the two descriptions of linear hulls in I.1.04 and Ex. I.1.3b. We can also define the convex hull C(S) (Fig. 2.3) as

$$\left\{\lambda_1 x_1 + \ldots + \lambda_k x_k \,\middle|\, x_i \in S, \lambda_i > 0, i \in \{1,\ldots,k\}, k \in N, \sum_{i=1}^{k} \lambda_i = 1\right\}.$$

Also, convex sets are those S with C(S) = S. Now, (Ex. 2b) a flat has H(S) = S and since $S \subseteq C(S) \subseteq H(S)$, a flat is always convex. Convex sets are of great practical importance, for instance in linear programming and control theory. We shall not develop it here: but note that intervals in R are convex.

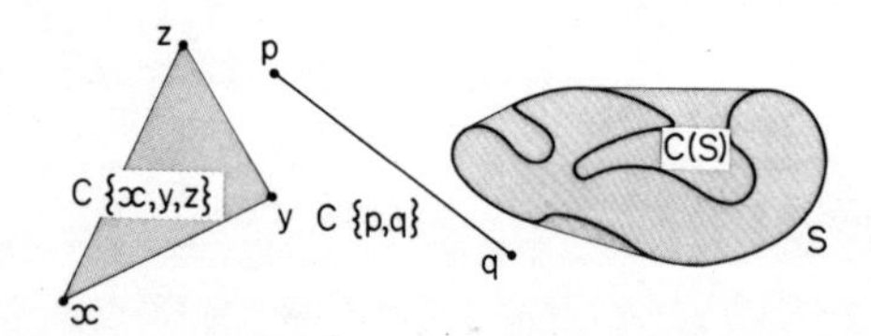

Fig. 2.3

Exercises II.2

1a) Prove from Def^n 2.01 that

$$\lambda(\mu x+(1-\mu)y)+(1-\lambda)z = \lambda\mu x+(1-\lambda\mu)\left(\frac{\lambda(1-\mu)}{1-\lambda\mu}\, y + \left(1 - \frac{(1-\mu)}{1-\lambda\mu}\right) z\right).$$

b) For any permutation

$$m : \{1,\ldots,k\} \to \{1,\ldots,k\} : i \mapsto m_i \text{ (cf. I.3.06)},$$

with $\lambda_1,\ldots,\lambda_k \in R$ s.t. $\lambda_1+\ldots+\lambda_k=1$, and $x_1,\ldots,x_k \in X$,

$$\text{then } \lambda_1 x_1 + (1-\lambda_1)\left(\frac{\lambda_2}{1-\lambda_1} x_2 + \ldots + \left(1 - \frac{\lambda_{k-1}}{1-\lambda_1-\ldots-\lambda_{k-2}}\right) x_k\right)\ldots\Bigg)$$

$$= \lambda_{m_1} x_{m_1} + (1-\lambda_{m_1})\left(\frac{\lambda_{m_2}}{-\lambda_{m_1}} x_{m_2} + \ldots + \left(1 - \frac{\lambda_{m_{k-1}}}{1-\lambda_{m_1}-\ldots-\lambda_{m_{k-2}}}\right) x_{m_k}\right)\ldots\Bigg)$$

2a) Prove that the affine hull $H(S)$ of $S \subseteq X$ (1.03) consists exactly of the set

$$\{\lambda_1 x_1 + \ldots + \lambda_k x_k \mid x_i \in S, i \in \{1,\ldots,k\}, k \in N\}.$$

b) Prove that S is an affine subspace of X if and only if $H(S) = S$. (cf. Ex.1.3c)

3) Suppose that $S = \{x_1,\ldots,x_k\}$, contained in an affine space X, does not satisfy an equation

$$x_i = \lambda_1 x_1 + \ldots + \lambda_{i-1} x_{i-1} + \lambda_{i+1} x_{i+1} + \ldots + \lambda_k x_k$$

for any i. Then using Def^n 1.07

$\dim H(S) = k - 1$, and

$H(S) = X$ if and only if $k = \dim X + 1$.

3. MAPS

One attraction of affine combinations is that they are "intrinsic to the space": one could argue that the idea of the midpoint of x and y is more basic than "x plus ½ the difference vector from x to y", which was our definition. It is certainly several thousand years older, and one can pinpoint the introduction of the more general $\mu x + \lambda y$ to Eudoxus's theory of proportions. We had the machinery of Chapter I to hand, however, so Def^n 1.01 was technically more convenient.

The structure-minded reader will find it a fruitful exercise to define an affine space as a set X with a map

$$\Lambda : X \times X \times R \to X$$

to be thought of as

$$(x,y,\lambda) \rightsquigarrow (1-\lambda)x + \lambda y$$

satisfying appropriate axioms, and construct the corresponding T and $\underset{\sim}{d}$. Notice (Fig. 3.1) that by Ex.1, starting with Def^{n}s 1.01, 2.01

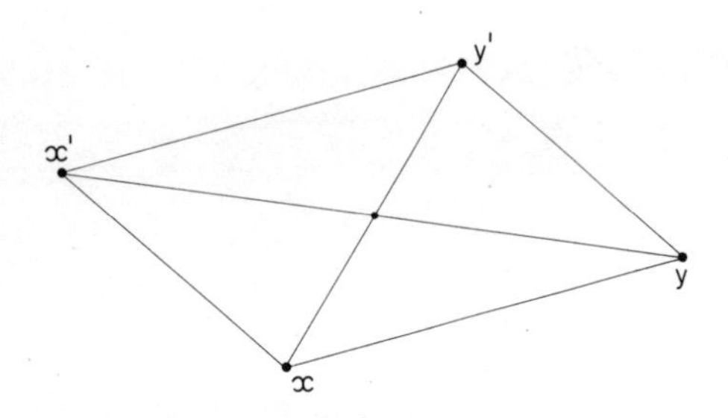

Fig. 3.1

$$\underset{\sim}{d}(x,y)=\underset{\sim}{d}(x',y') \iff \tfrac{1}{2}x + \tfrac{1}{2}y' = \tfrac{1}{2}x' + \tfrac{1}{2}y.$$

So starting from affine combinations, we could define

$$(x,y) \sim (x',y') \iff \tfrac{1}{2}x + \tfrac{1}{2}y' = \tfrac{1}{2}x' + \tfrac{1}{2}y,$$

and prove from the chosen axioms that ~ is an equivalence relation. Then T as the set of equivalence classes

$$[(x,y)] = \{(x',y') \mid (x,y) \sim (x',y')\}$$

and $\underset{\sim}{d}$ as the map

$$X \times X \to T : (x,y) \rightsquigarrow [(x,y)]$$

should have the structures of vector space (I.1.01) and difference map (1.01) if good axioms for Λ have been picked.

We leave this programme to the reader, but it motivates our next definition. With any kind of "set with structure" we are interested in maps from set to set that "respect the structure". With vector spaces it was linear maps; now, it is those that preserve affine combinations.

3.01 Definition

A map $A : (X,T) \to (Y,S)$ between affine spaces (or $P \to Q$ between convex sets in X,Y) is <u>affine</u> if for any $x, x' \in X$, $\lambda \in R$ (or $x, x' \in P$, $0 \leq \lambda \leq 1$) it satisfies

$$A((1-\lambda)x+\lambda x') = (1-\lambda)Ax + \lambda Ax' .$$

(We shall only want the convex sets case in Chapter IX., with P and Q as intervals in R : cf. Ex. 9.)

From the way that we built up the meaning of multiple combinations in §2, it is clear that this implies

$$A(\lambda_1 x_1 + \ldots + \lambda_k x_k) = \lambda_1 Ax_1 + \ldots + \lambda_k Ax_k$$

and (applying Ex. 2.2a) that

$$A(H(S)) = H(A(S)) .$$

Thus A preserves affine combinations and affine hulls: in particular, using Ex. 2.3b, $A : X \to Y$ carries flats to flats (such as lines to lines, or lines to points - why to nothing else?). Note that the map taking all of X to the same $y \in Y$ is a perfectly good affine map, just as the zero map between vector spaces is linear.

Affine maps may squash flat, but never bend.

3.02 Definition

An affine map $A : X \to Y$ takes all pairs of points in X separated by a given free vector $\underset{\sim}{t} \in T$ to pairs of points in Y all separated by the same vector in S, which we may call $\underset{\sim}{A}\underset{\sim}{t}$. (This is just a rephrasing of Ex. 2.) Ex. 3 checks that $\underset{\sim}{A}$ is a map $T \to S$ and is linear, so we may call it the linear part of A. Clearly for any $x_0 \in X$,

$$A(x) = A(x_0 + \underset{\sim}{d}(x_0, x)) = Ax_0 + \underset{\sim}{A}(\underset{\sim}{d}(x_0, x)),\ \forall\ x \in X,$$

so if we know the linear part of A and the image of any one point in X, we know A completely. A linear map, indeed, is its own linear part (Ex. 8.)

3.03 Definition

An affine map $A : X \to Y$ is an affine isomorphism if there is an affine map $B : Y \to X$ such that AB, BA are identity maps. An affine isomorphism $X \to X$ is an affine automorphism.

A translation of X is a map of the form $x \mapsto x + \underset{\sim}{t}$, for some free vector $\underset{\sim}{t}$. (One can add and scalar multiply translations just like their corresponding free vectors, and this gives yet another approach to defining an affine space.) Evidently every translation is an affine automorphism, and the translate of a subspace (Defn 1.04) is its image under a translation.

3.04 Definition

The image AX of an affine map $A : X \to Y$ is its set-theoretic image $\{Ax | x \in X\}$. Since X is (trivially) an affine subspace of itself, AX is a flat of Y.

The rank $r(A)$ of A is $\dim(AX) \leq \dim(Y)$. (Def^n 1.07)

The nullity $n(A)$ of A is the nullity of the linear part of $\underset{\sim}{A}$.

3.05 Components

Applying the equation after Def^n 3.02, it is clear that fixing origins O_X, O_Y for X and Y, and bases for T, S, we can write A as

$$A(x^1,\ldots,x^n) = A^i_j \begin{bmatrix} x^1 \\ \vdots \\ x^n \end{bmatrix} + \begin{bmatrix} a^1 \\ \vdots \\ a^m \end{bmatrix}.$$

Here $[A^i_j]$ is the matrix for $\underset{\sim}{A}$ given by the chosen coordinates and $(a^1,\ldots,a^m)$ are the coordinates of $A(O_X)$ in the chart used on Y. In individual components,

$$(Ax)^i = A^i_j x^j + a^i$$

where X is n-dimensional and Y is m-dimensional.

Exercises II.3

1a) Prove that $x + \frac{1}{2}\underset{\sim}{d}(x,y) = x' + \frac{1}{2}\underset{\sim}{d}(x',y)$ if and only if $\underset{\sim}{d}(x,y) = \underset{\sim}{d}(x',y')$.

(Hint: let $\underset{\sim}{d}(x',y') = \underset{\sim}{d}(x,y) + \underset{\sim}{t}$, and observe $\underset{\sim}{d}(x,y') = \underset{\sim}{d}(x,y)+\underset{\sim}{d}(y,x')+\underset{\sim}{d}(x',y')$.)

b) Deduce that $\underset{\sim}{d}(x,y) = \underset{\sim}{d}(x',y')$ if and only if $\frac{1}{2}x + \frac{1}{2}y' = \frac{1}{2}x' + \frac{1}{2}y$.

2) Deduce from Ex. 1b that if $A : (X,T) \to (Y,S)$ always satisfies $A(\frac{1}{2}x + \frac{1}{2}x') = \frac{1}{2}Ax + \frac{1}{2}Ax'$ (in particular, if A is affine), then $\underset{\sim}{d}(x,y) = \underset{\sim}{d}(x',y') \Rightarrow \underset{\sim}{d}(Ax,Ay) = \underset{\sim}{d}(Ax',Ay')$.

3) If $A : (X,T) \to (Y,S)$ is affine then:

a) $\underset{\sim}{A} = \{(\underset{\sim}{t},\underset{\sim}{s}) \mid \exists\, x \in X \text{ s.t. } \underset{\sim}{d}(Ax,A(x+\underset{\sim}{t})) = \underset{\sim}{s}\}$ is a mapping $S \to T$. (So $\underset{\sim}{A}$ satisfies Axioms Fi, Fii on p. 7. Use Ex. 2 for Fii.)

b) $\underset{\sim}{A}(\underset{\sim}{t}_1+\underset{\sim}{t}_2) = \underset{\sim}{A}\underset{\sim}{t}_1 + \underset{\sim}{A}\underset{\sim}{t}_2$ (choose $x \in X$ and consider $x + 2\underset{\sim}{t}_1$, $x + 2\underset{\sim}{t}_2$ and the point midway between them.)

c) $\underset{\sim}{A}(\lambda\underset{\sim}{t}) = \lambda\underset{\sim}{A}\underset{\sim}{t}$ (consider $A(x+\lambda\underset{\sim}{t})$).

4) If $A : X \to Y$ is affine then :

a) For any flat Y' of Y, $\overleftarrow{A}(Y')$ is a flat of X : in particular, for any $y \in Y$, $\overleftarrow{A}(\{y\})$ is a (perhaps empty) flat of X.

b) For any $y, y' \in AX \subseteq Y$, $\dim(\overleftarrow{A}\{y\}) = \dim(\overleftarrow{A}\{y'\}) = n(A)$.

c) $n(A) + r(A) = \dim X$.

5a) If $Y', Y'' \subseteq AX$ are parallel subspaces of Y then $\overleftarrow{A}Y'$ and $\overleftarrow{A}Y''$ are parallel subspaces of X.

b) Deduce that if $y, y' \in AX$ then $\overleftarrow{A}\{y\}$ and $\overleftarrow{A}\{y'\}$ are parallel flats of X.

6) An affine map is an affine isomorphism if and only if its linear part is an isomorphism, and hence if and only if it is a bijection.

7) Affine maps $A, A' : X \to Y$ have the same linear part if and only if $A' = T \circ A$, where T is a translation $Y \to Y$.

8) Let X, Y be vector spaces, alias $\tilde{X}, \tilde{Y}$ when considered in the natural way as affine spaces (with free vector spaces X, Y). Show that a map $M : X \to Y$ is linear if and only if $M(\underset{\sim}{0}_X) = \underset{\sim}{0}_Y$ and, considered as $\tilde{M} : \tilde{X} \to \tilde{Y}$, it is affine. Deduce that $\tilde{M}$ then coincides with its linear part M as a map between sets.

9a) Any affine map A between convex sets $P \subseteq X$, $Q \subseteq Y$ is the restriction of an affine map $\bar{A} : X \to Y$, and $\bar{A}$ is uniquely fixed by A if and only if $H(P) = X$.

b) If P, Q are intervals in R, and R has its natural affine structure with vector space R, then for any affine map $A : P \to Q$ there are unique numbers $a_1, a_2 \in R$ such that

$$A(p) = a_1 p + a_2 \quad \text{for all} \quad p \in P.$$

III Dual Spaces

"A duality of what is discriminated takes place in spite of the fact that object and subject cannot be defined."

Lankavatāra Sutra

1.01 Notation

Throughout this chapter X and Y will denote finite-dimensional real vector spaces, and n and m their respective dimensions.

1.02 Linear functionals

Just as the linear maps from X to itself have a special role and a special name, so do those from X to the field of scalars, R. They are called linear functionals on X, or dual or covariant vectors. (The term "covariant" is to distinguish them from the vectors in X, which are called contravariant. This is related to the "backwardness" of the formula for changing basis discussed in I.2.08; we shall look at it in more detail in 1.07.) The space L(X;R) of linear functionals on X forms a vector space (as does any L(X;Y); cf. I.2.01) which will generally be denoted by X* and called the dual space of X.

Geometrically, a linear functional may best be thought of by its "contours". The geographical function "H = height above sea-level" is very effectively specified on a surface by drawing lines of constant height - that is, by drawing the sets $\overleftarrow{\underset{\sim}{H}}(h)$ for various values of h. (Fig. 1.1). Similarly, a non-zero linear functional $\underset{\sim}{f}$ on an n-dimensional space will have contours that are lines for n = 2, planes for n = 3, (Fig. 1.2) and parallel affine hyperplanes in general. (They will be parallel by Ex. II. 2.3, and hyperplanes since $\underset{\sim}{f} \neq \underset{\sim}{0} \Rightarrow \underset{\sim}{f}X = R \Rightarrow r(\underset{\sim}{f}) = 1$

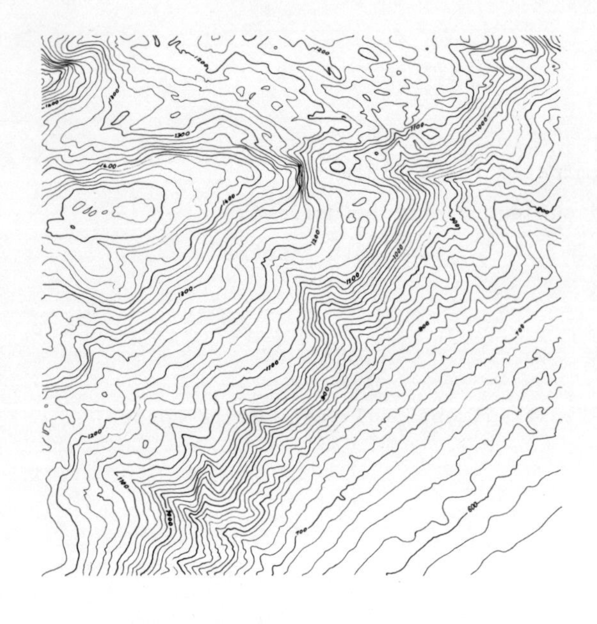

Fig. 1.1

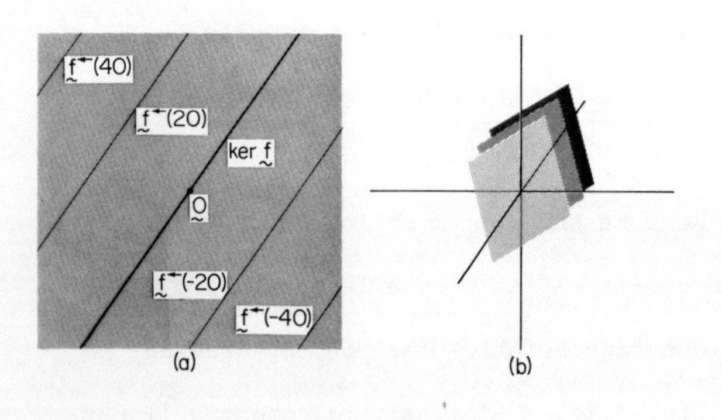

Fig. 1.2

$\Rightarrow n(\underset{\sim}{f}) = \dim(X)-1$ by Ex. II.4d.

1.03 Dual maps

From a linear map $\underset{\sim}{A} : X \to Y$, we do not get naturally any map $X^* \to Y^*$; a function $\underset{\sim}{A}$ defined on X cannot be expected to change a function $\underset{\sim}{f} \in X^*$ also defined on X to one defined on Y. However, we have a natural way to get a map the other way:

$$\underset{\sim}{A}^* : Y^* \to X^* : \underset{\sim}{f} \rightsquigarrow \underset{\sim}{f} \circ \underset{\sim}{A},$$

we call $\underset{\sim}{A}^*$ the <u>dual</u> map to $\underset{\sim}{A}$.

(This kind of reversal of direction is <u>also</u> called "contravariant", a habit that arose in a different part of mathematics entirely and conflicts with the usage for vectors, turning it from an oddity to a nuisance. However, both are too well entrenched to shift, so we shall be physicists and simply avoid this other usage for the word. But if you read mathematicians' books you must beware of it.)

1.04 Lemma

$\dim(X^*) = \dim X$.

Proof

Choose a basis $\beta = \underset{\sim}{b}_1, \ldots, \underset{\sim}{b}_n$ for X. Using the coordinates this gives us (I.2.07) we can define n functionals

$$\underset{\sim}{b}^i : X \to R : (a^1, \ldots, a^n) \rightsquigarrow a^i, \quad \text{for } i = 1, \ldots, n.$$

Any functional $\underset{\sim}{f}$, using β and the standard basis $\{\underset{\sim}{e}_1\}$ for $R = R^1$, must have a matrix, $[\underset{\sim}{f}]$, say. It follows that

$$\begin{aligned} [\underset{\sim}{f}] &= [f_1^1\ f_2^1 \ldots f_n^1] \\ &= f_1^1[1\ 0 \ldots 0] + f_2^1[0\ 1 \ldots 0] + \ldots + f_n^1[0\ 0 \ldots 1] \\ &= f_i^1[\underset{\sim}{b}^i] \\ &= [f_i^1\ \underset{\sim}{b}^i]. \end{aligned}$$

Now, $\underset{\sim}{f} = f_i^1\ \underset{\sim}{b}^i$ in a unique (why?) way, so $\beta^* = \{\underset{\sim}{b}^1, \ldots, \underset{\sim}{b}^n\}$ is a basis for X^*, giving $\underset{\sim}{f}$ coordinates $(f_1^1, \ldots, f_n^1) = (f_1, \ldots, f_n)$ for short, and so $\dim X^* = n = \dim X$. (This is in fact, just a special case of the general result (I.2.07) that $\dim(L(X;Y)) = \dim X \dim Y$, since X^* is just $L(X;R)$ and $\dim R = 1$.)

1.05 Remark

It is tempting to identify X^* with X, since using the basis $\underset{\sim}{b}^1, \ldots, \underset{\sim}{b}^n$ constructed in the proof of 1.04 we can set

$$\mathbf{B} : \beta \to \beta^* : \mathbf{b}_i \rightsquigarrow \mathbf{b}^i, \qquad \text{for } i = 1, \ldots, n$$

and by I.2.05, I.2.06 this determines an isomorphism $\mathbf{B} : X \to X^*$. However, this has great disadvantages, because the isomorphism depends very much on the choice of basis. Moreover, dual maps become confusing to talk about, because if X^* "is" just X, and Y^* "is" just Y, by virtue of isomorphisms $\mathbf{B}$, $\mathbf{B}'$ defined in this way, $\mathbf{A}^*$ goes from Y to X; you identify $\mathbf{A}^*$ with $\overleftarrow{\mathbf{B}}\mathbf{A}^*\mathbf{B}'$.

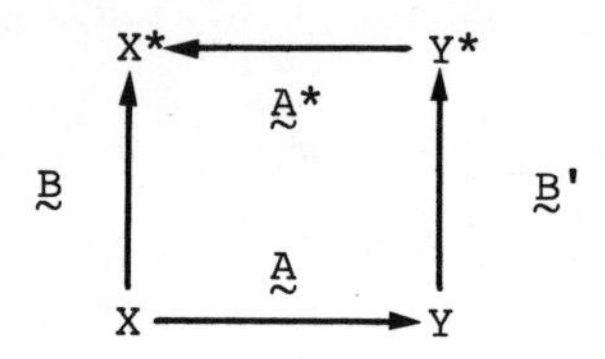

But if we choose a new basis $\frac{1}{2}\beta = \frac{1}{2}\mathbf{b}_1, \ldots, \frac{1}{2}\mathbf{b}_n$ for X, (and leave β' alone) we get a new basis for X^*, whose i-th member is a functional taking $\frac{1}{2}\mathbf{b}_i$, instead of $\mathbf{b}_i$, to 1. It thus takes $\mathbf{b}_i$ to 2 so the new basis $(\frac{1}{2}\beta)^*$ is $2\mathbf{b}^1, \ldots, 2\mathbf{b}^n$, and the new isomorphism $\mathbf{B}'' : X \to X^*$ defined by $\frac{1}{2}\mathbf{b}_i \rightsquigarrow 2\mathbf{b}^i$, $i = 1, \ldots, n$ is equal to $4\mathbf{B}$. Therefore

$$(\mathbf{B}'')^{\leftarrow} \mathbf{A}^*\mathbf{B}' = \tfrac{1}{4}\overleftarrow{\mathbf{B}}\mathbf{A}^*\mathbf{B}',$$

not at all the map just identified with A^*, though constructed in the same way. Moreover, identification of X^* with X is a particularly bad habit if carried over to infinite dimensions, where there may be <u>no</u> isomorphism, not just no natural one.

1.06 Dual basis

Although the <u>dual basis</u> $\beta^* = b^1, \ldots, b^n$ constructed for X^* in 1.04 should not be used to identify X^* with X, it can be used very effectively to simplify the algebra.

Given a vector $\underset{\sim}{x} \in X$ and a functional $\underset{\sim}{f} \in X^*$, with coordinates $(x^1, \ldots, x^n)$ and $(f_1, \ldots, f_n)$ according to β and β^*, we have

$$\begin{aligned} \underset{\sim}{f}(\underset{\sim}{x}) &= (f_i \underset{\sim}{b}^i)(\underset{\sim}{b}_j x^j) \\ &= f_i x^j (\underset{\sim}{b}_i(\underset{\sim}{b}_j)) \\ &= f_i x^j \delta^i_j \\ &= f_i x^i \end{aligned}$$

- a nice simple formula. So when we have a basis β chosen for X we usually choose the dual basis β^* for X^*. Notice that the dual basis $\mathcal{E}^* = \underset{\sim}{e}^1, \ldots, \underset{\sim}{e}^n$ to the standard basis $\mathcal{E}$ for R^n (cf. I.1.10) consists simply of the coordinate functions:-

$$\underset{\sim}{e}^i : R^n \to R : (x^1, \ldots, x^n) \rightsquigarrow x^i$$

Now, if we have bases $\beta = \underset{\sim}{b}_1, \ldots, \underset{\sim}{b}_n$ for X, $\beta' = \underset{\sim}{b}'_1, \ldots, \underset{\sim}{b}'_m$ for Y, giving an $m \times n$ matrix $A = [\underset{\sim}{A}]^{\beta'}_{\beta}$ for $\underset{\sim}{A} : X \to Y$, what is the matrix $[\underset{\sim}{A}^*]^{\beta^*}_{\beta'^*}$?

If $\underset{\sim}{f} = (f\ , \ldots, f_m)$ in "dual coordinates" on Y^*, then

$$A^*\underset{\sim}{f} = \underset{\sim}{A}^*(f_j \underset{\sim}{b}^j).$$

Hence,
$$\begin{aligned} (A^*\underset{\sim}{f})\underset{\sim}{b}_i &= (\underset{\sim}{A}^*(f_j \underset{\sim}{b}^j))\underset{\sim}{b}_i \\ &= f_j \underset{\sim}{b}^j (\underset{\sim}{A}\underset{\sim}{b}_i) \quad (\mathrm{def}^n) \\ &= f_j \underset{\sim}{b}^j (a^1_i, \ldots, a^n_i) \text{ since } \underset{\sim}{b}_i = (0, \ldots, \underset{\substack{\uparrow \\ i\text{-th place}}}{1}, \ldots, 0) \\ &= f_j a^j_i \quad (\mathrm{def}^n \text{ of } b^j), \end{aligned}$$

for any $\underset{\sim}{b}_i \in \beta$. Therefore in dual coordinates on X^*, $\underset{\sim}{A}^* \underset{\sim}{f}$ is $(a^j f_j, \ldots, a^j_n f_j)$.

This is exactly the result of applying the n × m matrix A^t, the <u>transpose</u> of A, obtained by switching rows and columns in A :-

$$\begin{bmatrix} a_1^1 & \cdot & \cdot & \cdot & \cdot & \cdot & \cdot \\ a_1^2 & a_2^2 & \cdot & \cdot & \cdot & \cdot & \cdot \\ \cdot & \cdot & \cdot & \cdot & \cdot & \cdot & \cdot \\ \cdot & \cdot & \cdot & \cdot & \cdot & \cdot & \cdot \\ \cdot & \cdot & \cdot & \cdot & a_m^m & \cdot & \cdot \end{bmatrix} \quad \text{becomes} \quad \begin{bmatrix} a_1^1 & a_1^2 & \cdot & \cdot & \cdot \\ \cdot & a_2^2 & \cdot & \cdot & \cdot \\ \cdot & \cdot & \cdot & \cdot & \cdot \\ \cdot & \cdot & \cdot & \cdot & \cdot \\ \cdot & \cdot & \cdot & \cdot & a_m^m \\ \cdot & \cdot & \cdot & \cdot & \cdot \\ \cdot & \cdot & \cdot & \cdot & \cdot \end{bmatrix}$$

In formulae, $[\underset{\sim}{A}^*] = [\underset{\sim}{A}]^t$

$$[\underset{\sim}{A}^*]^i_j = [\underset{\sim}{A}]^j_i \, .$$

Thus the use of dual bases nicely simplifies the finding of dual maps.

1.07 Change of basis

Since it is useful to have the basis of X* dual to that of X, when we change the basis of X from β to β' we want to change that of X* from β* to (β')*. Suppose then that as usual we are given the new basis vectors $(\underset{\sim}{b}'_i) = (b^j_i \underset{\sim}{b}_j) = (b^1_i, \dots, b^n_i)$ in terms of the old basis. To change the representation of a vector $\underset{\sim}{x}$ with old coordinates $(x^1, \dots, x^n)$, we have to work out

$$[\underset{\sim}{I}]^{\beta'}_{\beta} \begin{bmatrix} x^1 \\ \vdots \\ x^n \end{bmatrix} = ([\underset{\sim}{I}]^{\beta}_{\beta'})^{\leftarrow} \begin{bmatrix} x^1 \\ \vdots \\ x^n \end{bmatrix} = \begin{bmatrix} b^1_1 & \dots & b^1_n \\ \vdots & & \vdots \\ b^n_1 & \dots & b^n_n \end{bmatrix}^{\leftarrow} \begin{bmatrix} x^1 \\ \vdots \\ x^n \end{bmatrix} .$$

The i-th column of the matrix to be inverted is just the old coordinates of

$\underset{\sim}{b}_i$ (cf. I.2.08). If we have $\underset{\sim}{f} \in X^*$ represented by $(f_1, \ldots, f_n)$ in the coordinates dual to the old basis, what are its new coordinates? To get them we want the matrix $[\underset{\sim}{I}_{X^*}]^{\beta'^*}_{\beta^*}$, $= C^*$ for short. Evidently $\underset{\sim}{I}_{X^*} = (\underset{\sim}{I}_X)^*$, so by 1.06 the matrix C^* is exactly the transpose of the matrix $[\underset{\sim}{I}_X]^{\beta}_{\beta'}$, uninverted. So to find the new coordinates of f, just work out

$$(X^*,\beta^*) \xrightarrow{(I_X)^*} (X^*,\beta'^*)$$

$$(X,\beta) \xleftarrow[\underset{\sim}{I}_X]{} (X,\beta')$$

$$\begin{bmatrix} b^1_1 & \cdots & b^n_1 \\ \vdots & & \vdots \\ b^1_n & \cdots & b^n_n \end{bmatrix} \begin{bmatrix} f_1 \\ \vdots \\ f_n \end{bmatrix}$$

where the rows of the matrix are given by the old coordinates of the $\underset{\sim}{b}_i$'s.

This is what is meant by the statement that dual vectors "transform covariantly", since

$$f'_i = b^j_i f_j \qquad \text{compared with} \qquad \underset{\sim}{b}'_i = b^j_i \underset{\sim}{b}_j$$

shows that the dual vectors "co-vary" with the basis in transformation of their components. Contrariwise, we need the inverse matrix for the transformation of ordinary or "contravariant" vectors:

$$(x^i)' = \tilde{b}^i_j x^j, \qquad \text{where} \qquad \tilde{b}^i_j b^j_k = \delta^i_k,$$

as shown in I.2.08.

1.08 Notation

Consistently with what we have used so far, and with physical practice, lower indices for the components relative to a basis of a single object, such as in

$$\underset{\sim}{a} = (a_1, \ldots, a_n), \quad \underset{\sim}{a}^3 = (a_1^3, \ldots, a_n^3)$$

will indicate covariance, and upper indices such as in

$$\underset{\sim}{b} = (b^1, \ldots, b^n), \quad \underset{\sim}{b}^i_{jk} = (b^{i1}_{jk}, \ldots, b^{in}_{jk})$$

will refer to contravariance. (These examples emphasise that there may be more indices around, when the vector is one of a family labelled by these further indices - as with the $\underset{\sim}{b}_i$'s in a basis.)

Thus in general $a_i b^i$ will refer to the value $\underset{\sim}{a}(\underset{\sim}{b})$ of $\underset{\sim}{a}$ applied to $\underset{\sim}{b}$ (or $\underset{\sim}{b}(\underset{\sim}{a})$, via the identifications in the next section). The reader will notice that the numbering $\underset{\sim}{b}_1, \ldots, \underset{\sim}{b}_n$ or $\underset{\sim}{b}^1, \ldots, \underset{\sim}{b}^n$ of the vectors in a basis (which are <u>not</u> the components of an object) is done with indices the other way up. This is peculiar, but standard. It permits us to use the summation convention not only to represent $\underset{\sim}{a}(\underset{\sim}{b})$ but - as we used it in Chapter I - for linear combinations, like

$$(a^1, \ldots, a^n) = a^i \underset{\sim}{b}_i .$$

Since we shall normally suppress reference to the basis that we are using and work with n-tuples (or, for instance, $m \times n \times p \times q$ arrays) of numbers defined by the use of it, this should not cause too much confusion.

(Warning: it <u>is</u> in fact possible to regard the n vectors in a basis as the components of something called an <u>n-frame.</u> At that point the summation convention becomes more trouble than it's worth. We shall simply dodge this problem by only using "frame" in the traditional physicists' sense as short for "frame of reference". That is a particular choice of basis or coordinate system, two notions which we sometimes wish to separate (cf. II.1.08), neither of which is an object with variance at all.)

1.09 Double duals

Though there is no natural map $X \to X^*$ ("natural" meaning "independent of arbitrary choices"; this intuitive idea can be replaced by the beautiful and useful formalisation that is category theory, but we shall skip the formalities here) there is a very nice map

$$\theta : X \to (X^*)^* : \underset{\sim}{x} \rightsquigarrow [\underset{\sim}{f} \rightsquigarrow \underset{\sim}{f}(\underset{\sim}{x})]. \qquad \text{(cf. Ex.1).}$$

Now for any basis $\beta = \underset{\sim}{b}_1, \ldots, \underset{\sim}{b}_n$ for X with dual basis $\beta^* = \underset{\sim}{b}^1, \ldots, \underset{\sim}{b}^n$ for X^* and the basis $(\beta^*)^* = (b^1)^*, \ldots, (b^n)^*$ dual to that for $(X^*)^*$, the isomorphism $X \to (X^*)^*$ defined on bases by $\underset{\sim}{b}_i \rightsquigarrow (\underset{\sim}{b}^i)^*$ is exactly the map θ. For,

$$\begin{aligned}
(\underset{\sim}{b}^i)^* (\underset{\sim}{f}) &= (\underset{\sim}{b}^i)^* (f_1, \ldots, f_n) \\
&= f_i \\
&= (f_1\underset{\sim}{b}^1 + \ldots + f_n\underset{\sim}{b}^n)(\underset{\sim}{b}_i), \text{ since } \underset{\sim}{b}^j(\underset{\sim}{b}_i) = \delta^j_i \\
&= \underset{\sim}{f}(\underset{\sim}{b}_i) \\
&= (\theta(\underset{\sim}{b}_i))(\underset{\sim}{f}).
\end{aligned}$$

Thus, θ is an isomorphism, and so thoroughly "natural" that we can use it to identify $(X^*)^*$ with X, and $(\underset{\sim}{b}^i)^*$ with $\underset{\sim}{b}_i$, without ever creating difficulties for ourselves. We shall simply regard X and X^* as each other's duals, and forget about $(X^*)^*$; in fact this is why the word "dual" is used here at all. The practical-minded among us may find comfort in this identification. For, we started with abstract elements in X but dual vectors <u>had</u> a role, they attacked vectors by definition; now we have a role for vectors, they attack dual vectors!

CAUTION : The above argument rested firmly on the finite-dimensionality of X. We can always define θ, and it will always be injective, but without finite bases around it is <u>not</u> always surjective. This is sometimes not realised in physics texts, particularly earlier ones such as [Dirac].

Exercises III.1

1a) Prove that for any $\underset{\sim}{x} \in X$ there is a linear map

$$\overset{**}{\underset{\sim}{x}} : X^* \to R : \underset{\sim}{f} \rightsquigarrow \underset{\sim}{f}(\underset{\sim}{x}).$$

b) Prove that the map

$$\theta : X \to (X^*)^* : \underset{\sim}{x} \rightsquigarrow \overset{**}{\underset{\sim}{x}},$$

which by a) takes values in the right space $(X^*)^*$ and is thus well defined, is linear.

2) Prove, by considering matrices for the operators $\underset{\sim}{A}$ and $\underset{\sim}{A}^*$ in any basis and its dual and applying Ex. I.3.2a) that $\det \underset{\sim}{A}^* = \det \underset{\sim}{A}$.

IV Metric Vector Spaces

> "He who is wise sees near and far
> As the same,
> Does not despise the small
> Or value the great:
> Where all standards differ
> How can you compare?"
>
> Chuang Tzu.

1. METRICS

So far, we have worked with all non-zero vectors on an equal footing, unconcerned with the idea of their length except in illustration (as in I.3.14), or of angles between them. All the ideas we have considered have been independent of these concepts, and for instance either of the bases in Fig. 1.1 can be regarded as an equally good basis for the plane. Now, the notions of length and angle are among the most fruitful in geometry, and we need to use them in our theory of vector spaces. But this means adding a "length structure" to each vector space, and since it turns out that many are possible we must choose one - and define what we mean by one.

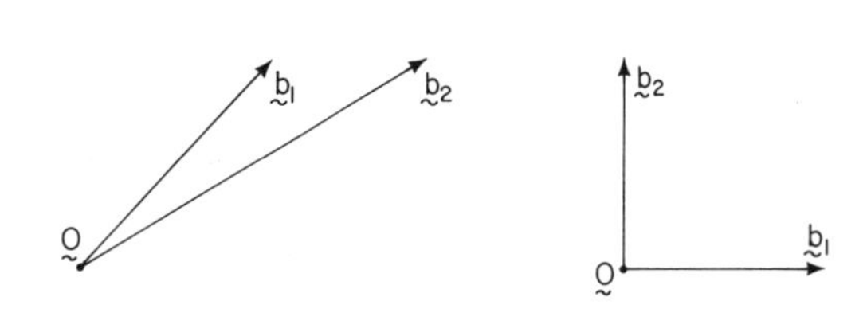

Fig. 1.1

To motivate this, let us look at R^2 with all its usual Euclidean geometry of lengths and angles, and consider two of its non-zero vectors $\underset{\sim}{v}$ and $\underset{\sim}{w}$, in coordinates (v^1, v^2) and (w^1, w^2). By Pythagoras's Theorem, the lengths $|\underset{\sim}{v}|$, $|\underset{\sim}{w}|$ of $\underset{\sim}{v}, \underset{\sim}{w}$ are

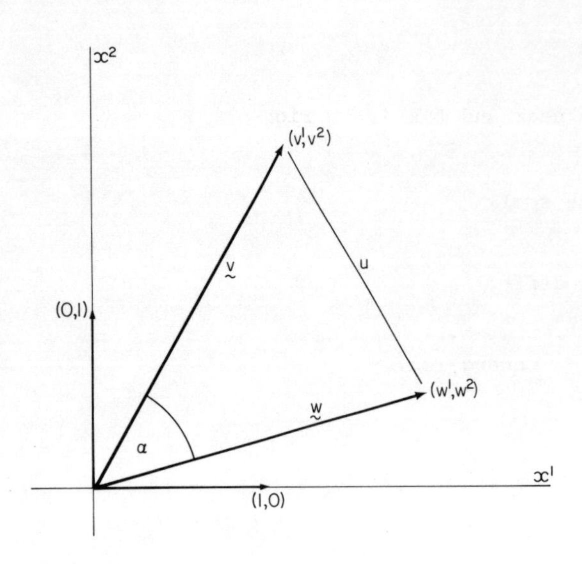

Fig. 1.2

$\sqrt{(v^1)^2 + (v^2)^2}$, $\sqrt{(w^1)^2 + (w^2)^2}$ respectively. The angle α may be found by the cosine formula for a triangle:

$$u^2 = |v|^2 + |w|^2 - 2|v|\,|w| \cos\alpha.$$

Applying Pythagoras, this gives

$$(v^1 - w^1)^2 + (v^2 - w^2)^2 = ((v^1)^2 + (v^2)^2) + ((w^1)^2 + (w^2)^2) - 2|v|\,|w| \cos\alpha.$$

Multiplying out and cancelling, we get

$$-2v^1w^1 - 2v^2w^2 = -2|v|\,|w| \cos\alpha.$$

So, $\quad v^1w^1 + v^2w^2 = |v|\,|w| \cos\alpha.$

The left-hand side of this involves coordinates, but the right hand side involves only Euclidean, coordinate-free, ideas of length and angle. Denoting $|v|\,|w| \cos\alpha$ for short by $v.w$, we can get both lengths and angles from it directly:

$$|v| = (v.v)^{\frac{1}{2}}, \quad \alpha = \cos^{\leftarrow} \frac{v.w}{|v|\,|w|}.$$

It has nice neat properties; $v.w = w.v$, and "w goes to $v.w$" is a linear functional for any v (Ex. 1). In coordinates it has a very simple formula

$$\underset{\sim}{v}.\underset{\sim}{w} = v^1w^1 + v^2w^2.$$

This depended for its proof only on the x^1 and x^2 axes being at right angles (so that we could apply Pythagoras) and the scales on them being right (the basis vectors (1,0) and (0,1) being actually of length 1). Also we can use it, itself, to define these conditions. For two non-zero vectors $\underset{\sim}{v},\underset{\sim}{w}$ with an angle α between them have

$$\begin{aligned} \underset{\sim}{v}.\underset{\sim}{w} = 0 &\Leftrightarrow |\underset{\sim}{v}|\ |\underset{\sim}{w}| \cos\alpha = 0 \\ &\Leftrightarrow \cos\alpha = 0, \text{ since } |\underset{\sim}{v}| \neq 0 \neq |\underset{\sim}{w}| \\ &\Leftrightarrow \quad \alpha = \pi/2 \end{aligned}$$

and a basis vector $\underset{\sim}{b}$ is of length 1 exactly when $\underset{\sim}{b}.\underset{\sim}{b} = 1$. So the argument establishing the formula works for any basis $\underset{\sim}{b}_1,\underset{\sim}{b}_2$ for R^2 provided that

$$\underset{\sim}{b}_i.\underset{\sim}{b}_j = \delta_{ij} .$$

In any such basis, $\underset{\sim}{v}.\underset{\sim}{w}$ will have the formula $v^1w^1 + v^2w^2$. So this "dot product" carries complete information about lengths and angles, and defines neatly in what bases it has a simple formula. It looks, then, like a good candidate for a "length structure": all that remains is to formalise it. So, on to the definition - generalising while we're at it, because we shall want a different sort of "length" for vectors in a "timelike direction" than for "spacelike" ones, when we come to spaces that model physical measurements (cf. 1.04). Moreover we shall find such generalised structures on, for example, the space of 2 × 2 matrices (cf. IX §6).

1.01 Definition

A bilinear form on a vector space X is a function

$$\underset{\sim}{F} : X \times X \to R$$

which is "linear in each variable separately". That is to say it satisfies

Bi) $\underset{\sim}{F}(\underset{\sim}{x} + \underset{\sim}{x}',\underset{\sim}{y}) = \underset{\sim}{F}(\underset{\sim}{x},\underset{\sim}{y}) + \underset{\sim}{F}(\underset{\sim}{x}',\underset{\sim}{y})$ & $\underset{\sim}{F}(\underset{\sim}{x},\underset{\sim}{y} + \underset{\sim}{y}') = \underset{\sim}{F}(\underset{\sim}{x},\underset{\sim}{y}) + \underset{\sim}{F}(\underset{\sim}{x},\underset{\sim}{y}')$

Bii) $\underset{\sim}{F}(\underset{\sim}{x}a,\underset{\sim}{y}) = a\underset{\sim}{F}(\underset{\sim}{x},\underset{\sim}{y}) = F(\underset{\sim}{x},\underset{\sim}{y}a)$.

The geometrical significance of a bilinear form depends on what further properties it has (the "dot product" discussed above is a bilinear form, but so is $(\underset{\sim}{v},\underset{\sim}{w}) \mapsto 0$, for instance. We need more conditions on a "length structure" than just bilinearity). A bilinear form in X is

(i) symmetric if $\underset{\sim}{F}(\underset{\sim}{x},\underset{\sim}{y}) = \underset{\sim}{F}(\underset{\sim}{y},\underset{\sim}{x})$ for all $\underset{\sim}{x},\underset{\sim}{y} \in X$.

(ii) anti-symmetric (or skew-symmetric) if $\underset{\sim}{F}(\underset{\sim}{x},\underset{\sim}{y}) = -\underset{\sim}{F}(\underset{\sim}{y},\underset{\sim}{x})$ for all $\underset{\sim}{x},\underset{\sim}{y} \in X$.

(iii) non-degenerate if "$\underset{\sim}{F}(\underset{\sim}{x},\underset{\sim}{y}) = 0$ for all $\underset{\sim}{y} \in X$" implies $\underset{\sim}{x} = \underset{\sim}{0}$.

(iv) positive definite if $\underset{\sim}{F}(\underset{\sim}{x},\underset{\sim}{x}) > 0$ for all $\underset{\sim}{x} \neq \underset{\sim}{0}$.

(v) negative definite if $\underset{\sim}{F}(\underset{\sim}{x},\underset{\sim}{x}) < 0$ for all $\underset{\sim}{x} \neq \underset{\sim}{0}$.

(vi) indefinite if not either positive or negative definite.

The most significant types of bilinear forms are among the non-degenerate ones. Specifically:

(vii) A metric tensor on X is a symmetric non-degenerate bilinear form. We will often follow physicists' practice in shortening this to just "metric", despite a certain risk (VI.1.02) of confusion.

(viii) An inner product on X is a positive or negative definite metric tensor. (cf. Ex. 3). We shall always take it to be positive unless otherwise indicated: there is no essential difference since a change of sign changes one to the other, without altering the geometry.

(ix) A symplectic structure on X is a skew-symmetric non-degenerate bilinear form.

(We shall not be concerned with symplectic forms here, but they play a central role in classical mechanics. See for instance [Abraham and Marsden], [Maclane (1)], or for a brief exposition [Maclane (2)].)

We denote the space of all linear forms on X by $L^2(X;R)$ or $L(X,X;R)$. (cf. Ex.4).

If S is a subspace of X, then $\underset{\sim}{F}$ is symmetric/anti-symmetric/.../symplectic on S according as the restriction

$$S \times S \to R : (\underset{\sim}{x},\underset{\sim}{y}) \rightsquigarrow \underset{\sim}{F}(\underset{\sim}{x},\underset{\sim}{y})$$

is symmetric/anti-symmetric/.../symplectic. (cf. Ex. 5)

It will often save writing to call a subspace on which a metric tensor is non-degenerate a non-degenerate subspace of X.

1.02 Definition

A metric vector space $(X,\underset{\sim}{G})$ is a vector space X with a metric tensor $\underset{\sim}{G} : X \times X \to R$. In particular:

An inner product space $(X,\underset{\sim}{G})$ is a vector space X with an inner product $\underset{\sim}{G} : X \times X \to R$.

For a given metric vector space $(X,\underset{\sim}{G})$ we shall often abbreviate $\underset{\sim}{G}(\underset{\sim}{x},\underset{\sim}{y})$ to $\underset{\sim}{x}\cdot\underset{\sim}{y}$ and $(X,\underset{\sim}{G})$ to X, when it is clear by context which metric tensor is involved.

We shall reserve the symbol $\underset{\sim}{G}$ exclusively to metric tensors (including inner products).

1.03 Definition

The standard inner product on R^n is defined by

$$(x^1, \ldots, x^n).(y^1, \ldots, y^n) = x^1y^1 + \ldots + x^ny^n = \sum_{i=1}^{n} x^iy^i.$$

Notice that we cannot use the summation convention here, since both sets of indices are upper; $\underset{\sim}{x}$ and $\underset{\sim}{y}$ are both contravariant vectors. The summation convention operates where we are combining something covariant with something contravariant, $\underset{\sim}{a}(\underset{\sim}{b}) = a_ib^i$ say, (cf. III. 1.08): an operation which depends only on general vector space definitions, and has this formula with respect to any basis and its dual. An inner product or metric tensor is extra structure. To give it a nice formula we must have a basis nice with respect to it, as we indicated at the beginning of the chapter. We examine this more precisely in §3.

The Lorentz metric on R^4 is defined by

$$(x^0,x^1,x^2,x^3).(y^0,y^1,y^2,y^3) = x^0y^0 - x^1y^1 - x^2y^2 - x^3y^3.$$

The indices run 0-3 instead of 1-4 by convention, for no particular reason except to make the odd coordinate out in the formula more distinctive. This metric originates in the physics discussed in Chapter 0 §3. For a single vector

$$\underset{\sim}{x} = (x^0,x^1,x^2,x^3)$$

it gives us

$$\underset{\sim}{x}\cdot\underset{\sim}{x} = (x^0)^2 - (x^1)^2 - (x^2)^2 - (x^3)^2,$$

a more systematic expression of the relativistically invariant quantity $t^2-x^2-y^2-z^2$

that we encountered before. The analogy with the Euclidean $|v||w|\cos\alpha$, for the dot product of two distinct vectors, can be elaborated using $\cosh\alpha$ instead. We explore some of the geometry behind this in Chapter IX §6.

CAUTION: some authors use $x^1y^1 + x^2y^2 + x^3y^3 - x^4y^4$ (essentially the negative of the metric above) as "the" Lorentz metric. And it is not customary in the journals to mention which has been chosen: you just have to work it out. We shall mention the differences made by this choice at the appropriate points.

The <u>determinant metric</u> on R^4 is defined by

$$\begin{aligned} \underset{\sim}{x}.\underset{\sim}{y} &= (x^1,x^2,x^3,x^4).(y^1,y^2,y^3,y^4) \\ &= \tfrac{1}{2}(x^1y^4 + x^4y^1) - \tfrac{1}{2}(x^3y^2 + x^2y^3). \end{aligned}$$

For this metric

$$\underset{\sim}{x}.\underset{\sim}{x} = \det \begin{bmatrix} x^1 & x^2 \\ x^3 & x^4 \end{bmatrix}.$$

The determinant of $n \times n$ matrices for $n \neq 2$ is not associated in this way with a metric tensor on R^{n^2}. But the particular case of $n = 2$ gives us, in Chapter IX, a short cut to an explicit example of indefinite geometry in Lie group theory that is important in itself.

1.04 Definition

In a vector space X with metric tensor $\underset{\sim}{G}$:-

The <u>length</u> $|\underset{\sim}{x}|_{\underset{\sim}{G}}$ of the vector $\underset{\sim}{x}$ is $\sqrt{\underset{\sim}{x}.\underset{\sim}{x}}$. (We shall suppress the $\underset{\sim}{G}$ if only one metric is in question.) Notice that with an indefinite metric a non-zero vector may have positive, zero or imaginary length. For example, in the Lorentz metric,

$$|(1,0,0,0)| = 1, \quad |(1,0,1,0)| = 0, \quad |(0,0,1,0)| = \sqrt{-1}.$$

For this reason $\underset{\sim}{x}.\underset{\sim}{x}$ is far more important than $|\underset{\sim}{x}|$ since the imaginary numbers are adventitious; they obscure the essentially real (rather than complex) structure in use.

In the situation of Chapter 0 §3, a vector labelled (1,0,0,0) or (-1,0,0,0) by some observer represents a separation purely in time from the origin, with no difference in spatial position - according to that observer. These vectors have positive Lorentz dot product with themselves. On the other hand a point labelled $(0,x^1,x^2,x^3)$ by someone, with a "purely spatial" separation from the origin according to this label, gives a negative number. Finally, possible "arrival" points for light flashes with "departure" at (0,0,0,0) or vice versa give zero, by the Principle of Relativity (as we say in Chapter 0 §3). We shall see (Ex. 3.5) that the sign of $\underset{\sim}{x}.\underset{\sim}{x}$ completely determines whether the spatial or temporal part of the separation $\underset{\sim}{x}$ can be eliminated by a suitable choice of coordinates. Borrowing language from this case even when not thinking of physics we shall call $\underset{\sim}{x}$

<u>timelike</u> if $\underset{\sim}{x}.\underset{\sim}{x} > 0$

<u>spacelike</u> if $\underset{\sim}{x}.\underset{\sim}{x} < 0$

<u>lightlike</u> or <u>null</u> if $\underset{\sim}{x}.\underset{\sim}{x} = 0$.

1.05 Examples

For the sake of the examples they provide, we introduce here the (non-standard because there are no standard ones) symbols H^2 for R^2 with the metric

$$(x^0,x^1).(y^0,y^1) = x^0y^0 - x^1y^1$$

and H^3 for R^3 with the metric

$$(x^0,x^1,x^2).(y^0,y^1,y^2) = x^0y^0 - x^1y^1 - x^2y^2.$$

In H^2 the null vectors are all those of the form (x,x) and $(x,-x)$. In H^3 the null vectors are those (x^0,x^1,x^2) with $(x^1)^2 + (x^2)^2 = (x^0)^2$. Fig. 1.3 shows

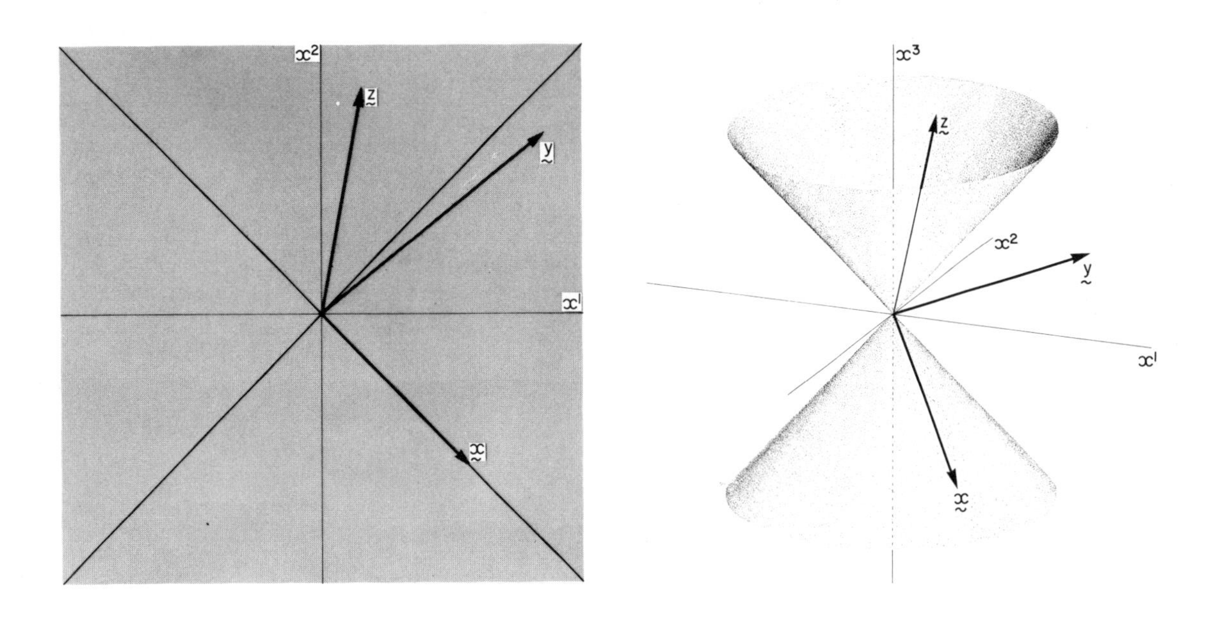

Fig. 1.3

$\underset{\sim}{x}$ as null, $\underset{\sim}{y}$ spacelike and $\underset{\sim}{z}$ timelike in each diagram. For R^4 with the Lorentz metric (which we shall call Lorentz space and denote by L^4) a similar picture is true but hard to draw.

The set $\{\underset{\sim}{x} | \underset{\sim}{x}.\underset{\sim}{x} = 0\}$ of null vectors is called the null cone or light cone of X: it is never a subspace of X with any non-degenerate indefinite metric. (Ex. 6.)

1.06 Definition

A norm on a vector space X is a function

$$X \to R : \underset{\sim}{x} \rightsquigarrow \|\underset{\sim}{x}\|$$

such that for all $\underset{\sim}{x}, \underset{\sim}{y}, \in X$ and $a \in R$,

Ni) $\|\underset{\sim}{x}\| = 0$ implies $\underset{\sim}{x} = \underset{\sim}{0}$.

Nii) $\|\underset{\sim}{x}a\| = |a| \; \|\underset{\sim}{x}\|$.

Niii) $\|\underset{\sim}{x}+\underset{\sim}{y}\| \leq \|\underset{\sim}{x}\| + \|\underset{\sim}{y}\|$.

A <u>partial</u> norm satisfies (Nii) but not necessarily (Niii), and only

N'i) $\|\underset{\sim}{x}\| \geq 0$ for all $\underset{\sim}{x} \in X$

instead of (Ni). (cf. Ex. 7a).

On an inner product space $(X,\underset{\sim}{G})$ we have a norm given exactly by length, $|\underset{\sim}{x}| = \|\underset{\sim}{x}\|_{\underset{\sim}{G}} = +\sqrt{\underset{\sim}{x}.\underset{\sim}{x}}$ (Ex. 7b) but for a general metric vector space $\sqrt{\underset{\sim}{x},\underset{\sim}{x}}$ need not be real, so that this does not define a function $X \to R$. We can however define, for a metric vector space $(X,\underset{\sim}{G}')$,

$$\|\underset{\sim}{x}\|_{\underset{\sim}{G}'} = +\sqrt{|\underset{\sim}{G}'(\underset{\sim}{x},\underset{\sim}{x})|}.$$

If $\underset{\sim}{G}'$ is an inner product this coincides with the length and we shall use $|\;|$ and $\|\;\|$ indifferently; in general $\|\;\|_{\underset{\sim}{G}'}$ is a partial norm. (Ex. 7c).

In any metric vector space $(X,\underset{\sim}{G})$ we shall abbreviate $\|\;\|_{\underset{\sim}{G}}$ to $\|\;\|$, when possible without confusion. We shall call $\|\underset{\sim}{x}\|$ the <u>size</u> of $\underset{\sim}{x}$, as against the length $|\underset{\sim}{x}|$.

A <u>unit</u> vector $\underset{\sim}{x}$ in a metric vector space is one such that $\|\underset{\sim}{x}\| = 1$.

Any non-null vector $\underset{\sim}{x}$ may be <u>normalised</u> to give the unit vector $\underset{\sim}{x}/\|\underset{\sim}{x}\|$ in the same direction.

<u>1.07 Lemma</u>

In any inner product space $(X,\underset{\sim}{G})$ we have, for any $\underset{\sim}{x},\underset{\sim}{y} \in X$

$$\underset{\sim}{x}\cdot\underset{\sim}{y} \le |\underset{\sim}{x}|\ |\underset{\sim}{y}|$$

with equality for $\underset{\sim}{x},\underset{\sim}{y}$ non-zero if and only if $\underset{\sim}{y} = \underset{\sim}{x}a$, for some $a \in R$ (when the two vectors are collinear.)

(This is obviously necessary to make possible the equation $\underset{\sim}{x},\underset{\sim}{y} = |\underset{\sim}{x}|\ |\underset{\sim}{y}| \cos\alpha$ of the remarks opening this chapter. It is called the Schwarz inequality and it is false for $\|\ \|_{\underset{\sim}{G}}$ when $\underset{\sim}{G}$ is indefinite.)

Proof

For any $a \in R$, $(\underset{\sim}{x}a - \underset{\sim}{y})\cdot(\underset{\sim}{x}a - y) = \underset{\sim}{x}a\cdot\underset{\sim}{x}a - \underset{\sim}{y}\cdot\underset{\sim}{x}a - \underset{\sim}{x}a\cdot\underset{\sim}{y} + \underset{\sim}{y}\cdot\underset{\sim}{y}$

$$= (\underset{\sim}{x}\cdot\underset{\sim}{x})a^2 - (2\underset{\sim}{x}\cdot\underset{\sim}{y})a + \underset{\sim}{y}\cdot\underset{\sim}{y} \quad .$$

Since $\underset{\sim}{G}$ is positive definite, $\underset{\sim}{z}\cdot\underset{\sim}{z} \ge 0$ for all $\underset{\sim}{z}$, and so in particular

$$(\underset{\sim}{x}\cdot\underset{\sim}{x})a^2 - (2\underset{\sim}{x}\cdot\underset{\sim}{y})a + \underset{\sim}{y}\cdot\underset{\sim}{y} \ge 0 \text{ for all } a.$$

Therefore the quadratic equation

$$(\underset{\sim}{x}\cdot\underset{\sim}{x})a^2 - (2\underset{\sim}{x}\cdot\underset{\sim}{y})a + \underset{\sim}{y}\cdot\underset{\sim}{y} = 0$$

in a cannot have distinct real roots, hence

$$(2\underset{\sim}{x}\cdot\underset{\sim}{y})^2 - 4(\underset{\sim}{x}\cdot\underset{\sim}{x})(\underset{\sim}{y}\cdot\underset{\sim}{y}) \le 0.$$

$$\therefore \quad (\underset{\sim}{x}\cdot\underset{\sim}{y})^2 \le (\underset{\sim}{x}\cdot\underset{\sim}{x})(\underset{\sim}{y}\cdot\underset{\sim}{y})$$

$$\therefore \quad |(\underset{\sim}{x}\cdot\underset{\sim}{y})| \le \sqrt{\underset{\sim}{x}\cdot\underset{\sim}{x}}\ \sqrt{\underset{\sim}{y}\cdot\underset{\sim}{y}}$$

$$\therefore \quad \underset{\sim}{x}\cdot\underset{\sim}{y} \le \|\underset{\sim}{x}\|\ \|\underset{\sim}{y}\|$$

In the case of equality, the equation has exactly one real root

$(a = (2\underset{\sim}{x}.\underset{\sim}{y})/(2\underset{\sim}{x}.\underset{\sim}{x}) = \|\underset{\sim}{y}\| / \|\underset{\sim}{x}\|)$ and for this value we have precisely

$$(\underset{\sim}{x}a - \underset{\sim}{y}).(\underset{\sim}{x}a - \underset{\sim}{y}) = 0.$$

Hence by the definiteness of $\underset{\sim}{G}$

$$\underset{\sim}{x}a - \underset{\sim}{y} = \underset{\sim}{0}$$
$$\underset{\sim}{x}a = \underset{\sim}{y}.$$ ■

1.08 Definition

Two vectors $\underset{\sim}{x},\underset{\sim}{y}$ in a metric vector space are <u>orthogonal</u> whenever

$$\underset{\sim}{x}.\underset{\sim}{y} = 0.$$

If the metric is an inner product, this coincides with the Euclidean idea of "at right angles" (for which orthogonal is just the Greek) but in an indefinite metric vector space a null vector is orthogonal to itself. Fig. 1.4 shows several pairs of vectors in H^2, with each matching pair orthogonal.

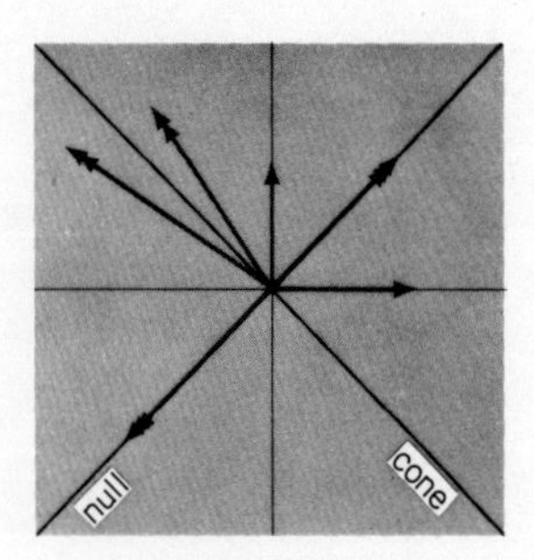

Fig. 1.4

For any $\underset{\sim}{x} \in X$, the set $\underset{\sim}{x}^{\perp}$ of vectors orthogonal to it is a subspace of X, since if $\underset{\sim}{x}.\underset{\sim}{y} = 0 = \underset{\sim}{x}.\underset{\sim}{y}'$ we have $\underset{\sim}{x}.(\underset{\sim}{y}+\underset{\sim}{y}') = \underset{\sim}{x}.\underset{\sim}{y}+\underset{\sim}{x}.\underset{\sim}{y}' = 0$, $\underset{\sim}{x}.(\underset{\sim}{y}a) = a(\underset{\sim}{x}.\underset{\sim}{y}) = 0$. ($\underset{\sim}{x}^{\perp}$ can be pronounced "$\underset{\sim}{x}$ perp.", from perpendicular.) This idea should be familiar for R^3 with the standard inner product (Fig. 1.5a); for H^3 it is illustrated in Fig. 1.5b-f. The plane in Fig. 1.5d is a good example of a degenerate subspace larger than a null line

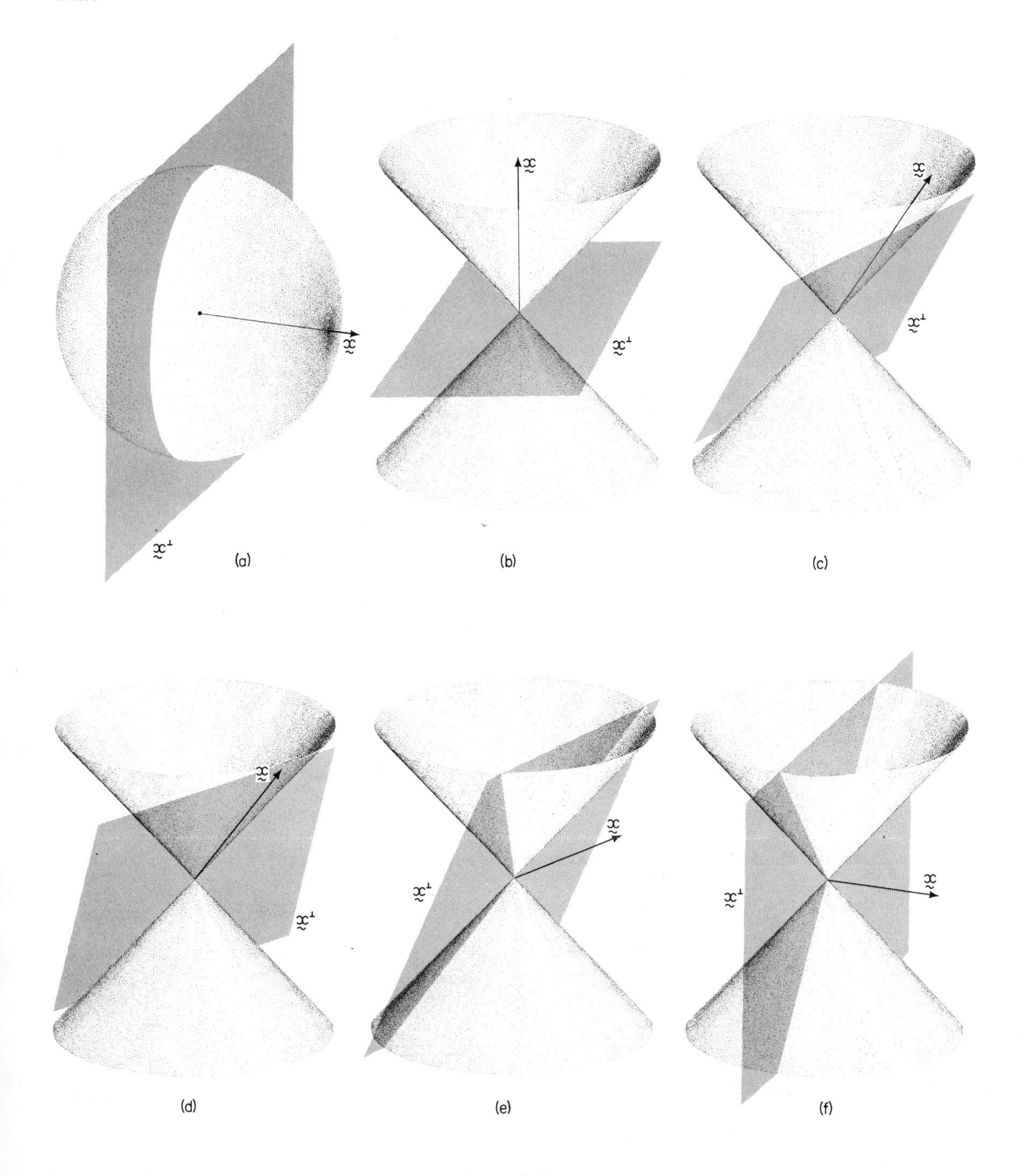

Fig. 1.5

In a similar way, the set of vectors $\underset{\sim}{y}$ for which $\underset{\sim}{x}\cdot\underset{\sim}{y}$ equals some given number a, is an affine subspace parallel to $\underset{\sim}{x}^{\perp}$. (In the inner product case, it is the set of vectors "with component $a/|\underset{\sim}{x}|$ in the $\underset{\sim}{x}$ direction" as in Fig. 1.6. Notice how this geometrical idea depends on orthogonality.)

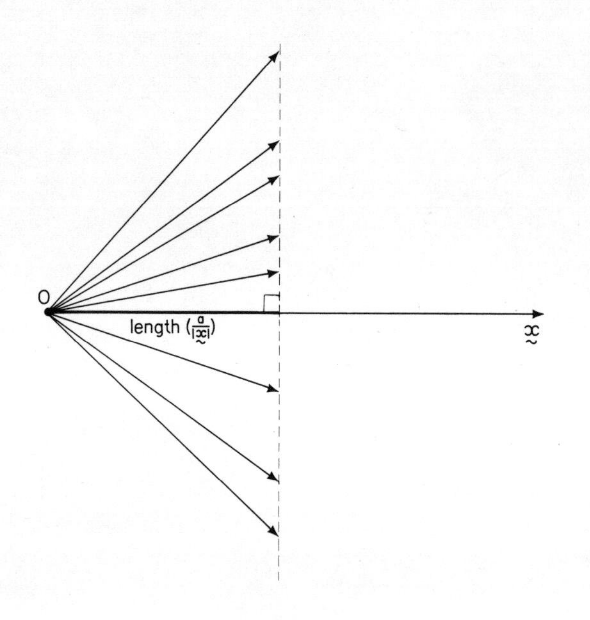

Fig. 1.6

Via orthogonality, then, we can go from vectors in X to parallel slicings of X. We have in fact found a transfer from X to X*, since these slices, for $\underset{\sim}{x} \in X$, are exactly the affine hyperplane "contours" (cf. III 1.02) of the linear functional

$$\underset{\sim}{x}^* : X \to R : \underset{\sim}{y} \rightsquigarrow \underset{\sim}{x}\cdot\underset{\sim}{y} \ .$$

Similarly, given a function $\underset{\sim}{f}$ we have a "gradient vector" for it: we can choose a vector $\underset{\sim}{x}$ in the unique direction orthogonal to $\ker\underset{\sim}{f}$, and with a length indicating how "steep" the functional is - how closely the contours are spaced. A metric tensor, then, gives us a geometrical way of changing from contravariant vectors to covariant ones and vice versa. As usual, the algebra gives us a grip on this (in the next theorem) which is useful in proofs and computations, but the geometry is the heart of the matter.

1.09 Theorem

For any non-degenerate bilinear form $\underset{\sim}{F}$ on a vector space X, the map

$$\underset{\sim}{F}_{\downarrow} : X \to X^*$$
$$\underset{\sim}{x} \rightsquigarrow \begin{pmatrix} \underset{\sim}{x}^* : X \to R \\ \underset{\sim}{y} \rightsquigarrow F(\underset{\sim}{x},\underset{\sim}{y}) \end{pmatrix}$$

is linear and an isomorphism.

Proof

For any $\underset{\sim}{x}$, $\underset{\sim}{x}',\underset{\sim}{y} \in X$, $a \in R$

$(\underset{\sim}{F}_{\downarrow}(\underset{\sim}{x}+\underset{\sim}{x}'))\underset{\sim}{y} = \underset{\sim}{F}((\underset{\sim}{x}+\underset{\sim}{x}'),\underset{\sim}{y}) = \underset{\sim}{F}(\underset{\sim}{x},\underset{\sim}{y}) + \underset{\sim}{F}(\underset{\sim}{x}',\underset{\sim}{y}) = \underset{\sim}{F}_{\downarrow}(\underset{\sim}{x})\underset{\sim}{y} + \underset{\sim}{F}_{\downarrow}(\underset{\sim}{x}')\underset{\sim}{y} = (\underset{\sim}{F}_{\downarrow}(\underset{\sim}{x})+\underset{\sim}{F}_{\downarrow}(\underset{\sim}{x}'))\underset{\sim}{y}$

So, $\underset{\sim}{F}_{\downarrow}(\underset{\sim}{x}+\underset{\sim}{x}') = \underset{\sim}{F}_{\downarrow}(\underset{\sim}{x}) + \underset{\sim}{F}_{\downarrow}(\underset{\sim}{x}')$.

And $\underset{\sim}{F}_{\downarrow}(\underset{\sim}{x}a)\underset{\sim}{y} = \underset{\sim}{F}(\underset{\sim}{x}a,\underset{\sim}{y}) = a\underset{\sim}{F}(\underset{\sim}{x},\underset{\sim}{y}) = a(\underset{\sim}{F}_{\downarrow}(\underset{\sim}{x})\underset{\sim}{y}) = (a\underset{\sim}{F}_{\downarrow}(\underset{\sim}{x}))\underset{\sim}{y}$.

So $\underset{\sim}{F}_{\downarrow}(\underset{\sim}{x}a) = (\underset{\sim}{F}_{\downarrow}(\underset{\sim}{x}))a$.

Hence $\underset{\sim}{F}_{\downarrow}$ is linear. Since $\underset{\sim}{F}$ is non-degenerate,

$$\begin{aligned} \underset{\sim}{F}_{\downarrow}(\underset{\sim}{x}) = \underset{\sim}{0} \quad &\Rightarrow \quad \underset{\sim}{F}_{\downarrow}(\underset{\sim}{x})\underset{\sim}{y} = 0 \text{ for all } \underset{\sim}{y} \\ &\Rightarrow \quad \underset{\sim}{F}(\underset{\sim}{x},\underset{\sim}{y}) = 0 \text{ for all } \underset{\sim}{y} \\ &\Rightarrow \quad \underset{\sim}{x} = \underset{\sim}{0}. \end{aligned}$$

Thus $\ker(\underset{\sim}{F}_{\downarrow}) = \{\underset{\sim}{0}\}$, so that $n(\underset{\sim}{F}_{\downarrow}) = 0$. (cf. I.2.09)

Hence by Theorem I.2.10 and III.1.04 we have

$\dim(\underset{\sim}{F}_{\downarrow}X) = r(\underset{\sim}{F}_{\downarrow}) = \dim X = \dim X^*$.

So, $\underset{\sim}{F}_{\downarrow}X = X^*$, since X^* is the only subspace of itself with the same dimension.

So $\underset{\sim}{F}_{\downarrow}$ is an injective (Ex. I.2.1) and surjective linear map, and hence an isomorphism by I.2.03. (Notice, once again, that finite dimension is crucial.

Continuity is involved in such infinite dimensional versions as are true.)

1.10 Notation

The inverse of the isomorphism $\underset{\sim}{F}_{\downarrow}$ will be denoted by $\underset{\sim}{F}_{\uparrow}$. In the sequel we shall make extensive use of $\underset{\sim}{G}_{\downarrow}$ and $\underset{\sim}{G}_{\uparrow}$ induced by a metric tensor $\underset{\sim}{G}$.

1.11 Lemma

A non-degenerate bilinear form $\underset{\sim}{F}$ on a vector space X induces a bilinear form $\underset{\sim}{F}^*$ on X* by

$$\underset{\sim}{F}^*(\underset{\sim}{f},\underset{\sim}{g}) = \underset{\sim}{F}(\underset{\sim}{F}_{\uparrow}(\underset{\sim}{f}),\ \underset{\sim}{F}_{\uparrow}(\underset{\sim}{g}))$$

(that is, change the functionals to vectors with $\underset{\sim}{F}_{\uparrow}$, and then apply $\underset{\sim}{F}$) which is also non-degenerate and is symmetric/anti-symmetric/.../indefinite (cf. 1.01) according as $\underset{\sim}{F}$ is.

Proof

We shall prove non-degeneracy, and leave the preservation of the other properties as Ex. 8.

If for some $\underset{\sim}{f} \in X^*$ we have $\underset{\sim}{F}^*(\underset{\sim}{f},\underset{\sim}{g}) = 0$ for all $\underset{\sim}{g} \in X^*$, this means

$$\underset{\sim}{F}(\underset{\sim}{F}_{\uparrow}(\underset{\sim}{f}),\ \underset{\sim}{F}_{\uparrow}(\underset{\sim}{g})) = 0 \text{ for all } \underset{\sim}{g} \in X^*.$$

$$\text{So, } \underset{\sim}{F}(\underset{\sim}{F}_{\uparrow}(\underset{\sim}{f}),\underset{\sim}{y}) = 0 \text{ for all } \underset{\sim}{y} \in X \ (\underset{\sim}{F}_{\uparrow} \text{ surjective}).$$

$$\text{Hence } \quad \underset{\sim}{F}_{\uparrow}(\underset{\sim}{f}) = \underset{\sim}{0} \qquad\qquad (\underset{\sim}{F} \text{ non-degenerate}),$$

and $\qquad \mathbf{f} = \mathbf{0} \qquad$ ($\mathbf{F}$ injective).

Thus $\mathbf{F}_{\uparrow}$ is non-degenerate. ■

1.12 Corollary

A metric tensor [respectively inner product] $\mathbf{G}$ on X induces a metric tensor [respectively inner product] $\mathbf{G}^*$ on X*. ■

Exercises IV.1

1) Prove by Euclidean geometry (no coordinates) that if for $\mathbf{v}, \mathbf{w}$ geometrical vectors (directed distances from O) in Euclidean space, with lengths v,w, we define

$$\mathbf{v}.\mathbf{w} \;=\; v\; w\; \cos\alpha$$

where α is the angle between them, then

a) $\mathbf{v}.\mathbf{w} = \mathbf{w}.\mathbf{v}$

b) $(\mathbf{v}a).\mathbf{w} = a(\mathbf{w}.\mathbf{v})$

c) $\mathbf{v}.(\mathbf{u}+\mathbf{w}) = \mathbf{v}.\mathbf{u} + \mathbf{v}.\mathbf{w}$.

2) There are 30 possible implications, such as "(iv) => (v)" among properties (i) - (vi) in 1.01. Which are true? Which properties always contradict each other?

3) The "dot product" $\mathbf{v}.\mathbf{w}$ defined in Ex. 1 is an inner product.

4) Addition and scalar multiplication of bilinear forms defined pointwise:-

$$(\mathbf{F}+\mathbf{F}')(\mathbf{x},\mathbf{y}) = \mathbf{F}(\mathbf{x},\mathbf{y}) + \mathbf{F}'(\mathbf{x},\mathbf{y}) \text{ for all } \mathbf{x},\mathbf{y} \in X$$

$$(\mathbf{F}a)(\mathbf{x},\mathbf{y}) = a(\mathbf{F}(\mathbf{x},\mathbf{y}))$$

make $L^2(X;R)$ a vector space.

5) Which of properties (i) - (iv) in 1.01 must hold for $\underset{\sim}{F}$ on any subspace of X if they hold for $\underset{\sim}{F}$ on X? (Test by looking at, for instance, line subspaces.)

6) In H^2 the null vectors do not form a subspace. Deduce with the aid of Theorem 3.05 that for no non-zero vector space with an indefinite metric tensor do the null vectors form a subspace.

7a) Prove from 1.05 Ni), Nii), Niii) that if $\| \ \|$ is a norm, then $\|\underset{\sim}{x}\| \geq 0$ for all $\underset{\sim}{x}$, so that $\| \ \|$ is also a partial norm.

b) Prove that if $\underset{\sim}{G}$ is an inner product, then $\| \ \|_{\underset{\sim}{G}}$ is a norm. (For Niii), expand $\|\underset{\sim}{x}+\underset{\sim}{y}\|_{\underset{\sim}{G}}$ and hence $(\|\underset{\sim}{x}+\underset{\sim}{y}\|_{\underset{\sim}{G}})^2$, using the bilinearity of $\underset{\sim}{G}$.)

c) If $\underset{\sim}{G}$ is an indefinite metric, $\| \ \|_{\underset{\sim}{G}}$ is a partial norm. Show that neither Ni) nor Niii) hold, by considering null vectors and sums of null vectors in, for example, H^2.

d) Show that if $\underset{\sim}{G}$ is any symmetric bilinear form (in particular a metric tensor), $\underset{\sim}{G}$ can be defined from $| \ |_{\underset{\sim}{G}}$ by means of the result that for all $\underset{\sim}{x},\underset{\sim}{y}$

$$\underset{\sim}{G}(\underset{\sim}{x},\underset{\sim}{y}) = \tfrac{1}{4}\underset{\sim}{G}(\underset{\sim}{x}+\underset{\sim}{y},\underset{\sim}{x}+\underset{\sim}{y}) - \tfrac{1}{4}\underset{\sim}{G}(\underset{\sim}{x}-\underset{\sim}{y},\underset{\sim}{x}-\underset{\sim}{y}) = \tfrac{1}{4}(|\underset{\sim}{x}+\underset{\sim}{y}|^2_{\underset{\sim}{G}} - |\underset{\sim}{x}-\underset{\sim}{y}|^2_{\underset{\sim}{G}}).$$

(This is known as the <u>polarisation identity</u>. It shows that the Euclidean, Lorentz and determinant <u>metric tensors</u> of 1.03 can be recovered from the corresponding $(\text{length})^2$ and determinant <u>functions</u>. In other words, a <u>quadratic</u> form determines a symmetric <u>bilinear</u> form. For, if we know $\underset{\sim}{G}(\underset{\sim}{x},\underset{\sim}{x})$ for all $\underset{\sim}{x} \in X$, the above identity shows how to find $\underset{\sim}{G}(\underset{\sim}{x},\underset{\sim}{y})$.)

8) Prove that F* on X* (Lemma 1.11) is symmetric, skew-symmetric, positive definite, negative definite or indefinite according as F itself has or has not each property.

2. MAPS

The isomorphism $\underset{\sim}{G}_{\downarrow}$ constructed above illustrates an unusual point about a metric tensor; usually, with any structure, we are interested only in functions that respect it (as linear maps do addition etc). While maps having $A\underset{\sim}{x}.A\underset{\sim}{y} = \underset{\sim}{x}.\underset{\sim}{y}$ in analogous fashion are important (see below, 2.07) they are not the only maps we allow here; we still work with the whole class of linear maps. Of equal importance with maps preserving the metric tensor are those constructed by means of it, such as $\underset{\sim}{G}_{\downarrow}$ and those we are about to define.

One of the most frequent operations with a vector in conventional three-dimensional space, from the moment it is introduced at school level, is to find its component in some direction, or in some plane. (If a particle is constrained to move in some sloping plane, the <u>in the plane</u> - components of gravity and other forces that it experiences suffice to determine its motion). This idea generalises to metric vector spaces:

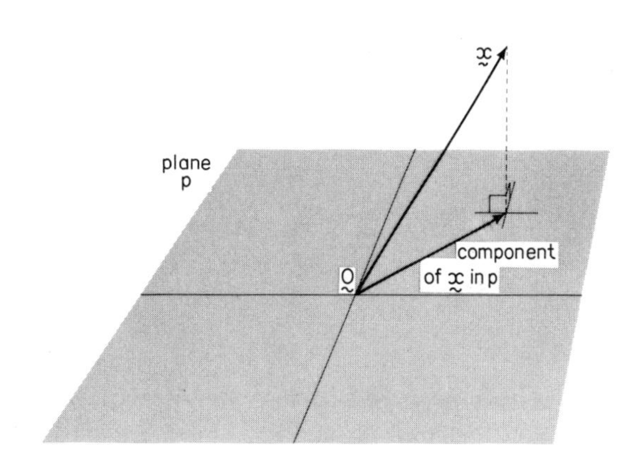

Fig. 2.1

2.01 Theorem

Let S be a non-degenerate subspace of a metric vector space X. Then there is a unique linear operator

$$\underset{\sim}{P} : X \to S$$

(called <u>orthogonal projection onto</u> S) such that $(\underset{\sim}{x} - \underset{\sim}{P}\underset{\sim}{x}).\underset{\sim}{y} = 0$ for all $\underset{\sim}{y} \in S$. (Essentially $\underset{\sim}{Px}$ is "the component of $\underset{\sim}{x}$ in S" and $\underset{\sim}{x} - \underset{\sim}{Px}$ is "the component of $\underset{\sim}{x}$ orthogonal to S" and $\underset{\sim}{x}$ is their sum.)

<u>Proof</u>

Existence:

We define $\underset{\sim}{Px}$ by exchanging $\underset{\sim}{x}$ for the functional "dot product with $\underset{\sim}{x}$", restrict that to a functional on S, and exchange the result for a vector <u>in</u> S, via the restriction of the metric. (Which is why we require the metric to be non-degenerate on S, thereby inducing an isomorphism $S \to S^*$.) Notice that although S is a subspace of X, S* is not a subspace of X* in a natural way: the dual of the inclusion $\underset{\sim}{i} : S \hookrightarrow X$ goes the other way. We have

$$\underset{\sim}{i}^* : X^* \to S^* : \underset{\sim}{f} \rightsquigarrow \underset{\sim}{f}|S = \underset{\sim}{f} \circ i.$$

Formally, if $\underset{\sim}{G}$ is the metric tensor on X and $\underset{\sim}{G}'$ the induced metric tensor then, on S,

$$\underset{\sim}{G}' : S \times S \to R : (\underset{\sim}{x},\underset{\sim}{y}) \rightsquigarrow \underset{\sim}{G}(\underset{\sim}{x},\underset{\sim}{y}).$$

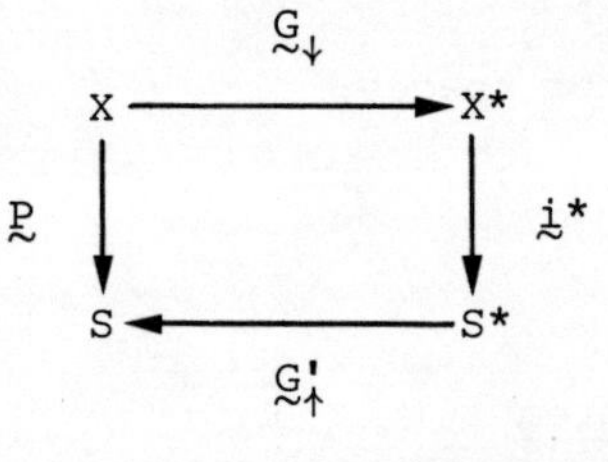

$$\begin{array}{ccc} X & \xrightarrow{\underset{\sim}{G}_\downarrow} & X^* \\ {\scriptstyle \underset{\sim}{P}}\downarrow & & \downarrow{\scriptstyle \underset{\sim}{i}^*} \\ S & \xleftarrow[\underset{\sim}{G}'_\uparrow]{} & S^* \end{array}$$

We set

$$\underset{\sim}{P} = (\underset{\sim}{G}'_\uparrow) \circ (\underset{\sim}{i}^*) \circ (\underset{\sim}{G}_\downarrow).$$

Then for any $\underset{\sim}{y} \in S$, $(\underset{\sim}{x} - \underset{\approx}{P}\underset{\sim}{x}).\underset{\sim}{y}$

$$= \underset{\sim}{x}.\underset{\sim}{y} - \underset{\sim}{G}_{\uparrow}'(\underset{\sim}{i}^*(\underset{\sim}{G}_{\downarrow}\underset{\sim}{x})).\underset{\sim}{y}$$

$$= \underset{\sim}{x}.\underset{\sim}{y} - \underset{\sim}{i}^*(\underset{\sim}{G}_{\downarrow}\underset{\sim}{x})\underset{\sim}{y} \quad (\underset{\sim}{y} \in S)$$

$$= \underset{\sim}{x}.\underset{\sim}{y} - (\underset{\sim}{G}_{\downarrow}\underset{\sim}{x})\underset{\sim}{y} \quad (\text{def}^n)$$

$$= \underset{\sim}{x}.\underset{\sim}{y} - \underset{\sim}{x}.\underset{\sim}{y} \quad (\text{def}^n)$$

$$= 0.$$

Uniqueness:

Suppose that we also have $\underset{\approx}{Q} : X \to S$ such that

$$(\underset{\sim}{x} - \underset{\approx}{Q}\underset{\sim}{x}).\underset{\sim}{y} = 0 \quad \text{for all } \underset{\sim}{y} \in S,\ \underset{\sim}{x} \in X.$$

Then by linearity of $\underset{\sim}{G}$,

$$(\underset{\sim}{x} - \underset{\approx}{P}\underset{\sim}{x} - (\underset{\sim}{x} - \underset{\approx}{Q}\underset{\sim}{x})).\underset{\sim}{y} = 0$$

$$(\underset{\approx}{Q}\underset{\sim}{x} - \underset{\approx}{P}\underset{\sim}{x}).\underset{\sim}{y} = 0 \quad \text{for all } \underset{\sim}{y} \in S,\ \underset{\sim}{x} \in X.$$

But $\underset{\approx}{P}\underset{\sim}{x}$, $\underset{\approx}{Q}\underset{\sim}{x}$, hence also $\underset{\approx}{Q}\underset{\sim}{x} - \underset{\approx}{P}\underset{\sim}{x}$, are in S, and $\underset{\sim}{G}$ is non-degenerate on S, hence $\underset{\approx}{P}\underset{\sim}{x} = \underset{\approx}{Q}\underset{\sim}{x}$ for all $\underset{\sim}{x} \in X$. ■

2.02 Corollary

The projection operator $\underset{\sim}{P}$ onto S is idempotent. (cf. I.3.02)

Proof

If $\underset{\sim}{y} = \underset{\approx}{P}\underset{\sim}{x}$ for some $\underset{\sim}{x}$, then $\underset{\sim}{y} \in S$. Moreover

$$(\underset{\sim}{y} - \underset{\sim}{P}\underset{\sim}{y}) \cdot \underset{\sim}{y}' = 0 \quad \text{for any } \underset{\sim}{y}' \in S.$$

But $(\underset{\sim}{y} - \underset{\sim}{P}\underset{\sim}{y}) \in S$, and $\underset{\sim}{G}$ is non-degenerate on S, so

$$\underset{\sim}{y} = \underset{\sim}{P}\underset{\sim}{y}$$

therefore $$\underset{\sim}{P}\underset{\sim}{x} = \underset{\sim}{P}(\underset{\sim}{P}\underset{\sim}{x}) \quad \text{for any } \underset{\sim}{x} \in X.$$ ■

It will be seen in the next section that a metric vector space possesses non-degenerate subspaces of all dimensions $0 < d \leq n$. Several orthogonal projections are illustrated in Fig. 2.2: (a) represents a typical projection in R^2 with the standard inner product, (b) and (c) projections in H^2.

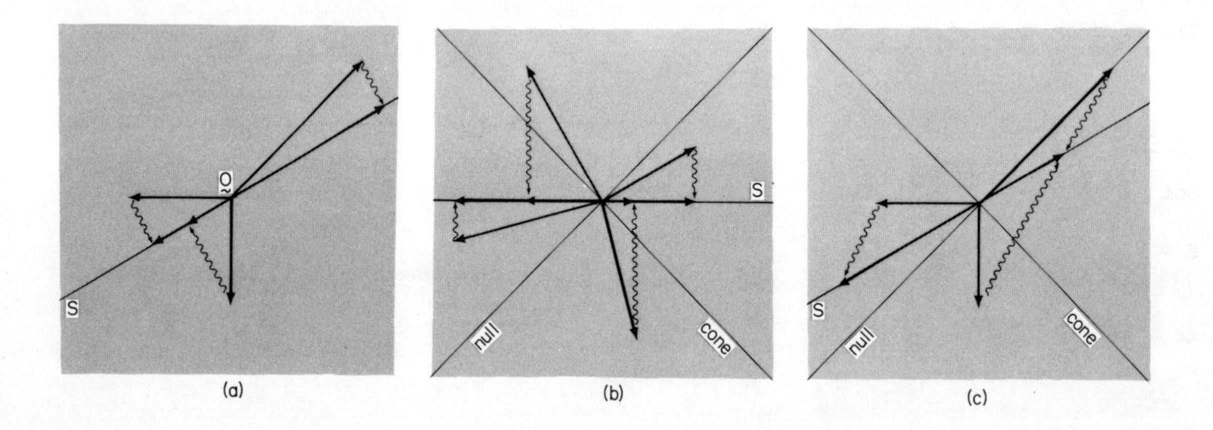

Fig. 2.2

Notice by comparing (a) and (c) how strongly the projection depends on the metric.

2.03 Definition

The kernel of the orthogonal projection onto a non-degenerate subspace S of X (cf. Ex. 9a) is called the orthogonal complement of S in X, and denoted by $S^{\perp}$. (We shall also call the orthogonal complement of the subspace spanned by one or more vectors simply the orthogonal complement of the vector or set of vectors; like $\underset{\sim}{x}^{\perp}$ in 1.08.)

2.04 Lemma

For any non-degenerate subspace S of X, each $\underset{\sim}{x} \in X$ can be expressed, in a unique way, as

$$\underset{\sim}{x} = \underset{\sim}{s} + \underset{\sim}{t},$$

where $\underset{\sim}{s} \in S$, and $\underset{\sim}{t} \in S^{\perp}$. (This gives an example of a direct sum $X = S \oplus S^{\perp}$, cf. Ex. VII.3.1 a-d.)

Proof

Set $\underset{\sim}{s} = \underset{\sim}{P}\underset{\sim}{x}$, $\underset{\sim}{t} = \underset{\sim}{x} - \underset{\sim}{P}\underset{\sim}{x}$, where $\underset{\sim}{P}$ is the orthogonal projection onto S. Then we have $\underset{\sim}{s} \in S$, $\underset{\sim}{t} \in S^{\perp}$ by the definition of $\underset{\sim}{P}$, and

$$\underset{\sim}{x} = \underset{\sim}{s} + \underset{\sim}{t}.$$

Uniqueness follows from that of $\underset{\sim}{P}$. ■

2.05 Corollary

If $\dim S = k$, $\dim X = n$, then $\dim S^{\perp} = n - k$.

Proof

Ex. 9b. ■

2.06 Corollary

If $\underset{\sim}{G}$ is non-degenerate on S, it is non-degenerate on $S^{\perp}$.

Proof

If $\underset{\sim}{x} \in S^{\perp}$, then $\underset{\sim}{x}.\underset{\sim}{s} = 0$ for all $\underset{\sim}{s} \in S$.

Therefore we have, for $\underset{\sim}{x} \in S^{\perp}$

$$\underset{\sim}{x}.\underset{\sim}{t} = 0 \text{ for all } \underset{\sim}{t} \in S^{\perp} \Rightarrow \underset{\sim}{x}.\underset{\sim}{t} + \underset{\sim}{x}.\underset{\sim}{s} = 0 \text{ for all } \underset{\sim}{s} \in S,\ \underset{\sim}{t} \in S^{\perp}$$

$$\Rightarrow \underset{\sim}{x}.(\underset{\sim}{s}+\underset{\sim}{t}) = 0 \quad \text{for all } \underset{\sim}{s} \in S,\ \underset{\sim}{t} \in S^{\perp}$$

$$\Rightarrow \underset{\sim}{x}.\underset{\sim}{y} = 0 \quad \text{for all } \underset{\sim}{y} \in X$$

$$\Rightarrow \underset{\sim}{x} = 0. \quad (\underset{\sim}{G} \text{ non-degenerate on } X)$$ ■

2.07 Definition

A linear map $\underset{\sim}{A} : X \to Y$ between metric vector spaces is an <u>isometry</u> if it is surjective and

$$(\underset{\sim}{A}\underset{\sim}{x}).(\underset{\sim}{A}\underset{\sim}{x}') = \underset{\sim}{x}.\underset{\sim}{x}' \text{ for all } \underset{\sim}{x},\underset{\sim}{x}' \in X.$$

(That is, if it preserves the metric, just as linearity means "preserving addition" with $\underset{\sim}{A}\underset{\sim}{x} + \underset{\sim}{A}\underset{\sim}{y} = \underset{\sim}{A}(\underset{\sim}{x}+\underset{\sim}{y})$ etc.) "Preserving lengths and angles" is evidently a strong condition: in particular it implies that $\underset{\sim}{A}$ is injective as well as surjective, and hence an isomorphism. (Ex. 2c) If $\underset{\sim}{A}$ preserves the metric but is not surjective, it is an isometry <u>into</u> Y. (Ex. 2d)

An operator on X which is an isometry is called unitary or orthogonal. (This term arose when attention was more on matrices than on the operators they describe. It will be accounted for at the end of §3 below. See also Ex. 3.7.)

In an inner product space, an orthogonal operator with positive determinant is simply a rotation since it preserves lengths and angles. The determinant condition ensures that the space is not "turned over". For example, if $\underset{\sim}{A}(x,y) = (-x,y)$ then $\underset{\sim}{A}$ is not a rotation but it is an orthogonal operator on R^2 with the standard inner product. Orthogonal operators must take unit/spacelike/timelike/null vectors to vectors of the same sort, and can (Ex. 3.7) take a vector $\underset{\sim}{v}$ to any $\underset{\sim}{w}$ with $\underset{\sim}{w}.\underset{\sim}{w} = \underset{\sim}{v}.\underset{\sim}{v}$. Fig. 2.3 shows the surfaces consisting of the possible end points for images of a vector $\underset{\sim}{x}$ under an orthogonal operator $\underset{\sim}{A}$, in various situations. For a more detailed discussion, see [Porteous], p. 427.

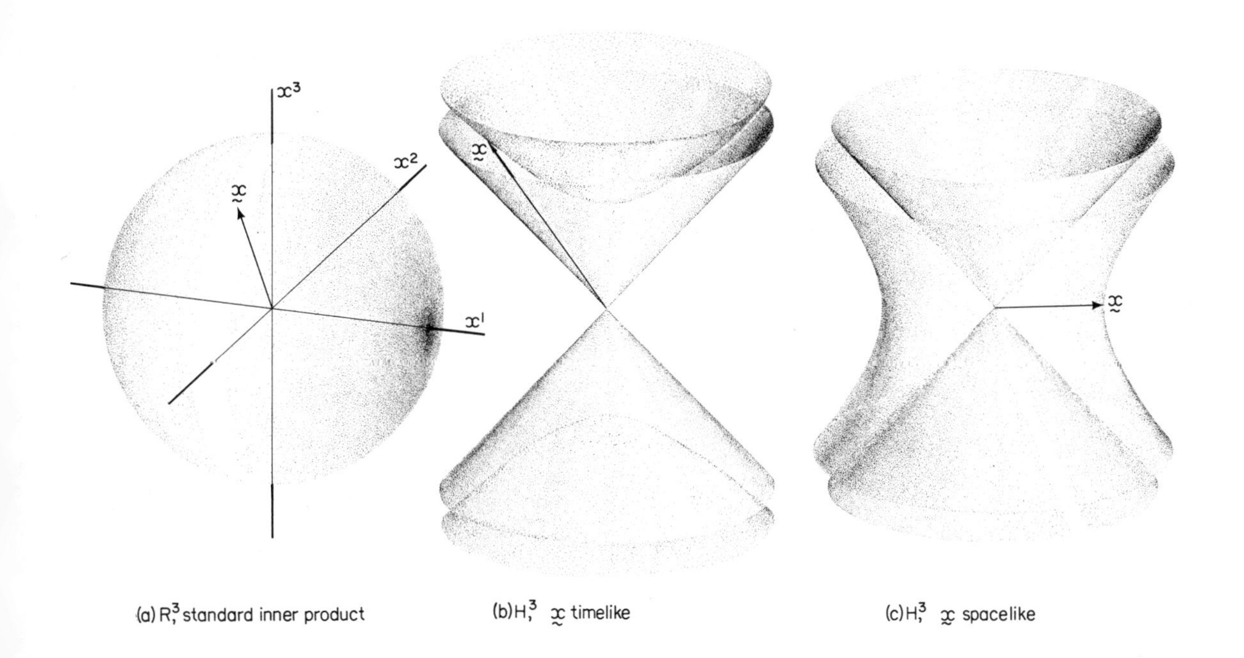

Fig. 2.3

For Lorentz space L^4, an orthogonal operator is sometimes called a Lorentz transformation, but this term is often reserved for a change of orthonormal basis (cf. next section) which involves the complications discussed in I.2.08 and III.1.07. It is thus a good idea to avoid using the same term for both, and we shall stick to the second usage. (In space, rotating an object and rotating your axes for its description are both practicable. On the other hand, if L^4 is thought of as spacetime, then it is obviously hard physically to "move it around" by an operator, whereas relabelling is just a matter of changing how you look at it - or who looks at it. A choice of who is "at rest" is exactly a choice of x^o-axis, since moving along this axis or parallel to it involves no change in "space" coordinates, only in time, which is what "at rest in a frame of reference" means: cf. XI.2.01.)

2.08 Definition

The adjoint $\underset{\sim}{A}^T$ of a linear operator $\underset{\sim}{A}$ on a metric vector space X is defined by the equation

$$\underset{\sim}{A}^T\underset{\sim}{x}\cdot\underset{\sim}{y} = \underset{\sim}{x}\cdot\underset{\sim}{A}\underset{\sim}{y} \quad \text{for all } \underset{\sim}{x},\underset{\sim}{y} \in X.$$

(The geometrical meaning of $\underset{\sim}{A}^T$ depends on the nature of $\underset{\sim}{A}$; if $\underset{\sim}{A}$ is orthogonal, for instance, $\underset{\sim}{A}^T$ coincides with $\underset{\sim}{A}^{\leftarrow}$ (Lemma 2.09). In other cases, such as $\underset{\sim}{A}$ a projection, $\underset{\sim}{A}^T$ coincides with $\underset{\sim}{A}$ (Lemma 2.11). These two situations are the important cases.) Obviously, $(\underset{\sim}{A}^T)^T = \underset{\sim}{A}$ by the symmetry of the metric.

The defining equation means exactly that

$$(\underset{\sim}{G}_{\downarrow}(\underset{\sim}{A}^T\underset{\sim}{x}))\underset{\sim}{y} = (\underset{\sim}{G}_{\downarrow}\underset{\sim}{x})\underset{\sim}{A}\underset{\sim}{y}.$$

So, $\quad (G_{\downarrow}(\underset{\sim}{A}^T\underset{\sim}{x}))\underset{\sim}{y} = (\underset{\sim}{A}^*(\underset{\sim}{G}_{\downarrow}\underset{\sim}{x}))\underset{\sim}{y}$ for all $\underset{\sim}{y} \in X$. (def^n of A*, III.1.03)

Hence we have, $\mathbf{G}_\downarrow(\mathbf{A}^T\mathbf{x}) = \mathbf{A}^*(\mathbf{G}_\downarrow\mathbf{x})$

$$\mathbf{A}^T\mathbf{x} = \mathbf{G}_\uparrow(\mathbf{A}^*(\mathbf{G}_\downarrow\mathbf{x})).$$

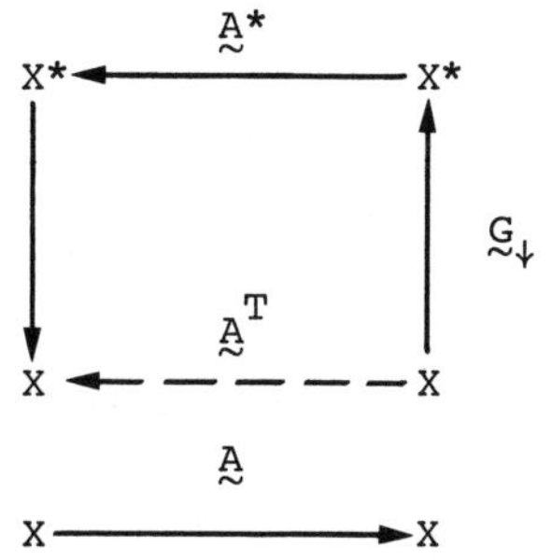

$$\begin{array}{ccc} X^* & \xleftarrow{\mathbf{A}^*} & X^* \\ \mathbf{G}_\uparrow \downarrow & & \uparrow \mathbf{G}_\downarrow \\ X & \xleftarrow{\mathbf{A}^T} & X \\ & & \\ X & \xrightarrow{\mathbf{A}} & X \end{array}$$

Therefore, $\mathbf{A}^T$ exists and is unique, being exactly the composite

$$\mathbf{A}^T = \mathbf{G}_\uparrow\mathbf{A}^*\mathbf{G}_\downarrow.$$

An operator $\mathbf{A}$ on X is <u>self-adjoint</u> if $\mathbf{A}^T = \mathbf{A}$, or equivalently if $\mathbf{Ax}.\mathbf{y} = \mathbf{x}.\mathbf{Ay}$ for all $\mathbf{x},\mathbf{y} \in X$.

Self-adjoint operators are very common and very useful (hence very important) in a great variety of contexts, particularly when the vector space in question is infinite-dimensional and a lot of tools useful in finite dimensions no longer apply. Self-adjointness has a very straightforward geometrical interpretation, which - since it is so intimately related to the nice form such operators can be given in coordinates - we leave till after the next section.

<u>2.09 Lemma</u>

An operator $\mathbf{A}$ on a metric vector space is orthogonal if and only if

$$\mathbf{A}^T\mathbf{A} = \mathbf{I}.$$

<u>Proof</u>

For any $\mathbf{x} \in X$,

$$\underset{\sim}{A}\underset{\sim}{x}.\underset{\sim}{A}\underset{\sim}{y} = \underset{\sim}{x}.\underset{\sim}{y} \text{ for all } \underset{\sim}{y} \in X \iff \underset{\sim}{A}^T(\underset{\sim}{A}\underset{\sim}{x}).\underset{\sim}{y} = \underset{\sim}{x}.\underset{\sim}{y} \text{ for all } \underset{\sim}{y} \in X \text{ (def}^n \text{ of } \underset{\sim}{A}^T)$$

$$\iff (\underset{\sim}{A}^T\underset{\sim}{A}\underset{\sim}{x} - \underset{\sim}{I}\underset{\sim}{x}).\underset{\sim}{y} = 0 \text{ for all } \underset{\sim}{y} \in X$$

$$\iff \underset{\sim}{A}^T\underset{\sim}{A}\underset{\sim}{x} - \underset{\sim}{I}\underset{\sim}{x} = \underset{\sim}{0} \text{ by non-degeneracy,}$$

hence $\underset{\sim}{A}\underset{\sim}{x}.\underset{\sim}{A}\underset{\sim}{y} = \underset{\sim}{x}.\underset{\sim}{y}$ for all $\underset{\sim}{x},\underset{\sim}{y} \in X \iff \underset{\sim}{A}^T\underset{\sim}{A} = \underset{\sim}{I}$. ■

2.10 Corollary

An operator $\underset{\sim}{A}$ on a metric vector space is orthogonal if and only if $\underset{\sim}{A}^T$ is orthogonal. ■

2.11 Lemma

Orthogonal projection $\underset{\sim}{P}$ onto a non-degenerate subspace S of a metric vector space X is a self-adjoint operator.

Proof

Let $\underset{\sim}{x} = \underset{\sim}{s} + \underset{\sim}{t}$, $\underset{\sim}{y} = \underset{\sim}{s}' + \underset{\sim}{t}'$, with $\underset{\sim}{s},\underset{\sim}{s}' \in S$, $\underset{\sim}{t},\underset{\sim}{t}' \in S^\perp$. (cf. 2.04)

Then

$\underset{\sim}{P}\underset{\sim}{x}.\underset{\sim}{y} = \underset{\sim}{P}(\underset{\sim}{s} + \underset{\sim}{t}).(\underset{\sim}{s}' + \underset{\sim}{t}') = \underset{\sim}{s}.(\underset{\sim}{s}' + \underset{\sim}{t}') = \underset{\sim}{s}.\underset{\sim}{s}' + \underset{\sim}{s}.\underset{\sim}{t}' = \underset{\sim}{s}.\underset{\sim}{s}'$

and similarly

$\underset{\sim}{x}.\underset{\sim}{P}\underset{\sim}{y} = \underset{\sim}{s}.\underset{\sim}{s}'$. ■

Exercises IV.2

1a) The orthogonal complement of a non-degenerate subspace S of X is exactly the set $\{\underset{\sim}{x} \in X \mid \underset{\sim}{x}.\underset{\sim}{y} = 0 \text{ for all } \underset{\sim}{y} \in Y\}$ of vectors orthogonal to all those in S.

b) If $\underset{\sim}{b}_1, \ldots, \underset{\sim}{b}_k$ is a basis for a non-degenerate subspace S, and $\underset{\sim}{b}'_1, \ldots, \underset{\sim}{b}'_\ell$ is a basis for $S^\perp$, show that between them they span X. Deduce their linear independence from the uniqueness in 2.04, and hence that they form a basis for X.

c) From (b), or from Theorem I.2.10, deduce that $\dim(S) + \dim(S^{\perp}) = \dim X$.

d) Suppose that an affine subspace S of X has $\underset{\sim}{d}(S\times S)$ a non-degenerate subspace V of the vector space T of X, for a given metric tensor on T. Then there is unique affine map $P : X \to S$ such that $(\underset{\sim}{d}(Px,x)).\underset{\sim}{y} = 0$ for any $x \in X$, $\underset{\sim}{y} \in V$, and that $P^2 = P$. (We shall call this also "orthogonal projection".) Show that P depends on the metric.

2a) In any metric vector space, for any operator $\underset{\sim}{A}$, $\det \underset{\sim}{A}^T = \det \underset{\sim}{A}$. (Consider the determinant of the matrix of $\underset{\sim}{A}^T = \underset{\sim}{G}_{\uparrow}\underset{\sim}{A}^*\underset{\sim}{G}_{\downarrow}$ in any basis, using Theorem I.3.08, and apply Chapter III, Ex. 2.)

b) Deduce via Lemma 2.09 that if $\underset{\sim}{A}$ is orthogonal then

$$\det \underset{\sim}{A} = \pm 1.$$

c) Deduce that any isometry is an isomorphism.

d) Deduce that any isometry into is injective and an isometry onto its image.

3) What are the pictures for H^2 corresponding to those in Fig. 2.3.b,c for H^3? Find the equations in coordinates of the surfaces shown and of the curves you draw.

4) From the polarisation identity (Ex. 1.7d)

$$\underset{\sim}{A}\underset{\sim}{x}.\underset{\sim}{A}\underset{\sim}{y} = \underset{\sim}{x}.\underset{\sim}{y},\quad \forall \underset{\sim}{x},\underset{\sim}{y} \quad \Leftrightarrow \quad \underset{\sim}{A}\underset{\sim}{x}.\underset{\sim}{A}\underset{\sim}{x} = \underset{\sim}{x}.\underset{\sim}{x},\quad \forall \underset{\sim}{x}.$$

So $\underset{\sim}{A}$ is orthogonal (preserving lengths <u>and</u> angles in the Euclidean case) if it preserves "dot squares" alone. Does preserving the size $\|\ \|_{\underset{\sim}{G}}$ imply orthogonality?

3. COORDINATES

3.01 Metrics

For a metric vector space $(X,\underset{\sim}{G})$, how do we write $\underset{\sim}{G}$, $\underset{\sim}{G}_{\uparrow}$ etc. in coordinates?

Choose a basis $\beta = \underset{\sim}{b}_1, \ldots, \underset{\sim}{b}_n$. Then we have coordinates for vectors. Suppose we know for any two basis vectors $\underset{\sim}{b}_i$, $\underset{\sim}{b}_j$ what $\underset{\sim}{G}(\underset{\sim}{b}_i,\underset{\sim}{b}_j)$ is. For two general vectors, $\underset{\sim}{x} = (x^1, \ldots, x^n)$, $\underset{\sim}{y} = (y^1, \ldots, y^n)$ in these coordinates, we have $\underset{\sim}{x} = x^i\underset{\sim}{b}_i$, $\underset{\sim}{y} = y^j\underset{\sim}{b}_j$. Now we use the bilinearity of $\underset{\sim}{G}$.

$$\begin{aligned}\underset{\sim}{G}(\underset{\sim}{x},\underset{\sim}{y}) &= \underset{\sim}{G}(x^i\underset{\sim}{b}_i,\ y^j\underset{\sim}{b}_j)\\ &= x^i\underset{\sim}{G}(\underset{\sim}{b}_i,\ y^j\underset{\sim}{b}_j) \qquad \text{(linearity in 1st variable)}\\ &= x^iy^j\underset{\sim}{G}(\underset{\sim}{b}_i,\underset{\sim}{b}_j)\ . \qquad \text{(linearity in 2nd variable)}\end{aligned}$$

Thus if we define

$$g_{ij} = \underset{\sim}{G}(\underset{\sim}{b}_i,\underset{\sim}{b}_j)$$

we have the formula

$$\underset{\sim}{G}(\underset{\sim}{x},\underset{\sim}{y}) = g_{ij}x^iy^j\ .$$

There are two lower indices for the components (Ex. 1a) g_{ij} of $\underset{\sim}{G}$, since $\underset{\sim}{G}$ is covariant "twice over": it is a map from two copies of X, where a covariant vector is a map from one.

Notice that whatever basis we choose, we have $g_{ij} = g_{ji}$ in the corresponding representation of $\underset{\sim}{G}$, since by definition a metric tensor must be symmetric.

3.02 Duality

Since $\underset{\sim}{G}_\downarrow$ is an isomorphism it takes a basis $\beta = \underset{\sim}{b}_1, \ldots, \underset{\sim}{b}_n$ for X to a basis $\underset{\sim}{G}_\downarrow\beta = \underset{\sim}{G}_\downarrow\underset{\sim}{b}_1, \ldots, G_\downarrow\underset{\sim}{b}_n$ for X*. Using these two bases, its matrix is of course just $[\underset{\sim}{G}_\downarrow]_\beta^{\underset{\sim}{G}_\downarrow\beta} = [\delta_{ij}]$, the identity matrix, which is nice and simple. However, the advantages of the dual basis β^* defined without reference to a metric (III.1.07) still apply, and it is in general handier to use β^*. In this basis, the j-th component of any $\underset{\sim}{f} \in X^*$ is the value of $\underset{\sim}{f}$ on $\underset{\sim}{b}_j$. In particular, for instance,

$$\begin{aligned}
\text{3rd component of } \underset{\sim}{G}_\downarrow(\underset{\sim}{x}) &= (\underset{\sim}{G}_\downarrow\underset{\sim}{x})\underset{\sim}{b}_3 \\
&= \underset{\sim}{G}(\underset{\sim}{x},\underset{\sim}{b}_3) \\
&= g_{ij}x^i y^j \text{ where } y^j = \begin{cases} 0, & i \neq 3 \\ 1, & i = 3 \end{cases} \\
&= g_{i3}x^i.
\end{aligned}$$

Similarly for the other components, so

$$\begin{aligned}
\underset{\sim}{G}_\downarrow(\underset{\sim}{x}) &= (g_{i1}x^1, \ldots, g_{in}x^n) \\
&= g_{ij}x^j\underset{\sim}{b}^i, \quad \text{where } \underset{\sim}{b}^1, \ldots, \underset{\sim}{b}^n \text{ is the dual basis,} \\
&= g_{ij}x^j \text{ for short.}
\end{aligned}$$

(We shall often shorten the symbols for indexed quantities in this fashion. Thus, as indicated in I.2.07 we may call a vector x^i instead of $(x^1,\ldots,x^n)$, and similarly a metric tensor just g_{ij}, etc.)

So the matrix $[\underset{\sim}{G}_\downarrow]_\beta^{\beta^*}$ is just

$$\begin{bmatrix} g_{11} & g_{12} & \cdots & g_{1n} \\ g_{21} & \cdot & & \cdot \\ \cdot & & \cdot & \cdot \\ \cdot & & & \cdot \\ \cdot & & \cdot & \\ g_{n1} & & & g_{nn} \end{bmatrix}$$

and the matrix $[\underset{\sim}{G}]^{\beta}_{\beta*}$ is its inverse, whose entry in the i-th row and j-th column we denote by g^{ij}. Then the g^{ij} are defined by the n equations

$$g^{ij}g_{jk} = \delta^i_k \quad (\text{or } g_{ki}g^{ij} = \delta^j_k, \text{ etc}).$$

Moreover, if we have $\underset{\sim}{x},\underset{\sim}{y} \in X^*$, x_i, y_i in dual coordinates to β, then (cf. 1.11, 1.12):-

$$\begin{aligned}
\underset{\sim}{G}^*(\underset{\sim}{x},\underset{\sim}{y}) &= \underset{\sim}{G}(\underset{\sim}{G}_{\uparrow}\underset{\sim}{x},\underset{\sim}{G}_{\uparrow}\underset{\sim}{y}) \\
&= \underset{\sim}{G}(g^{ik}x_k\underset{\sim}{b}_i,\ g^{j\ell}\ y_\ell\underset{\sim}{b}_j) \\
&= g_{ij}(g^{ik}x_k)(g^{j\ell}y_\ell) \\
&= (g_{ij}g^{j\ell})\ g^{ik}x_ky_\ell \\
&= \delta^\ell_i g^{ik}x_ky_\ell \\
&= g^{k\ell}x_ky_\ell\ , \text{ since obviously } g^{k\ell} = g^{\ell k}. \\
&= g^{ij}x_iy_j, \text{ changing dummy indices.}
\end{aligned}$$

So the components of $\underset{\sim}{G}^*$ in dual coordinates are exactly the g^{ij}'s.

When we are working in coordinates, it will obviously be a help to choose them so that these formulae become as simple as possible: that is, we want the matrix

$[g_{ij}]$ to have a nice simple form. To this end we define:

3.03 Definition

An orthogonal set in a metric vector space X is a subset S of X any two of whose members - $\underset{\sim}{x}, \underset{\sim}{y}$ say - are orthogonal and non-null: $\underset{\sim}{x}\cdot\underset{\sim}{x} \neq 0 \neq \underset{\sim}{y}\cdot\underset{\sim}{y}$, $\underset{\sim}{x}\cdot\underset{\sim}{y} = 0$. (This implies that S must be linearly independent; cf. Ex. 2)

An orthonormal set in X is an orthogonal set of unit vectors (cf. 1.06).

An orthonormal basis for X is a basis which is an orthonormal set.

3.04 Lemma

For $\beta = \underset{\sim}{b}_1, \ldots, \underset{\sim}{b}_n$, an orthonormal basis for X, in β-coordinates we have

$$g_{ij} = \pm\delta_{ij}.$$

Proof

$$g_{ij} = \underset{\sim}{b}_i \cdot \underset{\sim}{b}_j = \begin{cases} 0 & \text{if } i \neq j \quad (\beta \text{ orthogonal}) \\ \pm 1 & \text{if } i = j\ . \quad (\underset{\sim}{b}_1, \ldots, \underset{\sim}{b}_n \text{ unit vectors}) \end{cases}$$ ■

3.05 Theorem

Every metric vector space $(X, \underset{\sim}{G})$ possesses at least one orthonormal basis.

Proof

First we need a technical lemma:

3.06 Lemma

X possesses at least one non-null vector.

Proof

Suppose not. Then $\underset{\sim}{x}.\underset{\sim}{x} = 0$, all $\underset{\sim}{x} \in X$.

Hence $(\underset{\sim}{y} + \underset{\sim}{z}).(\underset{\sim}{y} + \underset{\sim}{z}) = 0$, all $\underset{\sim}{y},\underset{\sim}{z} \in X$.

So $\underset{\sim}{y}.\underset{\sim}{y} + 2\underset{\sim}{y}.\underset{\sim}{z} + \underset{\sim}{z}.\underset{\sim}{z} = 0$ all $\underset{\sim}{y},\underset{\sim}{z} \in X$. (since $\underset{\sim}{z}.\underset{\sim}{y} = \underset{\sim}{y}.\underset{\sim}{z}$)

$$\underset{\sim}{y}.\underset{\sim}{z} = -\tfrac{1}{2}(\underset{\sim}{y}.\underset{\sim}{y} + \underset{\sim}{z}.\underset{\sim}{z})$$
$$= 0.$$

Thus $\underset{\sim}{G}$ is the zero form, which is completely degenerate and hence not a metric as assumed.

Hence for $\underset{\sim}{G}$ a metric, X has at least one non-null vector. ■

Now by the lemma, choose $\underset{\sim}{x}_1 \in X$ such that $\underset{\sim}{x}_1.\underset{\sim}{x}_1 \neq 0$ and normalise by setting

$$\underset{\sim}{b}_1 = \underset{\sim}{x}_1 / \|\underset{\sim}{x}_1\|_{\underset{\sim}{G}}. \qquad \text{(cf. 1.06)}$$

Then suppose inductively that $n > k \geq 1$ and $\underset{\sim}{b}_1, \ldots, \underset{\sim}{b}_k$ is an orthonormal set, (the vector $\underset{\sim}{b}_1$ on its own is, so we have proved it for k=1). Let B_k be the subspace they span. $\underset{\sim}{G}$ is non-degenerate on B_k since if $\underset{\sim}{x} \in B_k$, $\underset{\sim}{x} = x^i\underset{\sim}{b}_i$ with $i = 1, \ldots, k$, so that if $\underset{\sim}{x}.\underset{\sim}{y} = 0$ for all $\underset{\sim}{y} \in B_k$, in particular $x^i = \underset{\sim}{x}.\underset{\sim}{b}_i = 0$ for each i. Hence $\underset{\sim}{x} = \underset{\sim}{0}$. Then $\dim(B_k^\perp) = n-k \neq 0$ and $\underset{\sim}{G}$ is non-degenerate on $B_k^\perp$ (2.05, 2.06). Hence, by the lemma, we may choose $\underset{\sim}{x}_{k+1} \in B_k^\perp$ non-null and set

$$\underset{\sim}{b}_{k+1} = \underset{\sim}{x}_{k+1} / \|\underset{\sim}{x}_{k+1}\|_{\underset{\sim}{G}},$$

a unit vector orthogonal to each of $\underset{\sim}{b}_1, \ldots, \underset{\sim}{b}_k$.

Inductively, this produces an orthonormal set of n vectors, which must (Ex. 2) be linearly independent and hence a basis for X.

3.07 Remark

For convenience we shall always order an orthonormal basis $\underset{\sim}{b}_1, \ldots, \underset{\sim}{b}_n$ so that

$$\underset{\sim}{b}_i \cdot \underset{\sim}{b}_i = \begin{cases} +1, & i \leq k \\ -1, & i > k \end{cases}$$

for some k, putting the "timelike" vectors first, as with the Lorentz metric (1.03). This gives us the standard formula

$$\underset{\sim}{x} \cdot \underset{\sim}{y} = x^1 y^1 + \ldots + x^k y^k - x^{k+1} y^{k+1} - \ldots - x^n y^n$$

which $\underset{\sim}{G}$ will have in any orthonormal basis, with not even k depending on the choice of basis:

3.08 Theorem

For any two orthonormal bases $\beta = \underset{\sim}{b}_1, \ldots, \underset{\sim}{b}_n$ and $\beta' = \underset{\sim}{b}'_1, \ldots, \underset{\sim}{b}'_n$ for a metric vector space $(X, \underset{\sim}{G})$, with

$$\underset{\sim}{b}_i \cdot \underset{\sim}{b}_i = \begin{cases} +1, & i \leq k \\ -1, & i > k \end{cases} \quad \text{and} \quad \underset{\sim}{b}'_j \cdot \underset{\sim}{b}'_j = \begin{cases} +1, & j \leq \ell \\ -1, & j > \ell \end{cases}$$

we have $k = \ell$.

Proof

If k = 0 or n then $\underset{\sim}{G}$ is definite and $k = \ell$, so suppose that $0 < k < n$. On the subspace N of X spanned by $\underset{\sim}{b}_{k+1}, \ldots, \underset{\sim}{b}_n$, $\underset{\sim}{G}$ is negative definite, since if $\underset{\sim}{x} \in N$

$$\begin{aligned} \underset{\sim}{G}(\underset{\sim}{x}, \underset{\sim}{x}) &= \underset{\sim}{G}(x^i \underset{\sim}{b}_{k+i}, x^i \underset{\sim}{b}_{k+i}) \\ &= \sum_{i=1}^{n-k} (-(x^i)^2). \end{aligned}$$

Consider any subspace W of X on which $\underset{\sim}{G}$ is positive definite, and choose a basis $\omega = \underset{\sim}{w}_1, \ldots, \underset{\sim}{w}_r$ for W. Then the set

$$P = \{\underset{\sim}{w}_1, \ldots, \underset{\sim}{w}_r, \underset{\sim}{b}_{k+1}, \ldots, \underset{\sim}{b}_n\}$$

is linearly independent. For suppose

$$a^1\underset{\sim}{w}_1 + \ldots + a^r\underset{\sim}{w}_r + a^{r+1}\underset{\sim}{b}_{k+1} + \ldots + a^{r+(n-k)}\underset{\sim}{b}_n = \underset{\sim}{0}.$$

Then we have $\qquad * \qquad a^i\underset{\sim}{w}_i = -a^{r+j}\underset{\sim}{b}_{k+j}$

and hence $\qquad ** \qquad (a^i\underset{\sim}{w}_i).(a^i\underset{\sim}{w}_i) = (-a^{r+j}\underset{\sim}{b}_{k+j}).(-a^{r+j}\underset{\sim}{b}_{k+j}).$

But $\underset{\sim}{G}$ is positive on W, negative on N, hence both sides of ** are zero, since LHS ≥ 0, RHS ≤ 0. Hence both sides of * are zero, by the definiteness of $\underset{\sim}{G}$ on N and W. Hence $a^i = 0$, $i = 1, \ldots, r + (n-k)$, by the linear independence of ω and β, so P is indeed independent.

P therefore must have $\leq n$ members, since an independent set cannot have more than dimX members. Hence

$$\dim W = r \leq k.$$

In particular, on the subspace spanned by $\{\underset{\sim}{b}'_1, \ldots, \underset{\sim}{b}'_\ell\}$ $\underset{\sim}{G}$ is positive definite, hence $\ell \leq k$.

But by a similar argument, $k \leq \ell$.

Hence $k = \ell$. ■

3.09 Corollary

The quantity $\sum_{i=1}^{n} g_{ii} = k(+1) + (n-k)(-1) = 2k-n$ is independent of which orthonormal basis is used to give $\underset{\sim}{G}$ in coordinates. ■

This quantity is usually called the signature of $\underset{\sim}{G}$; it specifies k, by $k = \frac{1}{2}(\sum_{i=1}^{n} g_{ii} + n)$, and is given by a shorter formula than k in terms of the components of $\underset{\sim}{G}$. (cf. Ex. 6)

3.10 Corollary ("Sylvester's Law of Inertia")

For any symmetric bilinear form $\underset{\sim}{F} : X \times X \to R$, there is a choice of basis for which $\underset{\sim}{F}$ has the form

$$\underset{\sim}{F}(x^1\underset{\sim}{b}_1 + \ldots + x^n\underset{\sim}{b}_n) = (x^1)^2 + \ldots + (x^k)^2 - (x^{k+1})^2 - \ldots (x^{k+\ell})^2, \; k+\ell \leq n.$$

Unless k or ℓ is zero, the subspace V^+ spanned by the basis vectors with $\underset{\sim}{b}_i \cdot \underset{\sim}{b}_i = +1$ depends on the choice of basis; so does the subspace V^- spanned by those with $\underset{\sim}{b}_i \cdot \underset{\sim}{b}_i = -1$. However, V^0, spanned by those with $\underset{\sim}{b}_i \cdot \underset{\sim}{b}_i = 0$, depends only on $\underset{\sim}{F}$, as do the dimensions k and ℓ.

Proof

The set $V^N = \{\underset{\sim}{x} \mid \underset{\sim}{F}(\underset{\sim}{x},\underset{\sim}{y}) = 0, \forall \underset{\sim}{y} \in X\}$ is a subspace of X since $\underset{\sim}{F}(\underset{\sim}{x},\underset{\sim}{y}) = \underset{\sim}{F}(\underset{\sim}{x}',\underset{\sim}{y}) = 0$ $\forall y$ implies, by bilinearity,

$$\underset{\sim}{F}(\underset{\sim}{x}+\underset{\sim}{x}',\underset{\sim}{y}) = \underset{\sim}{F}(\underset{\sim}{x},\underset{\sim}{y}) + \underset{\sim}{F}(\underset{\sim}{x}',y) = 0$$

and

$$\underset{\sim}{F}(\lambda\underset{\sim}{x},\underset{\sim}{y}) = \lambda\underset{\sim}{F}(\underset{\sim}{x},\underset{\sim}{y}) = 0.$$

Choose basis vectors for V^N and extend to a basis $\underset{\sim}{b}_1, \ldots, \underset{\sim}{b}_n$ for X, where

$\underset{\sim}{b}_{i+1}, \ldots, \underset{\sim}{b}_n$ are the basis vectors for V^N. Hence $\dim V^N = n-i$.

Denote by W the subspace spanned by $\underset{\sim}{b}_1, \ldots, \underset{\sim}{b}_i$. Now let $\underset{\sim}{w} \in W$.

Then: $\underset{\sim}{F}(\underset{\sim}{w},\underset{\sim}{v}) = 0 \ \forall \underset{\sim}{v} \in W$ implies that :

$$\underset{\sim}{F}(\underset{\sim}{w}, x^1\underset{\sim}{b}_1 + \ldots + x^i\underset{\sim}{b}_i) + x^{i+1}\ \underset{\sim}{F}(\underset{\sim}{w},\underset{\sim}{b}_{i+1}) + \ldots + x^n\underset{\sim}{F}(\underset{\sim}{w},\underset{\sim}{b}_n) = 0$$

for all $(x^1, \ldots, x^n)$.

Using the bilinearity of $\underset{\sim}{F}$ it follows that

$$\underset{\sim}{F}(\underset{\sim}{w}, x^1\underset{\sim}{b}_1 + \ldots + x^n\underset{\sim}{b}_n) = 0 \qquad \forall\ (x^1, \ldots, x^n)$$

$$\Rightarrow \underset{\sim}{F}(\underset{\sim}{w},\underset{\sim}{x}) = 0 \qquad \forall\ \underset{\sim}{x} \in X$$

$$\Rightarrow \underset{\sim}{w} \in V^N.$$

But $\underset{\sim}{w} \in W$ so $\underset{\sim}{w} = \underset{\sim}{0}$ because $\{\underset{\sim}{b}_1, \ldots, \underset{\sim}{b}_n\}$ is independent. Therefore $\underset{\sim}{F} \mid W \times W$ is non-degenerate and we can apply Theorem 3.05. Thus we replace $\underset{\sim}{b}_1, \ldots, \underset{\sim}{b}_n$ by the orthonormal basis resulting for W and we obtain the required form for $\underset{\sim}{F}$.

The independence of k and ℓ of all except $\underset{\sim}{F}$ itself follows by the same argument as Theorem 3.08. Evidently any other expression of $\underset{\sim}{F}$ in the form above has $\underset{\sim}{b}_{i+1}, \ldots, \underset{\sim}{b}_n \in V^N$, and being linearly independent and n-i in number these vectors span V^N. Accordingly the V^o given by any basis of the required kind has $V^o = V^N$, whose definition involves only $\underset{\sim}{F}$. ■

3.11 Lemma

The dual basis $\beta^* = \underset{\sim}{b}^1, \ldots, \underset{\sim}{b}^n$ to an orthonormal basis $\beta = \underset{\sim}{b}_1, \ldots, \underset{\sim}{b}_n$ for $(X,\underset{\sim}{G})$ is orthonormal in the dual metric $\underset{\sim}{G}^*$ on X^*.

Proof

β is orthonormal $\iff$ $\underset{\sim}{G}(\underset{\sim}{b}_i,\underset{\sim}{b}_j) = \pm\delta_{ij}$

$$\iff [g_{ij}] = \begin{bmatrix} 1 & & & & \\ & 1 & & 0 & \\ & & \ddots & & \\ & 0 & & -1 & \\ & & & & -1 \end{bmatrix} = M \text{ for short (up to order along diagonal)}$$

$$\iff [g^{ij}] = M \qquad \text{(using 3.01)}$$

$\iff$ β* is orthonormal. ■

3.12 Corollary

The signature of $\underset{\sim}{G}^*$ equals the signature of $\underset{\sim}{G}$. (Ex. 3) ■

3.13 Lemma

If $\underset{\sim}{A}$ is an operator on an inner product space $(X,\underset{\sim}{G})$, then (in the notations 2.08, III.1.06)

$$[\underset{\sim}{A}^T]^\beta_\beta = ([\underset{\sim}{A}]^\beta_\beta)^t$$

with respect to any orthonormal basis $\beta = \underset{\sim}{b}_1, \ldots, \underset{\sim}{b}_n$.

Proof

$(\underset{\sim}{G}_\downarrow \underset{\sim}{b}_i)\underset{\sim}{b}_j = \underset{\sim}{b}_i \cdot \underset{\sim}{b}_j = \delta_{ij} = \underset{\sim}{b}^i(\underset{\sim}{b}_j)$, hence $\underset{\sim}{G}_\downarrow(\underset{\sim}{b}_i) = \underset{\sim}{b}^i$.

Hence $[\underset{\sim}{G}_\downarrow]^{\beta^*}_\beta$ is the identity matrix I, hence so also is $[\underset{\sim}{G}_\uparrow]^\beta_{\beta^*}$.

Thus as matrices,

$$[A^T]^\beta_\beta = [\underset{\sim}{G}_\uparrow \underset{\sim}{A}^* \underset{\sim}{G}_\downarrow]^\beta_\beta = [\underset{\sim}{G}_\uparrow]^\beta_{\beta^*}\,[\underset{\sim}{A}^*]^{\beta^*}_{\beta^*}[\underset{\sim}{G}_\downarrow]^{\beta^*}_\beta = I[\underset{\sim}{A}^*]^{\beta^*}_{\beta^*}\,I = [\underset{\sim}{A}^*]^{\beta^*}_{\beta^*} = ([\underset{\sim}{A}]^\beta_\beta)^t \text{ by III.1.07.}$$

■

When $\underset{\sim}{G}$ is indefinite, the matrix of the adjoint is still closely related to the transpose. However, if a^i_j is a term giving the image of a timelike basis vector $\underset{\sim}{b}_j$ a component along a spacelike basis vector $\underset{\sim}{b}_i$, then we have a sign change in the i aspect from $\underset{\sim}{G}_\uparrow$ but not in the j aspect from $\underset{\sim}{G}_\downarrow$, so the sign of a^i_j changes. This is a vague statement: the precise one, of which 3.13 is a special case, follows.

3.14 Lemma

If $\underset{\sim}{A}$ is an operator on a metric vector space $(X,\underset{\sim}{G})$ then with respect to any orthonormal basis $\underset{\sim}{b}_1, \ldots, \underset{\sim}{b}_n$ we have

$$[\underset{\sim}{A}^T]^i_j = \left(\frac{g_{jj}}{g_{ii}}\right) [\underset{\sim}{A}]^j_i$$

for each i,j (with no summation).

Proof

Let $[\underset{\sim}{A}]^i_j = a^i_j$. We illustrate the case for i = 3.

$$\begin{aligned}(\underset{\sim}{A}\underset{\sim}{b}_j)\cdot\underset{\sim}{b}_3 &= (a^k_j\,\underset{\sim}{b}_k)\cdot\underset{\sim}{b}_3\\ &= a^k_j\,(\underset{\sim}{b}_k\cdot\underset{\sim}{b}_3)\\ &= a^3_j(\underset{\sim}{b}_3\cdot\underset{\sim}{b}_3) \text{ since } \underset{\sim}{b}_k\cdot\underset{\sim}{b}_3 = 0,\ k \neq 3\\ &= [\underset{\sim}{A}]^3_j\,(\underset{\sim}{b}_3\cdot\underset{\sim}{b}_3).\end{aligned}$$

Hence,

$$\begin{aligned}[\underset{\sim}{A}^T]^i_3 &= \frac{(\underset{\sim}{A}^T\underset{\sim}{b}_3)\cdot\underset{\sim}{b}_i}{\underset{\sim}{b}_i\cdot\underset{\sim}{b}_i}\\ &= \frac{\underset{\sim}{G}(\underset{\sim}{G}_\uparrow\underset{\sim}{A}^*\underset{\sim}{G}_\downarrow(\underset{\sim}{b}_3),\underset{\sim}{b}_i)}{\underset{\sim}{b}_i\cdot\underset{\sim}{b}_i}\end{aligned}$$

$$= \frac{(\mathbf{A}^*\mathbf{G}_{\downarrow}(\mathbf{b}_3))\mathbf{b}_i}{\mathbf{b}_i\cdot\mathbf{b}_i}$$

$$= \frac{\mathbf{G}_{\downarrow}\mathbf{b}_3(\mathbf{A}\mathbf{b}_i)}{\mathbf{b}\cdot\mathbf{b}_i}$$

$$= \frac{\mathbf{b}_3\cdot\mathbf{A}\mathbf{b}_i}{\mathbf{b}_i\cdot\mathbf{b}_i}$$

$$= \frac{a_i^3(\mathbf{b}_3\cdot\mathbf{b}_3)}{\mathbf{b}_i\cdot\mathbf{b}_i} \qquad \text{by the equation above}$$

$$= a_i^3\left(\frac{g_{33}}{g_{ii}}\right), \qquad \text{not summing over i.}$$

Thus if an operator $\mathbf{Q}$ on a metric vector space of dimension n and signature σ has a matrix, subdivided into

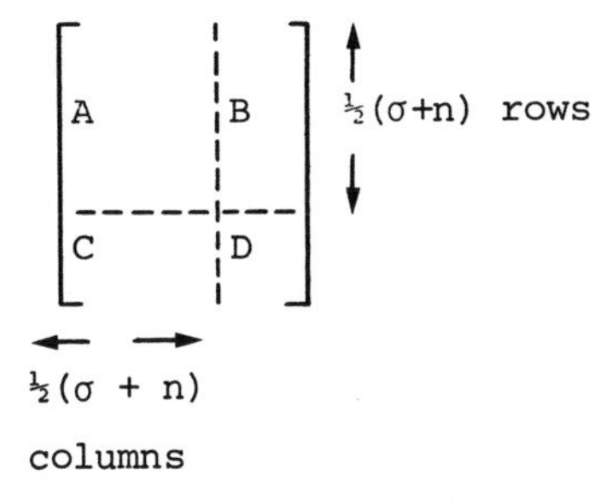

,

in an orthonormal basis with the first $\frac{1}{2}(\sigma + n)$ $\mathbf{b}_i$'s timelike, then the adjoint has the matrix

$$\left[\begin{array}{c|c} A^t & -C^t \\ \hline -B^t & D^t \end{array}\right].$$

<u>3.15 Corollary</u>

An operator $\mathbf{A}$ on a metric vector space is self-adjoint if and only if with respect to an orthonormal basis its matrix $[a_j^i]$ satisfies

$$a^i_j = \left(\frac{g_{jj}}{g_{ii}}\right) a^j_i$$

for each i,j (without summation). ■

With an inner product space, this means simply that $[a^i_j]$ is symmetrical about the diagonal; hence in this case self-adjoint operators are often called symmetric operators. With an indefinite metric the matrix is "symmetric" in some parts and "skew-symmetric" in others, and it is more natural to stick to "self-adjoint" (cf. 2.08).

3.16 Lemma

An operator $\underset{\sim}{A}$ on an inner product space is orthogonal if and only if with respect to an orthonormal basis it has a matrix whose columns [respectively, rows] regarded as column [respectively, row] vectors form an orthonormal set in the standard inner product. (This is also true for metric vector spaces, and little harder to prove (Ex. 4).)

Proof

$\underset{\sim}{A}$ is orthogonal $\Leftrightarrow$ $\underset{\sim}{A}^T\underset{\sim}{A} = \underset{\sim}{I}$ (2.09).

$$\Leftrightarrow [\underset{\sim}{A}^T]^k_i\,[\underset{\sim}{A}]^i_j = \delta^k_j$$

$$= \sum_{i=1}^{n} a^i_k\, a^i_j = \delta^k_j,$$ which is exactly the statement that the columns of $[\underset{\sim}{A}]$ are an orthonormal set.

For the "rows" version, apply Lemma 2.10. ■

3.17 Corollary

The rows of a matrix are an orthonormal set (in the standard inner product) if and only if the columns are. ■

This fact, rather impressive and magical at the "matrix theory" level, was the reason for the term "orthogonal" matrix, and hence orthogonal operators. Since the columns are exactly the coordinates of the images under $\underset{\sim}{A}$ of the vectors in the standard basis, which is orthonormal in the standard inner product, 3.16 amounts geometrically to the statement that $\underset{\sim}{A}$ preserves the metric for all vectors if it does so for basis vectors. This is true for any basis, by linearity, but not so simply stated in coordinates for a non-orthonormal basis.

Exercises IV.3

1a) Define a basis $\underset{\sim}{M}^{ij}$ for the vector space $L^2(X;R)$ in terms of that chosen for X such that

$$\underset{\sim}{G} = g_{ij}\underset{\sim}{M}^{ij}$$

where the g_{ij} are those defined in 3.01. (Thus the g_{ij} are components of $\underset{\sim}{G}$ in the vector space $L^2(X;R)$, as the a^i_j are for an operator $\underset{\sim}{A} \in L(X;X)$: cf. I.2.07. These two cases are specialisations of more general definitions given in Chapter V.)

b) If $\underset{\sim}{G}$ has components g_{ij} in the basis β, and a new basis consists of the vectors $\underset{\sim}{b}_i = (b^1_i, \ldots, b^n_i)$, i = 1, ..., n, in β-coordinates, derive the formula

$$g'_{k\ell} = g_{ij}b^i_k\, b^j_\ell$$

for the components of $\underset{\sim}{G}$ in β'-coordinates.

2) If $\{\underset{\sim}{b}_1, \ldots, \underset{\sim}{b}_r\}$ in $(X,\underset{\sim}{G})$ is an orthogonal set (so $\underset{\sim}{b}_i \cdot \underset{\sim}{b}_j = 0$ if and only if $i \neq j$), and $\underset{\sim}{x} = a^i \underset{\sim}{b}_i = \underset{\sim}{0}$, deduce that $\underset{\sim}{x} \cdot \underset{\sim}{b}_j = 0$ for $j = 1, \ldots, r$ and hence that each $a^i = 0$.

3) Deduce Corollary 3.12 from the proof of Lemma 3.11.

4) Prove Lemma 3.16 for metric vector spaces.

5) Show, by the method of proof of Theorem 3.05, that for any $\underset{\sim}{x}$ in 4-dimensional Lorentz space with $\underset{\sim}{x} \cdot \underset{\sim}{x} > 0$ [respectively $\underset{\sim}{x} \cdot \underset{\sim}{x} < 0$] there is a choice of coordinates giving $\underset{\sim}{x}$ the form $(t,0,0,0)$ [respectively $(0,x,0,0)$] and giving the metric the standard expression of Def^n 1.03.

6) Show that $\{(1,0,0,1), (0,1,-1,0), (1,0,0,-1), (0,1,1,0)\}$ is an orthonormal basis for R^4 with the determinant metric (cf. 1.03). Find the signature of this metric tensor (cf. 3.09).

7a) Show, by the method of proof of Theorem 3.05, that any unit timelike vector $\underset{\sim}{v}$ can be a member of an orthonormal basis $\underset{\sim}{v}, \underset{\sim}{b}_2, \ldots, \underset{\sim}{b}_n$.

b) Deduce that for any two unit timelike vectors $\underset{\sim}{v},\underset{\sim}{w}$ there is an orthogonal operator $\underset{\sim}{A}$ with $\underset{\sim}{A}\underset{\sim}{v} = \underset{\sim}{w}$. (Construct bases as in (a), use I.2.05 to find $\underset{\sim}{A}$ then establish its orthogonality.)

c) Prove the same thing for $\underset{\sim}{v},\underset{\sim}{w}$ unit spacelike vectors.

d) Prove the same thing for two null vectors. (Hint: an orthonormal basis indicates how to write a null vector as $\underset{\sim}{s}+\underset{\sim}{t}$, with spacelike $\underset{\sim}{s}$ and timelike $\underset{\sim}{t}$ separately moveable.)

e) Deduce that for any two vectors $\underset{\sim}{v},\underset{\sim}{w}$ there is an orthogonal operator $\underset{\sim}{A}$ with $\underset{\sim}{A}\underset{\sim}{v} = \underset{\sim}{w}$ if and only if $\underset{\sim}{v} \cdot \underset{\sim}{v} = \underset{\sim}{w} \cdot \underset{\sim}{w}$.

4. DIAGONALISING SYMMETRIC OPERATORS

It is convenient to have an orthonormal basis, so that the matrix of g_{ij}'s has a simple diagonal form. Likewise, if for some operator $\underset{\sim}{A}$ we can find a basis consisting entirely of eigenvectors of $\underset{\sim}{A}$, then $[\underset{\sim}{A}]$ will be very simple. For, if $\underset{\sim}{b}_i$ belongs to λ_i, (with $\lambda_1, \ldots, \lambda_n$ not necessarily distinct) we have

$$\begin{aligned}
\underset{\sim}{A}(x^1, \ldots, x^n) &= \underset{\sim}{A}(x^i \underset{\sim}{b}_i) \\
&= x^i (\underset{\sim}{A}\underset{\sim}{b}_i) \\
&= x^i (\lambda_i \underset{\sim}{b}_i) \qquad \text{(by I.3.07)} \\
&= (\lambda_1 x^1, \ldots, \lambda_n x^n)
\end{aligned}$$

so that the matrix of $\underset{\sim}{A}$ is just

$$\begin{bmatrix} \lambda_1 & & & \\ & \lambda_2 & & 0 \\ & & \ddots & \\ & 0 & & \lambda_n \end{bmatrix}$$

This is algebraically convenient and geometrically clear: it breaks $\underset{\sim}{A}$ down into scalar multiplication in various directions. This simplification, however is not worth it if the basis of eigenvectors is not orthonormal in the metric we are using: it is more helpful to have the metric in a simple form than to simplify an operator. The great advantage of <u>symmetric</u> operators, self-adjoint operators on an inner product space, is that we can have both, as we establish in the next few lemmas. The idea involved is as follows.

If we have an operator $\mathbf{A}$ for which there is an orthonormal basis of eigenvectors, which is in fact true if $\mathbf{A}$ is symmetric, the unit sphere $\{\mathbf{x} \mid \mathbf{x}.\mathbf{x} = 1\}$ is taken by $\mathbf{A}$ to an ellipsoid with the eigenvectors along the principal axes. Examples for two and three dimensions are shown in Fig. 4.1. There the $\rightarrow$ arrows represent unit eigenvectors, and the circle/sphere represents the unit sphere, carried by $\mathbf{A}$ to the $\twoheadrightarrow$ arrows and the ellipse/ellipsoid.

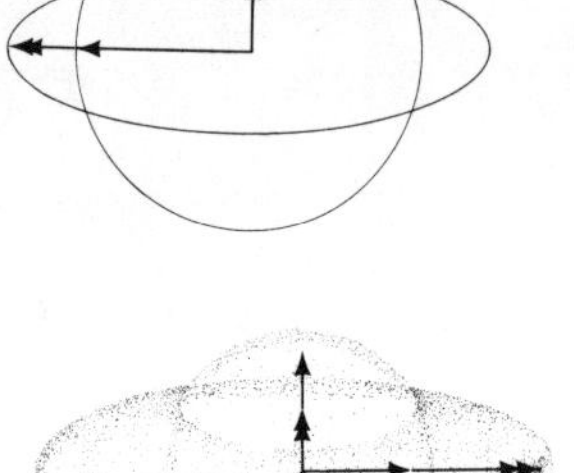

Fig. 4.1

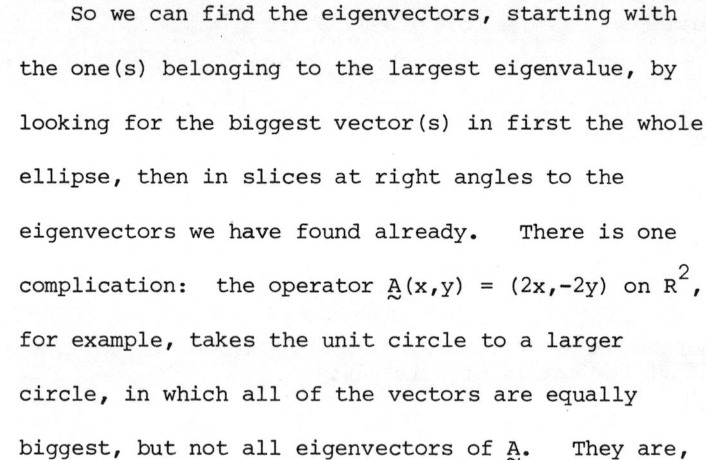

So we can find the eigenvectors, starting with the one(s) belonging to the largest eigenvalue, by looking for the biggest vector(s) in first the whole ellipse, then in slices at right angles to the eigenvectors we have found already. There is one complication: the operator $\mathbf{A}(x,y) = (2x,-2y)$ on R^2, for example, takes the unit circle to a larger circle, in which all of the vectors are equally biggest, but not all eigenvectors of $\mathbf{A}$. They are, however, eigenvectors of $\mathbf{A}^2$, and it turns out that by an algebraic trick (Lemma 4.03) the eigenvectors of $\mathbf{A}^2$ easily lead to those of $\mathbf{A}$.

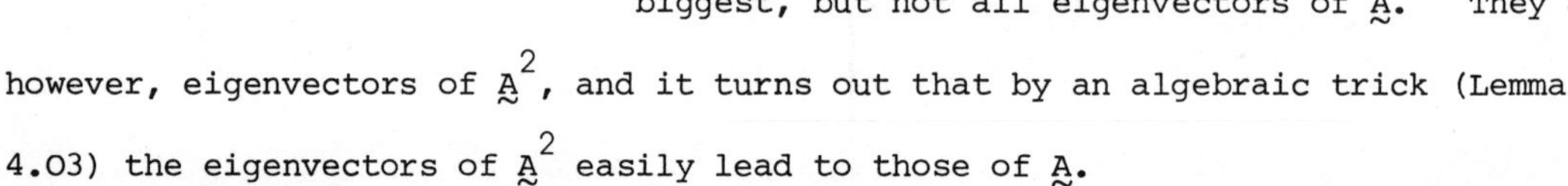

4.01 Definition

If $\mathbf{A}$ is any operator on an inner product space X, a vector $\mathbf{x}$ is <u>maximal</u> for $\mathbf{A}$ if $\mathbf{x}$ is a unit vector and

$$\mathbf{Ax}.\mathbf{Ax} \geq \mathbf{Ay}.\mathbf{Ay}$$

for all unit vectors $\mathbf{y} \in X$.

For X finite dimensional, it turns out that $\mathbf{A}$ must always have maximal vectors. For $\mathbf{x} \rightsquigarrow \mathbf{Ax}.\mathbf{Ax}$ is a continuous real-valued function on the set of unit vectors, which is closed and bounded in X (which can be taken as a copy of R^n here). Hence its maximum value exists and is attained, as proved below in Chapter VI 4.12. (This is essentially a topological fact, which is why we must defer its proof till we have the right machinery. The proof will not depend on the diagonalisability of symmetric operators, so we are not being circular.)

If $\underset{\sim}{x}$ is maximal for $\underset{\sim}{A}$, then $\|\underset{\sim}{A}\underset{\sim}{x}\| = \max\{\underset{\sim}{A}\underset{\sim}{y}\cdot\underset{\sim}{A}\underset{\sim}{y} \mid \underset{\sim}{y} \in X, \underset{\sim}{y}\cdot\underset{\sim}{y} = 1\}$ is often called the <u>norm</u> of $\underset{\sim}{A}$ and denoted by $\|\underset{\sim}{A}\|$. For <u>all</u> $\underset{\sim}{y}$, then, we have $\|\underset{\sim}{A}\underset{\sim}{y}\| \leq \|\underset{\sim}{A}\|\,\|\underset{\sim}{y}\|$. (Normalise $\underset{\sim}{y}$, apply the definition of $\|\underset{\sim}{A}\|$, and denormalise; Ex. 1.)

<u>4.02 Lemma</u>

If $\underset{\sim}{x}$ is a maximal vector of a symmetric operator $\underset{\sim}{A}$ on an inner product space X, then $\underset{\sim}{x}$ is an eigenvector of the operator $\underset{\sim}{A}^2$, belonging to the eigenvalue $\|\underset{\sim}{A}\|^2$.

<u>Proof</u>

$$
\begin{aligned}
\|\underset{\sim}{A}\|^2 = \|\underset{\sim}{A}\underset{\sim}{x}\|^2 &= \underset{\sim}{A}\underset{\sim}{x}\cdot\underset{\sim}{A}\underset{\sim}{x} \quad (\text{def}^n \text{ of } \|\underset{\sim}{A}\underset{\sim}{x}\|)\\
&= \underset{\sim}{A}^2\underset{\sim}{x}\cdot\underset{\sim}{x} \quad (\underset{\sim}{A} \text{ symmetric})\\
&\leq \|\underset{\sim}{A}^2\underset{\sim}{x}\|\,\|\underset{\sim}{x}\| \quad (\text{Lemma 1.07}) \quad \bigstar\\
&= \|\underset{\sim}{A}^2\underset{\sim}{x}\| \quad (\underset{\sim}{x} \text{ a unit vector})\\
&= \|\underset{\sim}{A}(\underset{\sim}{A}\underset{\sim}{x})\|\\
&\leq \|\underset{\sim}{A}\|\,\|\underset{\sim}{A}\underset{\sim}{x}\|\\
&\leq \|\underset{\sim}{A}\|\,(\|\underset{\sim}{A}\|\,\|\underset{\sim}{x}\|)\\
&= \|\underset{\sim}{A}\|^2. \quad (\underset{\sim}{x} \text{ a unit vector})
\end{aligned}
$$

So all of the inequalities must actually be equalities in this case, since they are squeezed in between equal quantities. But in the Schwarz inequality★, equality only holds if $\underset{\sim}{A}^2\underset{\sim}{x} = \underset{\sim}{x}a$ for some $a \in R$. Thus $\underset{\sim}{x}$ is an eigenvector of $\underset{\sim}{A}^2$, with eigenvalue

$$a = a(\underset{\sim}{x}.\underset{\sim}{x}) = (\underset{\sim}{x}a).\underset{\sim}{x} = (\underset{\sim}{A}^2\underset{\sim}{x}).\underset{\sim}{x} = \|\underset{\sim}{A}\|^2 .$$ ■

4.03 Lemma

A symmetric operator $\underset{\sim}{A}$ on an inner product space has an eigenvector belonging to an eigenvalue $+\|\underset{\sim}{A}\|$ or $-\|\underset{\sim}{A}\|$.

Proof

Take a maximal vector $\underset{\sim}{x}$ of $\underset{\sim}{A}$. Then by 4.02,

$$\underset{\sim}{A}^2\underset{\sim}{x} = \underset{\sim}{x}\|\underset{\sim}{A}\|^2,$$

$$\text{so,} \quad (\underset{\sim}{A}^2 - \|\underset{\sim}{A}\|^2\underset{\sim}{I})\underset{\sim}{x} = \underset{\sim}{0}.$$

Hence $(\underset{\sim}{A} + \|\underset{\sim}{A}\|\underset{\sim}{I})(\underset{\sim}{A} - \|\underset{\sim}{A}\|\underset{\sim}{I})\underset{\sim}{x} = \underset{\sim}{0}$.

Hence either $(\underset{\sim}{A} - \|\underset{\sim}{A}\|\underset{\sim}{I})\underset{\sim}{x} = \underset{\sim}{0}$, in which case $\underset{\sim}{x}$ is an eigenvector of $\underset{\sim}{A}$ belonging to the eigenvalue $+\|\underset{\sim}{A}\|$, or not, in which case $(\underset{\sim}{A} - \|\underset{\sim}{A}\|\underset{\sim}{I})\underset{\sim}{x}$ is an eigenvector of $\underset{\sim}{A}$ belonging to the eigenvalue $-\|\underset{\sim}{A}\|$. (cf. Ex. 2) ■

4.04 Lemma

If $\underset{\sim}{x}$ is an eigenvector of a self-adjoint operator $\underset{\sim}{A}$ on a metric vector space, then

$$\underset{\sim}{x}.\underset{\sim}{y} = 0 \Rightarrow \underset{\sim}{x}.\underset{\sim}{A}\underset{\sim}{y} = 0.$$

That is, $A(x^{\perp}) \subseteq x^{\perp}$, so that $y \rightsquigarrow Ay$ defines an operator on $x^{\perp}$: that induced by A. (This is not true for A not self-adjoint (Ex. 3), nor very useful if $x \in x^{\perp}$.)

Proof

If x belongs to the eigenvalue λ, then

$$x \cdot y = 0 \Rightarrow \lambda(x \cdot y) = 0 \Rightarrow (x\lambda) \cdot y = 0 \Rightarrow Ax \cdot y = 0 \Rightarrow x \cdot Ay = 0. \blacksquare$$

4.05 Theorem

If A is a symmetric operator on an inner product space X, then X has an orthonormal basis of eigenvectors of A.

Proof

Let dim X = n.

Find an eigenvector x_1 by Lemma 4.03, and set

$$b = x_1 / \|x_1\|$$

to give a unit eigenvector. Then by Lemma 4.04 we can consider the induced operator

$$A' : b^{\perp} \to b^{\perp} : x \rightsquigarrow Ax.$$

This is again symmetric, with respect to the restriction of the inner product. Hence we can find a unit eigenvector b_2 of A', which is then also an eigenvector of A and orthogonal to b_1 since it is in $b_1^{\perp}$. Inductively, we may continue this process, each time producing a unit eigenvector orthogonal to all the previous ones. Since we decrease the dimension of the space (to which we apply 4.03) by one each time, we produce exactly n eigenvectors before we run out of space. These are an orthonormal set of n vectors, hence independent and a basis. ■

4.06 Corollary

$\underset{\sim}{A}$ can be represented by a diagonal matrix with respect to an orthonormal basis. ■

4.07 Corollary

If μ is a root of multiplicity m of the characteristic equation

$$\det(\underset{\sim}{A} - \lambda \underset{\sim}{I}) = 0$$

then the eigenspace belonging to μ has dimension m. (This applies as a general result only if the metric is an inner product and $\underset{\sim}{A}$ symmetric : cf. Ex. 5.)

Proof

Represent $\underset{\sim}{A}$, via an orthonormal basis $\underset{\sim}{b}_1, \ldots, \underset{\sim}{b}_n$, by a diagonal matrix with entries the eigenvalues $\lambda_1, \ldots, \lambda_n$ of $\underset{\sim}{A}$ (not necessarily distinct or non-zero).

Then $\det(\underset{\sim}{A} - \lambda \underset{\sim}{I}) = (\lambda_1 - \lambda)(\lambda_2 - \lambda) \ldots (\lambda_n - \lambda)$

so that μ is a root with multiplicity m if and only if $\lambda_j = \mu$ for exactly m j's, $\lambda_{j_1} = \lambda_{j_2} = \ldots = \lambda_{j_m} = \mu$, say. Then the subspace spanned by $\underset{\sim}{b}_{j_1}, \ldots, \underset{\sim}{b}_{j_m}$ is exactly the eigenspace belonging to μ. ■

4.08 Corollary

All the roots of the characteristic equation of $\underset{\sim}{A}$ are real. ■

4.09 Corollary

In an inner product space $(X, \underset{\sim}{G})$ for any symmetric bilinear form $\underset{\sim}{h}$ on X we can find an orthonormal basis $\underset{\sim}{b}_1, \ldots, \underset{\sim}{b}_n$ for X such that

$$\underset{\sim}{h}(\underset{\sim}{b}_i, \underset{\sim}{b}_j) = 0 \text{ if } i \neq j.$$

Proof

For $\underset{\sim}{x} \in X$ define $\underset{\sim}{h}_{\underset{\sim}{x}} : X \to R : \underset{\sim}{y} \rightsquigarrow \underset{\sim}{h}(\underset{\sim}{x}, \underset{\sim}{y})$.

Next define $\underset{\sim}{A}_{\underset{\sim}{h}} : X \to X : \underset{\sim}{x} \rightsquigarrow \underset{\sim}{G}_{\uparrow}(\underset{\sim}{h}_{\underset{\sim}{x}})$. Then we have

$$\underset{\sim}{A}_{\underset{\sim}{h}}\underset{\sim}{x}\cdot\underset{\sim}{y} = (\underset{\sim}{G}_{\downarrow}(\underset{\sim}{A}_{\underset{\sim}{h}}\underset{\sim}{x}))\underset{\sim}{y} = \underset{\sim}{h}_{\underset{\sim}{x}}(\underset{\sim}{y}) = \underset{\sim}{h}(\underset{\sim}{x},\underset{\sim}{y}) = \underset{\sim}{h}(\underset{\sim}{y},\underset{\sim}{x}) = \underset{\sim}{h}_{\underset{\sim}{y}}(\underset{\sim}{x}) = \underset{\sim}{A}\underset{\sim}{y}\cdot\underset{\sim}{x} = \underset{\sim}{x}\cdot\underset{\sim}{A}\underset{\sim}{y},$$

since $\underset{\sim}{G}$ and $\underset{\sim}{h}$ are symmetric. Choosing by 4.05 an orthonormal basis $\underset{\sim}{b}_1, \ldots, \underset{\sim}{b}_n$ which diagonalises $\underset{\sim}{A}$ we have

$$\underset{\sim}{h}(\underset{\sim}{b}_i, \underset{\sim}{b}_j) = \underset{\sim}{A}\underset{\sim}{b}_i\cdot\underset{\sim}{b}_j = \lambda_i(\underset{\sim}{b}_i\cdot\underset{\sim}{b}_j) = 0 \quad \text{if } i \neq j.$$

(Unless $\underset{\sim}{h}$ is non-degenerate, some of the eigenvalues λ_i of $\underset{\sim}{A}$ will be zero.) ■

4.10 Definition

The $\underset{\sim}{b}_i$ of 4.09 are called <u>principal directions</u> of $\underset{\sim}{h}$. Notice that if $\det(\underset{\sim}{A}-\lambda\underset{\sim}{I})$ has coincident zeros, the directions are not uniquely defined. For if $\underset{\sim}{b}_1$ and $\underset{\sim}{b}_2$ both belong to λ, then so do $\underset{\sim}{b}_1' = \underset{\sim}{b}_1+\underset{\sim}{b}_2$ and $\underset{\sim}{b}_2' = \underset{\sim}{b}_1-\underset{\sim}{b}_2$. Thus $\underset{\sim}{b}_1'\sqrt{\tfrac{1}{2}}, \underset{\sim}{b}_2'\sqrt{\tfrac{1}{2}}, \underset{\sim}{b}_3, \ldots, \underset{\sim}{b}_n$ is an orthonormal basis diagonalising $\underset{\sim}{A}$ and $\underset{\sim}{h}$ equally well.

If they are completely indeterminate, that is if $\underset{\sim}{A}$ has one eigenvalue λ to which all of X belongs, $\underset{\sim}{h}$ is called <u>isotropic</u>.

4.11 Lemma

If $\underset{\sim}{h}$ is isotropic, then $\underset{\sim}{h} = \lambda\underset{\sim}{G}$ for some $\lambda \in R$ and the corresponding $\underset{\sim}{A} = \lambda\underset{\sim}{I}$. ■

4.12 Remark

Theorems 4.05, 4.09 are not (unlike for instance the existence of orthonormal bases, 3.05) true whether $\underset{\sim}{G}$ is definite or indefinite. For, if $\underset{\sim}{A} : H^2 \to H^2 : (x,y) \mapsto (x-y,x+y)$, $\underset{\sim}{A}$ is self-adjoint and $\underset{\sim}{h}((x,y),(x',y')) = \underset{\sim}{A}(x,y).(x',y') = xx' - x'y - xy' - yy'$ is correspondingly symmetric. But $\underset{\sim}{A}$ has no eigenvectors and $\underset{\sim}{h}$ has no principal directions. (Either look at the characteristic equation of $\underset{\sim}{A}$ (cf. I. 3.13) or draw pictures - preferably both; cf. also Ex. 5.) We shall need:

4.13 Lemma

If a self-adjoint linear operator $\underset{\sim}{A}$, on Lorentz space L^4 (cf. 1.03), has a timelike eigenvector $\underset{\sim}{v}$, then L^4 has an orthonormal basis of eigenvectors of $\underset{\sim}{A}$.

Proof

By 4.04, $\underset{\sim}{A}$ restricts to an operator on $\underset{\sim}{v}^{\perp}$, which is non-degenerate by 2.06. By the arguments of the proof of 3.08, $\underset{\sim}{G}$ is negative definite on $\underset{\sim}{v}^{\perp}$, and thus an inner product. Hence 4.05 applies and we have three spacelike orthonormal eigenvectors for $\underset{\sim}{A} \mid \underset{\sim}{v}^{\perp}$. These, with $\underset{\sim}{v}$, provide the required basis. ■

Exercises IV.4.

1) Prove the inequality $\|\underset{\sim}{A}\underset{\sim}{x}\| \leq \|\underset{\sim}{A}\| \, \|\underset{\sim}{x}\|$ of 4.01

2) Draw the pictures corresponding to the two possibilities involved in the proof of Lemma 4.03, in the case of the operator $\underset{\sim}{A}(x,y) = (2x,-2y)$ on R^2. What difference would it make if we factorised $(\underset{\sim}{A}^2 - \|\underset{\sim}{A}\|^2 \underset{\sim}{I})$ as $(\underset{\sim}{A} - \|\underset{\sim}{A}\| \underset{\sim}{I})(\underset{\sim}{A} + \|\underset{\sim}{A}\| \underset{\sim}{I})$?

3) Give an example of an operator $\underset{\sim}{A}$, on R^2 with the standard inner product, having an eigenvector $\underset{\sim}{x} = (0,1)$ and such that $\underset{\sim}{A}(\underset{\sim}{x}^{\perp}) \not\subseteq \underset{\sim}{x}^{\perp}$.

4) If $\underset{\sim}{x},\underset{\sim}{y}$ are eigenvectors of a symmetric operator belonging to eigenvalues λ,μ, with $\lambda \neq \mu$, then

$$\lambda(\underset{\sim}{x}\cdot\underset{\sim}{y}) = \mu(\underset{\sim}{x}\cdot\underset{\sim}{y})$$

Deduce that eigenvectors belonging to distinct eigenvalues are orthogonal. (cf. Ex. I. 3.7 for the non-symmetric case.)

5a) If $\underset{\sim}{A} : R^2 \to R^2$ has the matrix $\begin{bmatrix} 1 & 1 \\ -1 & 3 \end{bmatrix}$, show that $\det(\underset{\sim}{A}-\lambda\underset{\sim}{I}) = 0$ has 2 as a root with multiplicity 2, but the eigenspace belonging to 2 has dimension only 1. (Draw it!)

b) Show that $\underset{\sim}{A}$ is self-adjoint as an operator on H^2 (1.05).

V Tensors and Multilinear Forms

"In that One Void the two are not distinguished:
each contains complete within itself the ten thousand forms."
Seng-ts'an.

1. MULTILINEAR FORMS

Starting with a vector space X, we already have several spaces derived from it, such as the dual space X* and the spaces $L^2(X;R)$ and $L^2(X^*;R)$ of bilinear forms on X and X*. We shall now produce some more. Fortunately, rather than adding more spaces to an ad hoc list, all as different as, say X* and $L^2(X;R)$ seem from each other, our new construction gives a general framework in which all the spaces so far considered occur as special cases. (This gathering of apparently very different things into one grand structure where they appear as examples is common in mathematics - both because it is often a very powerful tool and because many mathematicians have great difficulty in remembering facts they can't deduce from a framework, like the atomic weight of copper or the date of the battle of Pondicherry. This deficiency is often what pushed them to the subject and away from chemistry or history in the first place, at school.)

We start by defining a generalisation of bilinear forms.

1.01 Definition

A function

$$\underset{\sim}{f} : X_1 \times X_2 \times \dots \times X_n \to Y$$

where $X_1, \ldots, X_n$, Y are vector spaces, is a multilinear mapping if

(i) $\underset{\sim}{f}(\underset{\sim}{x}_1,\ldots,\underset{\sim}{x}_i+\underset{\sim}{x}_i',\ldots,\underset{\sim}{x}_n) = \underset{\sim}{f}(\underset{\sim}{x}_1,\ldots,\underset{\sim}{x}_i,\ldots,\underset{\sim}{x}_n) + \underset{\sim}{f}(\underset{\sim}{x}_1,\ldots,\underset{\sim}{x}_i',\ldots,\underset{\sim}{x}_n)$

(ii) $\underset{\sim}{f}(\underset{\sim}{x}_1,\ldots,\underset{\sim}{x}_i a,\ldots,x_n) = (\underset{\sim}{f}(\underset{\sim}{x}_1,\ldots,\underset{\sim}{x}_i,\ldots,\underset{\sim}{x}_n))a$

for any $\underset{\sim}{x}_1 \in X_1,\ldots,\underset{\sim}{x}_i$ and $\underset{\sim}{x}_i' \in X_i,\ldots,\underset{\sim}{x}_n \in X_n$, $i \in \{1,\ldots,n\}$, $a \in R$.

The vector space (cf. Ex. 1) of all such functions is denoted by $L(X_1,\ldots,X_n;Y)$.

In particular, denote $L(\overbrace{X,\ldots,X}^{n \text{ times}};Y)$ by $L^n(X;Y)$.

If $\underset{\sim}{f} \in L^n(X;R)$, $\underset{\sim}{f}$ is called a multilinear form on X. (Notice that $L^1(X;R) = L(X;R) = X^*$, and that $L^2(X;R)$ has already been introduced (IV.1.01) under exactly this symbol.)

1.02 Examples

(i) Let X be a three-dimensional space and $\underset{\sim}{f}(\underset{\sim}{x}_1,\underset{\sim}{x}_2,\underset{\sim}{x}_3)$ be the volume of the parallelepiped determined by $\underset{\sim}{x}_1$, $\underset{\sim}{x}_2$ and $\underset{\sim}{x}_3$ (with negative volume if they are in "left-hand" order as in Fig. 1.1). Euclidean geometry will show that $\underset{\sim}{f}$ is multilinear on X (Ex. 2a).

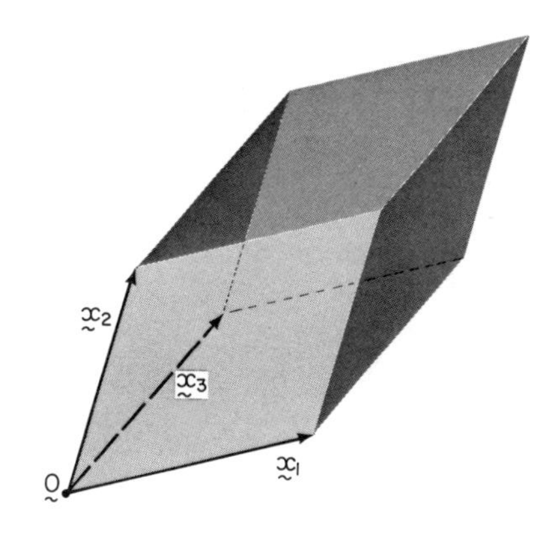

Fig. 1.1

(ii) If we define

$$\underset{\sim}{f} : L(W;X) \times L(X;Y) \times L(Y;Z) \to L(W;Z) : (\underset{\sim}{A},\underset{\sim}{B},\underset{\sim}{C}) \mapsto \underset{\sim}{CBA}$$

then $\underset{\sim}{f} \in L\Big(L(W;X), L(X;Y), L(Y;Z);\ L(W;Z)\Big)$.

(iii) If we define

$$\underset{\sim}{f} : X \times X^* \to R : (\underset{\sim}{x},\underset{\sim}{g}) \rightsquigarrow \underset{\sim}{g}(\underset{\sim}{x})$$

then $\underset{\sim}{f}$ is linear in the first variable by the linearity of each $\underset{\sim}{g} \in X^*$, in the second by the definition of addition and scalar multiplication in X^*. It is thus a (highly important) vector in $L(X,X^*;R)$.

(iv) If $\underset{\sim}{x}_1,\ldots,\underset{\sim}{x}_n$ are n specified vectors in X and we define

$$\underset{\sim}{f} : X^* \times X^* \times \ldots \times X^* \to R$$
$$(\underset{\sim}{g}_1, \underset{\sim}{g}_2, \ldots, \underset{\sim}{g}_n) \rightsquigarrow \underset{\sim}{g}_1(\underset{\sim}{x}_1)\underset{\sim}{g}_2(\underset{\sim}{x}_2) \ldots \underset{\sim}{g}_n(\underset{\sim}{x}_n)$$

then we have $\underset{\sim}{f} \in L^n(X^*;R)$, by a straightforward check.

Dually, if $\underset{\sim}{g}_1,\ldots,\underset{\sim}{g}_n$ are n specified linear functionals on X and we define

$$\underset{\sim}{g} : X \times X \times \ldots \times X \to R$$
$$(\underset{\sim}{x}_1, \underset{\sim}{x}_2, \ldots, \underset{\sim}{x}_n) \rightsquigarrow \underset{\sim}{g}_1(\underset{\sim}{x}_1)\underset{\sim}{g}_2(\underset{\sim}{x}_2) \ldots \underset{\sim}{g}_n(x_n)$$

then we have $\underset{\sim}{g} \in L^n(X;R)$. Notice that we may in this way "multiply" $\underset{\sim}{g}_1,\ldots,\underset{\sim}{g}_n$, but that we get a higher-order multilinear form by doing so, not just another functional. If we tried to define a "product functional" $\underset{\sim}{g}_1\underset{\sim}{g}_2 \ldots \underset{\sim}{g}_n$ by

$$(\underset{\sim}{g}_1\underset{\sim}{g}_2 \ldots \underset{\sim}{g}_n)\underset{\sim}{x} = \underset{\sim}{g}_1(\underset{\sim}{x})\underset{\sim}{g}_2(\underset{\sim}{x}) \ldots \underset{\sim}{g}_n(\underset{\sim}{x})$$

then we would have
$$\begin{aligned}(\underset{\sim}{g}_1\underset{\sim}{g}_2 \ldots \underset{\sim}{g}_n)\underset{\sim}{x}a &= a\underset{\sim}{g}_1(\underset{\sim}{x})\ a\underset{\sim}{g}_2(\underset{\sim}{x}) \ldots a\underset{\sim}{g}_n(\underset{\sim}{x}) \\ &= a^n\underset{\sim}{g}_1(\underset{\sim}{x})\ \underset{\sim}{g}_2(x) \ldots \underset{\sim}{g}_n(x) \\ &= a^n(\underset{\sim}{g}_1\underset{\sim}{g}_2 \ldots \underset{\sim}{g}_n)\underset{\sim}{x}\end{aligned}$$

which is clearly not linear, since $a^n \neq a$ in general, and so our "product" is not a functional. The $\underset{\sim}{g}$ we have defined lies in a higher space; we call it therefore the <u>tensor product</u> $\underset{\sim}{g}_1 \otimes \underset{\sim}{g}_2 \otimes \ldots \otimes \underset{\sim}{g}_n$ of $\underset{\sim}{g}_1, \ldots, \underset{\sim}{g}_n$ to distinguish it clearly from all the ordinary products, in many situations, which take two or more objects and give another of the same type. (The term <u>inner product</u> is never abbreviated to "product", for the same reason.)

1.03 Tensor products

We have a map

$$\begin{aligned} \otimes : X_1^* \times X_2^* \times \ldots \times X_n^* &\to L(X_1, \ldots, X_n; R) \\ (\underset{\sim}{g}_1, \underset{\sim}{g}_2, \ldots, \underset{\sim}{g}_n) &\rightsquigarrow \underset{\sim}{g}_1 \otimes \underset{\sim}{g}_2 \otimes \ldots \otimes \underset{\sim}{g}_n \end{aligned}$$

where $\underset{\sim}{g}_1 \otimes \underset{\sim}{g}_2 \otimes \ldots \otimes \underset{\sim}{g}_n$ is defined exactly as above, which is evidently multilinear (Ex. 3). It is not, however, surjective in general. This is most easily seen in an example:

$$\otimes : (R^2)^* \times (R^2)^* \to L^2(R^2;R)$$

takes a pair $\underset{\sim}{f}, \underset{\sim}{g}$ of linear functionals on R^2 to a bilinear form $(\underset{\sim}{x}, \underset{\sim}{y}) \rightsquigarrow \underset{\sim}{f}(\underset{\sim}{x})\underset{\sim}{g}(\underset{\sim}{y})$ on R^2. But such a form cannot for instance be non-degenerate. For if we choose a non-zero $\underset{\sim}{x} \in \ker \underset{\sim}{f}$, which is always possible (by I.2.10) since $\dim(\ker \underset{\sim}{f}) \geq 1$, then $\underset{\sim}{f}(\underset{\sim}{x})\underset{\sim}{g}(\underset{\sim}{y}) = 0$ for all $\underset{\sim}{y} \in R^2$.

However, any bilinear form can be expressed in terms of its effects on basis elements, in the manner of IV.3.01, for any basis $\underset{\sim}{b}_1, \underset{\sim}{b}_2$ of R^2;

$$\underset{\sim}{F}(\underset{\sim}{x}, \underset{\sim}{y}) = \underset{\sim}{F}((x^1, x^2), (y^1, y^2)) = \underset{\sim}{F}(x^1\underset{\sim}{b}_1 + x^2\underset{\sim}{b}_2, y^1\underset{\sim}{b}_1 + y^2\underset{\sim}{b}_2)$$

$$= x^1y^1\underset{\sim}{F}(\underset{\sim}{b}_1,\underset{\sim}{b}_1) + x^1y^2\underset{\sim}{F}(\underset{\sim}{b}_1,\underset{\sim}{b}_2) + x^2y^1\underset{\sim}{F}(\underset{\sim}{b}_2,\underset{\sim}{b}_1) + x^2y^2\underset{\sim}{F}(\underset{\sim}{b}_2,\underset{\sim}{b}_2) \text{ by bilinearity of } \underset{\sim}{F}$$

$$= x^iy^j\underset{\sim}{F}(\underset{\sim}{b}_i,\underset{\sim}{b}_j) \quad \text{using the summation convention.}$$

Setting $\underset{\sim}{F}(\underset{\sim}{b}_i,\underset{\sim}{b}_j) = f_{ij}$, we have

$$\underset{\sim}{F}(\underset{\sim}{x},\underset{\sim}{y}) = f_{ij}x^iy^j$$

$$= f_{ij}(\underset{\sim}{b}^i(\underset{\sim}{x}))(\underset{\sim}{b}^j(\underset{\sim}{y}))$$

$$= (f_{ij}\underset{\sim}{b}^i \otimes \underset{\sim}{b}^j)(\underset{\sim}{x},\underset{\sim}{y}).$$

Hence $\underset{\sim}{F} = f_{ij}\underset{\sim}{b}^i \otimes \underset{\sim}{b}^j$

so that the four tensor products $\underset{\sim}{b}^i \otimes \underset{\sim}{b}^j$ span the vector space $L^2(R^2;R)$. (Note that $\underset{\sim}{b}^1 \otimes \underset{\sim}{b}^2 \neq \underset{\sim}{b}^2 \otimes \underset{\sim}{b}^1$; tensor products do not commute.)

Thus the tensor products $f \otimes g$ in $L^2(R^2;R)$ do not constitute a vector space, since they are not closed under addition - we may have $\underset{\sim}{f} \otimes \underset{\sim}{g} + \underset{\sim}{f}' \otimes \underset{\sim}{g}' \neq \underset{\sim}{f}'' \otimes \underset{\sim}{g}''$ for any $\underset{\sim}{f}''$ and $\underset{\sim}{g}''$ in $(R^2)^*$. This is sad, as a vector space structure is too useful willingly to do without. However, if we formally put the sums in with them, subject to the linearity conditions we already have, we get a vector space which is essentially $L^2(R^2;R)$ back again. The construction involved is formal nonsense† of the type this book is dedicated to omitting, and the fact that the result is naturally isomorphic to a space we have already set up lets us avoid it with no ill consequences.

So: the natural notion of calling the set of all tensor products $\underset{\sim}{g}_1 \otimes \underset{\sim}{g}_2 \otimes \dots \otimes \underset{\sim}{g}_n$ the tensor product of the <u>spaces</u> $X_1^*, \dots, X_n^*$ is unsatisfactory,

---

† a respectable pure-mathematical technical term.

because this is not a vector space. However it sits naturally inside, and (Ex. 4a) spans, $L(X_1, \dots, X_n; R)$ which is a vector space. (Remember that $(\underset{\sim}{f}a)\otimes\underset{\sim}{g}$ is to be identified with, because it is the same function as, $\underset{\sim}{f}\otimes(\underset{\sim}{g}a)$.) This is then a good candidate for the tensor product: we set

$$X_1^* \otimes X_2^* \otimes \dots \otimes X_n^* = L(X_1, \dots, X_n; R).$$

This has the following two properties :-

Ti) $\boxtimes : X_1^* \times X_2^* \times \dots \times X_n^* \to X_1^* \otimes X_2^* \otimes \dots \otimes X_n^*$ is multilinear. (Ex. 4b)

Tii) If $\underset{\sim}{f} : X_1^* \times X_2^* \dots X_n^* \to Y$ is multilinear, then there is a unique linear map

$$\hat{\underset{\sim}{f}} : X_1^* \otimes X_2^* \otimes \dots \otimes X_n^* \to Y$$

such that $\underset{\sim}{f} = \hat{\underset{\sim}{f}}\circ\boxtimes$. (Ex. 4d)

Diagramatically:

$$\begin{array}{ccc} X_1^* \times X_2^* \times \dots \times X_n^* & \xrightarrow{\boxtimes} & X_1^* \otimes X_2^* \otimes \dots \otimes X_n^* \\ & \underset{\sim}{f} \searrow \quad \swarrow \hat{\underset{\sim}{f}} & \\ & Y & \end{array}$$ commutes.

These two properties between them pin down the tensor product completely, and permit us to define it for <u>any</u> set of vector spaces. Discarding whatever in the above is motivation rather than proof:

<u>1.04 Definition</u>

A <u>tensor product</u> of vector spaces $X_1, \dots, X_n$ is a space X together with a map

$$\otimes : X_1 \times X_2 \times \ldots \times X_n \to X$$

having properties Ti) and Tii).

1.05 Lemma

A tensor product of $X_1, \ldots, X_n$ always exists, and any two are isomorphic in a natural way.

Proof

Existence:

We have shown existence for the spaces $X_1^*, \ldots, X_n^*$, since $L(X_1, \ldots, X_n; R)$ does the job. But $X_i \cong (X_i^*)^*$ naturally, so $L(X_1^*, \ldots, X_n^*; R)$ will serve for $X_1, \ldots, X_n$.

Uniqueness:

If X , X', with maps $\otimes$, $\otimes'$ both have the properties Ti) and Tii), then:

By Ti) for $(X,\otimes)$ and Tii) for $(X',\otimes')$

there exists a unique $\underset{\sim}{\Psi}$ such that

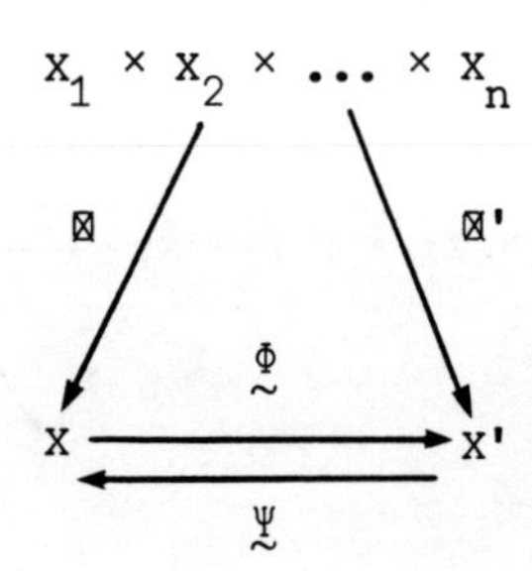

$$\underset{\sim}{\Psi}\otimes' = \otimes.$$

By Ti) for $(X',\otimes')$ and Tii) for $(X,\otimes)$

there exists a unique $\underset{\sim}{\Phi}$ such that

$$\underset{\sim}{\Phi}\otimes = \otimes'.$$

Hence $\underset{\sim}{\Psi}\underset{\sim}{\Phi}\otimes = \underset{\sim}{\Psi}\otimes' = \otimes = \underset{\sim}{I}_X\otimes$.

But by Ti) for $(X,\otimes)$, $\otimes = \underset{\sim}{I}\otimes$ is multilinear, hence by the uniqueness in Tii) for $(X,\otimes)$, this means that $\underset{\sim}{\Psi}\underset{\sim}{\Phi} = \underset{\sim}{I}_X$. Similarly, $\underset{\sim}{\Phi}\underset{\sim}{\Psi} = \underset{\sim}{I}_{X'}$, so that $X \cong X'$.

1.06 Language

We shall not go into the technical justification here, but the isomorphism of the theorem is in the strongest sense natural (like that of X with (X*)* but <u>not</u> with X*, unless X is allowed extra structure such as a metric); it is clear that it involves no arbitrary choices. As a consequence we may without confusion use it to regard all tensor products of $X_1, \ldots, X_n$ as essentially the same, and talk of <u>the</u> tensor product. We always denote this by $X_1 \otimes X_2 \otimes \ldots \otimes X_n$, and its elements will all be sums of scalar multiples of tensor products:

$$(\underset{\sim}{x}_1 \otimes \underset{\sim}{x}_2 \otimes \ldots \otimes \underset{\sim}{x}_n)a + (\underset{\sim}{x}'_1 \otimes \underset{\sim}{x}'_2 \otimes \ldots \otimes \underset{\sim}{x}'_n)a' + \ldots \text{ (finitely many terms)}$$

where $\underset{\sim}{x}_i, \underset{\sim}{x}'_i, \ldots \in X_i$, $a, a', \ldots \in R$, satisfying the equations

TA) $\underset{\sim}{x}_1 \otimes \ldots \otimes (\underset{\sim}{x}_i + \underset{\sim}{x}'_i) \otimes \ldots \otimes \underset{\sim}{x}_n = \underset{\sim}{x}_1 \otimes \ldots \otimes \underset{\sim}{x}_i \otimes \ldots \otimes \underset{\sim}{x}_n + \underset{\sim}{x}_1 \otimes \ldots \otimes \underset{\sim}{x}'_i \otimes \ldots \otimes \underset{\sim}{x}_n$

TS) $(\underset{\sim}{x}_1 a) \otimes \underset{\sim}{x}_2 \otimes \ldots \otimes \underset{\sim}{x}_n = \underset{\sim}{x}_1 \otimes (\underset{\sim}{x}_2 a) \otimes \ldots \otimes \underset{\sim}{x}_n = \underset{\sim}{x}_1 \otimes \underset{\sim}{x}_2 \otimes \ldots \otimes (\underset{\sim}{x}_n a) = (\underset{\sim}{x}_1 \otimes \underset{\sim}{x}_2 \otimes \ldots \otimes \underset{\sim}{x}_n)a.$

(cf. Ex. 5). This description of the elements could be used to set up the vector space $X_1 \otimes X_2 \otimes \ldots \otimes X_n$ directly, but this would involve more definitions. Since by Lemma 1.05 any construction giving something with properties Ti) and Tii) produces a result essentially the same as any other, we have chosen a quick one that uses only the tools ready to hand. From now on, the construction can be forgotten: Ti) and Tii) characterise the tensor product on spaces, TA) and TS) on vectors, and these between them suffice for all proofs and manipulations (reduced if need be to coordinate form).

Notice that an element of $X \otimes Y$ need not be of the form $\underset{\sim}{x} \otimes \underset{\sim}{y}$; it may be a sum of several such. It need not be a sum in a unique way. For, (Ex. 6a):

$$\underset{\sim}{x} \otimes \underset{\sim}{y} + \underset{\sim}{x}' \otimes \underset{\sim}{y} = (\underset{\sim}{x}\tfrac{1}{2} + \underset{\sim}{x}'\tfrac{1}{2}) \otimes (\underset{\sim}{y} + \underset{\sim}{y}') + ((\underset{\sim}{x} + \underset{\sim}{x}') \otimes (\underset{\sim}{y} + \underset{\sim}{y}'))\tfrac{1}{2}.$$

Vectors in a tensor product of spaces which can be expressed as a single tensor product of vectors are called simple tensors; those which can only be expressed as a sum, compound. Note that since the simple tensors span the tensor product, a linear map is entirely fixed by the values it takes on them.

1.07 Lemma

There is a natural isomorphism $X_1^* \otimes X_2^* \otimes \ldots \otimes X_n^* \cong (X_1 \otimes X_2 \otimes \ldots \otimes X_n)^*$.

Proof

Define $\Phi : (X_1 \otimes X_2 \otimes \ldots \otimes X_n)^* \to L(X_1, \ldots, X_n; R) = X_1^* \otimes X_2^* \otimes \ldots \otimes X_n^*$

$$\text{by } \Phi : f \rightsquigarrow f \circ \boxtimes$$

(where $\boxtimes : X_1 \times X_2 \times \ldots \times X_n \to X_1 \otimes X_2 \otimes \ldots \otimes X_n$ as above)

and $\Psi : L(X_1, \ldots, X_n; R) \to (X_1 \otimes X_2 \otimes \ldots \otimes X_n)^* : g \rightsquigarrow \hat{g}$.

Here $\hat{g}$ is uniquely given for each g by Tii).

Evidently Φ is linear, and Ψ is an inverse function for it, so that Φ is a bijection and hence an isomorphism (I.2.03). (Hence, of course, Ψ also is linear.)

Naturality we shall as usual take to follow from lack of special choices involved, since we do not want to go into the technicalities of category theory; they are indeed here technicalities only, and this isomorphism is another that may safely be used to identify two spaces. ■

This result, and our techniques for its proof, illustrate the usefulness of the tensor product: it swiftly reduces the theory of multilinear forms on a collection of spaces to that of linear functionals on a single space - their tensor product. We thus do not need to do all our work on functionals over again for multilinear forms.

1.08 Lemma

For any two vector spaces X_1, X_2 there is a natural isomorphism

$$L(X_1;X_2) \to X_1^* \otimes X_2$$

Proof

Define $\underset{\sim}{f} : X_1^* \times X_2 \to L(X_1;X_2)$.

$$(\underset{\sim}{g},\underset{\sim}{x}_2) \rightsquigarrow \left(\underset{\sim}{x}_1 \rightsquigarrow \underset{\sim}{x}_2(\underset{\sim}{f}(\underset{\sim}{x}_1)\right)$$

Then $\underset{\sim}{f}$ is multilinear (Ex. 7a) and so by Tii) induces a linear map

$$\hat{\underset{\sim}{f}} : X_1^* \otimes X_2 \to L(X_1,X_2) \quad \text{with} \quad \hat{\underset{\sim}{f}}\boxtimes = \underset{\sim}{f}.$$

Now $\hat{\underset{\sim}{f}}(\underset{\sim}{g}\otimes\underset{\sim}{x}_2) = \underset{\sim}{0} \;\Rightarrow\; \underset{\sim}{f}(\underset{\sim}{g},\underset{\sim}{x}_2) = \underset{\sim}{0}$

$\Rightarrow \underset{\sim}{x}_2(\underset{\sim}{g}(\underset{\sim}{x}_1)) = \underset{\sim}{0}$ for all $\underset{\sim}{x} \in X_1$

$\Rightarrow \underset{\sim}{x}_2 = \underset{\sim}{0}$ or $\underset{\sim}{g} = \underset{\sim}{0}$

$\Rightarrow \underset{\sim}{g}\otimes\underset{\sim}{x}_2 = \underset{\sim}{0}.$

Hence $\hat{\underset{\sim}{f}}$ is injective (Ex. 7b), and since

$$\begin{aligned} \dim(X_1^*\otimes X_2) &= \dim X_1^* \dim X_2 && \text{(Ex. 4c)} \\ &= \dim X_1 \dim X_2 && \text{(III.1.04)} \\ &= \dim(L(X_1;X_2)) && \text{(by I.2.07)} \end{aligned}$$

it is an isomorphism, as required. ■

This is a very important and very useful isomorphism (thanks again to naturality). It is far more often helpful to think of $L(X;Y)$ than of $X^*\otimes Y$. We shall generally identify the two, just as we identify $(X^*)^*$ and X.

1.09 Tensor products of maps

If we have linear maps $\underset{\sim}{A}_i : X_i \to Y_i$, $i = 1,\ldots,n$ then the composite map

$$X_1\times X_2\times\ldots\times X_n \xrightarrow{(\underset{\sim}{A}_1,\underset{\sim}{A}_2,\ldots,\underset{\sim}{A}_n)} Y_1\times Y_2\times\ldots\times Y_n \xrightarrow{\otimes} Y_1\otimes Y_2\otimes\ldots\otimes Y_n$$

$$(\underset{\sim}{x}_1,\underset{\sim}{x}_2,\ldots,\underset{\sim}{x}_n) \rightsquigarrow (\underset{\sim}{A}_1\underset{\sim}{x}_1,\underset{\sim}{A}_2\underset{\sim}{x}_2,\ldots,\underset{\sim}{A}_n\underset{\sim}{x}_n) \rightsquigarrow \underset{\sim}{A}_1\underset{\sim}{x}_1\otimes\underset{\sim}{A}_2\underset{\sim}{x}_2\otimes\ldots\otimes\underset{\sim}{A}_n\underset{\sim}{x}_n$$

is multilinear (check!) and hence induces by Tii) a unique linear map

$$\underset{\sim}{A}_1\otimes\underset{\sim}{A}_2\otimes\ldots\otimes\underset{\sim}{A}_n : X_1\otimes X_2\otimes\ldots\otimes X_n \to Y_1\otimes Y_2\otimes\ldots\otimes Y_n,$$

with, on simple elements,

$$\underset{\sim}{A}_1\otimes\underset{\sim}{A}_2\otimes\ldots\otimes\underset{\sim}{A}_n(\underset{\sim}{x}_1\otimes\underset{\sim}{x}_2\otimes\ldots\otimes\underset{\sim}{x}_n) = \underset{\sim}{A}_1\underset{\sim}{x}_1\otimes\underset{\sim}{A}_2\underset{\sim}{x}_2\otimes\ldots\otimes\underset{\sim}{A}_n\underset{\sim}{x}_n.$$

This is called the <u>tensor product</u> of the <u>maps</u> $\underset{\sim}{A}_1,\ldots,\underset{\sim}{A}_n$.

1.10 Notation

The most important cases of tensor products of spaces are of the type

$$\underbrace{X\otimes X\otimes\ldots\otimes X}_{k \text{ times}}\otimes\underbrace{X^*\otimes X^*\otimes\ldots\otimes X^*}_{h \text{ times}}$$

for some particular space X. This is denoted by X^k_h. Vectors in X^k_h are called <u>tensors</u> on X, <u>covariant of degree</u> h and <u>contravariant of degree</u> k, or <u>of type</u> $\binom{k}{h}$. We abbreviate X^k_0 to X^k, X^0_h to X_h.

Evidently, $X = X^1$ and $X^* = X_1$.

By convention, $X^0_0 = R$.

By Ex. 4c, $\dim(X^k_h) = (\dim X)^{k+h}$.

Sometimes such a space arises in a less tidy sequence, such as

$$X \otimes X \otimes X^* \otimes X \otimes X^* \otimes X^*$$

(for instance, as the tensor product of two tidy ones, $X \otimes X \otimes X^*$ and $X \otimes X^* \otimes X^*$; cf. Ex. 8a) and while it is legitimate (Ex. 8b) to reshuffle them if we wish, it may be inconvenient - $\underset{\sim}{x} \otimes \underset{\sim}{x}'$ could then mean one thing before the shuffle and another after. So, for example, the space

$$X \otimes X \otimes X \otimes X^* \otimes X^* \otimes X \otimes X \otimes X^* \otimes X \otimes X$$

will be denoted by $X^{3\ \ 2\ \ 2}_{\ \ 2\ \ 1}$. Its elements will then be covariant of degree 2+1=3, contravariant of degree 3+2+2, and of type $\left(^{3\ \ 2\ \ 2}_{\ \ 2\ \ 1}\right)$.

Since tensors on X are simply vectors in a space constructed from X, we shall denote them by symbols $\underset{\sim}{x}, \underset{\sim}{y}$ etc. of the same kind (modified by the habits we have already got into of $\underset{\sim}{f}, \underset{\sim}{g}$ etc. for functionals, $\underset{\sim}{G}$ for a metric tensor, etc., when convenient). Various notations such as bold sans-serif capitals are in use, dating from the days when tensors were thought mysterious and impressive, but this is unnecessary. Moreover, the borderland between those who never use anything but "indexed quantities" and those who reserve fancy type for much fancier objects is fast disappearing. So we shall not worry the typist.

1.11 Contraction

For any mixed tensor space, $X^{3\ 1}_{\ 1\ 2}$ for example, if we choose one copy of X* and one of X, say the 3rd X and the 2nd X* for definiteness here,

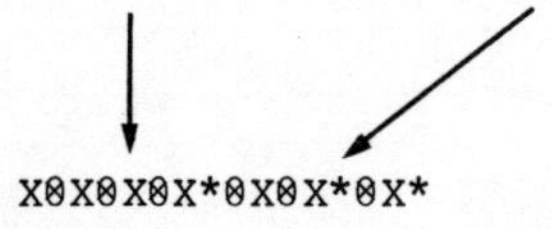

then we can define the corresponding linear contraction map

$$\underset{\sim}{C}^3_2 \ : \ X\otimes X\otimes X\otimes X^*\otimes X\otimes X^*\otimes X^* \to X\otimes X^*\otimes X\otimes X^*$$

on simple elements by

$$\underset{\sim}{C}^3_2(\underset{\sim}{x}_1\otimes\underset{\sim}{x}_2\otimes\underset{\sim}{x}_3\otimes\underset{\sim}{f}_1\otimes\underset{\sim}{x}_4\otimes\underset{\sim}{f}_2\otimes\underset{\sim}{f}_3) = (\underset{\sim}{x}_1\otimes\underset{\sim}{x}_2\otimes\underset{\sim}{f}_1\otimes\underset{\sim}{x}_4\otimes\underset{\sim}{f}_3)\underset{\sim}{f}_2(\underset{\sim}{x}_3)$$

(cf. Ex. 9). The image under this map of a tensor $\underset{\sim}{x}$ on X is called a contraction of $\underset{\sim}{x}$. (We can distinguish $\underset{\sim}{C}^3_2$ from a component (1.12) by the presence of the twiddle. We omit the suffixes when possible without ambiguity.)

A contraction map lowers both covariant and contravariant degree by one. If the original degrees are equal, successive contractions define a map right down to $X^0_0 = R$, but not uniquely. (There are k! possible total contractions $X^k_k \to R$, according to how we pair off the X's and the X*'s, and k! = 1 only if k = 0 or 1).

1.12 Components

By Ex. 4, if $\underset{\sim}{b}_1,\ldots,\underset{\sim}{b}_n$ is a basis for X then the set of all tensors of the form

$$\underset{\sim}{b}_i\otimes\underset{\sim}{b}_j\otimes\underset{\sim}{b}_k\otimes\underset{\sim}{b}^\ell\otimes\underset{\sim}{b}^m,$$

where i,j,k,ℓ,m are (not necessarily distinct) labels drawn from {1,2,...,n}, is a basis for X^3_2. (Ex. 4 is one of the chief examples of a careful check that is essential to do, but almost worthless merely to see done. Like a Zen exercise, you must experience it to gain anything. It should not be hard unless you have completely lost sight of what is going on, in which case you should return to the earlier chapters - or find a better book - rather than wish for this manipulation to be in the text.)

Thus, for any $\underset{\sim}{x} \in X^3_2$, $\underset{\sim}{x}$ is a unique linear combination (Ex. I.1.8)

$$\underset{\sim}{x} = (\underset{\sim}{b}_i \otimes \underset{\sim}{b}_j \otimes \underset{\sim}{b}_k \otimes \underset{\sim}{b}^{\ell} \otimes \underset{\sim}{b}^m) x^{ijk}_{\ell m}$$

and we may represent $\underset{\sim}{x}$ by its n^5 components $x^{ijk}_{\ell m}$. If we wish to change basis, to $\underset{\sim}{b}'_1, \ldots, \underset{\sim}{b}'_n$ say, then in the notation of I.2.08 and III.1.07, by precisely similar arguments, we have new components

$$x^{i'j'k'}_{\ell' m'} = \tilde{b}^{i'}_i \tilde{b}^{j'}_j \tilde{b}^{k'}_k b^{\ell}_{\ell'} b^m_{m'} x^{ijk}_{\ell m} \qquad \bigstar$$

where $\qquad \underset{\sim}{b}'_p = (b^1_p, b^2_p, \ldots, b^n_p)$ in the old coordinates,

for p (and so also i, j, etc.) $= 1, \ldots, n$, and

$$\tilde{b}^{i'}_i b^i_p = \delta^{i'}_p \text{ , etc.}$$

This is the traditional <u>definition</u> of a tensor of type $\binom{3}{2}$, as "a set of n^5 numbers that transform according to the equation$\bigstar$". As it stands, that is frankly meaningless; you can transform <u>any</u> set of n^5 numbers by that formula. A better expression of this approach is, for instance "a covariant tensor of order 3 is a rule which in any coordinate system allows us to construct n^3 numbers (components) x_{ijk}, each of which is specified by giving the indices i,j,k definite values from 1 to n such that the results in two different bases are related by the formula

$$x_{i'j'k'} = b^i_{i'} b^j_{j'} b^k_{k'} x_{ijk} \quad "$$

[Shilov], and so forth for other types. This is then logically satisfactory. The reader must decide whether for him it is more illuminating than the approach we have chosen.

We too say "and so forth for other types", since the completely general rule would have to replace i,j,k, ... by $i_1, \ldots, i_p$, so the formula would involve terms like $\tilde{b}^{i'_j}_{i_j}$. We shall only add that tensors of type, say, $\binom{3\ 1}{1\ 2}$ are represented by components labelled by expressions of the form $x^{ijk\ \ m}_{\ \ \ \ \ell\ np}$, transforming according to

$$x^{i'j'k'\ \ m'}_{\ \ \ \ \ \ \ell'\ n'p'} = (\tilde{b}^{i'}_{i}\tilde{b}^{j'}_{j}\tilde{b}^{k'}_{k} b^{\ell}_{\ell'}\tilde{b}^{m'}_{m} b^{n}_{n'} b^{p}_{p'}) x^{ijk\ \ m}_{\ \ \ \ \ell\ np}.$$

Notice that if $\underset{\sim}{v} = v^{i_1 \ldots i_k}_{j_1 \ldots j_h} (\underset{\sim}{b}_{i_1} \otimes \ldots \otimes \underset{\sim}{b}_{i_k} \otimes \underset{\sim}{b}^{j_1} \otimes \ldots \otimes \underset{\sim}{b}^{j_h}) \in X^k_h$,

and $\underset{\sim}{w} = w^{a_1 \ldots a_\ell}_{b_1 \ldots b_m} (\underset{\sim}{b}_{a_1} \otimes \ldots \otimes \underset{\sim}{b}^{b_m}) \in X^\ell_m$,

then $\underset{\sim}{v} \otimes \underset{\sim}{w} = v^{i_1 \ldots i_k}_{j_1 \ldots j_h} w^{a_1 \ldots a_\ell}_{b_1 \ldots b_m} (\underset{\sim}{b}_{i_1} \otimes \ldots \otimes \underset{\sim}{b}^{j_h} \otimes \underset{\sim}{b}_{a_1} \otimes \ldots \otimes \underset{\sim}{b}^{b_m}) \in X^{k\ \ell}_{h\ m} \cong X^{k+\ell}_{h+m}$.

So the components of $\underset{\sim}{v} \otimes \underset{\sim}{w}$ are simply all the possible products of components of $\underset{\sim}{v}$ and $\underset{\sim}{w}$. We shall sometimes follow the physics literature practice of referring to a tensor, $\underset{\sim}{x} \in X^{3\ 1}_{1\ 2}$, say, by its "typical component" $x^{ijk\ \ m}_{\ \ \ \ \ell\ np}$ when we have a fixed basis or chart in mind : this makes some room for confusion, but physics students need to get used to it.

Occasionally when we have names for the coordinates, rather than numbers (such as (x,y,z) not (x^1,x^2,x^3) on R^3, which is sometimes convenient in saving indices) we shall let the names stand for the numbers : like t^{xy}_{xz} instead of t^{12}_{13}. This has to be used with caution, owing to the summation convention - t^{xy}_{xz} means something quite different if x is a dummy index.

Contraction has a simple formula in coordinates. On a basis vector $\underset{\sim}{b}_i \otimes \underset{\sim}{b}_j \otimes \underset{\sim}{b}_k \otimes \underset{\sim}{b}^\ell \otimes \underset{\sim}{b}^m$ of $X \otimes X \otimes X \otimes X^* \otimes X^*$, for instance, the effect of "<u>contracting</u> <u>over</u> j <u>and</u> ℓ" (that is, applying $\underset{\sim}{C}^2_1$) is to take it to $\underset{\sim}{b}_i \otimes \underset{\sim}{b}_k \otimes \underset{\sim}{b}^m (\underset{\sim}{b}^\ell (\underset{\sim}{b}_j))$. Now,

$$\underset{\sim}{b}^\ell (\underset{\sim}{b}_j) = \delta^\ell_j$$

by definition. So the image in X^2_1 of

$$\underset{\sim}{x} = (\underset{\sim}{b}_i \otimes \underset{\sim}{b}_j \otimes \underset{\sim}{b}_k \otimes \underset{\sim}{b}^\ell \otimes \underset{\sim}{b}^m) x^{ijk}_{\ell m} \in X^3_2,$$

under contraction is a vector whose component along each basis vector $\underset{\sim}{b}_{i'} \otimes \underset{\sim}{b}_{k'} \otimes \underset{\sim}{b}^{m'}$ of X^2_1 is the sum of those $x^{ijk}_{\ell m}$'s having $j = \ell$ and $i = i'$, $k = k'$, $m = m'$. Thus $\underset{\sim}{C}^2_1 \underset{\sim}{x}$ has coordinates precisely $\delta^\ell_j x^{ijk}_{\ell m} = x^{ijk}_{jm}$, using the summation convention.

The naturality of the isomorphism $\hat{\underset{\sim}{f}}$ of Lemma 1.08 is illustrated by its form in coordinates. If we have $\underset{\sim}{a} \in X^*_1 \otimes X_2$, with

$$\underset{\sim}{a} = \underset{\sim}{b}^j \otimes \underset{\sim}{b}'_i a^i_j$$

with respect to bases $\underset{\sim}{b}_1, \dots, \underset{\sim}{b}_n$ for X_1, $\underset{\sim}{b}'_1, \dots, \underset{\sim}{b}'_m$ for X_2, then

$$\begin{aligned}(\hat{\underset{\sim}{f}}\underset{\sim}{a})\underset{\sim}{x} &= (\underset{\sim}{b}'_i (\underset{\sim}{b}^j (\underset{\sim}{x}))) a^i_j \\ &= (\underset{\sim}{b}'_i x^j) a^i_j \\ &= \underset{\sim}{b}'_i (a^j a^i_j).\end{aligned}$$

Hence, $\quad (\hat{\underset{\sim}{f}}\underset{\sim}{a})(x^1, \dots, x^n) = (a^1_j x^j, \; a^2_j x^j, \; \dots, \; a^n_j x^j)$

and the matrix of $\hat{\underset{\sim}{f}}\underset{\sim}{a}$ is exactly $[a^i_j]$. So whatever bases we choose for X_1 and X_2, they give the same representation for $\hat{\underset{\sim}{f}}\underset{\sim}{a}$ as for $\underset{\sim}{a}$.

Notice that $\hat{\underset{\sim}{f}}$ carries the contraction function $\underset{\sim}{C} : X^* \otimes X \to R$ to the trace function $\mathrm{tr} : L(X;X) \to R$, since $\underset{\sim}{C}\underset{\sim}{a} = a^i_i = \mathrm{tr}(\hat{\underset{\sim}{f}}\underset{\sim}{a})$ in any coordinate system. As $\hat{\underset{\sim}{f}}$ can safely be used to identify the two spaces, this gives a more intrinsic and coordinate-free way of thinking about the trace than we had in I.3.14, but it remains heavily algebraic.

(If $\underset{\sim}{A}$ is thought of as an "infinitesimal operator", using the differential structure of $L(X;X)$ then $\mathrm{tr}\underset{\sim}{A}$ becomes an "infinitesimal change of determinant". We shall discuss a precise formulation in a later volume in the context of Lie groups and their Lie algebras, or see [Porteous]. That allows a geometric interpretation of trace, exploited implicitly in IX §6 below, but not a universally applicable one.)

1.13 Tensors on metric spaces

Suppose that X has a metric tensor $\underset{\sim}{G}$. Then isomorphisms

$$\underset{\sim}{G}_{\downarrow} : X \to X^*, \qquad \underset{\sim}{G}_{\uparrow} : X^* \to X$$

give rise to isomorphisms between various of the spaces constructed from X. For example,

$$\underset{\sim}{I} \otimes \underset{\sim}{I} \otimes \underset{\sim}{G}_{\uparrow} \otimes \underset{\sim}{G}_{\downarrow} \otimes \underset{\sim}{G}_{\uparrow} \otimes \underset{\sim}{I} \otimes \underset{\sim}{G}_{\uparrow} : X \otimes X^* \otimes X^* \otimes X \otimes X^* \otimes X \otimes X^* \to X \otimes X^* \otimes X \otimes X^* \otimes X \otimes X \otimes X$$

and so forth. In general we have an isomorphism

$$X^k_h \cong X^{k'}_{h'},$$

preserving the order in which tensor products are taken, whenever

$$k + h = k' + h'.$$

If the metric has been fixed once and for all, these can be used, for instance, to make all tensors entirely contravariant. This might seem a simplification, but it is not. For example, velocity at a point arises naturally as a contravariant vector. The gradient of a potential at a point arises as a covariant one and the contours of the functional (cf. III.1.02 and VII.1.02) are the local linear approximation to those of the potential. Similar things happen for higher degrees. So it is better to keep dual objects distinguished, using the isomorphisms when convenient, rather than let them merge into the One Void: the goal of physics is not Nirvana.

The formulae for these isomorphisms come straight from those for $\underset{\sim}{G}_{\downarrow}$ and $\underset{\sim}{G}_{\uparrow}$ (IV.3.02). In general, let $\underset{\sim}{A} : X \to Y$ and $A' : X' \to Y'$ have matrices $[a^i_j]$ and $[a'^k_\ell]$ with respect to bases $\{\underset{\sim}{b}_1,\ldots,\underset{\sim}{b}_n\}$, $\{\underset{\sim}{b}'_1,\ldots,\underset{\sim}{b}'_{n'}\}$, $\{\underset{\sim}{c}_1,\ldots,\underset{\sim}{c}_m\}$ and $\{\underset{\sim}{c}'_1,\ldots,\underset{\sim}{c}'_{m'}\}$ for X, X', Y, Y' respectively. Then for $\underset{\sim}{x} \in X \otimes X'$ we have

$$\begin{aligned}
\underset{\sim}{A}\otimes\underset{\sim}{A}'(\underset{\sim}{x}) &= \underset{\sim}{A}\otimes\underset{\sim}{A}'(\underset{\sim}{b}_j \otimes \underset{\sim}{b}'_\ell x^{j\ell}) \\
&= (\underset{\sim}{A}\underset{\sim}{b}_j \otimes \underset{\sim}{A}'\underset{\sim}{b}'_\ell) x^{j\ell} \\
&= ((\underset{\sim}{c}_i a^i_j) \otimes (\underset{\sim}{c}'_k a'^k_\ell)) x^{j\ell} \\
&= (\underset{\sim}{c}_i \otimes \underset{\sim}{c}'_k) a^i_j a'^k_\ell x^{j\ell}.
\end{aligned}$$

So the $nn' \times mm'$ entries of $[\underset{\sim}{A}\otimes\underset{\sim}{A}']$ are just the multiples $a^i_j a'^k_\ell$, and so on for the higher orders. Thus, for instance, the isomorphism

$$\Theta = \underset{\sim}{I}\otimes\underset{\sim}{I}\otimes\underset{\sim}{G}_{\uparrow}\otimes\underset{\sim}{G}_{\downarrow}\otimes\underset{\sim}{G}_{\uparrow}\otimes\underset{\sim}{I}\otimes\underset{\sim}{G}_{\uparrow} \quad : \quad \underset{\sim}{x} \rightsquigarrow \underset{\sim}{y}$$

at the beginning of this section has the formula

$$y_{j\ \ \ell'}^{i\ k'\ \ m'np'} = g^{k'k}g_{\ell'\ell}g^{m'm}g^{p'p}x_{jk\ m\ p}^{i\ \ \ell\ n}.$$

Application of these isomorphisms is known as "raising and lowering indices", for obvious reasons. This lies behind our notations $\underset{\sim}{G}_{\uparrow}$ and $\underset{\sim}{G}_{\downarrow}$.

One of these isomorphisms is significant enough to merit special mention. It gives us the composite

$$\underset{\sim}{\Psi} : L(X;X) \xrightarrow{\hat{\underset{\sim}{f}}^{\leftarrow}} X^*\otimes X \xrightarrow{I\otimes G_{\downarrow}} X^*\otimes X^* = L^2(X;R). \quad \text{(cf. 1.03)}$$

Here $\hat{\underset{\sim}{f}}$ is as in Lemma 1.08, so we have an isomorphism between the space of operators and that of bilinear forms. If $\underset{\sim}{A} \in L(X,X)$ has matrix $[a_j^i]$ and $\underset{\sim}{F} = \underset{\sim}{\Psi}\underset{\sim}{A}$ has components $f_{k\ell}$, then we have the formula

$$f_{k\ell} = g_{ki}\, a_{\ell}^{i}\, .$$

In fact, $\underset{\sim}{\Psi}$ is most clearly represented otherwise by the formulation (Ex. 10a)

$$L(X;X) \to L^2(X;R) : \underset{\sim}{A} \rightsquigarrow [(\underset{\sim}{x},\underset{\sim}{y}) \rightsquigarrow \underset{\sim}{A}\underset{\sim}{x}.\underset{\sim}{y}].$$

In this form it is easy to prove (Ex. 10b) that $\underset{\sim}{A}$ is non-singular if and only if $\underset{\sim}{\Psi}\underset{\sim}{A}$ is non-degenerate, and that $\underset{\sim}{A}$ is self-adjoint if and only if $\underset{\sim}{\Psi}\underset{\sim}{A}$ is symmetric

This equivalence makes it seem that perhaps the separate proofs for the diagonalisation of symmetric operators (IV.4.06) and of symmetric bilinear forms (IV.3.05; only the "ortho<u>normal</u>" condition on the basis vectors requires non-degeneracy) were superfluous, and that one should be deducible from the other straight off. However, one involves a basis orthonormal with respect to $\underset{\sim}{G}$, the other a basis orthonormal with respect to $\underset{\sim}{\Psi A}$, which can be any bilinear form at all ($\underset{\sim}{\Psi}$ being surjective) so that the two are not closely related. Moreover, if $\underset{\sim}{G}$ is indefinite IV.4.06 is false (IV.4.12) but IV.3.05 remains true, so no close relation can be expected.

1.14 Geometry

The reader may have noticed a scarcity of pictures in this chapter. This is not because tensors are un-geometric. It is because they are so geometrically various. They include vectors, linear functionals, metric tensors, the "volume" form 1.02 (i) (cf. also Ex. 11) and nearly everything else we have looked at so far. That all of these wrap up in the same algebraic parcel is a great convenience, but it does mean that geometrical interpretations must attach to particular types of tensor, not to the tensor concept. We shall provide such interpretations, as far as possible, as we proceed.

Exercises V. 1

1) Define addition and scalar multiplication of multilinear maps, by analogy with IV. Ex. 1.4 for the bilinear case, and prove that $L(X_1,\ldots,X_n;Y)$ is then a vector space.

2a) Prove that $\underset{\sim}{f}$ as defined in 1.02(i) is multilinear, via Euclidean "base × height" arguments on the volume of parallelepipeds.

b) Prove from the definitions of addition and scalar multiplication of maps that $\underset{\sim}{f}$ as defined in 1.02(ii) is multilinear.

3) Prove from the definitions that the map $\otimes$ of 1.03 is multilinear.

4a) By choosing bases for $X_1,\ldots,X_n$, show that the set

$$\otimes(X_1^*\times X_2^*\times\ldots\times X_n^*) \quad \text{spans} \quad L(X_1,\ldots,X_n;R).$$

b) Check Ti) in 1.03.

c) Prove that the set of all possible tensor products of the form

$$\mathbf{b}_{i_1} \otimes \mathbf{b}_{i_2} \otimes \ldots \otimes \mathbf{b}_{i_n} ,$$

where each $\mathbf{b}_{i_j}$ is a vector in the basis chosen in part (a) for X_i, is a basis for $X_1^* \otimes X_2^* \otimes \ldots \otimes X_n^* = L(X_1, \ldots, X_2; R)$.

(Essentially, the argument is the same as for the case $X_1 = X_2 = R^2$ of 1.03.)

Deduce that $\dim(X_1^* \otimes X_2^* \otimes \ldots \otimes X_n^*) = \dim(X_1^*)\dim(X_2^*)\ldots\dim(X_n^*)$.

d) By examining its necessary values on basis elements, prove the existence and uniqueness of the map $\hat{\mathbf{f}}$ in Tii) of 1.03.

5) Prove that the tensor products of functionals defined as in 1.02 and 1.03 satisfy TA) and TS) of 1.06; deduce that the tensor products of vectors do likewise.

6a) Prove from equations TA) and TS) that

$$((\mathbf{x}\tfrac{1}{2} + \mathbf{x}'\tfrac{1}{2}) \otimes (\mathbf{y} + \mathbf{y}')) + ((\mathbf{x} - \mathbf{x}') \otimes (\mathbf{y} - \mathbf{y}'))\tfrac{1}{2} = \mathbf{x} \otimes \mathbf{y} + \mathbf{x}' \otimes \mathbf{y}'$$

b) Prove that if $\mathbf{x} \otimes \mathbf{y} = \mathbf{x}' \otimes \mathbf{y}'$, then $\mathbf{x} = a\mathbf{x}'$, $\mathbf{y}' = a\mathbf{y}$ for some $a \in R$.

7a) Prove that if we define $(\mathbf{f}(\mathbf{g}, \mathbf{x}_2))\mathbf{x}_1 = \mathbf{x}_2(\mathbf{g}(\mathbf{x}_1))$, $\mathbf{x}_i \in X_i$, then

$$\left.\begin{aligned} \mathbf{f}(\mathbf{g} + \mathbf{g}', \mathbf{x}_2) &= \mathbf{f}(\mathbf{g}, \mathbf{x}_2) + \mathbf{f}(\mathbf{g}', \mathbf{x}_2) \\ \mathbf{f}(\mathbf{g}, \mathbf{x}_2 + \mathbf{x}_2') &= \mathbf{f}(\mathbf{g}, \mathbf{x}_2) + \mathbf{f}(\mathbf{g}, \mathbf{x}_2') \\ \mathbf{f}(\mathbf{g}a, \mathbf{x}_2) &= (\mathbf{f}(\mathbf{g}, \mathbf{x}_2))a = \mathbf{f}(\mathbf{g}, \mathbf{x}_2 a) \end{aligned}\right\} \text{ as linear maps } X_1 \to X_2.$$

b) Show that any finite sum

$$\underset{\sim}{t} = \underset{\sim}{g}\otimes\underset{\sim}{x} + \underset{\sim}{g}'\otimes x' + \ldots$$

is equal to a similar expression with all the $\underset{\sim}{x}, \underset{\sim}{x}', \ldots$ linearly independent. Deduce that if $\hat{\underset{\sim}{f}}(\underset{\sim}{g}\otimes\underset{\sim}{x}) = \underset{\sim}{0} \Rightarrow \underset{\sim}{g}\otimes\underset{\sim}{x} = \underset{\sim}{0}$, then $\hat{\underset{\sim}{f}}(\underset{\sim}{t}) = \underset{\sim}{0} \Rightarrow \underset{\sim}{t} = \underset{\sim}{0}$, so that $\hat{\underset{\sim}{f}}$ is injective.

8a) Prove from Ti) and Tii) that

$$(X_1\otimes\ldots\otimes X_n)\otimes(Y_1\otimes\ldots\otimes Y_m) \cong (X_1\otimes\ldots\otimes X_n\otimes Y_1\otimes\ldots\otimes Y_m)$$

b) Prove that for any permutation m (Chapter I. 3.06) if we define

$$\underset{\sim}{M} : X_1\otimes X_2\otimes\ldots\otimes X_n \to X_{m_1}\otimes X_{m_2}\otimes\ldots\otimes X_{m_n}$$

on simple elements by

$$\underset{\sim}{M}(\underset{\sim}{x}_1\otimes\underset{\sim}{x}_2\otimes\ldots\otimes\underset{\sim}{x}_n) = \underset{\sim}{x}_{m_1}\otimes\underset{\sim}{x}_{m_2}\otimes\ldots\otimes\underset{\sim}{x}_{m_n}$$

then M is well defined and an isomorphism.

9) Check that contraction is well defined, in that tensors equal by TA) and TS) go to equal tensors.

10a) Prove that if $\underset{\sim}{\psi}$ is the composite isomorphism from 1.13

$$L(X;X) \to X^*\otimes X \xrightarrow[\underset{\sim}{I}\otimes G_\downarrow]{} X^*\otimes X^* = L^2(X;R)$$

then for any $\underset{\sim}{A} : X \to X$, we have $\underset{\sim}{\psi}\underset{\sim}{A}(\underset{\sim}{x},\underset{\sim}{y}) = \underset{\sim}{A}\underset{\sim}{x}\cdot\underset{\sim}{y}$.

b) Prove that $\underset{\sim}{\psi}\underset{\sim}{A}$ is non-degenerate [respectively, symmetric] if and only if $\underset{\sim}{A}$ is non-singular [respectively, self-adjoint].

11) A multilinear map $\underset{\sim}{f} : X\times\ldots\times X \to Y$ is <u>skew-symmetric</u> if

$$\underset{\sim}{f}(\ldots,\underset{\sim}{u},\ldots,\underset{\sim}{v},\ldots) = -\underset{\sim}{f}(\ldots,\underset{\sim}{v},\ldots,\underset{\sim}{u},\ldots)$$

whenever we fill in the empty spaces, for $\underset{\sim}{u},\underset{\sim}{v}$ in any positions. (A linear functional is regarded as skew-symmetric and symmetric, trivially.)

a) The set of skew-symmetric k-linear forms on x is a vector space. We denote it by $\Lambda^k X$. (That it is a subspace of $T^0_k X$ not $T^k_0 X$ is to do with the cultural barriers between mathematicians and physicists: $\Lambda^k X$ is largely used by mathematicians, thinking of the meanings for "co-" and "contravariant" that (III.1.03) we have chosen to avoid.)

b) If $(\underset{\sim}{b}^1,\underset{\sim}{b}^2,\underset{\sim}{b}^3)$ is a basis for X^*, then a basis for $\Lambda^2 X$ is $(\underset{\sim}{b}^1\otimes\underset{\sim}{b}^2-\underset{\sim}{b}^2\otimes\underset{\sim}{b}^1,\ \underset{\sim}{b}^1\otimes\underset{\sim}{b}^3-\underset{\sim}{b}^3\otimes\underset{\sim}{b}^1,\ \underset{\sim}{b}^2\otimes\underset{\sim}{b}^3-\underset{\sim}{b}^3\otimes\underset{\sim}{b}^2)$.

c) Find a general way of writing a basis for $\Lambda^k X$, where X^* has the basis $(\underset{\sim}{b}^1,\ldots,\underset{\sim}{b}^n)$. (The notation of I.3.06 should help.) Deduce that $\dim(\Lambda^k X) = \binom{k}{n}$, the number of combinations of k things chosen out of n. In particular $\dim(\Lambda^n X) = 1$ and $\dim(\Lambda^k X) = 0$ for $k>n$.

d) Since $\Lambda^n X$ is one-dimensional, for any $\underset{\sim}{A} \in L(X;X)$ the operator

$$\boxtimes^n \underset{\sim}{A}^* = \underbrace{\underset{\sim}{A}^*\otimes\ldots\otimes\underset{\sim}{A}^*}_{n\text{ times}} : X^*\otimes\ldots\otimes X^* \to X^*\otimes\ldots\otimes X^*$$

restricted to Λ^n (prove that we can so restrict it by showing that $\boxtimes^n\underset{\sim}{A}^*(\underset{\sim}{f})$ is skew-symmetric when $\underset{\sim}{f}$ is) is just scalar multiplication by some scalar $c(\underset{\sim}{A})$. Show that if $\underset{\sim}{b}_1,\ldots,\underset{\sim}{b}_n$ is any basis for X, $\underset{\sim}{f}\in\Lambda^n X$ non-zero, and $\underset{\sim}{A}$ an operator on X, then $c(\underset{\sim}{A}) = \underset{\sim}{f}(\underset{\sim}{A}\underset{\sim}{b}_1,\ldots,\underset{\sim}{A}\underset{\sim}{b}_n)/\underset{\sim}{f}(\underset{\sim}{b}_1,\ldots,\underset{\sim}{b}_n)$.

e) Find $c(\underset{\sim}{A})$ explicitly, and deduce that $c(\underset{\sim}{A}) = \det \underset{\sim}{A}$. Thus $\det \underset{\sim}{A}$ is "what $\underset{\sim}{A}$ does to skew-symmetric n-linear forms."

f) Why is "skew-symmetric n-linear form" the natural notion of "volume measure" on an n-dimensional vector space? (cf. 1.02(i), I.3.05 and Ex. 2)

12) If we have non-zero $\underset{\sim}{f} \in \Lambda^n X$, $\underset{\sim}{g} \in \Lambda^n Y$, for X,Y n-dimensional and $\underset{\sim}{A} : X \to Y$, define $\det(\underset{\sim}{g}\underset{\sim}{A}/\underset{\sim}{f}) = \underset{\sim}{g}(\underset{\sim}{A}\underset{\sim}{b}_1,\ldots,\underset{\sim}{A}\underset{\sim}{b}_n)/\underset{\sim}{f}(\underset{\sim}{b}_1,\ldots,\underset{\sim}{b}_n)$, where $\beta = (\underset{\sim}{b}_1,\ldots,\underset{\sim}{b}_n)$ is any basis for X.

a) Show that this definition is independent of β.

b) How could the definition be made without reference to a basis?

c) Show that if we choose bases $\underset{\sim}{b}_1,\ldots,\underset{\sim}{b}_n$ for X, $\underset{\sim}{c}_1,\ldots,\underset{\sim}{c}_n$ for Y such that $\underset{\sim}{f}(\underset{\sim}{b}_1,\ldots,\underset{\sim}{b}_n) = 1 = \underset{\sim}{g}(\underset{\sim}{c}_1,\ldots,\underset{\sim}{c}_n)$, then $\det(\underset{\sim}{g}\underset{\sim}{A}/\underset{\sim}{f})$ is given by the usual formula from the corresponding matrix for $\underset{\sim}{A}$.

VI Topological Vector Spaces

"That which gives things their suchness
Cannot be delimited by things.
So when we speak of "limits" we remain confined
To limited things."

Chuang Tzu.

1. CONTINUITY

When we use logarithms for practical calculations, we rarely know exactly the numbers with which we are working; never, if they result from any physical operation other than counting. However if the data are about right, so is the answer. To increase the accuracy of the answer, we must increase that of the data (and perhaps, to use this accuracy, refer to log tables that go to more figures). In fact for any required degree of accuracy in the final answer, we can find the degree of accuracy in our data which we would need in order to guarantee it - whether or not we can actually _get_ data that accurate. The same holds for most calculations, particularly by computer. Errors may build up, but sufficiently accurate data will produce an answer accurate to as many places as required. (The other side of this coin is summarised in the computer jargon GIGO - "Garbage In, Garbage Out".)

On the other hand, suppose our calculation aims to predict what is going to happen to a spherical ball B of mixed U_{235} and U_{238} in a certain ratio λ : and that for this shape and ratio, theory says that critical mass is exactly $9\frac{1}{4}$ kg. Assume we have found the mass of B to three significant figures as 9.25 kg. Now 9.25 kg _is_ the mass "to three significant figures" but that means exactly that it could be up to 5×10^{-3} kg more or less than $9\frac{1}{4}$ kg precisely. And depending on where in that range it is, we have either a bomb or a melting lump of metal, and we cannot calculate which from our measurements. If we knew the mass more accurately as 9.250 kg, to four significant figures, we would have the same problem. Around the critical mass, no degree of accuracy in our knowledge of the mass (even ignoring the fact that at a really accurate level everything becomes probabilistic anyway), will guarantee that the energy output is within the laboratory rather than the kiloton range. The accuracy of our computed answer breaks down in spectacular fashion. The function

$$f : R \rightarrow R$$

where $f(x)$ is the energy output in the next minute of such a ball of uranium of mass x, is discontinuous at the critical mass. Our useful general ability to guarantee any required level of accuracy in the answer by reducing possible errors in the data far enough does not apply here. Around any mass definitely less than the critical one (by however little) we can get an answer close to what happens if we can reduce our measurement error to less than that little; similarly with a definitely greater mass than critical. This is not possible around the critical mass itself. Thus f is continuous everywhere except at the critical mass, both in the intuitive sense and according to the following definition.

1.01 Definition

A function $f : R \rightarrow R$ is continuous at $x \in R$ if for any positive number ε (however small) there exists a positive number δ such that if

$$|y - x| < \delta \quad \text{then} \quad |f(y) - f(x)| < \varepsilon.$$

(Notice the requirement that δ must not be zero; zero would always work, since

$$|x - y| = 0 \Rightarrow x - y = 0 \Rightarrow x = y \Rightarrow f(x) = f(y) \Rightarrow |f(x) - f(y)|<\varepsilon,$$

but from where could we obtain infinitely accurate data? It is in fact a theorem that to get them would take infinite energy.)

The use of ε and δ in this context are among the most standard notations in all of mathematics (to the point where the word "epsilontics" has been coined for complicated continuity proofs). Which symbol is used where, can be remembered by the observation that ε is the maximum allowable error in the εnd result of applying f; that condition we can satisfy by making the error in the δata less than δ. The fun of the game lies in the diversity of continuous functions for which δ depends intricately on ε.

If for some choice of ε (and hence for all smaller choices) no such δ exists, f is by definition <u>discontinuous</u> at x. There may be such a δ for <u>some</u> ε (such as $\varepsilon > \frac{3}{4}$ for f as indicated by the graph in Fig. 1.1) but continuity requires that for <u>each</u> ε there must be a δ. (Cf. also Ex. 1)

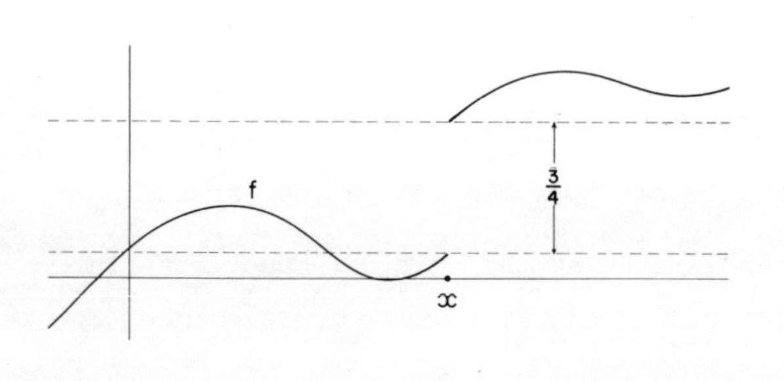

Fig. 1.1

This definition generalises immediately to a much wider context, with the help of a further term:

1.02 Definition

A <u>metric</u> (or <u>distance function</u>) on a set X is a function

$$d : X \times X \to R$$

satisfying

i) $d(x,y) = d(y,x)$

ii) $d(x,y) = 0$ if and only if $x = y$

iii) $d(x,z) \leq d(x,y) + d(y,z)$.

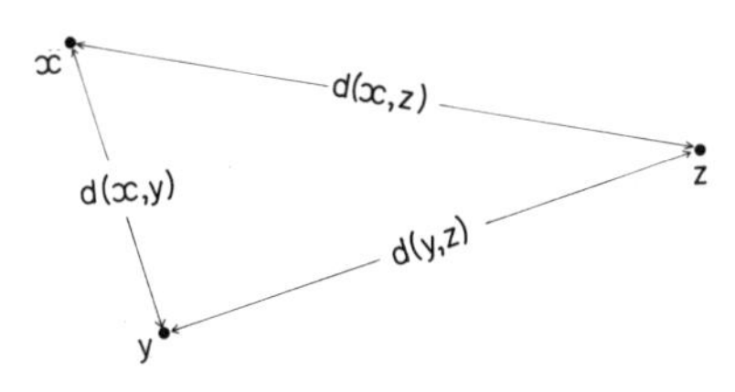

Fig. 1.2

Plainly these are reasonable properties for a "distance from x to y" function. Condition (iii) is called the <u>triangle inequality</u>, since the lengths of the sides of plane triangles give the most familiar examples. Notice the differences from II.1.01.

The pair (X,d) is a <u>metric space</u>: as usual for a set-plus-structure we shall often denote it by just X where we can do so without confusion.

Every set X has the <u>trivial metric</u> : $d(x,x) = 0$, $d(x,y) = 1$ if $x \neq y$

Whereas a metric may be defined on any set, a metric <u>tensor</u> is defined only on a vector space. Part of the connection between the two is indicated in Ex. 2, but some of it must wait until we consider manifolds.

A function $\rho : X \times X \to R$ that satisfies (i) and (iii), but, instead of (ii), only

(ii)* $\rho(x,x) = 0$ for all x, but $\rho(x,y)$ may be zero for $x \neq y$

is a <u>semimetric</u> or <u>pseudometric</u>, and (X,ρ) is a <u>semimetric</u> or <u>pseudometric space</u>. Thus, every metric is a semimetric. Moreover, every semimetric has a non-negative image in R, as may be seen by putting $x = z$ in (iii) and using (i).

CAUTION: Normally it causes no confusion when "metric tensor" is abbreviated to "metric". However, (Ex. 2) a definite metric tensor gives rise to a metric in the sense just defined. In this context it is safer (but not usual) to refer to the latter as a nontensorial metric, to emphasise the distinction. A similar situation occurs on Riemannian manifolds (cf. IX.3.10 and 4.03).

1.03 Definition

A function $f : X \to Y$ between metric spaces (X,d) and (Y,d') is continuous at $x \in X$ if for any $0 < \varepsilon \in R$ there exists $0 < \delta \in R$ such that if $d(x,y) < \delta$ then $d'(f(x),f(y)) < \varepsilon$.

If f is continuous at all $x \in S$, where $S \subseteq X$, f is continuous on S. If $S = X$, we just call f continuous.

(Notice that this coincides with 1.01 when R is given the natural metric $d(x,y) = |x - y|$.)

Now, another way of phrasing 1.03 is to say that the image under f of the set $\{y | d(x,y) < \delta\}$, called the open ball $B(x,\delta)$ of radius δ around x, is inside $B(f(x),\varepsilon)$, similarly defined and named. The reason for the word "ball" is obvious from Fig. 1.4. There it is illustrated for maps $f : R \to R$, $g : R^2 \to R^2$ and $h : R^2 \to R^3$ with the usual notion of distance. (We shall see later that other notions can be important.) "Open" refers to the fact that all the points in a ball $B(x,\delta)$ are strictly inside it, in the following sense. If $y \in B(x,\delta)$, so that $d(x,y) = r < \delta$, then by the triangle inequality all points in $B(y, \frac{\delta-r}{2})$ are in $B(x,\delta)$ too. Hence y is completely surrounded by points in $B(x,\delta)$ (Fig. 1.3). So $B(x,\delta)$ has no points in it of what it is natural to call its boundary (cf. 1.04

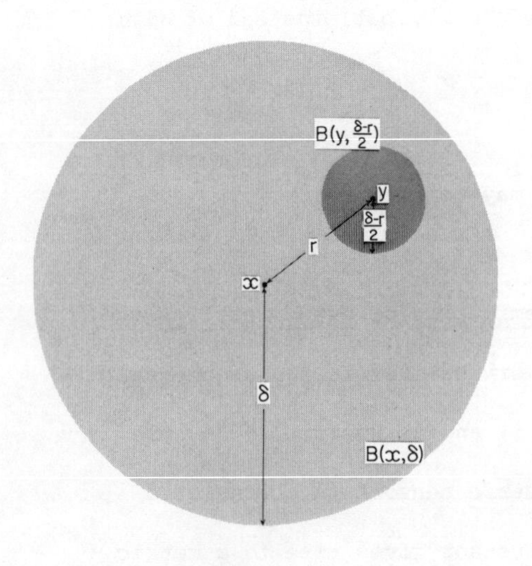

Fig. 1.3

below). By association of ideas this "not including a boundary" is thought of as being "unfenced" and hence "open". (See also Ex. 3c; this is motivation from English usage, however, which is always warped by making it precise enough for mathematics - a sort of uncertainty principle, perhaps. The reader would do well to forget the motivation in favour of the defined meaning as soon as he can : a crutch is useful to get you walking, but once your leg has healed it stops you running, and you should throw it away.) In the course of defining this language precisely, we extend it to other sets also:

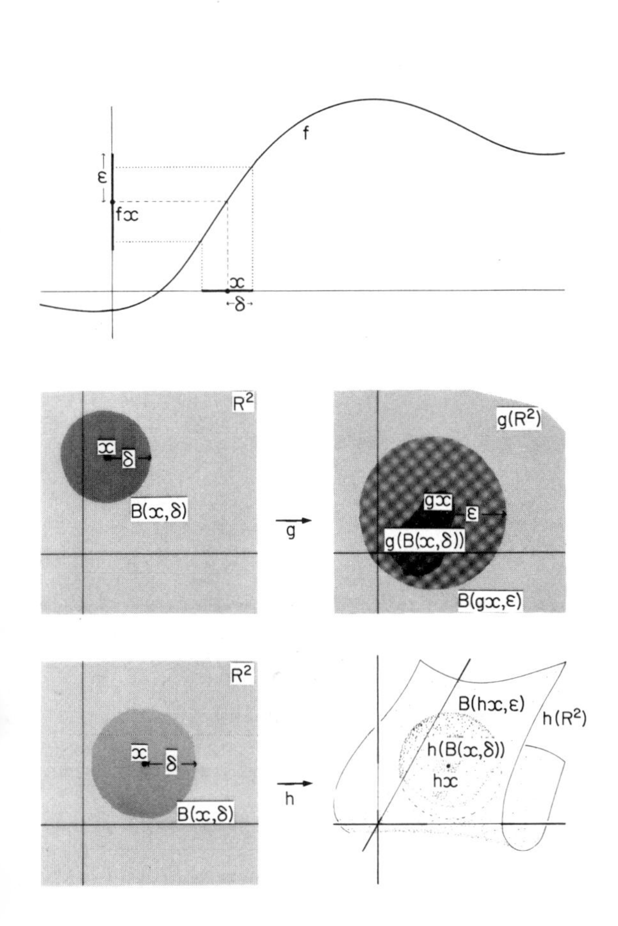

Fig. 1.4

<u>1.04 Definition</u>

A <u>boundary point</u>, or <u>point of closure</u>, of a set S in a metric space X is a point x such that for any $0 < \delta \in R$, $B(x,\delta)$ contains both points in S and points not in S.

The <u>boundary</u> ∂S of the set S is the set of boundary points of S.

The set S is <u>open</u> if any boundary points it has are not contained in it, <u>closed</u> if all boundary points it has are contained in it.

(Notice that since $\emptyset$ has no points, no $B(x,\delta)$ for any x or δ contains points, so it has no boundary points. Since it contains all the boundary points

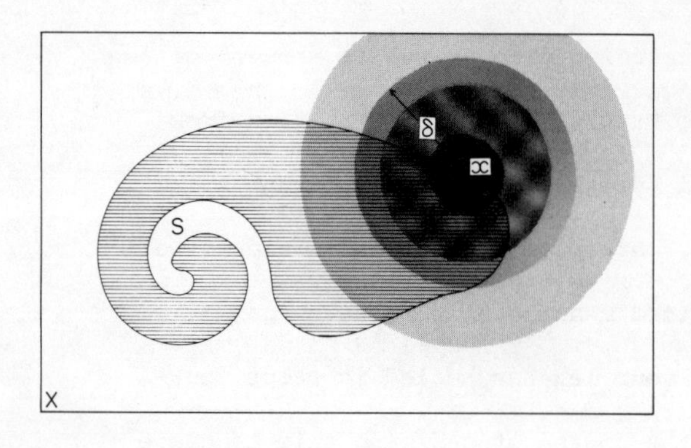

Fig. 1.5

it has, ∅ is closed; since it contains no boundary points, it is open. By a similar argument, the whole space X is both open and closed. This is one point where the crutch of common usage is a hindrance more than a help.)

The closure $\bar{S}$ of S is the set $S \cup \partial S$. (cf. Ex. 3f)

If you have not met these terms before, you should do Ex. 3 before going much further, to learn what they mean in practice. The definitions alone cannot give you the flavour, and as we are not writing a topology book we cannot roll them fully around the tongue in the text.

Now, we sometimes want to talk about continuity when we do not have a natural choice of metric, and to suppose we had would confuse things thoroughly. Most notably this occurs on indefinite metric vector spaces (compare and contrast Ex. 2b and Ex. 2c). It is precisely this false supposition of a metric in models of spacetime which still sustains a lot of innumerate or semi-numerate "Philosophers Of Science" in their belief in a twins "paradox" (cf. Chapter 0.5.3). To avoid this confusion it is convenient to have a "continuity structure" separate from particular choices of metric. This kind of structure is called a topology, and we shall define it in a moment in 1.07. Moreover, just as leaving out bases can greatly clarify some parts of linear algebra, the separation of continuity from specific metrics proved such a powerful tool that topologies have become as central to modern mathematics and physics as vector spaces. Before the definition, we shall prove a lemma which says essentially that the roundness of the balls $B(x,\delta)$ used is irrelevant to the definition of continuity; all that matters is their openness.

1.05 Lemma

A function $f : X \to Y$ between two metric spaces is continuous at $x \in X$ if and only if for any open set V containing $f(x)$, there is an open set U containing x such that $f(U) \subseteq V$.

Proof

(i) Suppose f is continuous at x.

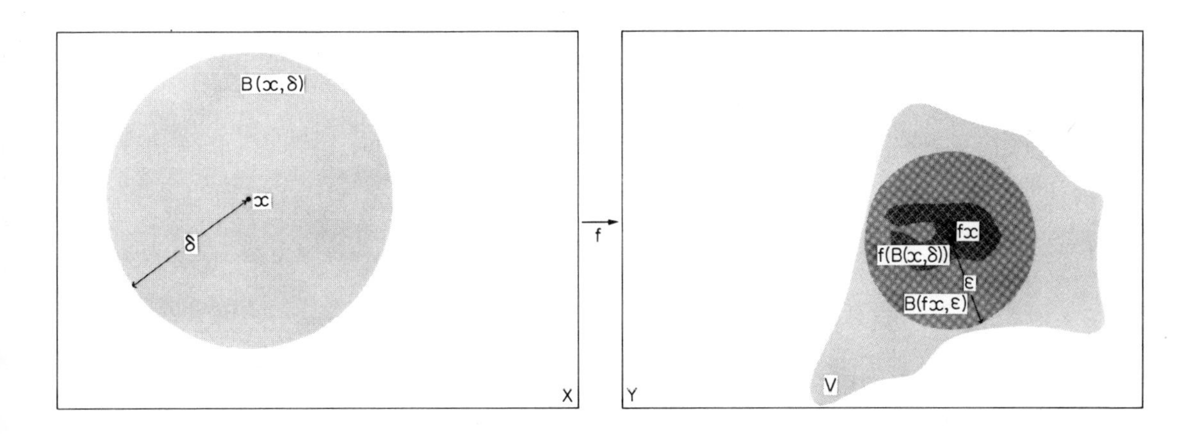

Fig. 1.6

Then since V is open and $f(x) \in V$, there exists $\varepsilon \in R$ such that $B(f(x),\varepsilon) \subseteq V$. (Ex. 3b).

Now, by continuity of f there exists $\delta \in R$ such that $d(x,y)<\delta \Rightarrow d(f(x),f(y))<\varepsilon$. Hence $f(B(x,\delta)) \subseteq B(f(x),\varepsilon) \subseteq V$, and $B(x,\delta)$ is open (Ex. 4a), so if we set $U = B(x,\delta)$ we are done.

(ii) Suppose for each open set V containing $f(x)$, there is an open set U containing x such that $f(U) \subseteq V$.

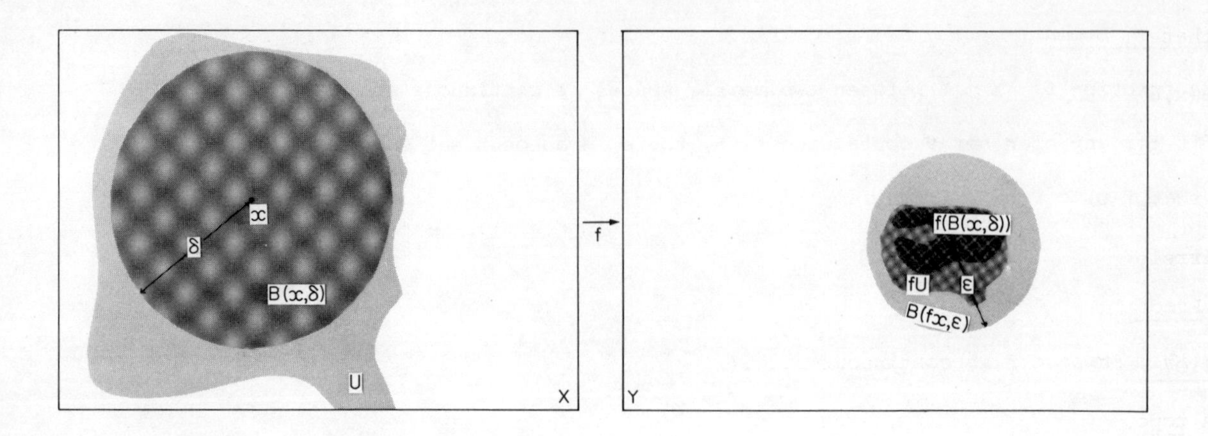

Fig. 1.7

Then in particular, since each $B(f(x),\varepsilon)$ is open (Ex. 3a), there is an open set U such that $f(U) \subseteq B(x,\varepsilon)$, with $x \in U$. Hence by Ex. 3b there exists $\delta \in R$ such that $B(x,\delta) \subseteq U$. But then

$$f(B(x,\delta)) \subseteq f(U) \subseteq B(f(x),\varepsilon).$$

Since this can be done for any ε, f is continuous at x. ■

1.06 Corollary

A map $f : X \to Y$ between metric spaces is continuous if and only if $\overleftarrow{f}(V)$ is open for each open set V in Y.

Proof

Suppose f is continuous

Then it is continuous at each $x \in X$, and in particular at each $x \in \overleftarrow{f}(V)$. Now, since V is open there exists by the above some open set U for each such x, such that $f(U) \subseteq V$. Hence by Ex. 3b applied to U, for each x we have some δ such that $B(x,\delta) \subseteq U \subseteq \overset{\leftarrow}{f}(V)$. Now by Ex. 3b applied to $\overset{\leftarrow}{f}(V)$, $\overset{\leftarrow}{f}(V)$ must be open.

Similarly for the converse. ■

We are now able to capture the essential aspects of continuity, with no irrelevancies, in the following definitions.

1.07 Definition

A topology on a set X is a specification of which subsets of X are to be considered open. More set-theoretically, it is a family $\mathcal{T}$ of subsets of X, called the open sets of the topology, satisfying the axioms

OA) $\emptyset \in \mathcal{T}$ and $X \in \mathcal{T}$.

OB) For any finite family $\{U_i \mid i = 1,\ldots,n\}$ of open sets, $\bigcap_{i=1}^{n} U_i$ is open.

OC) For any family (finite or infinite) $\{U_\alpha \mid \alpha \in A\}$ of open sets, $\bigcup_{\alpha\in A} U_\alpha$ is open.

The topology is Hausdorff (pronounced "housed orff" and named after the German mathematician F. Hausdorff (1868-1942)) if it satisfies one extra axiom:

OD) For any two distinct points $x,y \in X$, there exists open sets $U,V \in \mathcal{T}$ such that $x \in U$, $y \in V$, and $U \cap V = \emptyset$.

(Since we have agreed that open sets can be of any shape, not just round balls, OD can be remembered by Fig. 1.8, which relates an English meaning to the German pronunciation.) This is such a useful

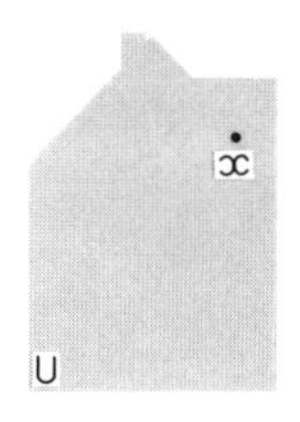

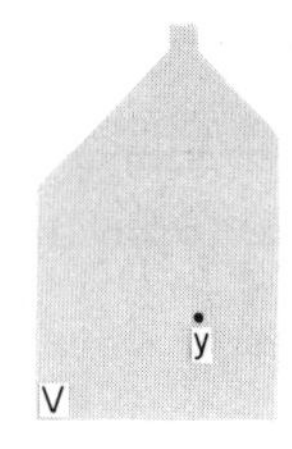

Fig. 1.8

condition that by "topology" we shall always mean "Hausdorff topology" unless otherwise stated.

The set X with the topology $\mathcal{T}$ is the topological space $(X,\mathcal{T})$, as usual denoted by just X if only one topology has been mentioned for the set. (It is not unheard of to give a set as many as ten topologies at a time. We shall content ourselves throughout with one or two.)

If X is a metric [respectively, pseudometric] space, then the metric [respectively, pseudometric] topology on X is the topology consisting of the open sets defined in 1.04 (cf. Ex. 4a). A metric topology is always Hausdorff (Ex. 4a), a pseudometric topology is not - which severely limits its usefulness. In general, many other metrics serve equally well to define a given metric topology by giving rise to the same open sets. We shall see this for vector spaces in §3.

The set R of real numbers will always in this book be assumed to have the usual metric topology, given by the metric $d(x,y) = |x - y|$.

If $\mathcal{T}$ is the metric topology corresponding to some metric on X then the topological space $(X,\mathcal{T})$ is metrisable. It can be useful arbitrarily to pick a metric to give a handle on $\mathcal{T}$ in computations, even when there is no natural choice, just as an arbitrary basis can be convenient when computing with a vector space. In particular, metrisability guarantees that X is Hausdorff, which is handy.

We can make the following extensions of the definitions in 1.04, showing that the underlying concepts are of a topological rather than a metric nature.

1.08 Definition

A neighbourhood of a point x in X, generally denoted by $N(x)$ or some variation of it, is an open set containing x. The role of open balls is taken over by neighbourhoods as we go to topology. (That this take-over sacrifices nothing as far as continuity is concerned is the essence of Lemma 1.05.)

A boundary point of a set S in a topological space X is a point x such that for any neighbourhood $N(x)$ of x, $N(x)$ contains both some points in S and some points not in S. (Fig. 1.9)

The *boundary* ∂S of S is the set of boundary points of S.

A set S is *closed* if it contains all its boundary points, or equivalently (Ex. 5a) if $X \backslash S$ is one of the open sets of the topology. It is important to note that a set may be neither open nor closed (cf. Ex. 3g).

The *closure* $\bar{S}$ of a set S is the set $S \cup \partial S$. (cf. Ex. 5b).

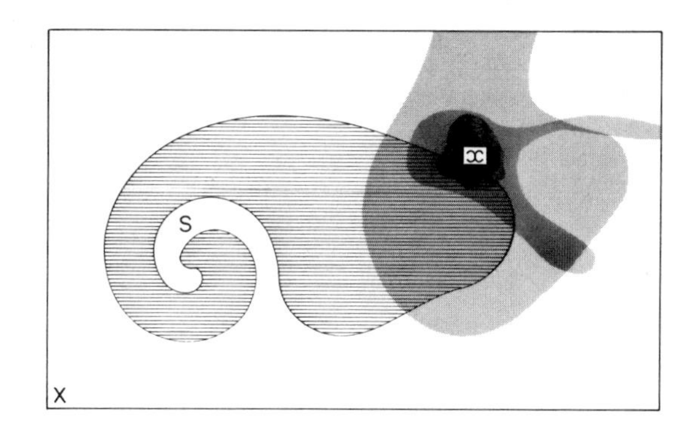

Fig. 1.9

We now have the framework in which we can define continuity in full generality, uncluttered by metrics.

1.09 Definition

A map $f : (X,T) \to (Y,\Sigma)$ between topological spaces is *continuous* if

$$V \in \Sigma \implies \overleftarrow{f}(V) \in T.$$

This is just the reformulation we reached in 1.06. (cf. also Ex. 5c)

1.10 Lemma

If $f : (X,T) \to (Y,\Sigma)$ and $g : (Y,\Sigma) \to (Z,\P)$ are continuous maps, then so is $g \circ f : (X,T) \to (Z,\P) : x \rightsquigarrow g(f(x))$.

Proof

$V \in \P \implies \overleftarrow{g}(V) \in \Sigma$ (g continuous)

$\Rightarrow \quad \overleftarrow{f}(\overleftarrow{g}(V)) \in T \quad$ (f continuous)

$\Rightarrow \quad \overleftarrow{(g \circ f)}(V) \in T \quad$ (same set).

The shortness of this proof illustrates the power of the topological viewpoint. Proving the same thing for the more limited case of continuous maps between metric spaces is actually harder, with assorted ε's and δ's (write it out and see how messy it looks!) but that result is implied by this one and 1.06.

1.11 Definition

A map $f : X \to Y$ between topological spaces is a homeomorphism if it is continuous, bijective and its inverse is also continuous. (cf. Ex. 7) (There is obviously a homeomorphism between the shapes of Fig. 1.10, though it cannot preserve distances: we cannot have $d(f(x), f(y)) = d(x,y)$ in general. This is another reason why continuity is most naturally considered in topological rather than metric terms, and the reason for the name "india-rubber geometry" for topology.) If there is a homeomorphism between two spaces they are homeomorphic.

Fig. 1.10

1.12 Lemma

If T, Σ are topologies on X, and the identity map $I_X : X \to X : x \rightsquigarrow x$ is a homeomorphism, then $T = \Sigma$. (This saves a lot of work when showing that different definitions give the same topology.)

Proof

$V \in T \Rightarrow \overleftarrow{I_X}(V) \in \Sigma \Rightarrow V \in \Sigma$ (I_X continuous)

$U \in \Sigma \Rightarrow (\overleftarrow{I_X})^{\leftarrow}(U) \in T \Rightarrow U \in T$ ($\overleftarrow{I_X}$, which is I_X again anyway, continuous.)

So $T = \Sigma$ as sets, and hence as topologies.

Exercises VI.1

1) Show that if, for some function $f : R \to R$, at some $x \in R$ we have continuity of f at x by virtue of the <u>same</u> choice of δ for each ε, then f is constant between $x - \delta$ and $x + \delta$.

2a) Using the Schwarz inequality (IV.1.07) show that for any inner product space $(X,\underset{\sim}{G})$ and vectors $\underset{\sim}{x},\underset{\sim}{y} \in X$,

$$\underset{\sim}{G}(\underset{\sim}{x} + \underset{\sim}{y}, \underset{\sim}{x} + \underset{\sim}{y}) \leq (\|\underset{\sim}{x}\|_{\underset{\sim}{G}} + \|\underset{\sim}{y}\|_{\underset{\sim}{G}})^2.$$

b) Show that for any inner product space $(X,\underset{\sim}{G})$ the function

$$d_{\underset{\sim}{G}} : X \times X \to R : (\underset{\sim}{x},\underset{\sim}{y}) \rightsquigarrow \|\underset{\sim}{x} - \underset{\sim}{y}\|_{\underset{\sim}{G}}$$

is a (non-tensorial) metric on X. (For the triangle inequality, apply (a) to

$$\|(\underset{\sim}{x} - \underset{\sim}{y}) + (\underset{\sim}{y} - \underset{\sim}{z})\|_{\underset{\sim}{G}} \quad .)$$

c) Show that if $\underset{\sim}{G}$ is an indefinite metric tensor, then $d_{\underset{\sim}{G}}$ is not even a semimetric (consider, for example, the vectors (0,0), (1,0) and (1,1) in H^2).

3a) Show that each open ball $B(x,\delta)$ in a metric space is indeed open by Def^n 1.04. (Hint : Fig. 1.4.)

b) Show that $S \subseteq X$ is open in the metric space X if and only if for each point

$x \in S$ there exists some $0 < \delta \in R$ such that $B(x,\delta) \subseteq S$.

c) Show that $S \subseteq X$ is open if and only if $X \backslash S$ is closed. (If the boundary between two countries is a fortified wall - Hadrian's Wall, say, or the Great Wall of China - then the country A that includes the wall is closed to invasion by B, while B is open to attacks from A. In topology, A and B cannot have a wall each.)

d) Show that $\partial B(x,\delta) = \{y | d(x,y) = \delta\}$, and that this set, called the <u>sphere</u> $S(x,\delta)$ of radius δ, centre x, is closed.

e) Show that the <u>closed ball</u> of radius δ, centre x and denoted by

$$\bar{B}(x,\delta) = \{y | d(x,y) \le \delta\}$$

is the closure of the open ball $B(x,\delta)$.

f) The closure $\bar{S}$ of any set $S \subseteq X$ is closed, so justifying the term "closure". (cf. Ex. 5b for the general case.)

g) The set $\{x | 0 < x \le 1\}$ is neither open nor closed as a subset of R with the usual metric.

h) Show that R itself is both open and closed as a subset of R.

i) Show that $R \times \{0\}$ is closed but not open in the plane $R \times R$ with the metric $d((x,y),(x',y')) = +\sqrt{|(x - x')^2 + (y - y')^2|}$.

4a) Show that a metric topology does indeed satisfy OA - OD, and a pseudometric topology satisfies OA - OC.

b) If x,y are distinct points in a pseudometric space X such that $d(x,y) = 0$, then any set open in the pseudometric topology that contains one contains the other, so that X is not Hausdorff.

c) The intersection of the infinite set $\{B(0,1 + 1/n) | n \in N\}$ of open balls in R is not open. Hence, OB cannot usefully be strengthened.

5a) A set S in a topological space X contains all its boundary points if and only if the set $X\backslash S$ of points of X not in S is open.

b) The closure $\bar{S}$ of any set S in a topological space X is closed. (cf. Ex. 3f for metrisable spaces.)

c) A map $f : X \to Y$ between topological spaces is continuous if and only if for every closed set $C \subseteq Y$, $\overleftarrow{f}(C) \subseteq X$ is also closed.

6a) In a Hausdorff topology, each set $\{x\}$ containing only one point is closed.

b) The continuous map, $f : R \to R : x \rightsquigarrow x^2$, takes the open set $U = \,]-1,1[$ to a set $f(U)$ which is neither open nor closed. (Note; for reasons of space we shall not go into the proofs that the elementary functions - polynomials, log, sin, etc. - are continuous. The work is not in proving continuity, but in defining the functions themselves sufficiently precisely to prove anything at all. This is done in any elementary analysis book.)

7) If X is the set $]0,1] \subset R$ and Y is the unit circle $\{(x,y)\,|\,x^2 + y^2 = 1\}$ in R^2, both with the usual metric topology given by Euclidean distance, then the map

$$X \to Y : (\sin 2\pi x, \cos 2\pi x)$$

which wraps X once round Y is a continuous bijection but not a homeomorphism.

2. LIMITS

The equipment that we have set up is very powerful. Notice that we have now two new kinds of object (metric and topological spaces) and allowable maps between them, to place alongside vector and affine spaces, with linear and affine maps. However the rule for allowing maps - continuity - is a little surprising. Instead of preserving, for instance, addition <u>forwards</u> as a linear map must do, to be continuous a map must preserve openness <u>backwards</u> ($\overleftarrow{f}$(open set) must be open) but not necessarily forwards. (cf. Ex. 1.6b. If it does carry open sets to open sets, f itself is called open.) This grew naturally out of our considerations of computability, but those were not the original motivation. The interest was more in something that is preserved forwards: the limit of a sequence.

2.01 Definition

A mapping $S : N \to X : i \rightsquigarrow S(i)$, where X is any set, is called a sequence of points in X. ($S(i)$ is often written x_i, for reasons of tradition and convenience. As usual, N denotes the natural numbers. We shall also continue to denote a neighbourhood of a point x by $N(x)$.)

A sequence S of points x_i in a topological space X has the point x as a limit if every neighbourhood $N(x)$ of x contains x_i for all but finitely many $i \in N$. (Hence after passing some x_m where $m = \max\{i \mid x_i \notin N(x)\}$, which is required to be a finite set and hence has a maximum, S stays inside $N(x)$.) If X is Hausdorff, S can have at most one limit (Ex. 1a) and we speak of the limit of S; it need not have any limit in general (Ex. 1b).

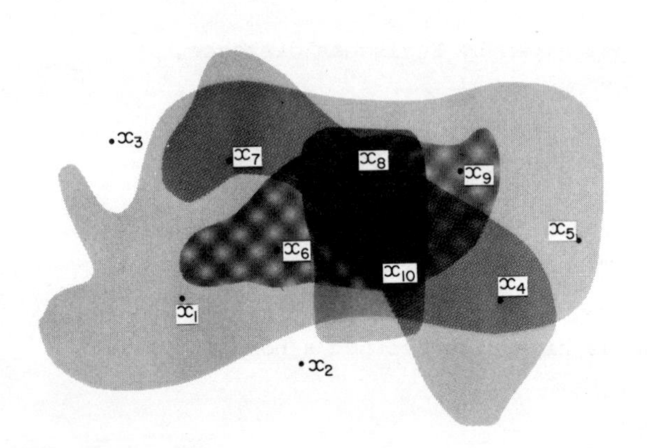

Fig. 2.1

If S has the limit x, then S is convergent and converges to x (cf. Ex. 1d). We write for short that

$$\operatorname{Lim}(S) = \lim_{i\to\infty} x_i = x.$$

If S does not converge, we may still have a convergent subsequence S' of S. (Formally S' is given in the form $S' = S \circ J$, where J is any order-preserving injective map $J : N \to N$. This just codifies the obvious notion, and guarantees that S' also will be an infinite sequence). We may have several subsequences of S converging to different points (Ex. 1b), but if S itself converges, so do all its subsequences, and to the same point. (Ex. 1c).

2.02 Lemma

A function $f : X \to Y$ between topological spaces, where (X,T) is metrisable, is continuous if and only if it preserves limits; formally, if and only if for any sequence of points in X

$$\lim_{i\to\infty} x_i = x \Rightarrow \lim_{i\to\infty} (f(x_i)) \text{ exists and is } f(x).$$

Proof

(i) If f is continuous, for any neighbourhood $N(f(x))$ of $f(x)$ there is a neighbourhood $N'(x)$ of x such that $f(N'(x)) \subseteq N(f(x))$ (Lemma 1.05, rephrased.) So since for any sequence

$$f(x_i) \notin N(f(x)) \Rightarrow f(x_i) \notin f(N'(x)) \Rightarrow x_i \notin N'(x),$$

we have

$$A = \{i | f(x_i) \notin N(f(x))\} \subseteq \{i | x_i \notin N'(x)\} = B.$$

If $\lim_{i\to\infty} x_i = x$, B must be finite by definition and hence so must A. Thus f preserves limits.

(ii) If f is not continuous at some x, then for some neighbourhood $N(f(x))$ of $f(x)$ every neighbourhood

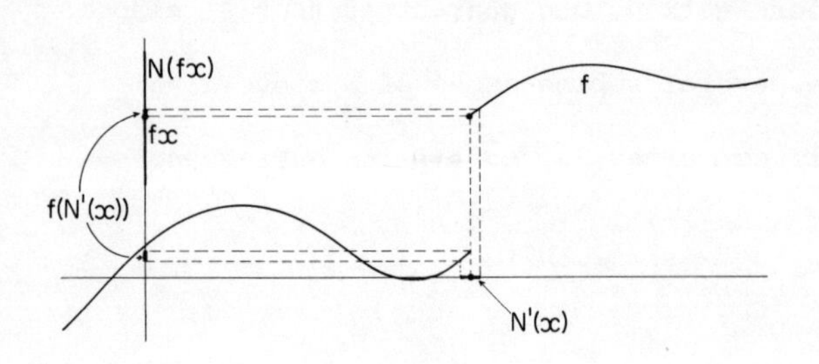

Fig. 2.2

$N'(x)$ of x contains points y such that $f(y) \notin N(f(x))$ (Fig. 2.2). Choosing any metric d on X such that the corresponding topology is T, we take a sequence

$$N_i(x) = B(x, 1/i)$$

of open balls in this metric, which are all neighbourhoods of x. Hence in each N_i we can choose y_i such that $f(y_i) \notin N(f(x))$. Now clearly the y_i converge to x (details, Ex. 2), but the sequence $f(y_i)$ stays outside $N(f(x))$ and cannot therefore converge to $f(x)$: so f does not preserve limits.

Hence if f <u>does</u> preserve limits, it must be continuous. ■

Notice that continuity <u>always</u> implies preserving limits: only the converse depended on metrisability of X. For full generality we could have replaced preserving limits by preserving the operation of closure. However, in the sequel we shall be dealing only with metrisable topologies. (We just don't want to confuse ourselves by a choice of metric; in spacetime that would turn out to depend on a choice of basis - whereas the topology which we shall use does not.) It will be safe throughout this book therefore to think of a topology as a minimum structure allowing us to take limits, and continuity as the preservation of this structure.

In analysis, this view of the nature of topologies and continuity is the most central; several kinds of convergence are juggled in the average infinite-dimensional proof.

<u>Exercises VI.2</u>

1a) Show that if x,y are both limits of a sequence x_i of points in a topological space X, any neighbourhoods N(x), N(y) of x,y have x_i in common for all but finitely many i. Deduce that if X is Hausdorff, then x = y.

b) The sequence $S : N \to R : i \rightsquigarrow (-1)^i$ has no limit, in the sense of Def^n 2.01. Find convergent subsequences of S converging to different points.

c) Prove, for any sequence S in a topological space X, that if $\lim_{i\to\infty} S(i) = x$, all subsequences of S converge to x.

d) You may be meeting topologies explicitly for the first time here, but you will have been taught a definition of convergence for a sequence (recall that a sequence, unlike a series, involves no adding up). Either

(i) Show that this is equivalent to Def^n 2.01 in the case of R with the usual metric topology or

(ii) show that it is not by producing a sequence that converges by one definition and not by the other.

In case (ii) destroy (or sell to an enemy) the text using the other definition: it is still mentally in the confusion about continuity that was only cleared up around the end of last century. Any definition <u>not</u> equivalent to 2.01 is known by bitter experience to bring chaos in its train.

2) Show that in a metric space X;

a) Every open ball $B(x,\delta)$ must contain an open ball of the form $B(x,1/n)$ for some $n \in N$.

b) Every neighbourhood N(x) of a point x must contain at least one of the open balls $B(x,1/n)$, $n \in N$, and hence all but finitely many of them.

c) Deduce that the sequence y_i in the proof of Lemma 2.02 converges to x.

3. THE USUAL TOPOLOGY

There is only one useful topology on any finite-dimensional affine or vector space, but a great many ways to define it. From a coordinate and limit point of view, it is described very simply. A sequence of vectors in R^n converges if and only if each sequence of j-th coordinates does. The j-th coordinate of the limit is then the limit of the j-th coordinates, as one would hope and expect. However, it is not transparently obvious that this means the same in all coordinate systems. The easiest proof that it does is to give a coordinate-free definition and show that it reduces to this form in any coordinates. We may approach such a definition as follows.

Another viewpoint on the nature of a topology on a set X (and a very powerful one when formalised) is as a rule for which functions on X are to be considered continuous. For example, if _all_ functions on X are continuous, all subsets of X must be open; this is called the _discrete_ topology, and is useful surprisingly often. Now, for X a vector space the mildest requirement that we can reasonably make (in the absence of any extra structure on X), and still expect to relate the topology to the linear structure, is that at least all linear functionals $X \to R$ should be continuous. In finite dimensions it turns out that this is enough to define the usual topology; in infinite dimensions it defines _a_ topology, but no one topology is "usual".

3.01 Definition

The _weak topology_ on a vector space V is the smallest (Ex. 1a) family T of subsets of V such that

Wi) T is a topology

Wii) For any linear functional $\underset{\sim}{f} : V \to R$ and open set $U \subseteq R$, $f^{\leftarrow}(U) \in T$.

Open sets in R are exactly unions of sets of open intervals (Ex. 1b,c). Hence the sets of the form $f^{\leftarrow}(U)$ in V are the unions of sets of infinite slabs (Fig. 3.1, a,b) which do not include their boundary hyperplanes. (These latter are lines and planes for $V = R^2, R^3$ respectively.) Lacking infinite space, we show only the heart of each slab.

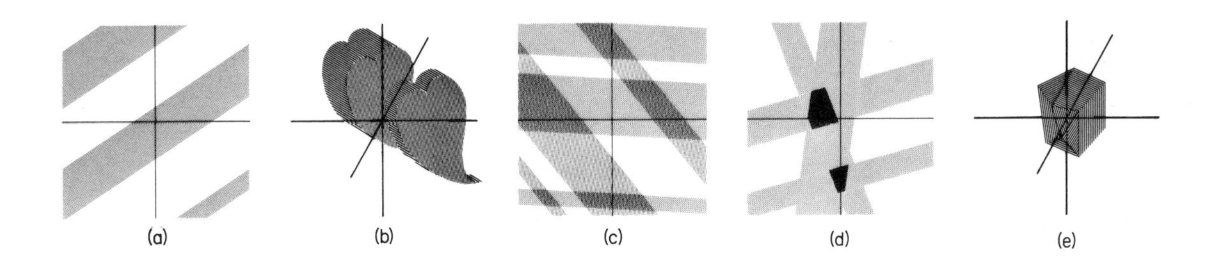

Fig. 3.1

But to satisfy Wi) we must also include finite intersections of such slabs or sets of slabs (Fig. 3.1 c,d,e), which gives us chunks of all flat-sided shapes and sizes, and infinite unions of such chunks, by which we can build up rounded figures (Ex. 2).

Now the condition that all linear functionals be continuous is a large one at first sight; to check the truth of it for a topology, we can reduce the work considerably via the following:

3.02 Lemma

For any topological space X, the sum $\sum_{i=1}^{n} f_i$ of a finite set of continuous functions $f_1,\ldots,f_n : X \to R$ is again continuous.

Proof

Represent the usual topology on R by the usual metric. (1.07). For any $x \in X$, and any positive $\varepsilon_1,\ldots,\varepsilon_n \in R$, there exist by hypothesis open sets $U_1,\ldots,U_n \subseteq X$ containing x such that

$$f_i(U_i) \subseteq B(f_i(x), \varepsilon_i) =]f_i(x) - \varepsilon_i, f_i(x) + \varepsilon_i[, \quad i = 1,\ldots,n,$$

since $B(f_i(x), \varepsilon)$ is an open set.

So for any $0 < \varepsilon \in R$ we can set each $\varepsilon_i = \varepsilon/n$, find corresponding U_i and define

$$U = \bigcap_{i=1}^{n} U_i.$$

This is again an open set by OB, containing x because each U_i does. Moreover,

$y \in U \Rightarrow y \in U_i$, for each $i = 1,\ldots,n$

$\Rightarrow f_i(y) \in B(f_i(x), \varepsilon_i)$ for each i

$\Rightarrow |f_i(y) - f_i(x)| < \varepsilon/n$ for each i

$\Rightarrow \sum_{i=1}^{n} |f_i(y) - f_i(x)| < \varepsilon$

$\Rightarrow |\sum_{i=1}^{n} f_i(g) - \sum_{i=1}^{n} f_i(x)| < \varepsilon$ (Ex. 3)

$\Rightarrow (\sum_{i=1}^{n} f_i)y \in B((\sum_{i=1}^{n} f_i)x, \varepsilon)$.

Thus $(\sum_{i=1}^{n} f_i)(U) \subseteq B((\sum_{i=1}^{n} f_i)x, \varepsilon)$, as required for continuity. ■

3.03 Corollary

For any basis $\underset{\sim}{b}_1,\ldots,\underset{\sim}{b}_n$ of a finite-dimensional vector space V, with some topology T, we have all $\underset{\sim}{f} \in V^*$ continuous if and only if the vectors $\underset{\sim}{b}^1,\ldots,\underset{\sim}{b}^n$ of the dual basis are continuous.

Proof

(i) If all covariant vectors are continuous, that includes $\underset{\sim}{b}^1,\ldots,\underset{\sim}{b}^n$.

(ii) Any $\underset{\sim}{f} \in V^*$ is a linear combination of $\underset{\sim}{b}^1,\ldots,\underset{\sim}{b}^n$; that is, exactly a sum of scalar multiples of them. Since any scalar multiple of a continuous function is continuous (Ex. 4), if the $\underset{\sim}{b}^i$'s are continuous then so is $\underset{\sim}{f}$. ■

This means that all the open sets of the weak topology are also open in the <u>open box topology</u> for any choice of coordinates:
we could replace Wii) by

Wii*) For every open set
$U \subseteq R$, each $(\underset{\sim}{b}^i)^{\leftarrow}(U) \in T$.

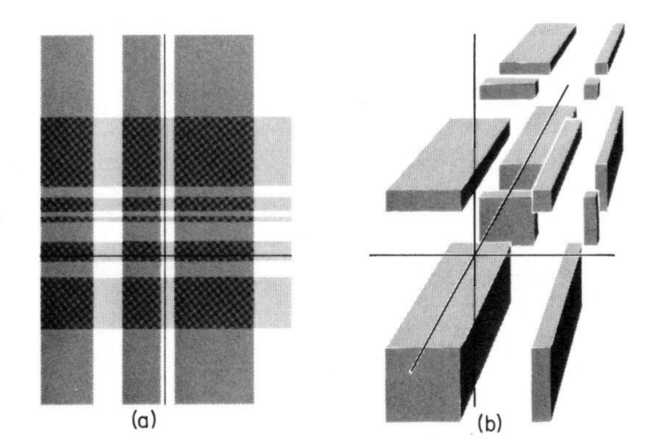

Fig. 3.2

That would replace Fig. 3.1 c,d,e by pictures like Fig. 3.2 a,b without changing the topology.

We have shown Wii) and Wii*) to be equivalent. Equivalence follows for the definitions "The smallest family of subsets of X such that Wi) and Wii) hold" and "The smallest family such that Wi) and Wii*) hold". Therefore the two topologies must be the same, and we can build up precisely the same collection of open sets by taking infinite unions of open boxes as by taking unions of the more arbitrary chunks of Fig. 3.1 c.d.e.

The same topology, then, goes under two names, depending on the choice of definition, and we could find others (cf. Ex. 5). Since they all refer to the same thing, we shall agree to call it the <u>usual</u> topology on a finite-dimensional vector space. We may use other names to specify that we are about to invoke a particular definition useful for the computation or argument we are about to develop. Since we have proved the open-box topology for <u>any</u> choice of coordinates to be the same as the weak topology, it is worth pointing out explicitly that we have proved:

3.04 Theorem

The open box topology is invariant with respect to change of basis. ■

Similarly we have

3.05 Theorem

If X is an affine space with vector space T, then the topology defined on X by choosing a chart C_a (cf. II.1.08) on X and setting

$$S \subseteq X \text{ is open in } X \iff C_a(S) \text{ is open in } R^n \text{ with the usual topology,}$$

does not depend on the choice of chart. (We call this also the usual topology on an affine space.) ■

3.06 Theorem

If V,W are finite-dimensional vector spaces, then all linear maps $\underset{\sim}{A} : V \to W$ are continuous in the usual topology. (We have only defined this to be true for W = R.)

Proof

Choose a basis $\underset{\sim}{b}_1,\ldots,\underset{\sim}{b}_n$ for W, and consider an arbitrary open box

$$U = \left\{ (v^1,\ldots,v^n) \in V \;\middle|\; v^1 \in]a_1,b_1[,\ v^2 \in]a_2,b_2[,\ \ldots,\ v^n \in]a_n,b_n[\right\}$$

$$= \bigcap_{i=1}^{n} (b^i)^{\leftarrow}(I_i),$$

where the $I_i =]a_i,b_i[$ are open intervals in R. Now the n maps

$$\underset{\sim}{b}^i \circ \underset{\sim}{A} : V \to R$$

are linear (being composites of linear maps) and have range R. So by Wii) the sets $(\underset{\sim}{b}^i \circ \underset{\sim}{A})^{\leftarrow}(I_i)$ are open in V. Therefore so also is their intersection. However,

$$\underset{\sim}{v} \in \bigcap_{i=1}^{n} (\underset{\sim}{b}^i \circ \underset{\sim}{A})^{\leftarrow}(I_i) \iff \underset{\sim}{b}^i \circ \underset{\sim}{A}(\underset{\sim}{v}) \in I_i, \text{ each } i$$

$$\iff \underset{\sim}{A}(\underset{\sim}{v}) \in (\underset{\sim}{b}^i)^{\leftarrow}(I_i), \text{ each } i$$

$$\iff \underset{\sim}{A}(\underset{\sim}{v}) \in U.$$

So $\underset{\sim}{A}^{\leftarrow}(U) = \bigcap_{i=1}^{n} (\underset{\sim}{b}^i \circ \underset{\sim}{A})^{\leftarrow}(I_i)$, which we have just shown to be open.

Hence for U an open box, $\underset{\sim}{A}$ satisfies 1.08, and since by Ex. 2 an arbitrary open set U' is a union $\bigcup_\alpha U_\alpha$ of open boxes $\underset{\sim}{A}$ satisfies 1.08 completely. Finally, $\underset{\sim}{A}^{\leftarrow}(U')$ is a union $\bigcup_\alpha (\underset{\sim}{A}^{\leftarrow}(U_\alpha))$ of open sets and hence again it is open. Thus $\underset{\sim}{A}$ is continuous. ■

3.07 Corollary

If X,Y are finite-dimensional affine spaces, all affine maps $X \to Y$ are continuous in the usual topology. ■

The open box topology, viewed as a topology on $R^n = R \times R \times \ldots \times R$, is a special case of the following useful tool. We have seen how often products of sets are convenient; here we add some extra structure:

3.08 Definition

Given topological spaces $X_1,\ldots,X_n$, the product topology on the set $X_1 \times X_2 \times \ldots \times X_n$ is defined to be the collection of all unions of sets of the form

$$U_1 \times \ldots \times U_n \subseteq X_1 \times \ldots \times X_n$$

where each U_i is open in X_i. (cf. Ex. 6a)

The open box topology on R^n illustrates this so well that further pictures should not be necessary. With this device, we can prove very easily the following geometrically useful result.

3.09 Lemma

If $\underset{\sim}{F} : V \times V \to R$ is any bilinear form on a finite-dimensional vector space V, then

$$f : V \to R : \underset{\sim}{v} \rightsquigarrow \underset{\sim}{F}(\underset{\sim}{v},\underset{\sim}{v})$$

is continuous.

(This function is called the quadratic form corresponding to $\underset{\sim}{F}$ because for all $\lambda \in R$, $\underset{\sim}{v} \in V$ we have $f(\lambda\underset{\sim}{v}) = \lambda^2 f(\underset{\sim}{v})$.)

Proof

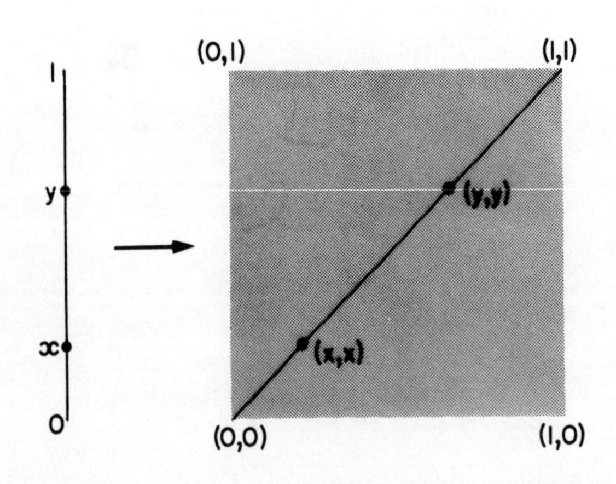

Fig. 3.3

f is the composite of the diagonal map

$$\text{Diag}: V \to V \times V : \underset{\sim}{v} \rightsquigarrow (\underset{\sim}{v},\underset{\sim}{v})$$

(Fig. 3.3, with [0,1] instead of V, explains the name) and $\underset{\sim}{F}$, which in turn is the composite (cf. V. 1.03) of the tensor product

$$\otimes : V \times V \to V \otimes V$$

and a linear map $\underset{\sim}{\hat{F}} : V\otimes V \to R$.

Now $\underset{\sim}{\hat{F}}$ is continuous by the definition of the topology on $V\otimes V$, $\otimes$ is continuous as a special case of Ex. 6b. Also Diag is continuous because $(\text{Diag})^{\leftarrow}(U_1 \times U_2)$, where U_1, U_2 are open sets in V, is exactly $U_1 \cap U_2$ which is again open (continuity follows as in Theorem 3.06). Hence by Lemma 1.09 f is continuous. ■

Exercises VI.3

1a) If $\{T_k | k \in K\}$ is a (non-empty, perhaps infinite) family of topologies satisfying 3.01 Wii), show that $\bigcap_{k\in K} T_k$ satisfies Wi), Wii). Deduce that if any T satisfies Wi), Wii) then there is a smallest (contained in all others) such T, and hence that the weak topology exists.

b) The open intervals $]a,b[= \{x | a < x < b\} \subseteq R$ are indeed open in the sense of Def^n 1.04.

c) Any open set U in R is the union of a set of open intervals. (Hint: use the open intervals that, by the metric definition of open, surround each $x \in U$.)

2a) Express the open ball $B(0,1) = \{(x,y) | x^2 + y^2 < 1\}$ in R^2 with the standard metric as a union of rectangular slabs. (Hint, Fig. 3.4)

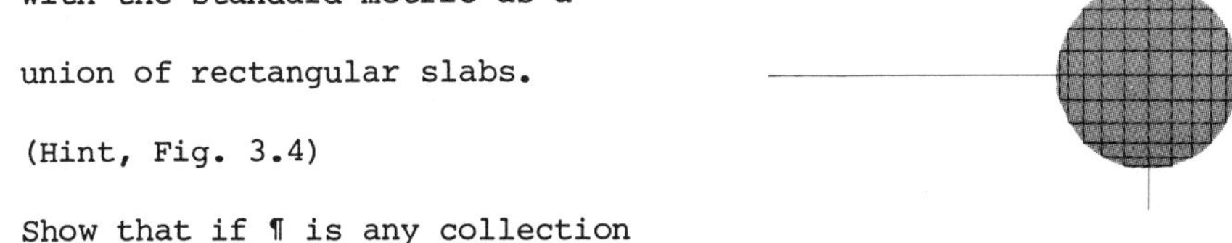

Fig. 3.4

b) Show that if ¶ is any collection of subsets of a set X, the set $\tilde{¶}$ of arbitrary unions of finite intersections of sets in ¶, together with ∅ and X, satisfies OA - OC and is thus a topology for X. (¶ is then called a <u>sub-basis</u> for the topology $\tilde{¶}$, which is <u>generated</u> by ¶.)

c) Show that the topology generated by ¶ is the smallest topology in which all the sets of ¶ are open, in the sense that if T is another topology such that ¶ $\subseteq$ T we have $\tilde{¶} \subseteq$ T.

d) Show that the product topology (Def^n 3.08) generated by the products of open sets U_i, is the smallest topology which makes the n projections

$$\pi_i : X_1 \times X_2 \times \ldots \times X_n \to X_i : (x_1, x_2, \ldots, x_n) \rightsquigarrow x_i$$

continuous.

3) Show by induction from the triangle inequality, $|a + b| \le |a| + |b|$ in the case of real numbers, that for any finite set $a_1, \ldots, a_n$, $b_1, \ldots, b_n \in R$ we have

$$\left| \sum_{i=1}^{n} a_i - \sum_{i=1}^{n} b_j \right| \le \sum_{i=1}^{n} |a_i - b_i|.$$

4) If X is a topological space, $f : X \to R$ is continuous, and $\lambda \in R$, show that for any $x \in X$, $0 < \varepsilon \in R$ there is an open neighbourhood $N(x)$ of x such that

$$f(N(x)) \subseteq B(f(x), \varepsilon/|\lambda|).$$

Deduce that the function, $\lambda f : X \to R : x \rightsquigarrow \lambda(f(x))$, is continuous.

5a) Show that for any choice of basis on a vector space X the function

$$d_s : X \times X \to R$$

$$((x^1, \ldots, x^n), (y^1, \ldots, y^n)) \rightsquigarrow \max \{|x^1 - y^1|, \ldots, |x^n - y^n|\}$$

is a (nontensorial) metric, that the corresponding open balls are open boxes in this basis, and that the corresponding topology is the usual topology.

(d_s is called the square metric because of the shape of its open balls in R^2 with the standard basis.)

b) Show by using the square metric that a sequence $\underset{\sim}{x}_i = (x^1(i),\ldots,x^n(i))$ in a vector space X converges in the usual topology if and only if each coordinate does, and that $\lim_{i\to\infty} \underset{\sim}{x}_i = (\lim_{i\to\infty} x^1(i),\ldots,\lim_{i\to\infty} x^n(i))$.

c) Show that the diamond metric d_d on a vector space X

$$d_d : X \times X \to R$$

$$((x^1,\ldots,x^n),(y^1,\ldots,y^n)) \rightsquigarrow |x^1 - y^1| + \ldots + |x^n - y^n|$$

is indeed a metric, draw $B((0,0),1) \subseteq R^2$ with this metric, and use Lemma 1.10 to prove that d_s and d_d give the same topology. (Notice that this "mutual inclusion" argument is much less work than expressing open sets in one directly as unions of explicitly defined open sets in the other, in the manner of Ex. 2a).

d) Show that the Euclidean metric

$$d_e : R^n \times R^n \to R$$

$$((x^1,\ldots,x^n),(y^1,\ldots,y^n)) \rightsquigarrow +\sqrt{[(x^1-y^1)^2 + \ldots + (x^n-y^n)^2]}$$

is a metric, draw $B((0,0),1) \subseteq R^2$ with this metric, and show that it gives the usual topology.

(The diamond and square metrics are much the most useful metrisations of the usual topology, since they do not involve square roots. And in spacetime unlike space, even the Euclidean metric is not independent of choice of orthonormal basis; it varies with the choice of "timelike" basis vector.

So being neither invariant nor easy to do sums with, it is of little use.)

6a) The product topology defined in 3.08 is Hausdorff if all the X_i are Hausdorff spaces. Is the converse true?

b) Prove that $\otimes : X_1 \times X_2 \times \ldots \times X_n \to X_1 \otimes X_2 \otimes \ldots \otimes X_n$ is continuous, using the product topology on its domain and the usual vector space topology on its image.

7a) If a metric space (X,d) has the metric topology and $X \times X$ the corresponding product topology, show that in the usual topology on R, $d : X \times X \to R$ is continuous.

b) Deduce by 2.02 that for $i \rightsquigarrow x_i$, $i \rightsquigarrow y_i$ sequences in X

$$\lim_{n\to\infty} d(x_n, y_n) = d(\lim_{n\to\infty} x_n, \lim_{n\to\infty} y_n)$$

whenever the limits on the right exist.

8a) Show that if $\| \ \|$ is a norm (IV.1.06) on X, $d(\underset{\sim}{x},\underset{\sim}{y}) = \|\underset{\sim}{x}-\underset{\sim}{y}\|$ defines a metric.

b) Show that each of the metrics of Ex. 5 is given in this way by a norm, $\|\underset{\sim}{x}\| = d(\underset{\sim}{0},\underset{\sim}{x})$.

c) Show from Axioms IV.1.06 that if $\| \ \|$, $\| \ \|'$ are norms on finite-dimensional X, there exist $\lambda, \lambda' > 0$ such that for any $\underset{\sim}{x} \in X$, $\lambda \|\underset{\sim}{x}\| \leq \|\underset{\sim}{x}\|' \leq \lambda' \|\underset{\sim}{x}\|$. (Pick a basis and show that $\|a^i b_i\| \leq |a^i| \ \|\underset{\sim}{b}_i\|$.) Deduce that the metric on X given by any norm defines the usual topology.

4. COMPACTNESS AND COMPLETENESS

We have already (IV.4.01) had to make use of an essentially topological argument, in proving that we could diagonalise symmetric operators. (In other books you may find proofs which _look_ purely algebraic, involving the complex numbers. In fact they require the so-called Fundamental Theorem of Algebra, which is actually a topological result, depending crucially on the completeness of the complex plane.) To prove the existence of maximal vectors, around which the proof IV.4.05 revolved, we must first look a little more closely at the topological properties of the real numbers, which we then extend to real vector spaces.

The first notion we need is that R is complete. There are many different ways of defining and proving this property. To prove it one must by one or another method construct the real numbers from the rationals, or even the integers, which would be out of place here. We shall therefore take it as an axiom, in a form in which it is clear that its failure would do such violence to our intuition of continuity (which it is one of the purposes of the real number system to express) that the real numbers would have been forgotten long ago. If this leaves you still wanting a proof, consult an analysis book that constructs the real numbers by Cauchy sequences, Dedekind cuts, or whatever.

4.01 Completeness axiom.

The Intermediate Value Theorem is true of the real numbers.

The Intermediate Value Theorem (not Axiom, because most books prefer to start from a less comprehensible equivalent statement and prove it from that) says that if a function

$$f : [0,1] \to R$$

is continuous, and for some $v \in R$ we have

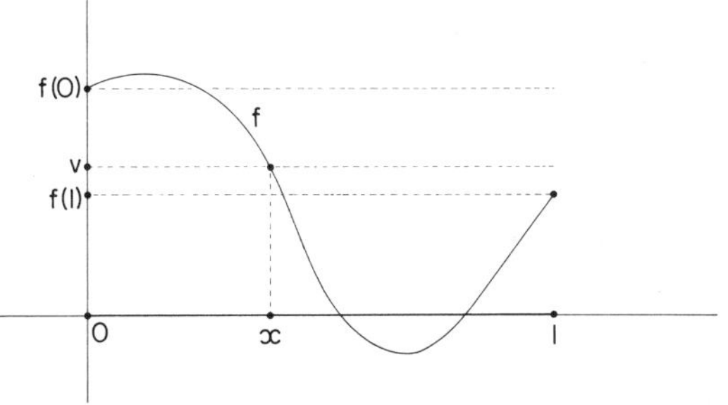

Fig. 4.1

$$f(1) < v < f(0)$$

$$\text{or} \quad f(0) < v < f(1)$$

then we must have at least one $x \in [0,1]$ such that $f(x) = v$ (the "intermediate value" in question). That is, the graph of f must cross the level v somewhere. (There <u>are</u>, incidentally, a few mathematicians who refuse to believe this, or rather say it has no meaning in their terms. But then, a really pure mathematician can disbelieve anything.)

We can tidy the statement a little. If we have an f which does <u>not</u> take the value v, then we can get a new function

$$\tilde{f} : [0,1] \to R : x \rightsquigarrow \frac{f(x)-v}{|f(x)-v|}$$

which is still continuous (Ex. 1d), and takes only the values +1 and -1, with <u>no</u> intermediate values at all. Thus the Intermediate Value Theorem above is equivalent to the following statement, which is the form it is most convenient to use:

There exists no continuous map $f : [0,1] \to R$ taking only the values -1, +1, with $f(0) = 1$, $f(1) = -1$.

Notice that this is <u>not</u> true if we replace [0,1] by the set Q of <u>rational</u> numbers x such that $0 \le x \le 1$, for if we define

$$f : Q \to R : x \rightsquigarrow \begin{cases} -1 \text{ if } x^2 < \frac{1}{2} \\ +1 \text{ if } x^2 > \frac{1}{2} \end{cases}$$

the "point of discontinuity" $1/\sqrt{2}$, is missing from Q because it is irrational, so <u>on its domain of definition</u> f is continuous. Hence the name "completeness"; the assertion is that the real unit interval, and hence the real line, does not have "missing points" of this kind. ("Complete" is defined more generally in the Appendix.)

We must now prove one of the equivalent statements, as a necessary tool to reach our main goal of exploring the complementary notion of compactness.

4.02 Lemma

If we have a sequence $J_1, J_2, J_3 \ldots$ of closed (cf. Ex. 2) subintervals $J_i = [j_i,k_i] \subseteq [0,1]$ of the unit interval, such that $J_{i+1} \subseteq J_i$ for each i (Fig. 4.2), then

Fig. 4.2

$$\bigcap_{i\in N} J_i \neq \emptyset.$$

That is to say, there is at least one point $x \in [0,1]$ which is in every J_i.

Proof

Suppose not.

Then for each point $x \in [0,1]$ there is at least one i (and hence all greater i) such that $x \notin J_i$. Therefore either $x < j_i$ or $x > k_i$; x is either to the left or to the right of the whole interval J_i. Define then

$$f : [0,1] \to R : x \leadsto \begin{cases} -1 \text{ if } x < j_i, \text{ for some } i \\ +1 \text{ if } x > k_i, \text{ for some } i. \end{cases}$$

Now f is well defined. (Since if x was to the left of some interval J_i and to the right of another interval J_i' we could not have either $J_i \subseteq J_i'$ or $J_i' \subseteq J_i$, contradicting the given fact that always $J_m \subseteq J_n$ when $m > n$, and one of i,i' must be the greater.) It is also continuous at each $x \in [0,1]$, since if $x < j_i$, say, so that $f(x) = -1$, we have

$$y \in B(x, \tfrac{1}{2}(j_i-x)) \implies y < j_i$$

so that $f(B(x,\frac{1}{2}(j_i-x))) = \{-1\} \subseteq B(-1,\varepsilon)$ for any positive ε; similarly for $x > k_j$. Now if 0 is not to the left of any J_i, it is in each J_i, and hence in $\bigcap_{i\in N} J_i$, which is not therefore empty. So if it is empty, $f(0) = -1$. Similarly $f(1) = +1$. Thus if the supposition that no point is in every J_i is true, we have a function which contradicts the Intermediate Value Theorem. ■

We now come to one of the characteristic properties of compact spaces - sometimes taken as a definition of compactness. We shall defer our more limited definition a little longer.

4.03 Theorem

If $S : N \to [0,1] : i \rightsquigarrow x_i$ is any sequence of points in the unit interval, then S has at least one convergent subsequence.

Proof

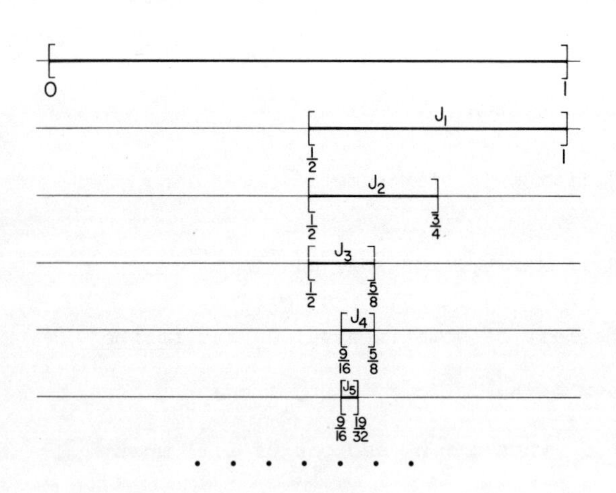

Fig. 4.3

If $x_i \in [0,\frac{1}{2}]$ for only a finite set of values of i, we must have $x_i \in [\frac{1}{2},1]$ for an infinite set, and vice versa, since N is infinite. So we can choose a half-interval J_1 from $[0,\frac{1}{2}]$ and $[\frac{1}{2},1]$ in which S takes values infinitely many times (if it does so in both, let us agree to take the left one). Next, by the same argument we can choose a closed half J_2 of J_1 in which S takes values x_i for infinitely many i, and so on. (Notice that we do not say "takes infinitely many values"; if

$x_i = \frac{1}{4}$ for all i, S converges by taking just one value - infinitely many times.) Thus we get a sequence

$$[0,1] \supseteq J_1 \supseteq J_2 \supseteq J_3 \supseteq \ \cdots$$

of closed intervals, as in Fig. 4.3, which must then by 4.02 have a point x in common. Moreover, in each J_i we know that S takes values infinitely many times, so we can choose a subsequence S' of S by

$$x'_j = \text{first } x_i \text{ after } x'_{j-1} \text{ to be inside } J_j$$

and still have an infinite sequence. But now each $x'_j \in J_j \subseteq J_k$ if $k < j$, so S' takes values at most j-1 times outside any J_j.

Now every open ball $B(x,\varepsilon)$ around x must contain at least one of the J_i, because

$$\begin{aligned} y \in J_i \quad &\Rightarrow \quad |x-y| \le \text{length } (J_i), \text{ since } x \in J_i \\ &\Rightarrow \quad |x-y| \le 1/2^i \\ &\Rightarrow \quad |x-y| < \varepsilon, \text{ if we choose } i \text{ large enough.} \end{aligned}$$

Hence $B(x,\varepsilon)$ contains x'_i for all but finitely many i. Since every neighbourhood of x contains some $B(x,\varepsilon)$, this extends to all neighbourhoods $N(x)$ of x and we have

$$\lim_{i\to\infty} x'_i = x$$

so that S' is a convergent subsequence of S. ■

Notice that the choice of convergent subsequence was not necessarily unique. Indeed, from any sequence of all the rational numbers from 0 to 1 (such as is used to prove that they are countable) we can choose a subsequence converging to any chosen one of the uncountable set of real numbers in [0,1]. (If you don't know about uncountable infinities, ignore this remark.)

4.04 Corollary

The same is true for a sequence in any closed interval [a,b].

Proof

Consider $\phi : [0,1] \rightarrow [a,b] : x \rightsquigarrow (b-a)x+a$, and its inverse

$\overleftarrow{\phi} : [a,b] \rightarrow [0,1] : x \rightsquigarrow (x-a)/(b-a)$.

These are affine maps, hence continuous by 3.07.

If S is a sequence in [a,b], $\overleftarrow{\phi}\circ S$ is a sequence in [0,1], which has a convergent subsequence S' by the theorem, with limit x, say. Then $\phi\circ S'$ is a subsequence of S, and by 2.02 we have

$$\lim_{i\to\infty} (\phi\circ S') = \phi(x).$$ ■

This property, of any sequence having a convergent subsequence is one of the several equivalent definitions of compactness. We shall not need compactness in full generality, however; we want it for rather more limited purposes than the usual mathematics text. Therefore since there is a nice geometrical characterisation of compact sets in finite-dimensional vector or affine spaces we shall consider it only for such embedded sets, not abstractly.

Notice that two characteristics are necessary for the unit interval [0,1] to have the convergent subsequence property: it is closed topologically, and it is bounded. That is (giving the definition a number, since it is so important):

4.05 Definition

A set $S \subseteq R$ is bounded if we can find a bound $b \in R$ such that

$$x \in S \Rightarrow |x| \leq b.$$

If a set $S \subseteq R$ is not closed, there is some boundary point x of S not in S: a sequence of points in S (and hence all its subsequences) can converge to x and thus not converge to any point in S. If S is not bounded, we can choose a sequence x_i in S such that $|x_n| > n$ for each $n \in N$, so that the sequence and all its subsequences "go to infinity" and cannot converge to any real number, let alone one in S.

In fact topologically (and even differentially) there is very little to choose between not being closed and not being bounded; Fig. 4.4 shows the graph of

$$f : \{x | 0 < x < 1\} \to R : x \rightsquigarrow \frac{2x-1}{2x(1-x)} .$$

This is a nice (analytic) homeomorphism from the open unit interval (which is bounded but not closed) to the whole real line (which is closed but not bounded).

If however we take a set C in R which is closed and bounded, the image of any continuous function from it to R will again be so (Ex. 3), even if we do not insist that the function be continuous (or even defined) outside C. This is a nice characteristic of the set and intrinsic to the set's topology: unlike the open

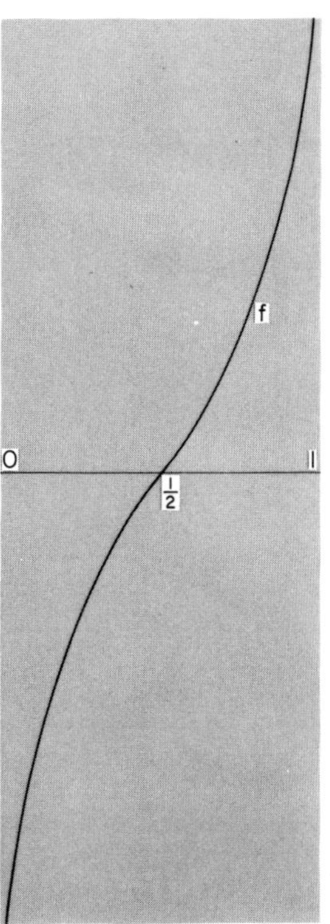

Fig. 4.4

interval it cannot be spread out continuously over infinite length. It is thus a sort of intrinsic "smallness" or "finiteness" property, for which the universal name has become compact.

For sets in a general finite-dimensional vector space, a very similar idea holds. Once again, we define it invariantly, and then reduce it to coordinates.

4.06 Definition

A set C in a finite-dimensional vector space X is compact if

(i) It is closed in the usual topology.

(ii) For any linear functional $\underset{\sim}{f} \in X^*$, $\underset{\sim}{f}(C) \subseteq R$ is bounded. (This obviously reduces to 4.05 if X = R).

4.07 Lemma

Choose any basis $\underset{\sim}{b}_1,\ldots,\underset{\sim}{b}_n$ for X. Then $\underset{\sim}{b}^1(C),\ldots,\underset{\sim}{b}^n(C) \subseteq R$ are bounded if and only if all $\underset{\sim}{f}(C)$ are, for $\underset{\sim}{f} \in X^*$.

Proof

Exactly in the style of that of 3.03. (Ex. 4a). ■

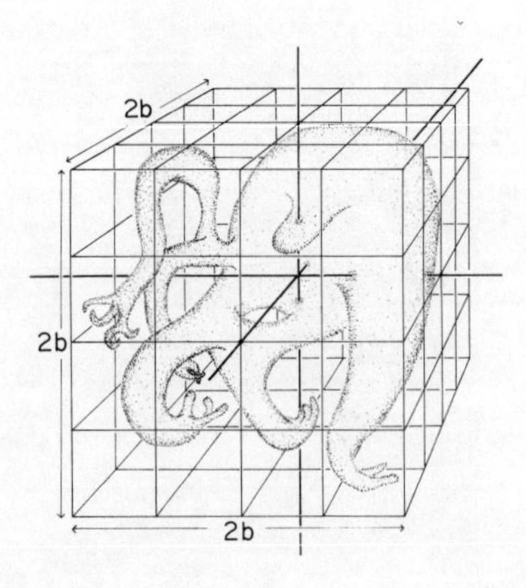

Fig. 4.5

4.8 Corollary

A set $S \subseteq X$ is bounded if and only if the values of the coordinates of points in S are all of modulus less than some $b \in R$. That is, S is completely inside some box of side 2b (illustrated in Fig. 4.5 for R^3). We then say S is bounded by b with respect to these coordinates.

We are now ready for our main theorem about compactness, except for a technical point which it is simpler not to dodge: in the continuous function with no intermediate value on the rational unit interval we constructed above, we took it for granted that we knew what "continuous on" a subset S of R meant, even if the function was not defined on the rest of R. In metric terms, this is so obvious as to be hardly worth mentioning - if we have a subset S of a metric space (X,d) we get an induced metric on S by restricting d to $S \times S \subseteq X \times X$. Then (cf. Ch. 0. §2) pairs of points in S retain the distances they had as points of X and we carry on as before. This metric is used explicitly in Ex. 3, for example. However, it is not always very appropriate or convenient: if we always took the induced metric for surfaces in R^3 for instance, we would say that the distance from London to Sydney was 7,900 miles, for example, whereas the more useful distance is the one within the surface of 11,760. Thus for a subset we may want a different metric, but we usually want the induced notion of continuity. Hence we define

4.09 Definition

If S is a subset of a topological space X, the induced topology on S is the collection of sets $\{ S \cap U \mid U \text{ open in } X\}$.

These sets are called open in S; this does not mean that they are necessarily open in X, if S is not. For example, if $X = R^2$ and $S = \{(x,y) \mid x^2 = y\}$ (Fig. 4.6) the intersection of S with the open disc $U = \{(x,y) \mid x^2 + y^2 < 2\}$ is neither open in X (since it does not contain a neighbourhood in the space X of, for example, (0,0)) nor closed in X,

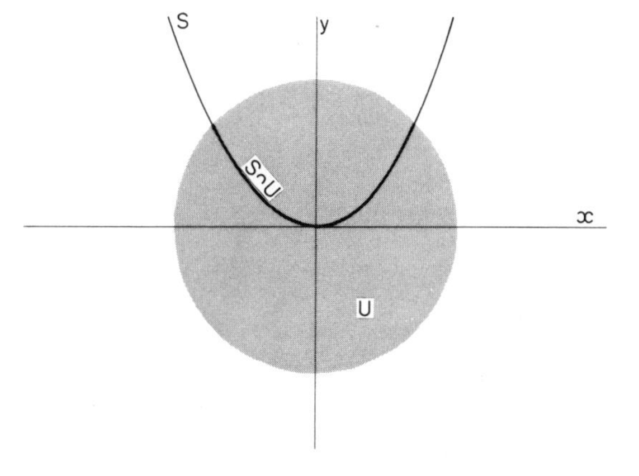

Fig. 4.6

since it does not contain (1,1) or (-1,1) which are boundary points of it. But by definition it is open in S. (cf. Ex. 5.)

With this topology a subset S is called a subspace of X. When we need to distinguish other than by context we call it a topological subspace as distinct from the vector or affine subspaces we have had before. (Notice that the example of S in Fig. 4.6 is neither a vector nor an affine subspace of R^2.) Sadly for the science-fiction fan, none of these kinds of subspace can be dodged into for faster-than-light travel - they are all just subsets of what we started with, with restrictions of the structure we started with.

With that technicality out of the way (if you are still uneasy about it, do Ex. 5) we can state and prove one of the main properties of compact sets, for which we have already found a use:

4.10 Theorem

If C is a compact set, in any finite-dimensional vector or affine space† X, and $f : C \to R$ is a continuous function with respect to the induced topology on C, then f(C) is bounded and closed.

Proof

We shall assume X a vector space (the proof transfers at once to affine spaces).

Choose a basis, and let C be bounded with respect to these coordinates by b.

First we prove that C has the convergent subsequence property.

Let S be any sequence of points $\underset{\sim}{c}_i$ in C, and write $\underset{\sim}{c}_i$ in coordinates as $(c^1(i),\ldots,c^n(i))$ - bringing the i's into brackets to emphasise that they have nothing to do with variance. Then we can choose a subsequence $\tilde{c}^1(j)$ of the

---

† If we had given a more abstract definition of "compact", this condition of being in X would be unnecessary. We are just saving effort and space.

sequence $c^1(i)$ of first coordinates converging to (say) x^1 by 4.04, since $c^1(i)$ is a sequence in the closed interval $[-b,b]$. We then define $\tilde{S}^1$ as the subsequence of S that has $\tilde{c}^1(j)$ as its sequence of first coordinates. Next, consider the sequence $\tilde{c}^2(j)$ of second coordinates of values of $\tilde{S}^1$, choose a convergent subsequence and define $\tilde{S}^2$ as the subsequence of $\tilde{S}$ having the chosen subsequence as second coordinates - converging to x^2 say. The first coordinates still converge to x^1, because they are a subsequence of a convergent sequence (cf. Ex. 2.1c).

Repeating this process n times, we get a subsequence $\tilde{S}=(\tilde{c}^1(i),\tilde{c}^2(i),\ldots,\tilde{c}^n(i))$ of S, with each sequence of j-th coordinates converging to some x^j. Hence, by Ex. 3.5b, $\tilde{S}$ converges to $(x^1,\ldots,x^n)$, which must therefore be in C since C contains its boundary points - that is, all the points in X to which sequences in C <u>can</u> converge are in C. By Ex. 5f, $\tilde{S}$ converges in the topology on C.

(Notice that our argument depended on finite-dimensionality: the limiting result of choosing a subsequence an infinite number of times might leave us with no points).

Now, if f(C) is not bounded we can choose a sequence x_i in f(C) that "goes to infinity" with all its subsequences. But for each x_i, since it is in f(C) we can choose a $\underset{\sim}{c}_i$ in C such that $f(\underset{\sim}{c}_i) = x_i$. This gives us a sequence in C, which must have a subsequence converging to a point $\underset{\sim}{c}$, say, in C, so that (restricting to the values of i in the subsequence)

$$\lim_{i\to\infty} x_i = \lim_{i\to\infty} f(\underset{\sim}{c}_i) = f(\lim_{i\to\infty} \underset{\sim}{c}_i) = f(\underset{\sim}{c}), \quad \text{by 2.02.}$$

Thus we have found a convergent subsequence of x_i, contrary to assumption. Therefore f(C) must be bounded.

Similarly, if x is a boundary point of f(C), choose a sequence x_i in f(C) converging to x, a sequence $\underset{\sim}{c}_i$ in C such that $f(\underset{\sim}{c}_i) = x_i$, and a convergent subsequence $\underset{\sim}{c}'_i$ of $\underset{\sim}{c}_i$, with limit $\underset{\sim}{c} \in C$. Then if $x'_i = f(\underset{\sim}{c}'_i)$, we know x'_i still converges to x by Ex. 2.1c, and

$$x = \lim_{i\to\infty} x_i' = \lim_{i\to\infty} f(\underset{\sim}{c}_i') = f(\lim_{i\to\infty} \underset{\sim}{c}_i') = f(\underset{\sim}{c})$$

so $x \in f(C)$. Thus $f(C)$ is closed. ■

4.11 Corollary

If $f : C \to R$ is a continuous function on a compact set, then f has a maximum on C; that is, not only do the values of f stay below a certain level, but there is some $\underset{\sim}{c} \in C$ such that for all $\underset{\sim}{c}' \in C$,

$$f(\underset{\sim}{c}') \leq f(\underset{\sim}{c}).$$

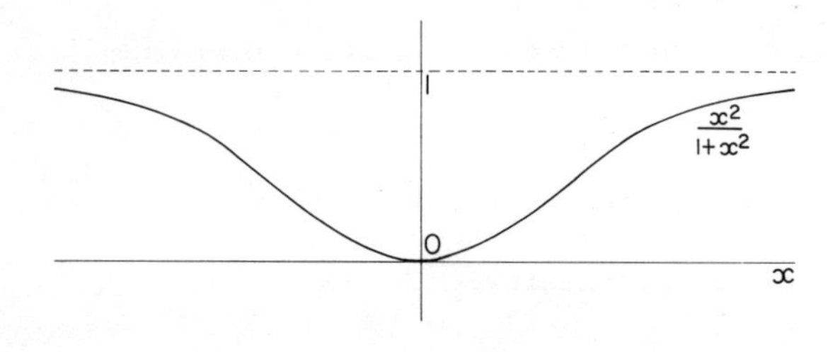

Fig. 4.7

(Fig. 4.7 shows that the function $x \rightsquigarrow x^2/(1+x^2)$ defined on all of R does not have this property.)

Proof

We have shown that $f(C)$ is bounded, say by k. The existence within $[-k,k]$ of a "top end" of the set $f(C)$ is another use of completeness: the set of rationals q in $[0,1]$ with $q^2 < \frac{1}{2}$ has no meaningful "top end" within the rationals. Precisely, we want some $x \in [-k,k]$ such that

(i) $x \geq y$ for all $y \in f(C)$.

(ii) We can find no finite length ε by which x is "above" $f(C)$.

That is, every $B(x,\varepsilon)$ contains at least one point of $f(C)$.

That is exactly the requirement that x lies in the closure of $f(C)$.

The proof of existence of x (Ex. 6) is precisely similar to the proof of 4.02. Condition (ii) means that x is either in $f(C)$ nor a boundary point of $f(C)$. Now, since $f(C)$ is closed, we have

$$x = f(\mathbf{c})$$

for some $\mathbf{c} \in C$, so we have found a $\mathbf{c}$ such that

$$\mathbf{c}' \in C \Rightarrow f(\mathbf{c}') \leq f(\mathbf{c})$$

as required. ■

4.12 Corollary

Let $\mathbf{A}$ be any operator on a finite-dimensional inner product space X. The function $\mathbf{x} \rightsquigarrow \mathbf{Ax}.\mathbf{Ax}$ on the unit sphere S in X has a maximum value, $\|\mathbf{A}\|$, which is attained by at least one vector $\mathbf{x} \in S$. That is, there exist maximal vectors for $\mathbf{A}$. (cf. Chapter IV. 4.01)

Proof

S is evidently bounded by 1 in orthonormal coordinates. By Ex. 1.6a the set $\{1\} \subseteq R$ is closed, hence since the function $\| \ \| : \mathbf{x} \rightsquigarrow \sqrt{\mathbf{x}.\mathbf{x}}$ is continuous by 3.09, the set $S = (\| \ \|)^{\leftarrow}(\{ \})$ is closed by Ex. 1.5c.

S is therefore compact, and since the quadratic form $\mathbf{x} \rightsquigarrow \mathbf{Ax}.\mathbf{Ax}$ is also continuous by 3.09, the result follows.

■

Remark

If you are still unconvinced that topological reasoning is necessary to prove that we can diagonalise symmetric operators, do Ex. 7.

Compactness is an extremely powerful tool, and one that a mathematician learns to use as readily as his fingers: "by compactness" is often an acceptable substitute for an argument in full, since everyone is so familiar with how compactness proofs go, and what can and cannot be done with them. It is *so* powerful, in fact, that mathematicians feel very naked facing the world without it, whereas physicists have to; for example the contours in Fig. IV. 2,3b,c, and the Lorentz group, are non-compact. Moreover, no remotely reasonable spacetime can be compact. This is in sharp contrast to "pure" differential geometry, which is largely conducted on nice compact manifolds, with nice compact groups in the background. In consequence, we shall not explore compactness further, since we shall have fewer opportunities to use it than "pure mathematics" texts in the same area. It remains however one of the central notions of topology, and the reader should seize every opportunity to get better acquainted with it. It is less useful, because less often applicable, in physics than in mathematics, but still essential as we have just seen. Th^m IV.4.05, for instance, is a tool we could not do without in what follows.

<u>Exercises VI.4</u>

a) If the function $g : [0,1] \to R$ is continuous, with $g(x) \neq 0$ for any $x \in [0,1]$, show that $1/g : [0,1] \to R : x \rightsquigarrow 1/g(x)$ is also continuous.

b) If $g : [0,1] \to R$ is continuous, show that $|g| : X \to R : x \rightsquigarrow |g(x)|$ is also continuous.

c) If $g,g' : [0,1] \to R$ are continuous, show that $gg' : [0,1] \to R : x \rightsquigarrow g(x)g'(x)$ is continuous.

d) If $f(x) \neq v$ for any $x \in [0,1]$, where $f : [0,1] \to R$ is continuous, show that if

$$\tilde{f}(x) = \frac{f(x)-v}{|f(x)-v|} = (f(x) - v)\frac{1}{|f(x)-v|},$$

then $\tilde{f}$ is continuous and $\tilde{f}(x) = -1$ or $+1$ according as $f(x) < v$ or $f(x) > v$.

2) The intersection of the infinite family

$$U_n = \{x | 0 < x < 1/2^n\}, \quad n \in N$$

of open intervals is empty, so that the "closed" condition in 4.02 is essential.

3a) Show that any closed bounded set C in R is contained in some closed interval [a,b], and deduce from 4.04 and the closedness of C that any sequence taking values in C has a convergent subsequence with its limit in C.

b) If f is a continuous map $C \to R$ (where continuity is defined on C by Def^n 1.05, using the same metric $d(x,y) = |x - y|$ on C as on R), and S is a sequence taking values in f(C), show by using a) and 2.02 that S must have a convergent subsequence with its limit in f(C).

c) Show that if f(C) were not closed, or not bounded, there would exist sequences taking values in f(C) but not having any subsequence converging to any point in f(C).

d) Deduce that f(C) is closed and bounded. (Notice that this is a proof of a special case of Th^m 4.10, by essentially the same method.)

4a) Write out the proof of Lemma 4.07.

b) Define "bounded" and "compact" for sets in a finite-dimensional affine space, and show in the manner of 4.07, 4.08 how these definitions may be expressed in coordinates.

5a) If a topology on X is given by a metric, prove that the induced topology on $S \subseteq X$ is given by the induced metric (so that we really are just transferring to topology the obvious notion in the metric case).

b) Prove that if T is open in S in the induced metric from X, and S is open in the metric sense on X, then T is open in X.

c) Prove that if T is closed in S and S is closed in X, then T is closed in X, again using metrics.

d) Repeat (b) and (c) using topologies and induced topologies instead of metrics.

e) Prove that if $f : X \to Y$ is a continuous function, X, Y topological spaces, and S a subspace of X, then $f|S$ is continuous.

f) Prove that a sequence in a subspace S of X converges to $x \in S$ in the induced topology on S if and only if it converges to x as a sequence in X.

6a) If S is a subset of the closed interval $[a,b]$ and there is no $x \in [a,b]$ such that

(i) $y \in S \Rightarrow y \leq x$

(ii) $x \in \bar{S}$ (cf. 1.04, 1.08 for closure)

construct a continuous function $f : [a,b] \to R$ with only the values -1, $+1$ such that $f(a) = -1$, $f(b) = 1$.

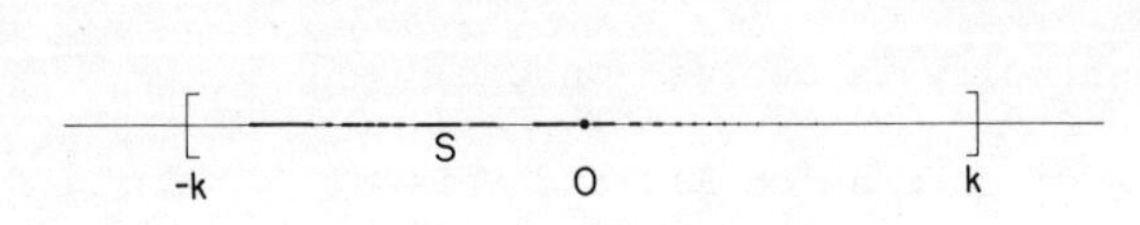

Fig. 4.8

b) Deduce that such an x must exist. (x is called the <u>supremum</u>, sup S, of the set S.

c) Deduce that S has an <u>infimum</u> $\inf S \in \bar{S}$ such that $x \in S \Rightarrow x \geq \inf S$.

7) All the definitions of Chapters I-V could be made with any other field (cf. Ex. I.1.10) substituted for R, such as the complex or rational numbers (though complex-valued inner products take a little care). Show that for the <u>rational</u> vector space $Q \times Q$, where Q represents the field of rational numbers and all scalars are to be rational numbers, with the obvious addition and scalar multiplication the operator represented by the matrix

$$\begin{bmatrix} 1 & 1 \\ 1 & 2 \end{bmatrix}$$

has no maximal vectors and no eigenvectors, though the operator on R^2 represented by the same matrix has.

8a) Show that if $0 < \lambda < 1$ and $m > n$, then $0 < \lambda^m < \lambda^n < 1$.

b) Deduce that if $y = \inf\{\lambda^n \mid n \in N\}$ (cf. Ex. 6c) then

$$S : N \to R : i \rightsquigarrow \lambda^i \text{ converges to } y.$$

c) Deduce by 2.02 that $T : N \to R : i \rightsquigarrow \lambda^{i+1}$ converges to λy, hence that $\lambda y = y$.

d) Deduce that S converges to 0.

VII Differentiation and Manifolds

"There are nine and sixty ways of constructing tribal lays,
And every single one of them is right."

Rudyard Kipling.

Throughout this chapter X,X' will be affine spaces of (finite) dimensions n, m respectively, with difference functions $\underset{\sim}{d}$,$\underset{\sim}{d}'$ and vector spaces T,T'.

1. DIFFERENTIATION

Differentiating a function f:R → R gives another function $\frac{df}{dx}$: R → R, whose value at x ∈ R is (Fig. 1.1a) the slope of the tangent at (x,f(x)) to the graph of f. Thus differentiation is an operator on a set of functions R → R. (Actually it is a linear operator and the functions form an infinite-dimensional vector space, with the usual addition and scalar multiplication.) This is a little misleading when we go to higher dimensions. For a map g : R^2 → R, we need two numbers to specify the "tangent" to its graph - a plane tangent to a surface, Fig. 1.1b - over each point (x,y) ∈ R^2. Then

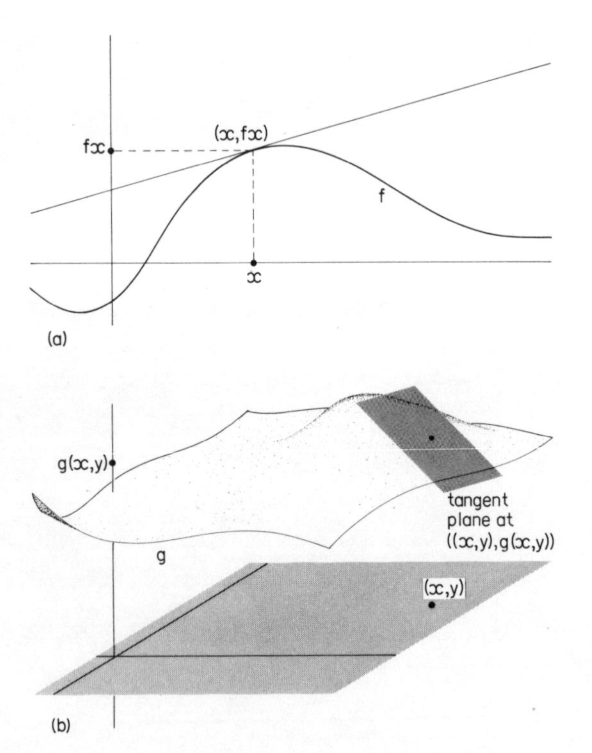

Fig. 1.1

differentiating can no longer give another function of the same kind, $R^2 \to R$. (It looks more like a function $R^2 \to R^2$.) Writing this down in coordinates involves "partial derivatives" like $\frac{\partial g}{\partial x}$, and for higher dimensional domain and image, rather many of them. To disentangle what these are actually doing, let us look at the geometry involved in differentiating maps between affine spaces.

The tangent line in Fig. 1.1a and the tangent plane in Fig. 1.1b are "flat approximations" at $(x,f(x))$ and $(x,y,g(x,y))$ to the graphs of f and g respectively. Differentiating at a point x means substituting for the given map f the one whose graph is this flat approximation. Or rather, since this would be the affine map approximating f, the linear part of it. This is the interesting part - we already know that x goes to f(x), and the value on any point combines with the linear part to specify an affine map completely (cf. Chapter 2 §2).

Just as for functions on the real line, we find the derivative of f at x by looking at the value of f on $(x + \Delta x)$, seeing how much this differs from f(x), and going to the limit as $\Delta x \to 0$. Now, however, we have more ways in which to move away from x - classified by the tangent vectors at x - and more directions in which its image can differ from f(x) - classified by the tangent vectors at f(x). Thus we have a map from the tangent space at x to the tangent space at f(x). These tangent spaces were defined in Chapter II 1.02.

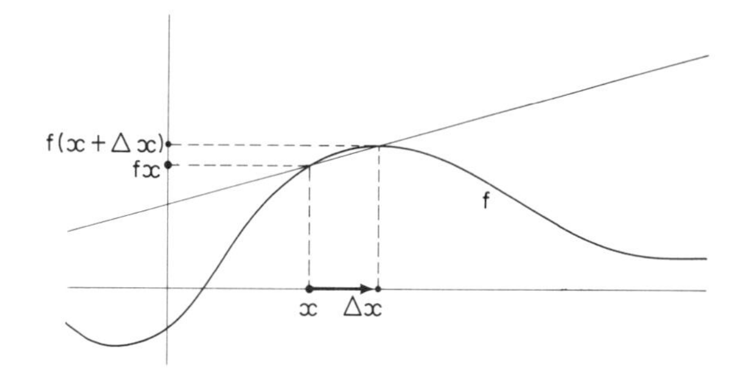

Fig. 1.2

In the following definition the usual topology (Ch. VI. §3) is used to provide neighbourhoods (Ch. VI. 1.08).

Before coming to technical details, it is worth recalling that by Ex. II. 1.3.

$$x + (\underset{\sim}{t}_1 + \underset{\sim}{t}_2) = (x + \underset{\sim}{t}_1) + \underset{\sim}{t}_2,$$

so we may unambiguously write both as $x + \underset{\sim}{t}_1 + \underset{\sim}{t}_2$.

1.01 Definition

If $f : X \to X'$ is a map (not necessarily affine) between affine spaces, a derivative of f at $x \in X$ is a linear map

$$\underset{\sim}{D}_x f : T_x X \to T_{f(x)} X'$$

such that for any neighbourhood N in $L(T_x X; T_{f(x)} X')$ of the zero linear map, there is a neighbourhood N' of $\underset{\sim}{0} \in T_x X$ such that if $\underset{\sim}{t} \in N'$ then

$$\underset{\sim}{d}'(f(x+\underset{\sim}{t}), f(x)) = \underset{\sim}{d}'_{f(x)}(\underset{\sim}{D}_x f(\underset{\sim}{t}) + \underset{\sim}{A}(\underset{\sim}{t}))$$

for some $\underset{\sim}{A} \in N$. (That is, if we get close enough to x, the correction term to make the flat approximation $\underset{\sim}{D}_x f$ agree with f is given by an arbitrary small - close to zero - map. Thus the correction term $\underset{\sim}{A}(\underset{\sim}{t})$ itself, being the image of a small vector by a small map, is "second order small" and vanishes in the limit.) This is illustrated for a map $f : R \to R^2$ in Fig. 1.3; notice that this time the image, not the graph of f is shown.

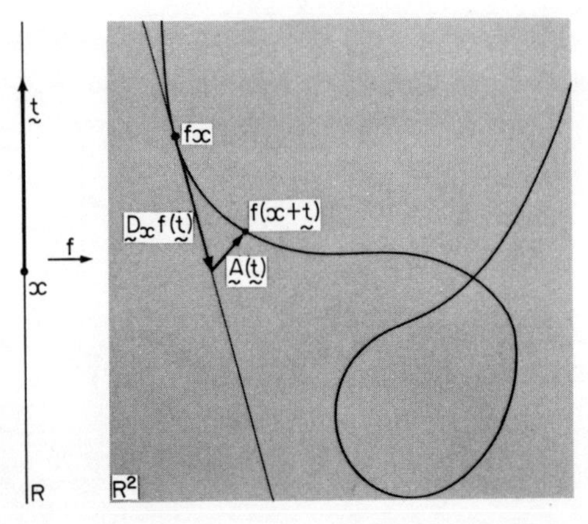

Fig. 1.3

If f has a derivative at x it is unique (Ex. 3) so we shall refer to the derivative $\underset{\sim}{D}_x f$ of f at x, and say f is differentiable at x; if f is differentiable wherever it is defined, we just say f is <u>differentiable</u>.

The derivative (if it exists) is given by the formula (Ex. 3b)

$$\bigstar \qquad \underset{\sim}{D}_x f(\underset{\sim}{t}) = \lim_{h\to 0} \overleftarrow{\underset{\sim}{d}'_{f(x)}} \left(\frac{\underset{\sim}{d}'(f(x), f(x+h\underset{\sim}{t}))}{h} \right) .$$

(The $\overleftarrow{\underset{\sim}{d}'_{f(x)}}$ is necessary to get a tangent vector in $T_{f(x)}X'$, not a free vector in T'.) Or if X' is a vector space, using the canonical affine structure the formula gives

$$\underset{\sim}{D}_x f(\underset{\sim}{t}) = \lim_{h\to 0} \overleftarrow{\underset{\sim}{d}'_{f(x)}} \left(\frac{f(x+h\underset{\sim}{t}) - f(x)}{h} \right) .$$

If X = X' = R, as in elementary calculus, any linear map between them may be described by its slope, alias the number it multiplies points in X by, alias value it takes on the basis vector 1. Leaving out the binding map, as is common, the formula thus reduces to

$$\underset{\sim}{D}_x f(1) = \lim_{h\to 0} \left(\frac{f(x+h.1) - f(x)}{h} \right)$$

$$= \lim_{\Delta x\to 0} \frac{f(x+\Delta x) - f(x)}{\Delta x}$$

relabelling h = h.1 as Δx. This is the classical expression for the derivative in the calculus of one variable.

The geometrically precise expression $\bigstar$ above is thus close in spirit to the elementary "x + Δx" approach though it uses a slightly more general definition of limit (Ex. 1a) than that of Ch. VI. It would make a simpler definition but for the fact that this limit may exist even when $\underset{\sim}{D}_x f$ does not (Ex. 2), so we have to be a little careful. However we shall generally require differentiability as a condition before we start, so from there on we can identify the derivative with this limit map, and not worry.

1.02 Higher derivatives

If a map $f : X \to X'$ is differentiable, this gives us a map

$$\hat{\underset{\sim}{D}}f : X \to L(T;T') : x \rightsquigarrow \underset{\sim}{d}'_{f(x)} \underset{\sim}{D}_x f \overleftarrow{\underset{\sim}{d}_x}$$

by forgetting to which point each tangent space is attached. If $\hat{\underset{\sim}{D}}f$ is continuous, we say f is continuously differentiable, or C^1; if $\hat{\underset{\sim}{D}}f$ is continuously differentiable we say f is C^2, and so on. If f is C^k for all finite k, we say f is C^∞, or smooth. (Sometimes C^0 is used for just "continuous", but whenever we say C^k without fixing k we will assume $k \geq 1$.)

Notice that $\hat{\underset{\sim}{D}}^2 f = \hat{\underset{\sim}{D}}(\hat{\underset{\sim}{D}}f)$ takes values in $L(T;L(T;T'))$ which is naturally isomorphic (Ex. 4) to $L^2(T;T')$, and $\hat{\underset{\sim}{D}}^k f$ similarly takes values in $L^k(T;T') \cong T^* \otimes \ldots \otimes T^* \otimes T'$ (V.1.07, 1.08). In particular when $T' = R$, $L^k(T;T') = T^* \otimes \ldots \otimes T^*$ by definition (V.1.03). Thus tensor quantities arise naturally from differentiation, even if we start with just scalar functions. We shall explore this more fully on manifolds (under "covariant differentiation"). There one is forced not to forget the distinctness of the points at which the tangent spaces are attached, which makes the structure involved clearer. For the moment, notice that the derivative at x of $f : X \to R$ gives a linear functional $T_x X \to T_{f(x)} R \cong R$ whose contours (cf. III.1.02) are exactly the local flat approximations in the tangent space to the contours of f in X.

Notice also, if you have previously done this material a different way, that the directional derivative of f in the direction of a tangent vector $\underset{\sim}{t}$ is simply $\underset{\sim}{D}_x f(\underset{\sim}{t})$, and in this setting hardly needs a special name.

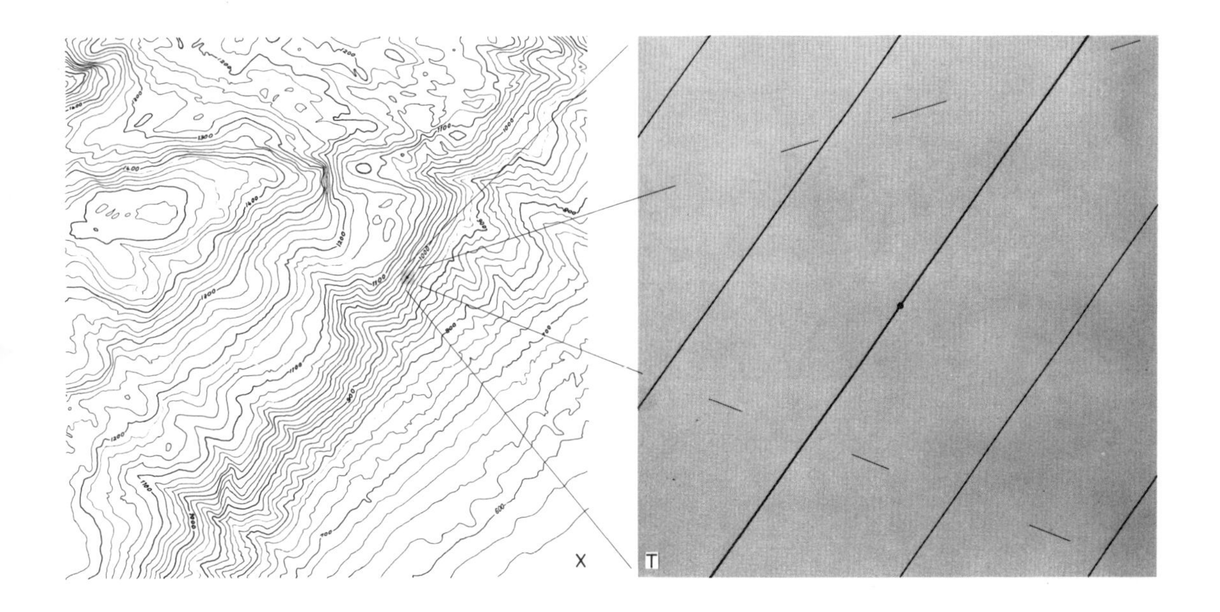

Fig. 1.4

1.03 Partial derivatives

These are just the components of <u>the</u> derivative, once we have chosen charts (C and C', say) for X and X' (cf. II.1.08). This fixes bases β_x, $\beta'_{f(x)}$ for $T_x X$ and $T_{f(x)} X'$ respectively so the linear map $\mathbf{D}_x f$ is represented by a matrix in the usual way. The partial derivatives are its entries, computed as follows.

We use the chart on X' to represent f as $(f^1,\ldots,f^m)$ where f^i is the composite

$$f^i = \mathbf{e}^i \circ C' \circ f : X \to X' \to R^m \to R \quad \text{and} \quad \mathbf{e}^i : (x^1,\ldots,x^m) \rightsquigarrow x^i.$$

If $\beta'_{f(x)}$ is $(\mathbf{c}_1,\ldots,\mathbf{c}_m)$, this means that $f(x) = x'_0 + f^i(x)\mathbf{c}_i$ where x'_0 is the origin, labelled $(0,\ldots,0)$, according to the chart C'. If then $\beta_x = (\mathbf{b}_1,\ldots,\mathbf{b}_n)$

$$\mathbf{D}_x f(\mathbf{b}_j) = \overleftarrow{\mathbf{d}}'_{f(x)} \left(\lim_{h\to 0} \frac{\mathbf{d}'(f(x), f(x+h\mathbf{b}_j))}{h} \right) \qquad \text{by 1.01.}$$

The components of this vector in $T_{f(x)}X'$ with respect to $\beta'_{f(x)}$ are exactly those of its image by $\underset{\sim}{d}'_{f(x)}$ in T', with respect to β'. (That is how $\beta'_{f(x)}$ is defined from β'.) So, since

$$\frac{\underset{\sim}{d}'(f(x),f(x+h\underset{\sim}{b}_j))}{h} = \frac{\underset{\sim}{d}'(x'+f^i(x)\underset{\sim}{c}_i, x'+f^i(x+h\underset{\sim}{b}_j)\underset{\sim}{c}_i)}{h}$$

$$= \frac{f^i(x+h\underset{\sim}{b}_j)\underset{\sim}{c}_i - f^i(x)\underset{\sim}{c}_i}{h}$$

$$= \frac{(f^i(x+h\underset{\sim}{b}_j) - f^i(x))\underset{\sim}{c}_i}{h}$$

and $\underset{\sim}{d}'_{f(x)}$ (being linear and hence continuous) preserves limits, we see that if $\underset{\sim}{D}_x f(\underset{\sim}{b}_j) = f^i_j(x)\underset{\sim}{c}_i$ we have

$$f^i_j(x) = \lim_{h\to 0} \frac{f^i(x+h\underset{\sim}{b}_j) - f^i(x)}{h}$$

These matrix entry functions f^i_j are normally denoted by $\frac{\partial f^i}{\partial x^j}$, or $\partial_j f^i$ for short, where $x^1,\ldots,x^n$ are the coordinates on X given by the chart C. Notice that if we identify the point $x \in X$ with its label $(x^1,\ldots,x^n) \in R^n$, as is common, the equation above becomes

$$\partial_j f^i(x) = \frac{\partial f^i}{\partial x^j}(x) = \lim_{h\to 0} \frac{f^i(x^1,\ldots,x^j+h,\ldots,x^n) - f^i(x^1,\ldots,x^j,\ldots,x^n)}{h}.$$

If the limit in this equation (written either way) exists, we shall call it $\partial_j f^i$ and "partial derivative" even if $\underset{\sim}{D}_x f$ does not exist, cf. Ex. 7.1c.

The matrix $[\partial_j f^i(x)]$, or less abbreviatedly

$$\begin{bmatrix} \frac{\partial f^1}{\partial x^1}(x) & \cdots & \frac{\partial f^1}{\partial x^n}(x) \\ \vdots & & \vdots \\ \frac{\partial f^m}{\partial x^1}(x) & \cdots & \frac{\partial f^m}{\partial x^n}(x) \end{bmatrix}$$

representing $\underset{\sim}{D}_x f$, is the celebrated <u>Jacobian matrix</u> of the map f at x. If X' is R itself, then $(f^1,\ldots,f^m)$ collapses effectively to f, so we have partial derivatives $\partial_j f$ and a matrix

$$[\partial_1 f,\ \partial_2 f,\ \ldots,\ \partial_n f].$$

If on the other hand X is R, while X' is some other space, the derivatives are no longer "partial" as x has only one direction to change in. The entries are usually given a different notation, $\frac{df^i}{dx}(x)$, or (less ambiguously) $\frac{df^i}{dt}(x)$, for $x \in R$. The Jacobian matrix takes the form

$$\begin{bmatrix} \frac{df^1}{dt}(x) \\ \vdots \\ \frac{df^m}{dt}(x) \end{bmatrix} .$$

As usual, the entries in such a "column matrix" are just the components of the vector to which the single basis vector of the domain is carried. In this case the single basis vector is the unit vector $\underset{\sim}{e}_x = \overleftarrow{\underset{\sim}{d}}_x(\underset{\sim}{e}_1)$ ($\underset{\sim}{e}_1$ being the ordered 1-tuple (1), considered as the single standard basis vector for $R = R^1$ as a real vector space. (cf. I.1.10) Its image, which determines $\underset{\sim}{D}_x f$ completely, is denoted by $f^*(x)$. This is the only * we attach to a symbol not previously denoting a vector, which should help the reader to remember that $f^*(x)$, unlike $f(x)$, is a vector.

Notice again that the <u>vector</u> $f^*(x)$ has the same components, relative to $\beta'_{f(x)}$, as the <u>map</u> $D_{\underset{\sim}{x}} f : T_x R \to T_{f(x)} X'$, relative to $\{\underset{\sim}{e}_x\}$ and $\beta'_{f(x)}$. This can encourage confusion when working in coordinates, particularly when X' is also R and $\frac{df}{dt}(x)$ is the "slope of tangent" mentioned at the beginning of the chapter. We have three entities - the linear map $D_{\underset{\sim}{x}} f$, the number $\frac{df}{dt}(x) \in R$ which is the unique entry in the 1×1 matrix representing $D_{\underset{\sim}{x}} f$, and the vector $f^*(x)$ - which are geometrically quite distinct though "componentwise" indistinguishable. By $\frac{df}{dt}$, with no (x), we will mean the function $R \to R : x \rightsquigarrow \frac{df}{dt}(x)$. When we use its value at x we always write it as

$$\frac{df}{dt}(x), \quad \text{not } \frac{d}{dt}(f(x)) \quad \text{or worse} \quad \frac{d}{dx}(f(x))$$

as f(x) is just a number, and cannot be differentiated. For a function of x like, say

$$f : x \rightsquigarrow \int_a^b g(x,s)h(x,0,s)ds, \quad \text{where } g : R^2 \to R,\ h : R^3 \to R$$

we write the differentiated function as

$$\frac{d}{dt} \int_a^b g(\ ,s)h(\ ,0,s)ds \quad \text{and its value at 0 as}$$

$$\left(\frac{d}{dt} \int_a^b g(\ ,s)h(\ ,0,s)ds)\ (0)\right] \quad \text{rather than} \quad \frac{d}{dt} \int_a^b g(0,s)h(0,0,s)ds,$$

which would be ambiguous.

Similar rules will apply to the "covariant" differential operator introduced in the next chapter.

The <u>Jacobian determinant</u> of f at x is the determinant of the Jacobian matrix (only defined when m=n). Like the matrix, this depends on choice of coordinates (since we could change them at one end but not the other, det is only invariant for operators (I.3.12)) but whether it is zero does not. This is very useful. For example, if f is C^1, then if $\underset{\sim}{D}_x f$ is non-singular for some x_0 its determinant is non-zero, hence by continuity of det and $\hat{\underset{\sim}{D}}f$ it must be non-zero in some neighbourhood of x. (Why? If you are not clear, prove this in detail by putting the definitions carefully together.) So we have $\underset{\sim}{D}_x f$ an isomorphism not just at $x = x_0$ but for all x in some neighbourhood of x. This "spreading out" from a point to a neighbourhood is a typical, powerful trick of differential topology; it gives us in particular the following very useful theorem.

1.04 Theorem (Inverse Function Theorem)

If f is C^k, for any k, $\underset{\sim}{D}_x f$ is an isomorphism (so we want n=m) if and only if there are neighbourhoods N of x, N' of f(x) such that f(N) = N' and we have a local C^k inverse $\overleftarrow{f} : N' \to N$. (That is, $\overleftarrow{f} \circ f = I_N$, $f \circ \overleftarrow{f} = I_{N'}$.)

We leave the proof of this result to the Appendix, since we there erect, anyway, machinery which permits a very efficient proof. An understanding of the proof is not in any way essential to an understanding of the result, which is not easy to doubt once understood.

1.05 Corollary

If $f : X \to X'$ is C^1 and $\underset{\sim}{D}_x f$ is injective, then there is a neighbourhood N of x such that $f|N$ is injective. (Thus we want dim X $\leq$ dim X' for either to be possible.)

This is a sufficient but not a necessary condition (Ex. 7).

Proof

Let dim X = n, dim X' = m.

Choose a basis $\underset{\sim}{b}_1, \ldots, \underset{\sim}{b}_n$ for $T_x X$. If $\underset{\sim}{D}_x f$ is injective $(\underset{\sim}{D}_x f)\underset{\sim}{b}_1, \ldots, (\underset{\sim}{D}_x f)\underset{\sim}{b}_n$ are linearly independent, so we can extend them to a basis

$\beta = (\underset{\sim}{D}_x f)\underset{\sim}{b}_1, \ldots, (\underset{\sim}{D}_x f)\underset{\sim}{b}_n, \underset{\sim}{c}_1, \ldots, \underset{\sim}{c}_{m-n}$ for $T_{f(x)}X'$. Define an affine map

$$A : X' \to R^n : (f(x) + a^i(\underset{\sim}{D}_x f)\underset{\sim}{b}_i + h^j \underset{\sim}{c}_j) \rightsquigarrow (a^1, \ldots, a^n)$$

using the chart on X' induced by β and the choice of f(x) as origin. Then clearly $\underset{\sim}{D}_x(A \circ f)$ is injective too, being $\underset{\sim}{D}_x f$ composed with the linear part of A which takes non-zero image vectors of $\underset{\sim}{D}_x f$ to non-zero vectors by construction. Hence it is an isomorphism, since $\dim(T_{A(f(x))}R^n) = \dim T_x X$. By the Theorem there exist neighbourhoods N, N' of x, A(f(x)) and $\phi : N' \to N$ such that $\phi \circ (A \circ f|N) = I_N$. That is $(\phi \circ A) \circ (f|N) = I_N$,so f|N is injective. ■

Both these results are <u>local</u>, asserting things only on neighbourhoods which may be very small, not <u>global</u>. They do not assert that f is invertible or injective as a whole map, even if $\underset{\sim}{D}_x f$ is invertible or injective for all x. (For example, the map $R^2 \to R^2$ taking (x,y) to the point - using complex labels - e^{x+iy} is locally invertible and injective everywhere, but takes infinitely many points to every point in R^2 except (0,0). Work out what is happening in this example if you are not familiar with it; it is illuminating.)

Both results amount to saying that the linear approximation $\underset{\sim}{D}_x f$ is worth making. That is, since $\underset{\sim}{D}_x f$ is supposed to be "arbitrarily close to" f in a sufficiently small neighbourhood of x, the properties of being injective or an isomorphism carry over. When the algebraic condition of injectivity on $\underset{\sim}{D}_x f$ fails, more elaborate approximations (Taylor expansions) are needed for a good local description of f. Recent results give straightforward algebraic criteria on the first k higher derivatives at x, which are "almost always" satisfied for some k, and guarantee that the approximation is locally <u>perfect</u> up to a smooth change of coordinates. For an elementary introduction, see [Poston and Stewart].

Exercises VII.1

1a) Suppose we have Hausdorff topological spaces X,Y and any map (not necessarily continuous or everywhere defined) $f : X \to Y$. Now define

$\lim_{x\to p} (f(x)) = q$ if and only if for any neighbourhood $N(q)$ of q we can find a neighbourhood $N(p)$ of p such that, if $x \in N(p)$ and $f(x)$ is defined, then $f(x) \in N(q)$. (Draw a picture!)

Show that if x_i is a sequence in X with $\lim_{i\to\infty} x_i = p$, $f(x_i)$ defined for infinitely many $i \in N$, and $\lim_{x\to p}(f(x))$ exists, then

$\lim_{i\to\infty} f(x_i) = f(p)$, if $f(p)$ is defined.

b) If X is the set of natural numbers 1,2,3,... together with one extra element which we label ∞ (it can _be_ anything - for instance this book) find a topology on X which makes Def^n VI.2.01 a special case of the one above.

2) Consider $f : R^2 \to R$ such that

$$f(x,y) = \begin{cases} |x| \exp\left(\dfrac{(y-2x^2)^2}{4x^4((y-2x^2)^2 - x^4)}\right) & \text{if } x^2 < y < 3x^2 \\ 0 \text{ otherwise.} & \end{cases}$$

a) Draw a picture of f.

b) Show that for any vector $\underset{\sim}{v} \in R^2$, $\lim_{h\to 0} \dfrac{f(h\underset{\sim}{v})-f(\underset{\sim}{0})}{h}$ exists and is zero, using the definition in Ex. 1a) of "limit" and the usual topologies on R^2 and R.

c) Show that if $\underset{\sim}{v}_i = (1/i, 2/i^2)$ we have $\lim_{i\to\infty} \underset{\sim}{v}_i = \underset{\sim}{0}$, but that in the notation of 1.01 if $x = 0$ we have $\underset{\sim}{d}'(f(x + \underset{\sim}{v}_i), f(x)) = 1/i$.

d) Find a neighbourhood N of the zero map $R^2 \to R$ such that

$$\underset{\sim}{A} \in N \Rightarrow \underset{\sim}{A}(1/i, 2/i^2) \neq 1/i.$$

e) Deduce that f has no derivative at (0,0). (If we put $f(x,y) = 1$ for $x^2 < y < 3x^2$, 0 otherwise, f would still have all partial derivatives $\partial_j f(0,0)$ without even being continuous at (0,0). We need the more complicated function above as a counterexample later.)

3) Show that if a map $f : X \to X'$ between affine spaces has a derivative $\underset{\sim}{D}_x f$ at $x \in X$,

a) $\underset{\sim}{D}_x f$ is unique (so if a linear map $\underset{\sim}{D}'_x f$ also satisfies the definition, $\underset{\sim}{D}'_x f = \underset{\sim}{D}_x f$)

b) $$\underset{\sim}{D}_x f(\underset{\sim}{t}) = \lim_{h\to 0} \underset{\sim}{d}'_{f(x)} \left(\frac{\underset{\sim}{d}'(f(x), f(x+h\underset{\sim}{t}))}{h}\right).$$

Note that as $h\to 0$ we are forcing the linear map $\underset{\sim}{A}$ in the definition 1.01 towards the zero map.

(Hint: to get a quantitative grip on the neighbourhoods involved and make possible a proof by epsilontics, choose norms arbitrarily - any norm will give, by Ex. VI.4.8, the usual topology in finite dimensions - on T and T', take the corresponding norm (V.4.01) on L(T;T'), and express the limits in these terms.)

c) Construct an example in the style of Ex. 2 to show that Theorems 1.04, 1.05 become false if we substitute $\tilde{\underset{\sim}{D}}_x f$, where $\tilde{\underset{\sim}{D}}_x f(\underset{\sim}{t}) = \lim_{h\to 0} \left(\frac{\underset{\sim}{d}'(f(x+h\underset{\sim}{t}), f(x))}{h}\right)$, for $\underset{\sim}{D}_x f$. (Thus we need the existence of $\underset{\sim}{D}_x f$, not just $\tilde{\underset{\sim}{D}}_x f$.)

d) If f is differentiable at x, it is continuous at x. (Hint: otherwise not even $\tilde{\underset{\sim}{D}}_x f$ could exist.)

e) If f is an affine map, then $\hat{\underset{\sim}{D}}_x f$ is the linear part of f. (In particular, we may treat a linear map as being its own $\hat{\text{d}}$erivative : cf. Ex. II.3.8.)

4) If $\underset{\sim}{A}$ is a linear map $T \to L(T;T')$, define

$$\underset{\sim}{A}' : T \times T \to T' : (\underset{\sim}{x},\underset{\sim}{y}) \rightsquigarrow (\underset{\sim}{A}(\underset{\sim}{x}))\underset{\sim}{y} \quad \text{and prove:}$$

a) $\underset{\sim}{A}'$ is bilinear.

b) The map $L(T,L(T;T')) \to L^2(T;T') : \underset{\sim}{A} \rightsquigarrow \underset{\sim}{A}'$ is a vector space isomorphism.

c) Similarly, prove that $L(T;L(T;\ldots;L(T;T')\ldots)) \cong L^k(T;T')$.

5a) If functions f,g defined in a neighbourhood of x in an affine space X, taking values in a vector space Y, are differentiable at x, then so is f + g, and we have, for any $a \in R$,

$$\underset{\sim}{D}_x(f + g) = \underset{\sim}{D}_x f + \underset{\sim}{D}_x g, \qquad \underset{\sim}{D}_x(af) = a(\underset{\sim}{D}_x f)$$

as linear maps.

Thus $\underset{\sim}{D}$ is a linear map from the (infinite-dimensional) vector space of differentiable maps $X \to Y$ to the space of maps $X \to L(T;Y)$, where T is the vector space of X.

b) If f,g are functions $X \to R$, X affine, show from the definitions that $\underset{\sim}{D}_x(fg) = (\underset{\sim}{D}_x f)g + f(\underset{\sim}{D}_x g)$, where $(fg)(x) = f(x)g(x)$, treating $\underset{\sim}{D}_x f$, $\underset{\sim}{D}_x g$ and $\underset{\sim}{D}_x(fg)$ as taking values in R. (Insert the appropriate freeing maps if desired.) In other terms

$$d(fg) = \frac{df}{dt}g + f\frac{dg}{dt}.$$

This fact, in one notation or another, should have been familiar from school onwards. Its usual name is Leibniz's rule, though some books, for example [Misner, Thorne and Wheeler] call it and its generalisations to tensors the chain rule - a name we reserve to its more usual meaning (Ex. 6).

6) Show that if the maps $f : X \to Y$, $g : Y \to Z$ between affine spaces are differentiable at $x \in X$, $f(x) \in Y$ respectively, then $g \circ f$ is differentiable at x and

$$\underset{\sim}{D}_x (g \circ f) = (\underset{\sim}{D}_{f(x)} g) \circ (\underset{\sim}{D}_x f).$$

(This is known as the chain rule for differentiation.)

Deduce that if f and g are C^k, then so is $g \circ f$.

7a) Use the function $x \rightsquigarrow x^3$ to show that the "if" of 1.05 cannot be strengthened to "if and only if".

b) Show that $f : R \to R : x \rightsquigarrow \begin{cases} 2x^2 \sin(1/x) + x & \text{if } x \neq 0 \\ 0 & \text{if } x = 0 \end{cases}$

is differentiable everywhere (draw it!) but not C^1.

c) Show that f has no inverse in any neighbourhood of 0, though $\underset{\sim}{D}_0 f$ is an isomorphism. (This illustrates why C^1 is so much more powerful a condition than just "differentiable" : the difference is much more than between C^1 and C^∞.)

2. MANIFOLDS

In constructing charts on affine spaces (II.1.08) we remarked that on, for example, the earth we could not do so globally. (That is, all over the globe - hence the word. In general we use it to mean "all over the manifold we are considering", which may for instance be the whole of spacetime.) We can however do it locally.

Around any point on the earth we have no trouble in drawing charts of the immediate locality - it is only when we try to cover the whole earth that we are forced into complications like Fig. 2.1. The same applies to any smooth closed surface in R^3; locally we can choose coordinates and make it look like a piece of R^2 (Fig. 2.2), globally not. Now, all the definitions of the last section depended only on having a function f defined in some neighbourhood of the point x we were interested in, since in the course of taking limits we eventually disregarded everything outside _any_ particular neighbourhood. (At this point, strictly speaking, we should write out all the definitions again with f defined on an open set in an affine space, instead of the whole space. What we shall actually do is to talk as though this rewriting has been done.) So a local resemblance to an affine space is all we need to set up the differential calculus. The existence of such a resemblance is exactly what we require in defining a manifold.

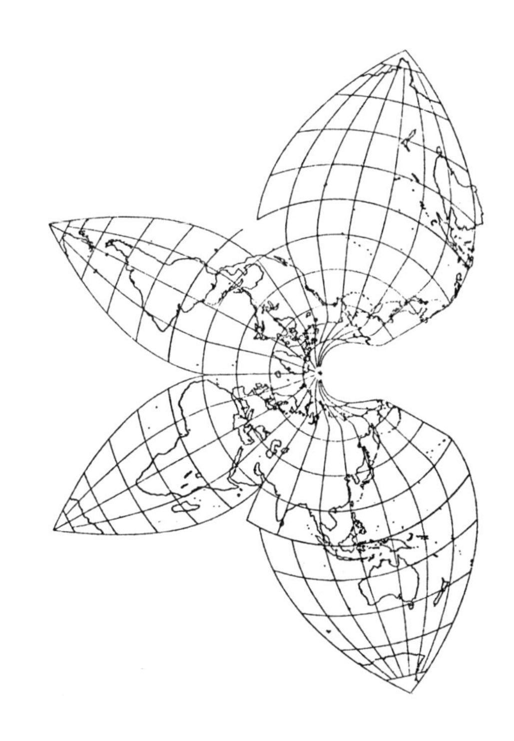

Fig. 2.1

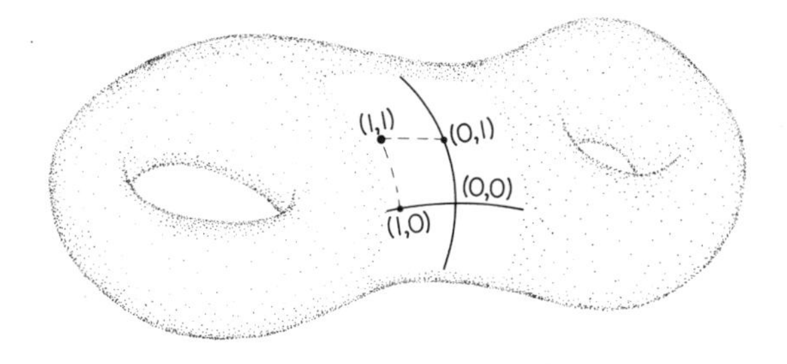

Fig. 2.2

2.01 Definition

A C^k manifold modelled on an affine space X (sometimes, in particular, R^n) is a Hausdorff topological space M together with a collection of open sets $\{U_a | a \in A\}$ in M and corresponding maps $\phi_a : U_a \to X$, such that

Mi) $\bigcup_{a \in A} U_a = M$.

Mii) Each ϕ_a defines a homeomorphism $U_a \to \phi_a(U_a)$.

Miii) If $U_a \cap U_b \neq \emptyset$, then the composites $\phi_a \circ \overleftarrow{\phi_b}$, $\phi_b \circ \overleftarrow{\phi_a}$, on the sets $\phi_b(U_a)$, $\phi_a(U_b)$ on which they are defined (Fig. 2.3), are C^k. (We deduce from Mii they are homeomorphisms; we are requiring them to be differentiable k times as well.)

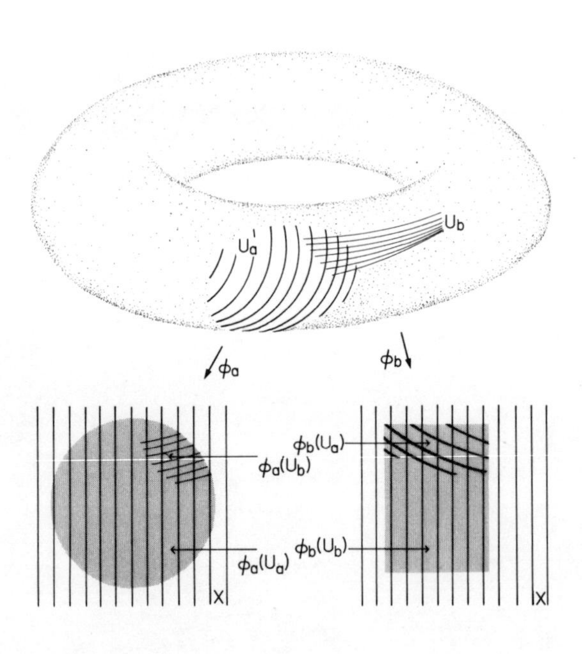

Fig. 2.3

The pairs (U_a, ϕ_a) are called charts on M, and the set $\{(U_a, \phi_a) | a \in A\}$ of all of them is an atlas. Ex. 1 is concerned with some specific examples of manifolds and atlases.

A new chart (U, ϕ), beyond those we have specified in defining M, is called admissible if for all $a \in A$, the maps $\phi \circ \overleftarrow{\phi_a}$ and $\phi_a \circ \overleftarrow{\phi}$ are C^k wherever they are defined. M is not changed in any significant way if we enlarge the family $\{(U_a, \phi_a) | a \in A\}$ by adding admissible charts, and we shall feel free to do so.

It will often be convenient, for a particular $x \in M$, to consider a chart

$\phi : u \to R^n$ with $\phi(x) = (0,\ldots,0)$ (Ex. 1g); a chart around x will always mean this, unless otherwise stated.

To shorten the statements of definitions and theorems we shall generally confine ourselves to C^∞, or smooth, manifolds; very little is lost by this. By "manifold" we mean "smooth manifold" unless otherwise stated.

The dimension dim(M) of M is the dimension of the affine space it is modelled on. We often call M an n-manifold if we want to specify its dimension. (Thus, a 2-manifold, or surface, is a manifold modelled on the plane. It need not be the "surface of" anything.)

The axioms Mi) - Miii) are natural enough; Mi) just says that no point in M is "uncharted", Mii) that the charts are topologically uncomplicated, relative to the topology on M, and Miii) that they are differentially nice (C^k) relative to each other. We cannot ask that they be C^k individually, because we do not yet have a notion of differentiation of maps defined on M; the one we are about to define depends precisely on the compatibility of the charts, which give a "differential structure" on M. (It is possible for a given topological manifold - something satisfying just Mi) and Mii) - to have many essentially different differential structures. For example the seven-dimensional sphere S^7 has 28, and the thirty-one dimensional sphere S^{31} has over 16 million. Certain topological manifolds admit none at all.) On the other hand we must exclude situations like Fig. 2.4 in which a map $f : R \to M$ gives a differentiable map $R \to X$ when composed with one chart, ϕ_b, but not when composed with another. If that kind of thing can happen we cannot hope to give differentiability of f itself a meaning independent of our choice of chart. Having in the affine setting separated differentiability from choice of charts so effectively, this would be a shame. Axiom Miii) is exactly sufficient to stop it happening. (Cf. Ex. 2, one of the most essential exercises in this book. Do 2a and 2b or at least be quite sure you understand what they say, before reading further.) This lets us make

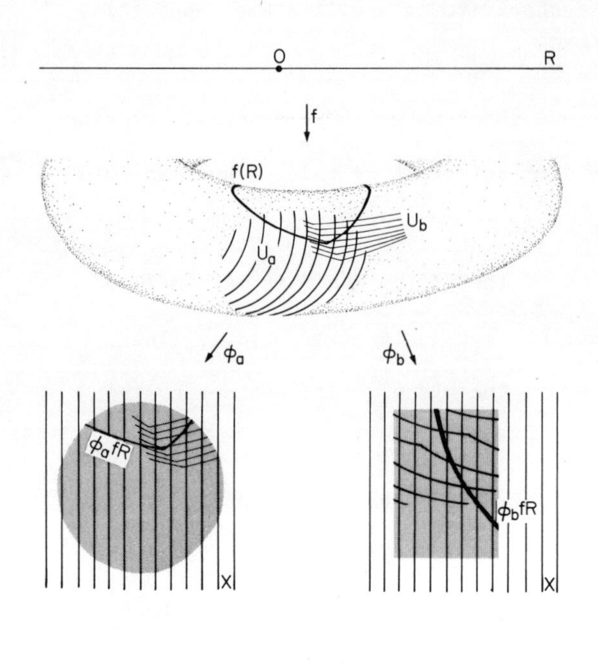

Fig. 2.4

2.02 Definition

A map $f : M \to N$ between smooth manifolds is differentiable, [respectively C^k] at $x \in M$ if for some charts (U,ϕ) on M, (V,ψ) on N, with $x \in U$, $f(x) \in V$, the map $\psi \circ f \circ \overleftarrow{\phi}$ (Fig. 2.5) is differentiable [respectively C^k] at $\phi(x)$. (Ex. 2 guarantees that this definition is independent of our choice of charts.)

A homeomorphism $f : M \to N$ between C^k manifolds is a C^k diffeomorphism if both f and $\overleftarrow{f}$ are C^k. (Note, if $f : R \to R$ has $f(x) = x^3$, then f is a homeomorphism and C^∞, but $\overleftarrow{f}$ is not differentiable at 0. So the C^k condition on $\overleftarrow{f}$ does not follow from f being C^k.) If there is a diffeomorphism between two manifolds they are diffeomorphic.

2.03 Tangent spaces

Now, differentiability of f ought reasonably to mean the existence of a derivative for f itself, not just for various maps $\psi \circ f \circ \overleftarrow{\phi}$, and so it will - once we have said what a derivative is now supposed to be.

Clearly, we shall want $\underset{\sim}{D}_x f$ to be, as before, the linear part of a flat approximation to f at x. For the linearity to be definable, $\underset{\sim}{D}_x f$ must therefore be a map between vector spaces, attached as before to the points x and f(x). Thus we need to attach tangent spaces to points in a manifold, like the ones we have been using attached to points in an affine space.

As with tribal lays, there are very many approaches to constructing tangent

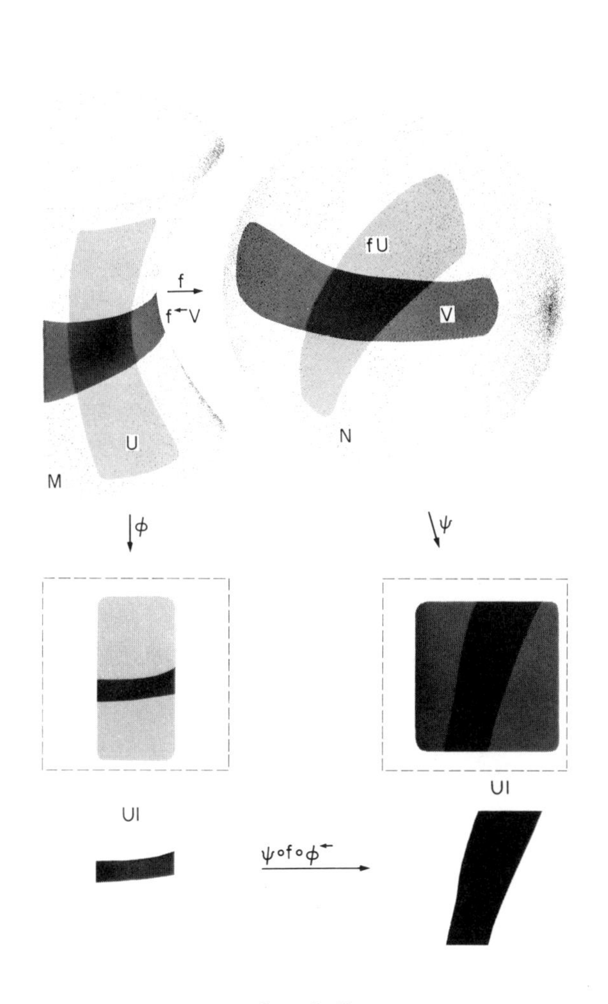

Fig. 2.5

spaces, which are all right ways. That is, they all give naturally isomorphic results (in the strong, technical sense of the word "natural") and they all illuminate one or another aspect of what is going on. We shall sketch the most geometrical, and give as Ex. 3 a particular formal construction chosen (i) because it is the one that requires no more machinery than we have already to hand, and (ii) because it is the rigorous replacement of the traditional "definition" (Ex. 4), which a physicist must be at home with to be able to understand his more unreconstructed colleagues and older books.

It is a (non-trivial) fact that any finite-dimensional manifold can be mapped smoothly and injectively into R^n, for some sufficiently high n. Establishing the lowest value of n that is sufficient for a given manifold is one of the major preoccupations of the differential topologists. For example, among 2-manifolds the sphere and torus can sit nicely in R^3, but the Klein bottle (Fig. 2.6) always has a self-intersection in three dimensions and cannot be mapped continuously and injectively into a less than four-dimensional Euclidean space. (This object, by

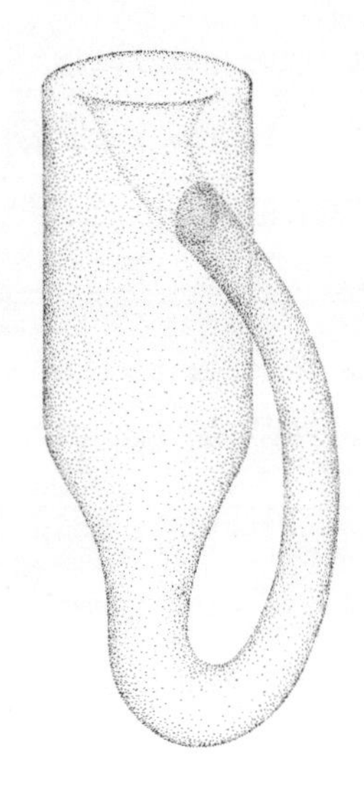

Fig. 2.6

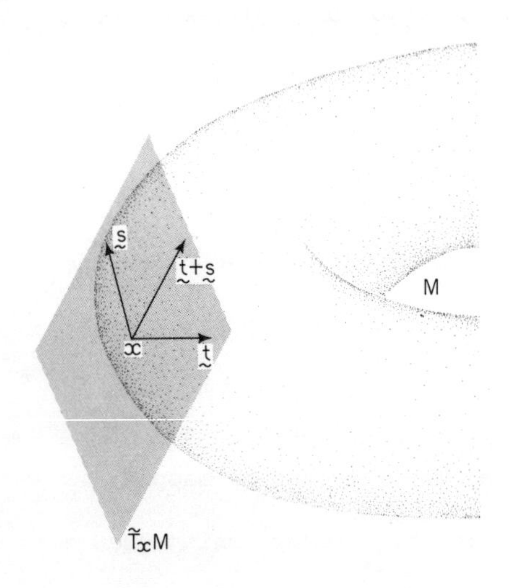

Fig. 2.7

the way, was originally called in unfrivolous 19th Century fashion the Kleinsche Fläche - Klein surface - but this was mistaken by an English translator for Kleinsche Flasche - Klein bottle - and the error took hold so strongly that now the Germans too call it a Flasche.) In general, if dim(M) = m, we may need n up to 2m+1, but never more, if we are concerned only with the differential structure on M (and not, for instance, with a metric as well); this is the Whitney Embedding Theorem, see [Guillemin and Pollack]. All we need here is that for some n, M can be thought of as a subset of R^n in a nice way (like the manifolds of Ex. 1) because we can then define the tangent space to M at $x \in M$ to be the affine subspace $\tilde{T}_x M$ of R^n that is geometrically tangent to M at x. Or rather, since we want a vector space, we use the tangent space $T_x(\tilde{T}_x M)$ in the affine space sense we already have. We shall call this $T_x M$. By Ex. 5 it is canonically isomorphic to the space defined in detail in Ex. 3 (and hence, also, independent of the embedding) and gives us a nice picture of it.

We shall often use this kind of picture in drawing illustrations, though the definition in Ex. 3 is more convenient for formal proofs and calculations. This is just like the interplay between geometric thinking and algebraic proof we have used for vector spaces. We do not give the strict definition of "embedding" here, which involves some technicalities to disallow pictures like Fig. 2.8, since we shall use the embedded picture throughout for illustration only, not proofs.

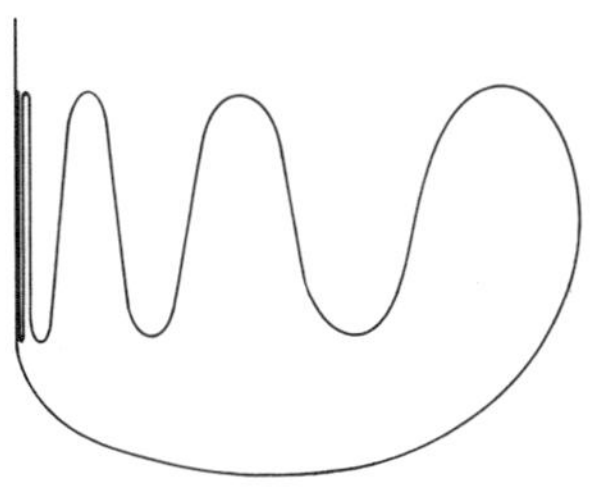

Fig. 2.8

We can also embed manifolds in each other. If M is embedded in N we may call it a <u>submanifold</u> of N. If then dim M = dim N-1, so that each tangent space T_xM is a hyperplane in T_xN (I.1.09), M is a <u>hypersurface</u> in N.

We can now once again interpret differentiability at x of a map f, now between manifolds $f : M \to N$, as the existence of a derivative - that is, a linear map

$$\underset{\sim}{D}_x f : T_xM \to T_{f(x)}N$$

which locally approximates f. The formal details are in Ex. 6. The idea is simply to look at regions of M and N around x and f(x) which are small enough to mistake via charts for pieces of affine space, and transfer to manifolds the affine space notion of derivative. In a similar way we can define higher derivatives, with $\underset{\sim}{D}_x^k f \in L^k(T_xM;T_{f(x)}N)$.

Notice that, unlike the affine space situation where each T_xX had a canonical isomorphism $\underset{\sim}{d}_x$ with the vector space T of X, for a manifold M modelled on X and $x \in M$ there is no such natural choice of isomorphism $T_xM \to T$. Any chart (U,ϕ) with $x \in U$ gives <u>an</u> isomorphism

$$\underset{\sim}{d}_{\phi(x)} \circ \underset{\sim}{D}_x \phi : T_xM \to T_{\phi(x)}X \to T,$$

but any other isomorphism $T_xM \to T$ can equally be realised by an admissible chart.

(Why? Prove it by composing an affine map with ϕ.) Thus we cannot reasonably identify T_xM with T, any more than we identify T with its dual T^*. In consequence, we cannot identify the tangent spaces T_xM, T_yM at different points in M with each other. That is, we cannot "forget to which point each tangent space is attached." Further implications of this will appear in the next section.

Exercises VII.2

1a) If $S^2 = \{(x^1,x^2,x^3) \in R^3 \mid (x^1)^2 + (x^2)^2 + (x^3)^2 = 1\}$, and

$$U_{i+} = \{(x^1,x^2,x^3) \in S^2 \mid x^i > 0\}, \quad i = 1,2,3$$

$$U_{i-} = \{(x^1,x^2,x^3) \in S^2 \mid x^i < 0\}, \quad i = 1,2,3$$

are the six open hemispheres obtained by slicing through S^2 with coordinate planes (draw them!) show that the six "flattening" maps, such as

$$\phi_{1+} : U_{1+} \to R^2 : (x^1,x^2,x^3) \rightsquigarrow (x^2,x^3)$$

and

$$\phi_{2-} : U_{2-} \to R^2 : (x^1,x^2,x^3) \rightsquigarrow (x^1,x^3)$$

constitute an atlas for S^2 making it a smooth manifold.

b) Show that the surface $\{\underset{\sim}{x} \mid \underset{\sim}{x}\cdot\underset{\sim}{x} = 1\}$ of Fig. IV.2.3b is a 2-manifold, with one chart for each component. Show that $\{\underset{\sim}{x} \mid \underset{\sim}{x}\cdot\underset{\sim}{x} = -1\}$ of Fig. IV.2.3c is also a 2-manifold, by finding an atlas for it. Generalise this to show that in any metric vector space, $\{\underset{\sim}{x} \mid \underset{\sim}{x}\cdot\underset{\sim}{x} = a\}$ is a manifold whenever $a \neq 0$.

Find atlases making the following into smooth manifolds:

c) The sets of positions of a unit rod in the plane, and in R^3.

d) The set of positions of a unit circle in R^3.

e) The set of all possible circles in R^3.

f) The set of ellipses in R^3 with one focus at $(0,0,0)$.

N.B.: None of these can be covered by a single chart (though to prove this rigorously is non-trivial). Case f is the first abstract (non-embedded) manifold ever considered: the space of Keplerian orbits around a body centred at the origin. Space engineers use various atlases, and regret deeply the lack of a single, smooth, unredundant complete way to define the coordinates or "elements" of an orbit.

g) Show that for a chart $\phi : U \to R^n$ and $x \in U$, there is an affine map $A : R^n \to R^n$ such that $A \circ \phi$ is an admissible chart taking x to $(0,\ldots,0)$.

2) Let M be a C^k manifold modelled on an affine space X, and (U_a, ϕ_a), (U_b, ϕ_b) charts on M with $U_a \cap U_b \neq \emptyset$.

a) If U is an open subset of an affine space Y, and f a map $U \to M$, use Ex. 1.6 and the Inverse Function Theorem (1.04) to show that for any $x \in U$ with $f(x) \in U_a \cap U_b$ (so that both sides are defined) we have, for all $j = 1,\ldots,k$,

$$\phi_a \circ f \text{ is } C^j \text{ at } x \iff \phi_b \circ f \text{ is } C^j \text{ at } x.$$

b) If W is an open subset of M, and f is a map from U to an affine space Z, show that for any $x \in W \cap (U_a \cap U_b)$ and for all $j = 1,\ldots,k$,

$$f \circ \overleftarrow{\phi_a} \text{ is } C^j \text{ at } x \iff f \circ \overleftarrow{\phi_b} \text{ is } C^j \text{ at } x.$$

c) Deduce that in the situation of Def^n 1.02, if (U',ϕ') and (V',ψ') are also charts on M,N, with $x \in U'$, $f(x) \in V'$, then

$$\psi' \circ f \circ \phi' \text{ is } C^k \text{ at } x \iff \psi \circ f \circ \phi \text{ is } C^k \text{ at } x.$$

3) Let (U,ϕ), (U',ϕ') be charts on a smooth manifold M, modelled on an affine space X with vector space T, $u \in U \cap U'$, and $\underset{\sim}{t}, \underset{\sim}{t}' \in T$. Define the

relation ~ by

$$(U,\phi,\underset{\sim}{t}) \sim (U',\phi',\underset{\sim}{t}') \iff \underset{\sim}{\hat{D}}_{\phi(u)}(\phi'\circ\overleftarrow{\phi})\underset{\sim}{t} = \underset{\sim}{t}'.$$

a) Show that ~ is an equivalence relation on the set of such triples.

b) Show that if $(U,\phi,\underset{\sim}{t}) \sim (U',\phi',\underset{\sim}{t}')$ and $(U,\phi,\underset{\sim}{s}) \sim (U',\phi',\underset{\sim}{s}')$, then

$$(U,\phi,\underset{\sim}{t}+\underset{\sim}{s}) \sim (U',\phi',\underset{\sim}{t}'+\underset{\sim}{s}')$$

and $(U,\phi,\underset{\sim}{t}a) \sim (U',\phi',\underset{\sim}{t}'a)$ for all $a \in R$.

Hence we have a well defined addition and scalar multiplication on the set T_uM of ~ equivalence classes, making it a vector space. Then T_uM is the <u>tangent space to</u> M <u>at</u> u and the ~ equivalence classes are <u>tangent vectors</u> to M at u.

4) If X in Ex. 3 is R^n, we may write ϕ,ϕ' in the form $\phi(u) = (x^1(u),\ldots,x^n(u))$, $\phi'(u) = (x'^1(u),\ldots,x'^n(u))$. By a standard abuse of language, if $(x^1,\ldots,x^n) = \underset{\sim}{x}$, we also write $\phi'(\overleftarrow{\phi}(\underset{\sim}{x})) = (x'^1(\underset{\sim}{x}),\ldots,x'^n(\underset{\sim}{x}))$. Then if $\underset{\sim}{t} = (t^1,\ldots,t^n)$, $\underset{\sim}{t}' = (t'^1,\ldots,t'^n)$, show that

$$(U,\phi,\underset{\sim}{t}) \sim (U',\phi',\underset{\sim}{t}') \iff t'^i = t^j\frac{\partial x'^i}{\partial x^j} \qquad (\bigstar)$$

by applying 1.03.

The traditional description of a vector in a differential context is "a set of n numbers that transform according to★". Two sets of n numbers, associated with two different charts, represent the same vector if they are related by the formula★.

(Warning: in the really confused books, you are told that a vector is a set of <u>functions</u> that transform according to★. What they mean is a vector <u>field</u>, which we come to shortly. Sometimes they say "quantities", which is at least vague enough not to be wrong.)

5a) If $i : M \to R^n$ is the inclusion used in 2.03, use Theorem 1.04 to show that if (U,ϕ) and (U',ϕ') are charts on M with $x \in U \cap U'$, then

$$(U,\phi,\underset{\sim}{t}) \sim (U',\phi',\underset{\sim}{t}') \quad \Leftrightarrow \quad \underset{\sim}{D}_{\phi(x)}(i\circ\overleftarrow{\phi})\underset{\sim}{t} = \underset{\sim}{D}_{\phi'(x)}(i\circ\overleftarrow{\phi'})\underset{\sim}{t}'.$$

Thus each tangent vector, in the sense of Ex. 4, is uniquely represented by a single vector in T_xR^n. The image of any member of it in T_xR^n is the same.

b) Show that these representing vectors form a subspace, Y say, of T_xR^n, and that the function, from the version of T_xM defined in Ex. 3 to Y, taking each tangent vector to its representative in Y, is a vector space isomorphism.

c) Give a precise definition for the geometrical notion of "tangent affine subspace" used in 2.03 (for instance as the union of the set of straight lines in R^n tangent at x to curves in M, which defines an affine subspace Y, cf. VIII. §1), and show that Y coincides with T_xM as defined in 2.03.

6a) Show that if $f : M \to N$ is differentiable at $u \in U \cap U'$, and (V,ψ) is a chart on N with $f(u) \in V$, then

$$(U,\phi,\underset{\sim}{t}) \sim (U',\phi',\underset{\sim}{t}') \Rightarrow \underset{\sim}{D}_{\phi(u)}(\psi\circ f\circ\overleftarrow{\phi})\underset{\sim}{t} = \underset{\sim}{D}_{\phi'(u)}(\psi\circ f\circ\overleftarrow{\phi'})\underset{\sim}{t}'$$

$$\Rightarrow (V,\psi,\underset{\sim}{D}_{\phi(u)}(\psi\circ f\circ\overleftarrow{\phi})\underset{\sim}{t}) \sim (V,\psi,\underset{\sim}{D}_{\phi'(u)}(\psi\circ f\circ\overleftarrow{\phi'})t')$$

so that f induces a well defined map

$$\underset{\sim}{D}_u f : T_uM \to T_{f(u)}N$$

taking the ~ equivalence class of $(U,\phi,\underset{\sim}{t})$ to that of $(V,\psi,\underset{\sim}{D}_{\phi(u)}(\psi\circ f\circ\overleftarrow{\phi})\underset{\sim}{t})$, and prove that $\underset{\sim}{D}_u f$ is linear.

b) We can now take the derivative at $u \in U$ of the chart $\phi : U \to X$ since U (being open in M) and X now both have differential structures. Show that $\underset{\sim}{D}_x\phi$ just takes each $\underset{\sim}{t} \in T_xM$ to its representative in $T_{\phi(x)}X$.

7) Suppose that $f : M \to M'$ is a C^k map between manifolds and that for some $x \in M$ the derivative $\underset{\sim}{D}_x f$ is injective. Let $\dim M = m \leq n = \dim M'$.

a) Deduce from 1.05 that x has a neighbourhood N such that $f|N$ is injective.

b) Construct a chart $\phi : U \to R^n$ around $f(x)$ such that
$\phi(f(N)) = \phi(U) \cap \{(x^1,\ldots,x^n) \mid x^{m+1} = x^{m+2} = \ldots = x^n = 0\}$.
(For $A : R^n \to R^m$ as constructed in proving 1.05, use A as a projection $R^n \to R^m$ and move an (n-m) - dimensional subspace through each $f(y)$ to a new origin.)

c) Show that if $B : R^n \to R^m : (x^1,\ldots,x^m,\ldots,x^n) \rightsquigarrow (x^1,\ldots,x^m)$, then $B\circ\phi\circ f$ is a chart map admissible on M'.

d) Deduce that x and f(x) have C^k charts around them which give f the local coordinate form

$$(x^1,\ldots,x^m) \rightsquigarrow (y^1,\ldots,y^m,0,\ldots,0).$$

8) Suppose $f : M \to M'$ is a C^k map between manifolds and that for some $x \in M$ the derivative $\underset{\sim}{D}_x f$ is surjective with $\dim M = m > n = \dim M'$.

a) Show similarly to Ex. 7 that x and f(x) have charts around them giving f the local form

$$(x^1,\ldots,x^{m-n},x^{m-n+1},\ldots,x^m) \rightsquigarrow (x^{m-n+1},\ldots,x^m).$$

(Hint: construct for some neighbourhood N of x a function

$F : N \to M' \times R^k : y \rightsquigarrow (f(y),?)$ such that $\underset{\sim}{D}_x F$ is bijective, and use 1.04.)

b) Deduce that if for some $p \in M'$ <u>every</u> $x \in \overleftarrow{f}(p)$ has $\underset{\sim}{D}_x f$ surjective, then a chart giving coordinates $(x^1,\ldots,x^{m-n})$ on $\overleftarrow{f}(p)$ may be constructed around each $x \in \overleftarrow{f}(p)$. Prove that these make $\overleftarrow{f}(p)$ into a C^k manifold by satisfying Mi) - Miii).

c) Deduce in particular that if $f : R^n \to R$ is C^∞ and has $\underset{\sim}{D}_x f \neq \underset{\sim}{0}$ for every $x \in \overleftarrow{f}(1)$, then $\overleftarrow{f}(1)$ has the structure of a smooth (n-1)-manifold. Construct such functions f to deduce with less work than in Ex.1 that the sets there given in a,b,c are manifolds.

3. BUNDLES AND FIELDS

From elementary vector analysis, the idea is familiar of a "vector field". That is, a choice of vector at each point in R^3, varying smoothly from point to point. Transferred to general manifolds, this will mean a choice of tangent vector, obviously; but how do we interpret smoothness? Plainly, it must be as smoothness of the map ($x \rightsquigarrow$ chosen vector at x). This means that we need a differential structure on the set of <u>all</u> tangent vectors $\bigcup_{x\in M} T_x M$, denoted by TM. (Each subset $T_x M$ of course has already a differential structure, being a finite-dimensional vector - and hence affine - space.) We can do this in the embedded picture (Fig. 3.1), labelling the vectors by their beginning and end points $(x, x+\underset{\sim}{t})$ and considering the differentiability of the vector field in terms of the resulting

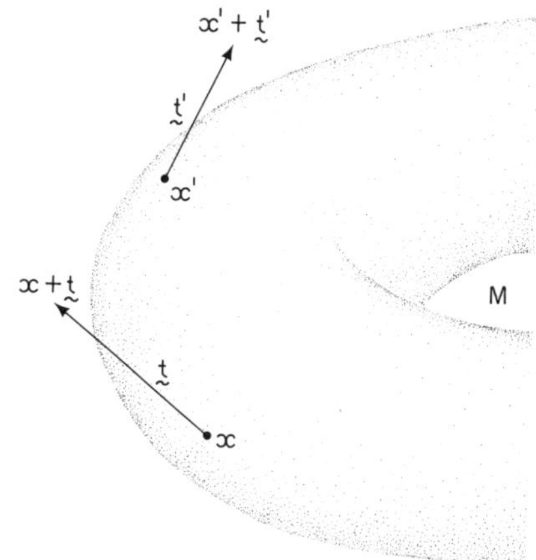

Fig. 3.1

map $M \to R^n \times R^n = R^{2n}$. This is intuitive, but cumbersome; it is more convenient to handle TM directly, via the construction of the tangent spaces by means of the charts. The results are essentially the same.

One minor technical point is needed here. In the last paragraph, in order to differentiate a map taking values in $R^n \times R^n$ we used the obvious identification with R^{2n}, which is an affine space, so that our definitions applied. There is an equally obvious affine space [respectively, vector space] structure on the set-theoretical product $X \times Y$ of any two affine [respectively, vector] spaces X and Y. (We could have introduced these in Chapters I and II, but not so easily have explained their usefulness.) The details are collected in Ex. 1.

3.01 Theorem

If M is an n-manifold modelled on an affine space X, with atlas $\{(U_a,\phi_a) \mid a \in A\}$ we set

$$TM = \bigcup_{x\in M} T_x M, \qquad TU_a = \bigcup_{x\in U_a} T_x M \subseteq TM.$$

Then, $\{(TU_a, \underset{\sim}{D}\phi_a) \mid a \in A\}$ is an atlas making TM a 2n-manifold modelled on $X \times X$, where $\underset{\sim}{D}\phi_a : TU_a \to TR^n$ is defined by $\underset{\sim}{D}\phi_a \mid T_x M = \underset{\sim}{D}_x \phi_a$. (We are just taking all the derivatives at once to make one big map.)

Proof

(The diagram must regrettably be for M a 1-manifold, since otherwise TM is at least four-dimensional! Recall Fig. II.1.4.)

We must confirm the axioms Mi) - Miii) of Definition 2.01; dim(TM) = 2n by Ex. 1b,c.

Mi) $\displaystyle\bigcup_{a\in A} (TU_a) = \bigcup_{a\in A} \Big(\bigcup_{x\in U_a} T_x M\Big) = \bigcup_{x\in M} T_x M = TM.$

Mii) We fix the topology on TM, similarly to VI.3.01 (the weak topology) by taking as open the smallest family of sets that makes all the $\underset{\sim}{D}\phi_a$'s continuous. (This we must have, if we hope for homeomorphisms!) That is, we take the family of finite intersections, and arbitrary unions of these, of sets of form $(\underset{\sim}{D}\phi_a)^{\leftarrow}(W)$, for W open in $X \times X$. (cf. Ex. 1e)

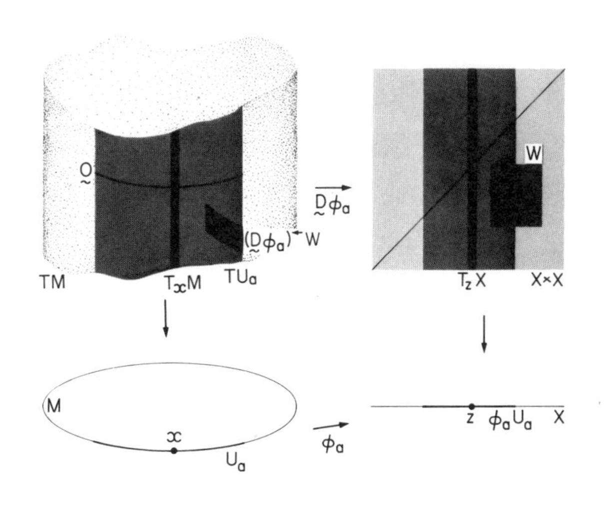

Fig. 3.2

Now we know that each $\underset{\sim}{D}\phi_a$ is injective for ϕ_a is bijective: If $\underset{\sim}{t},\underset{\sim}{t}'$ are in the tangent spaces at $x \neq x'$, say, they are mapped by $\underset{\sim}{D}\phi_a$ into the disjoint tangent spaces at $\phi_a(x) \neq \phi_a(x')$. If they are in the same tangent space at x say, use the fact that $\underset{\sim}{D}\phi_a|T_xM = \underset{\sim}{D}_x\phi_a$ is an isomorphism and hence injective. Similarly $\underset{\sim}{D}\phi_a$ is surjective. We have just picked the topology on TM to make each $\underset{\sim}{D}\phi_a$ continuous, so to prove Mii) it remains only to check that each $(\underset{\sim}{D}\phi_a)^{\leftarrow}$, which exists since $\underset{\sim}{D}\phi_a$ is bijective, is continuous. This means by Definition VI.1.09 that

$$U \text{ open in TM} \implies ((\underset{\sim}{D}\phi_a)^{\leftarrow})^{\leftarrow}(U) \text{ open in } X \times X.$$

(Note that $((\underset{\sim}{D}\phi_a)^{\leftarrow})^{\leftarrow}(U) = \underset{\sim}{D}\phi_a\,(U \cap U_a)$, since $\underset{\sim}{D}\phi_a$ need not be defined on all of U.)

This follows if we prove it for U of the form $(\underset{\sim}{D}\phi_b)^{\leftarrow}(W)$, with W open in $X \times X$, since any open set, by definition, is a union of finite intersections of such (Ex. 3a). Hence we want

$$W \text{ open in } X \times X \implies ((\underset{\sim}{D}\phi_a)^{\leftarrow})^{\leftarrow}((\underset{\sim}{D}\phi_b)^{\leftarrow}(W)) \text{ open in } X \times X.$$

That is, W open in X × X => $(\underset{\sim}{D}\phi_b \circ (\underset{\sim}{D}\phi_a)^{\leftarrow})^{\leftarrow}(W)$ open in X × X (cf. Ex. 3b). Hence, we want $\underset{\sim}{D}\phi_b \circ (\underset{\sim}{D}\phi_a)^{\leftarrow}$, equal by the Chain Rule to $\underset{\sim}{D}(\phi_b \circ \phi_a^{\leftarrow})$, to be continuous. But this follows at once from the requirement on M that each $\phi_b \circ \phi_a^{\leftarrow}$ be continuously differentiable, so we are done.

(We have dropped the temporary expedient (1.02) of forgetting where tangent spaces are attached, so that $\underset{\sim}{D}(\phi_b \circ \phi_a^{\leftarrow})$ is properly seen as a map $\phi_a(U_b) \times X \to \phi_b(U_a) \times X$, linear on tangent spaces $T_x = \{x\} \times X$, not a map $\phi_a(U_b) \to L(T;T)$. It is immediate that continuity in this view is equivalent to the earlier definition. Notice again the crucial nature of the continuity requirement (cf. Ex. 1.7, Ex. 3c). This is why we have not even given a name to things satisfying Mi) and Mii) but with the $\phi_b \circ \phi_a^{\leftarrow}$ only differentiable, since in the absence of this theorem they are of little use beyond what comes from satisfying Mi) and Mii) with no differential conditions at all.)

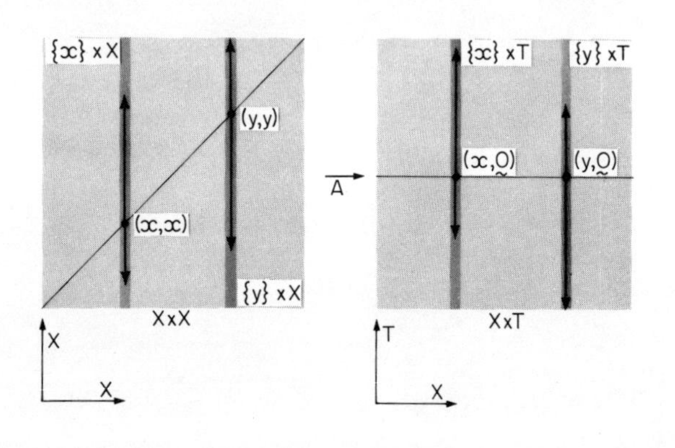

Fig. 3.3

Miii) Define a map

$$A: X \times X \to X \times T : (x,y) \rightsquigarrow (x, \underset{\sim}{d}(x,y)).$$

This is an affine map, hence its derivative everywhere is just its linear part $\underset{\sim}{A}$; trivially, it is C^{∞}. It nicely disentangles "affine space directions" in X×X, seen as a union of tangent spaces, from "tangent vector directions" (Fig. 3.3), so simplifying the algebra. The point is that in Fig. 3.3(a) a "horizontal" movement changes the vector, for instance from zero to non-zero, but not in Fig. 3.3(b).

We know by Ex. 1f that the derivative of

$$A\circ(\underset{\sim}{D}(\phi_b\circ\phi_a^{\leftarrow}))\circ A^{\leftarrow} : \phi_a(U_b)\times T \to \phi_b(U_a)\times T$$

$$(q,\underset{\sim}{t}) \rightsquigarrow ((\phi_b\circ\phi_a^{\leftarrow})q,\ \hat{\underset{\sim}{D}}_q(\phi_b\circ\phi_a^{\leftarrow})\underset{\sim}{t})$$

$$= (p,\underset{\sim}{s}) \quad \text{for short}$$

(why does this map have the expression given?) at $(q,\underset{\sim}{t})$ is exactly

$$\underset{\sim}{D}_q(\phi_b\circ\phi_a^{\leftarrow}) \oplus \underset{\sim}{D}_{\underset{\sim}{t}}(\underset{\sim}{D}_q(\phi_b\circ\phi_a^{\leftarrow})) : T_qX \oplus T_{\underset{\sim}{t}}(T_qX) \to T_pX \oplus T_{\underset{\sim}{s}}(T_pX).$$

Since $\underset{\sim}{D}_q(\phi_b\circ\phi_a^{\leftarrow})$ is linear, we can identify this with

$$\underset{\sim}{D}_q(\phi_b\circ\phi_a^{\leftarrow}) \oplus \underset{\sim}{D}_q(\phi_b\circ\phi_a^{\leftarrow}) : T_qX \oplus T_qX \to T_pX \oplus T_pX.$$

We then have by the Chain Rule that

$$\underset{\sim}{D}_{(q,\underset{\sim}{t})}(\underset{\sim}{D}(\phi_b)\circ(\underset{\sim}{D}(\phi_a))^{\leftarrow}) = \underset{\sim}{D}_{(q,\underset{\sim}{t})}(\underset{\sim}{D}(\phi_b\circ\phi_a^{\leftarrow}))$$

$$= \underset{\sim}{D}_{(q,\underset{\sim}{t})}(A^{\leftarrow}\circ(A\circ(\phi_b\circ\phi_a^{\leftarrow})\circ A^{\leftarrow})\circ A)$$

$$= \underset{\sim}{A}^{\leftarrow}\circ(\underset{\sim}{D}_q(\phi_b\circ\phi_a^{\leftarrow}) \oplus \underset{\sim}{D}_q(\phi_b\circ\phi_a^{\leftarrow}))\circ\underset{\sim}{A}$$

so that the differentiability, and its continuity, of $\underset{\sim}{D}(\phi_b)\circ(\underset{\sim}{D}(\phi_a))^{\leftarrow}$ follows from that of $\phi_b\circ\phi_a^{\leftarrow}$; similarly for higher derivatives. (Recall that we are assuming all manifolds to be C^∞ unless otherwise stated.) ■

3.02 Language

Notice the very different roles played in this proof by $\underset{\sim}{D}_q$ and $\underset{\sim}{D}$:

For any map $f : M \to N$ between manifolds we define $\underset{\sim}{D}f : TM \to TN$ by the requirement that for any $q \in M$, $\underset{\sim}{D}_q f : T_qM \to T_{f(q)}N$ is just the restriction of $\underset{\sim}{D}f$ to T_qM.

However, $\underset{\sim}{D}_q f$ is a linear map between vector spaces while $\underset{\sim}{D}f$ is a map between manifolds, which for the case above we had to prove was again differentiable. (In general this does not follow from the differentiability of f. In fact f is C^1 exactly when $\underset{\sim}{D}f$ exists and is continuous, C^2 when $\underset{\sim}{D}f$ is C^1, and so on - which gives a cleaner definition. The proof that this is equivalent to the chart definition (2.02) is purely a mechanical check.) We shall therefore call them by different names: Whereas the derivative of f, at a point $q \in M$, is the linear map $\underset{\sim}{D}_q f$ approximating f at q, the differential of f is the map $\underset{\sim}{D}f$ between the manifolds TM and TN. In this we are neither following nor departing from standard usage, because there is none - various authors use the words variously. So do not expect the distinction always to be made in the same way elsewhere.

One special case is worth a special symbol. If f is a real-valued function, then $\underset{\sim}{D}_q f$ is a linear map $T_q M \to T_{f(q)} R$. Composed with the isomorphism $\underset{\sim}{d}_{f(q)} : T_{f(q)} R \to R$, this gives us a linear map $T_q M \to R$, whose geometrical meaning is as explained in 1.02 and Fig. 1.4 for affine spaces. The map $TM \to R$ obtained by combining all of these we denote by df (this usage is standard) as distinct from the map $\underset{\sim}{D}f : TM \to TR$. The map df is highly important: it is often called the gradient of f. We shall extend our use of the word "differential" to include df also by a mild abuse of language. Notice that although $(df)_q$ is a (covariant) vector at q, we do not put a ~ under it. We have chosen this inconsistency to avoid confusion with the $\underset{\sim}{d}$, $\underset{\sim}{d}_x$ etc that we use for affine structures.

3.03 Definition

The tangent bundle on a manifold M is the manifold TM together with the map (trivially a C^∞ map) taking each tangent vector down to its point of attachment:

$$\begin{array}{c} TM \\ \Pi \downarrow \\ M \end{array} \qquad \text{where } \Pi(T_q M) = \{q\}.$$

Locally, Π looks like the projection of a product, as in Fig. 3.4; globally it may do so, as in Fig. 3.2, but it often does not, as we shall see.

The <u>bundle of tensors of type</u> $\binom{k}{h}$ (cf. V.1.10) <u>on</u> M is defined by taking the tensor product

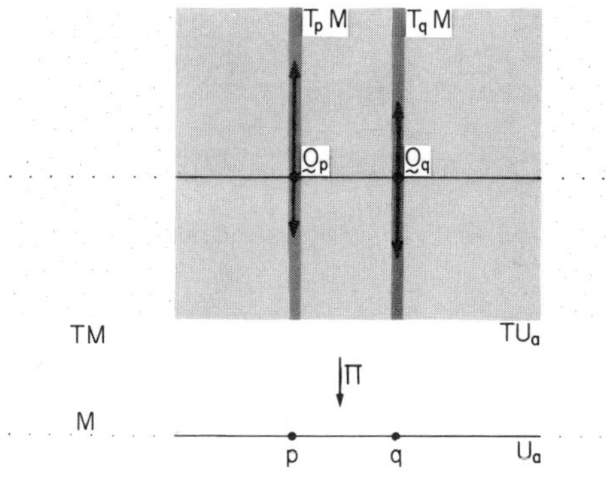

$$(T_xM)^k_h = \underbrace{T_xM\otimes\ldots\otimes T_xM}_{k\text{ times}}\otimes\underbrace{(T_xM)^*\otimes\ldots\otimes(T_xM)^*}_{h\text{ times}}$$

Fig. 3.4

over each point $x \in M$, and joining the separate spaces up into a manifold called T^k_hM. (The formal details of this construction are technically laborious but contains no ideas not in the proof of 3.01. The main challenge is to find a sufficiently succinct notation to get formulae for the maps involved that do not spread over more than two lines. This is a highly worthwhile exercise, and left as such in Ex. 4.) The bundle is then T^k_hM together with the map Π^k_h taking each tensor in $(T_xM)^k_h$ to x, as for the tangent vectors above. (We may for brevity refer to the bundle as simply T^k_hM, not the pair (T^k_hM,Π^k_h), but the map is always to be understood as part of the structure.) Tensor bundles of type $({}^k_h{}^m_n)$, etc, are defined similarly.

We make similar abbreviations to those in Ch. V.1.10, of T^0_hM to T_hM, T^k_0M to T^kM, T_1M to T^*M, $(T_xM)^*$ to T^*_xM, Π^0_1 to Π etc. By convention T^0_0M is just the product manifold $M\times R$ together with the projection $\Pi^0_0 : T^0_0M \to M : (x,r) \rightsquigarrow x$ as bundle map, so that $(\Pi^0_0)^{\leftarrow}(x) = (T_xM)^0_0$. We have a natural bijection $f \rightsquigarrow (x \rightsquigarrow (x,f(x)))$ between functions $M \to R$ and $\binom{0}{0}$ - tensor fields $M \to T^0_0M$ and we generally identify the two ideas.

For any of these bundles the vector space $(T_xM)^h_k$ (sometimes denoted by $(T^h_kM)_x$, according to taste), for a particular $x \in M$, is the <u>fibre</u> at, or over, x.

This word is suggested by Fig. 3.4 where the fibres are one-dimensional, but applies regardless of dimension.

<u>3.04 Definition</u>

A C^r <u>tensor field of</u> type $\binom{k}{h}$ on a manifold M is a C^r <u>section</u> of the bundle T^k_hM; that is, a C^r map $\underset{\sim}{v} : M \to T^k_hM$ such that $\Pi^k_h \circ \underset{\sim}{v} = I_M$, (Fig. 3.5). This is precisely "choice of a tensor at each point of M" in the manner of the beginning of this section. (We shall be concerned so invariably with C^∞, or <u>smooth</u>, fields that we shall take any tensor field to be C^∞ unless otherwise indicated, as we do for manifolds.)

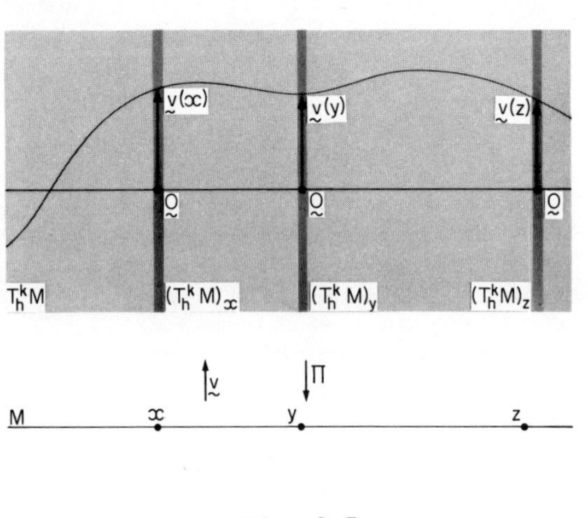

Fig. 3.5

We shall denote the (∞-dimensional) vector space of <u>all</u> $\binom{k}{h}$ - tensor fields on M by T^k_hM, omitting 0's etc, as in 3.03.

Notice that we use a symbol of the same kind (twiddled lower case, like $\underset{\sim}{v}$) for a tensor field as for a single tensor. The context should make clear what is meant, even when the overwhelming weight of tradition makes us abbreviate "tensor field" to just "tensor" in particular cases, such as the Einstein tensor. We shall do so as rarely as we can help, but with no further apology.

Sometimes, particularly when the value of $\underset{\sim}{v}$, at $p \in M$ is a function (for instance $\underset{\sim}{v}$ of type $\binom{0}{1}$, $\underset{\sim}{v}(x) : TM \to R$, and $\underset{\sim}{v}$ of type $\binom{0}{2}$, $\underset{\sim}{v}(x) : T_xM \times T_xM \to R$, are linear and bilinear maps respectively) we want an alternative way to write $\underset{\sim}{v}(x)$. This is to avoid expressions like $\underset{\sim}{v}(x)(\underset{\sim}{t})$ or $\underset{\sim}{v}(x)(\underset{\sim}{s},\underset{\sim}{t})$. We introduce the notation $\underset{\sim}{v}_x$ for $\underset{\sim}{v}(x)$ to achieve this. If $\underset{\sim}{v}$ has a complicated expression like $a^k_i b^j_h \underset{\sim}{f}^i$ we may write $a^k_i(x) b^j_k(x) (\underset{\sim}{f}^i)_x$ or put brackets round the lot, writing $(a^k_i b^j_k \underset{\sim}{f}^i)_x$.

We deviate from lower case in a similar way to our previous usage for single tensors (where, for example, a bilinear form in $L^2(X;R) = X^0_2$ was denoted by $\underset{\sim}{F}$(cf IV.1.01, V.1.03, V.1.10)), and from twiddles in an instance discussed in the next section.

In particular, we have:-

A contravariant vector field is a section of TM. That is a smooth choice of a tangent vector at each point. This is illustrated in the embedded case by Fig. 3.6. Contravariant vector fields are sometimes called tangent vector fields

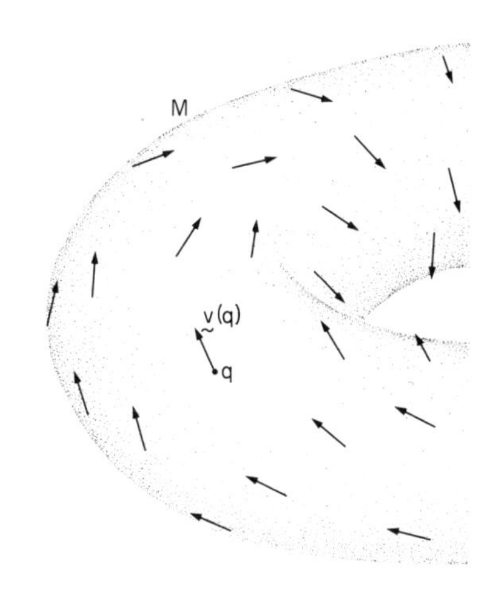

Fig. 3.6

A covariant vector field is a section of T*M. In particular the gradient df of a smooth function f is always a covariant vector field (cf. 3.02), but the converse is untrue. A linear functional $\underset{\sim}{f}$ in T^*_xM often called a cotangent vector at x and covariant vector fields are sometimes called cotangent vector fields or one-forms.

If we have a $\binom{k}{h}$-tensor field $\underset{\sim}{t}$ and an $\binom{i}{j}$-tensor field $\underset{\sim}{s}$ on M, then in each $(T^k_hM)_x \otimes (T^i_jM)_x \cong (T^{k\ i}_{h\ j}M)_x \cong (T^{k+i}_{h+j}M)_x$ we have an element $\underset{\sim}{t}_x \otimes \underset{\sim}{s}_x$. These define sections $x \rightsquigarrow \underset{\sim}{t}_x \otimes \underset{\sim}{s}_x$ of the $\binom{k\ i}{h\ j}$ and $\binom{k+i}{h+j}$-tensor bundles which - subject to the trivial check of being C^∞ if $\underset{\sim}{t}$ and $\underset{\sim}{s}$ are - give us $\binom{k\ i}{h\ j}$ and $\binom{k+i}{h+j}$-tensor fields on M. In particular, if $\underset{\sim}{t}$ is of type $\binom{0}{0}$, that is just a function $t : M \to R$ then $\underset{\sim}{t} \otimes \underset{\sim}{s}$ is just $t\underset{\sim}{s}$ with $(t\underset{\sim}{s})_x$ the scalar multiple $t(x)\underset{\sim}{s}(x)$.

For mixed tensors we can define various contraction maps

$$\underset{\sim}{C}^j_i : T^k_h M \to T^{k-1}_{h-1} M, \quad j \leq k,\ i \leq h,$$

fibre by fibre, exactly as in Ch. V.1.11. We have correspondingly for tensor fields the maps

$$T^k_h M \to T^{k-1}_{h-1} M : \underset{\sim}{t} \rightsquigarrow \underset{\sim}{C}^j_i \circ \underset{\sim}{t},$$

which we shall also denote by the symbols $\underset{\sim}{C}^j_i$.

A metric tensor field is a section $\underset{\sim}{G}$ of T_2M such that for each $x \in M$, $\underset{\sim}{G}(x)$ is a metric tensor (IV.1.01(vii)) on TM. $\underset{\sim}{G}$ is a Riemannian structure on M if each $\underset{\sim}{G}(x)$ is an inner product, a pseudo-Riemannian structure in the indefinite case. In particular, if M is a 4-manifold and the signature (IV.3.09) of $\underset{\sim}{G}(x)$ is everywhere -2, then $\underset{\sim}{G}$ is a Lorentz structure. A manifold with one of these structures is called a Riemannian, pseudo-Riemannian or Lorentz manifold accordingly. A Lorentz manifold is often called a spacetime. The definitions of timelike, spacelike and null vectors (IV.1.04) extend in the obvious way to tangent vectors and fields on a pseudo-Riemannian manifold.

3.05 Definition

Taking R^n as an affine space (and hence a manifold modelled on R^n by way of the single identity chart (I_{R^n}, R^n)) with difference function $\underset{\sim}{d}(\underset{\sim}{x},\underset{\sim}{y}) = \underset{\sim}{y}-\underset{\sim}{x}$, we define the standard Riemannian structure on R^n from the standard inner product (IV.1.03) by the equation

$$\underset{\sim}{G}(x)(\underset{\sim}{t},\underset{\sim}{t}') = (\underset{\sim}{d}_x\underset{\sim}{t}).(\underset{\sim}{d}_x\underset{\sim}{t}')$$

(where $\underset{\sim}{d}_x$ is as defined in II.1.02) for each point $x \in R^n$. We usually abbreviate this to $\underset{\sim}{t}.\underset{\sim}{t}'$. Call R^n with this metric tensor field <u>Euclidean n-space</u> and denote it by E^n to distinguish it from its vector space which we still call R^n.

We define the <u>standard Lorentz structure</u> on R^4 similarly. Call the result <u>Minkowski space</u> M^4, as distinct from its vector space L^4 (IV.1.05). Note the distinction. Lorentz space L^4 is a <u>vector</u> space with a metric tensor, while Minkowski space M^4 is an <u>affine</u> space with a (constant) metric tensor field. (The affine space R^4 can have a geometry given by an interesting non-constant C^∞ metric tensor field but a metric tensor on the vector space R^4 is a single bilinear form.)

These metric tensor fields are <u>constant</u>, in the sense of being given on the vector space and transferred to the tangent spaces by means of the canonical isomorphisms $\underset{\sim}{d}_x$. Only affine spaces have such $\underset{\sim}{d}_x$'s as part of their structure, and hence only on affine spaces can "constant tensors" be defined in this absolute sense except for three trivial cases:

(i) A tensor field of type $\binom{0}{0}$ is just a function, and can obviously be a constant one.

(ii) The <u>zero</u> tensor field $\underset{\sim}{0}$ of any type $\binom{h}{k}$, with $\underset{\sim}{0}(x) = \underset{\sim}{0} \in (T^k_h M)_x$,

(iii) The <u>identity</u> $\binom{1}{1}$-tensor field $\underset{\sim}{I}_x : T_x M \to T_x M$, its dual, and their scalar multiples and tensor powers (like $3\underset{\sim}{I}\otimes\underset{\sim}{I}\otimes\underset{\sim}{I}^*$) may reasonably be called constant, since in <u>any</u> chart their components are constant.

(If M has a metric tensor field $\underset{\sim}{G}$, there is a form of constancy relative to $\underset{\sim}{G}$ for other fields, which we define in VIII.7.10. The three cases (i)-(iii) above are constant relative to any $\underset{\sim}{G}$.)

If a manifold M is embedded in R^n, the standard Riemannian structure on R^n can be applied to pairs of vectors tangent to M at any point x, and this obviously defines an <u>induced</u> Riemannian structure on M. The same may hold but it does <u>not</u> hold <u>necessarily</u> for pseudo-Riemannian structures. For the tangent space T_xM may be a degenerate subspace of T_xR^n with the given metric tensor (cf. IV, 1.01 and 1.09, and Figs. 3.7 and IV.1.5.). An important case where it does hold appears

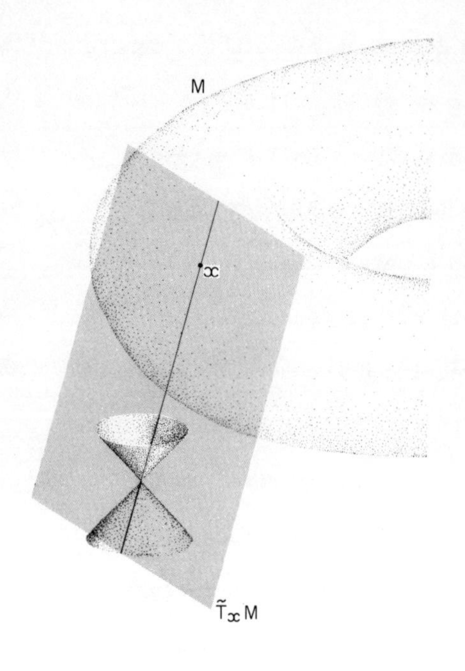

Fig. 3.7

in Ch. VIII. Ex. 1.4.

It is a fact that any metric tensor field, on any manifold, can be so induced by an embedding in some R^n with a constant metric tensor, but this is the consequence of deep, comparatively recent, techniques, not a classical result. It is far beyond the scope of this book. Moreover, n may need to be very large. For example, a Riemannian 2-manifold may need up to 10 dimensions for this, a spacetime may need up to 87 spacelike and 3 timelike dimensions, [Clarke]. (These numbers are not known to be best possible; this would require specific examples with proofs that no smaller flat space would hold them, even harder than the proof that <u>any</u> spacetime fits in 90 flat dimensions.) We shall see (X 1.08) a "flat" metric on the torus induced by a four-dimensional embedding : it is not hard to show that <u>no</u> three-dimensional embedding can induce it.

Exercises VII.3

1a) If S,T are vector spaces, prove that the definitions

$$\left.\begin{array}{c}(\underset{\sim}{s},\underset{\sim}{t}) + (\underset{\sim}{s}',\underset{\sim}{t}') = (\underset{\sim}{s} + \underset{\sim}{s}', \underset{\sim}{t} + \underset{\sim}{t}') \\ (\underset{\sim}{s},\underset{\sim}{t})a = (\underset{\sim}{s}a,\underset{\sim}{t}a)\end{array}\right\} \text{ for } \underset{\sim}{s},\underset{\sim}{s}' \in S,\ \underset{\sim}{t},\underset{\sim}{t}' \in T,\ a \in R$$

make the set $S \times T$ of ordered pairs $(\underset{\sim}{s},\underset{\sim}{t})$ into a vector space, the <u>product</u> or <u>direct sum</u> of S and T, denoted by $S \oplus T$. (The abstract definitions of "product" and "sum" applicable here coincide for vector spaces.) We often identify $(\underset{\sim}{s},\underset{\sim}{0})$ with $\underset{\sim}{s}$, $(\underset{\sim}{0},\underset{\sim}{t})$ with $\underset{\sim}{t}$, and hence write $(\underset{\sim}{s},\underset{\sim}{t})$ as $\underset{\sim}{s} + \underset{\sim}{t}$.

b) Show that $\dim(S\oplus T) = \dim S + \dim T$, by considering bases.

c) If X,Y are affine spaces with vector spaces S,T and difference functions $\underset{\sim}{d}_X, \underset{\sim}{d}_Y$ respectively, prove that the map

$$\underset{\sim}{d}((x,y),(x',y')) \to \underset{\sim}{d}_X(x,x') + \underset{\sim}{d}_Y(y,y')$$

from $(X \times Y) \times (X \times Y)$ to $S\oplus T$ is a difference function making $X \times Y$ an affine space with vector space $S\oplus T$. This affine space is called the <u>product</u> of X and Y, and denoted still by $X \times Y$. (The ideas of "sum" and "product" do <u>not</u> coincide here, precisely because there is no natural way of identifying $x \in X$ with any particular $(x,y) \in X \times Y$. The full treatment of these ideas is an (elementary) part of category theory.)

d) Show that if subspaces S,T of a vector space X have the property that any $\underset{\sim}{x} \in X$ can be written as a sum

$$\underset{\sim}{x} = \underset{\sim}{s}+\underset{\sim}{t}, \text{ with } \underset{\sim}{s} \in S,\ \underset{\sim}{t} \in T \text{ unique, so that}$$

$$\underset{\sim}{s}+\underset{\sim}{t} = \underset{\sim}{s}'+\underset{\sim}{t}' \Rightarrow \underset{\sim}{s} = \underset{\sim}{s}',\ \underset{\sim}{t} = \underset{\sim}{t}' \quad \text{for } \underset{\sim}{s},\underset{\sim}{s}' \in S,\ \underset{\sim}{t},\underset{\sim}{t}' \in T$$

then there is an isomorphism

$$S\oplus T \xrightarrow{\ \cong\ } X : (\underset{\sim}{s},\underset{\sim}{t}) \rightsquigarrow \underset{\sim}{s}+\underset{\sim}{t}.$$

Thus for instance Ex. I. 3.8 shows that if $\underset{\sim}{P}$ is a projection, $X \cong \underset{\sim}{P}(x) \oplus \ker \underset{\sim}{P}$. Lemma IV. 2.04 is a special case of this.

e) Show that if X and Y have the weak topology given by their affine space structures, the product topology on $X \times Y$ coincides with the weak topology given by the product affine space structure.

f) Prove that if X, X' are affine spaces, the map

$$T_{(x,x')} X \times X' \to (T_x X) \oplus (T_{x'} X')$$

$$((x,x'),(y,y')) \rightsquigarrow ((x,y),(x',y'))$$

is an (obviously natural) vector space isomorphism. Show that if $f : X \to Z$, $f' : X' \to Z'$ are maps between affine spaces the map

$$f \times f' : X \times X' \to Z \times Z'$$

$$(x,x') \rightsquigarrow (f(x), f'(x'))$$

is differentiable [respectively, C^∞, C^k] at (x,x') if and only if f and f' are differentiable [respectively, C^∞, C^k] at x and x', and that $\underset{\sim}{D}_{(x,x')}(f \times f')$, where it exists, is the map

$$T_{(x,x')}(X \times X) \cong T_x X \oplus T_{x'} X' \xrightarrow{\underset{\sim}{D}_x f \oplus \underset{\sim}{D}_{x'} f'} T_{f(x)} Z \oplus T_{f'(x')} Z' \cong T_{f \times f'(x,x')}(Z \times Z')$$

where the isomorphisms are as just defined.

2a) If M, N are manifolds modelled on affine spaces X, Y with atlases $\{(U_a \phi_a) | a \in A\}$, $\{(V_b, \psi_b) | b \in B\}$, then if $M \times N$ has the product topology and we define

$$\theta_{ab} : U_a \times V_b \to X \times Y : (u,v) \rightsquigarrow (\phi_a(u), \psi_b(v))$$

apply Ex. 1 to show that $\{(U_a \times U_b, \theta_{ab}) | (a,b) \in A \times B\}$ is an atlas making

$M \times N$ a manifold modelled on $X \times Y$, with $\dim(M \times N) = \dim M + \dim N$.

b) Show that for any $x \in N$, the map $M \to M \times N : y \rightsquigarrow (y,x)$ is smooth.

c) Construct a natural isomorphism $T_{(p,q)} M \times N \to T_p M \oplus T_q N$.

d) If $M = N = S^1$, the unit circle $\{(x,y) \in R^2 | x^2 + y^2 = 1\}$, give charts making S^1 a manifold modelled on the real line and construct a diffeomorphism from $M \times N$ to a torus. (Consider the torus obtained by rotating the circle $\{(x,y,z) \in R^3 | y = 0, (x-2)^2 + z^2 = 1\}$ round the z-axis.)

e) If M,N have metric tensor fields $\underset{\sim}{G}^M$, $\underset{\sim}{G}^N$, show that the <u>product</u> tensor field $\underset{\sim}{G}^{M\times N}$ defined by

$$\underset{\sim}{G}^{M\times N}_{(p,q)}(\underset{\sim}{s}+\underset{\sim}{t},\underset{\sim}{u}+\underset{\sim}{v}) = \underset{\sim}{G}^M_p(\underset{\sim}{s},\underset{\sim}{u}) + \underset{\sim}{G}^N_q(\underset{\sim}{t},\underset{\sim}{v})$$

where $\underset{\sim}{s}+\underset{\sim}{t}$, $\underset{\sim}{u}+\underset{\sim}{v}$ are the decompositions given by c) using the identification mentioned in Ex. 1a), is a metric tensor field on $M\times N$, positive definite if both $\underset{\sim}{G}^M$ and $\underset{\sim}{G}^N$ are.

3a) Show that if Y is a topological space in which every open set is a union of finite intersections of members of a family $\mathcal{V}$ of open subsets of Y, and $f : X \to Y$ is a map from a topological space X, then f is continuous if and only if $f^{\leftarrow}(V)$ is open for every $V \in \mathcal{V}$.

b) Show, by considering what it means to be a member of each set, that

$$(x,y) \in ((\underset{\sim}{D}\phi_a)^{\leftarrow})^{\leftarrow}((\underset{\sim}{D}\phi_b)^{\leftarrow}(W) \iff (x,y) \in (\underset{\sim}{D}\phi_b \circ (\underset{\sim}{D}\phi_a)^{\leftarrow})^{\leftarrow}(W)$$

so that the two sets are equal.

c) Find an atlas of charts on R satisfying Mi) - Miii) except that the $\phi_a^{\leftarrow} \circ \phi_b$, though differentiable, are not <u>continuously</u> so. (Charts on R are just real-valued functions; try an atlas consisting of (I_R, R) and f, where

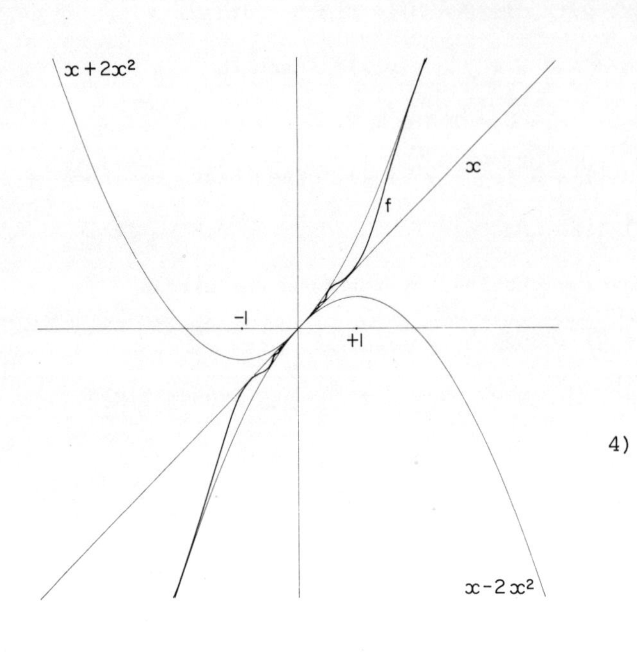

Fig. 3.8

$f(x) = x(1 + |x| + x\sin 1/x)$ (Fig. 3.8).)

Show that Theorem 3.01 fails for R with this atlas, in that the $\underset{\sim}{D}\phi_a$ are not homeomorphisms and the topology that they induce on TR is not even Hausdorff.

4) Show that if M is an n-manifold modelled on an affine space X with vector space T, T_h^kM is an $(n + n^{k+h})$-manifold modelled on $X \times (T_h^k)$, using the charts constructed in 3.01 on TM to construct those on T_h^kM.

4. COMPONENTS

It is at this point that the distinction between natural operations such as taking tensor products, and non-natural ones involving a choice of basis, becomes more than a matter of style. The former can be done smoothly, all over the manifold at once, as we have just seen. A choice of basis often cannot. (This is in contrast to the situation for a single vector space, where we always could choose a basis, and the question was whether it helped us.) A smooth choice of basis for each tangent space means, obviously, a set $\{\underset{\sim}{t}_1,\ldots,\underset{\sim}{t}_n\}$ of smooth vector fields with the property that for each x, $\{\underset{\sim}{t}_1(x),\ldots,\underset{\sim}{t}_n(x)\}$ is a basis for T_xM. In particular, each $\underset{\sim}{t}_i$ must have $\underset{\sim}{t}_i(x) \neq \underset{\sim}{0}$ for all x. Now for the tangent bundle of S^1 this is possible (drawn two ways in Fig. 4.1a) but for the Möbius strip bundle over S^1 (Fig. 4.1b), with the same property of looking like the projection

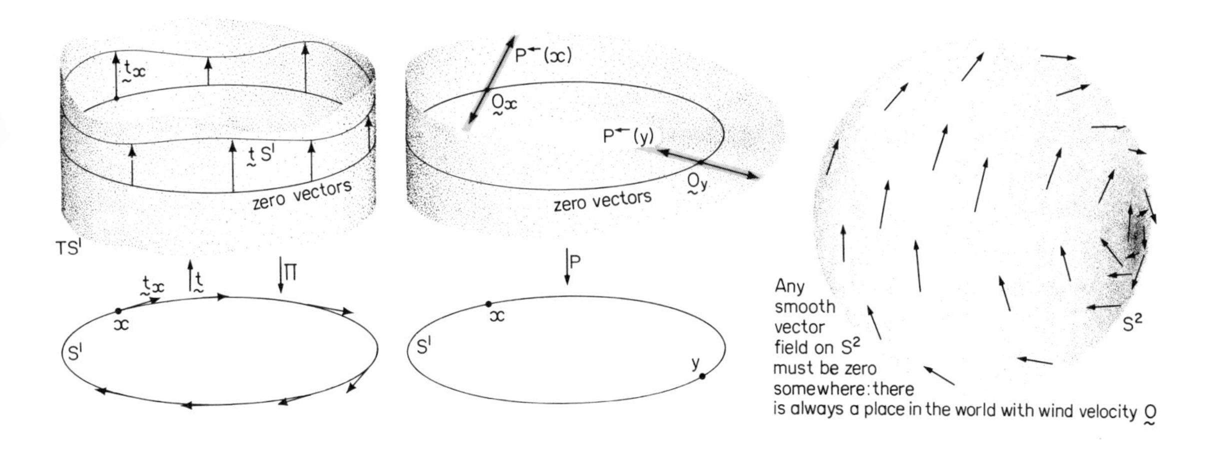

Fig. 4.1

$U \times R \to U$ over small open sets U in S^1, it is not. (This "local product structure" is the main defining property of a "fibre bundle" in general. Since we shall only be concerned hereafter with specific bundles constructed from TM we shall not go into the technicalities of this definition; this example, however, should be clear.) In the case of the tangent bundle to S^2, the non-existence of globally non-zero vector fields (Fig. 4.1c) is known as the Hairy Ball Theorem. (The name is due to the consequence that, for S^2 embedded in R^3, no smooth - or even continuous - choice of non-zero vectors $\underset{\sim}{t}(x)$ in T_xR^3 for each $x \in S^2$ can have all the $\underset{\sim}{t}(x)$ tangent to S^2. If a hair is attached to each point $x \in S^2$, and we take $\underset{\sim}{t}(x)$ as the unit vector along the hair at x, this implies that the coat of hair cannot be everywhere continuously combed flat to S^2. The result also applies to coconuts and dogs, insofar as they are topologically spheres.) The algebraic topology needed to prove the Hairy Ball Theorem is outside the scope of this book, but not very hard; the reader should consult Volume 1 of [Spivak].

A fortiori, there is no smooth choice of basis for $T_x S^2$ at every point $x \in S^2$. The possibility of such a choice is in fact quite a rare one (manifolds for which we can do it are called parallelisable); for instance among compact 2-manifolds it can be done only for the Klein bottle and torus, and among spheres only for S^1, S^3 and S^7.

However, if M is modelled on R^n, which we may suppose without loss of generality (why?), a particular chart $\phi : U \to R^n$ gives coordinate labels $(x^1(u),\ldots,x^n(u))$ to points $u \in U$. Take the standard basis for each tangent space $T_x R^n$ to be $\overleftarrow{d}_x(\mathcal{E})$ (cf. I.1.10), which by a minor abuse of language we shall also denote by $\mathcal{E} = e_1,\ldots,e_n$. Then ϕ gives us (Fig. 4.2) an obvious choice of basis

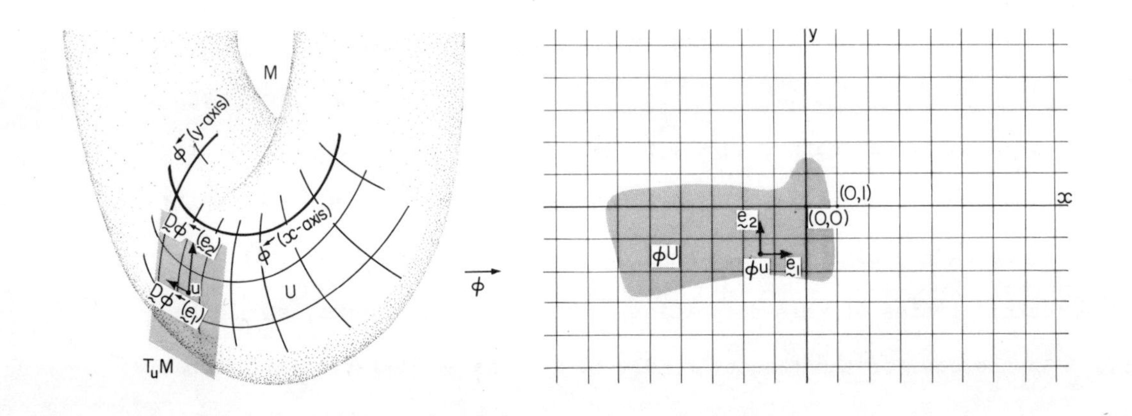

Fig. 4.2

$(D_u\phi)^{\leftarrow}(\mathcal{E}) = (D_u\phi)^{\leftarrow} e_1,\ldots, (D_u\phi)^{\leftarrow} e_n$ (depending of course on the chart, and not defined for tangent spaces outside U) and a corresponding dual basis for $T^*_u M$. These basis vectors have standard symbols, far more convenient than $(D_u\phi)^{\leftarrow}(e_i)$ and $((D_u\phi)^{\leftarrow}(e_i))^*$ and far more suggestive, which we shall now introduce. A little surprisingly the notation for the dual basis is the simpler to explain, and we shall do this first.

4.01 Covariant vectors

The dual basis $\mathcal{E}^*$ to the standard basis $\mathcal{E}$ for R^n consists of the coordinate functions e^i (cf. III.1.06) and hence the dual basis to $(\underset{\sim}{D}_u\phi)^{\leftarrow}(\mathcal{E})$ consists of the composite linear maps $\underset{\sim}{e}^i \circ \underset{\sim}{D}_u\phi$, $i = 1,\ldots,n$. But since the $\underset{\sim}{e}^i$ are linear, $\underset{\sim}{D}_{\phi(u)}(\underset{\sim}{e}^i) = \underset{\sim}{e}^i$ (cf. Ex. 1.3e), and thus

$$\begin{aligned}\underset{\sim}{e}^i \circ \underset{\sim}{D}_u\phi &= \underset{\sim}{D}_{\phi(u)}\underset{\sim}{e}^i \circ \underset{\sim}{D}_u\phi \\ &= \underset{\sim}{D}_u(\underset{\sim}{e}^i \circ \phi) \quad \text{by the Chain Rule (Ex. 1.6)} \\ &= \underset{\sim}{D}_u x^i\end{aligned}$$

since $\phi(u) = (x^1(u),\ldots,x^n(u))$ means exactly that $\underset{\sim}{e}^i \circ \phi = x^i$. Strictly, we are interested in maps $T_uM \to R$, not $T_uM \to TR$, so in the notation of 3.02 the i-th vector in this basis is dx^i. Doing this for each i and each $u \in U$ gives us vector fields $dx^1,\ldots,dx^n$ on U such that any covariant vector field can be written locally - that is, within the part U of M to which the "local choice of coordinates" ϕ applies - as a linear combination

$$\underset{\sim}{v} = v_1dx^1 + \ldots + v_ndx^n = v_idx^i,$$

with the v_i real-valued functions. (Expressed in this way, a covariant vector field is often called a Pfaffian in older books.) "In coordinates", then,

$$\underset{\sim}{v} = (v_1,\ldots,v_n)$$

or v_i for short, with respect to the chart ϕ.

If we are using, for example, (x,y,z) or (r,θ) as labels via the chart, instead of (x^1,x^2,x^3) or (x^1,x^2), we shall correspondingly call these basis covectors dx, dy, dz or dr, dθ and write $\underset{\sim}{v}$ as (v_x, v_y, v_z) or (v_r, v_θ), (cf. V. 1.12).

4.02 Contravariant vectors

If $\underset{\sim}{t}$ is a tangent vector at $u \in M$, we can "differentiate" a function f with respect to $\underset{\sim}{t}$ by taking the directional derivative $\underset{\sim}{D}_u f(\underset{\sim}{t})$, (cf. 1.02). If $\underset{\sim}{t}$ is one of the basis vectors $(\underset{\sim}{D}_u \phi)^{\leftarrow}(\underset{\sim}{e}_i)$ (= $\underset{\sim}{b}_i$ for short) we are interested in, then we agree as in 1.03 to denote $df(\underset{\sim}{b}_i)$ by $\frac{\partial f}{\partial x^i}$ or $\partial_i f$. For any tangent vector $\underset{\sim}{t}$ we have

$$\begin{aligned} \underset{\sim}{D}_u f(\underset{\sim}{t}) &= \underset{\sim}{D}_u f(t^i \underset{\sim}{b}_i) \\ &= (t^i \partial_i) f, \end{aligned}$$

by linearity. Thus we can identify $\underset{\sim}{t}$ with the linear map

$$\partial_{\underset{\sim}{t}} : f \rightsquigarrow df(\underset{\sim}{t})$$

since the correspondence

$$\underset{\sim}{t} \rightsquigarrow \partial_{\underset{\sim}{t}}$$

is both linear and natural, (and injective, since $\partial_{\underset{\sim}{t}} \neq \partial_{\underset{\sim}{t}'}$ for $\underset{\sim}{t}' \neq \underset{\sim}{t}$). Having done so, we have the $\partial_i = \frac{\partial}{\partial x^i}$ as a basis for $T_u M$; by a routine check (Ex. 3) this is precisely the basis to which $dx^1, \ldots, dx^n$ is the dual. (As with the dx^i, we have the ∂_i as fields on U.) Clearly, the ∂_i have their indices rightly placed for contravariant basis vectors; whether the i in $\frac{\partial}{\partial x^i}$ is "up" or "down" is debatable, but we shall regard it as "down" for the sake of the summation convention.

We shall carry this identification to the point of discarding the temporary notation $\partial_{\underset{\sim}{t}}$ just introduced, and simply write $\underset{\sim}{t}(f)$ for $df(\underset{\sim}{t})$: in coordinates

$$\underset{\sim}{t} = t^i \partial_i, \qquad \underset{\sim}{t}(f) = t^i \partial_i f.$$

(Notice that we are not writing ∂_i as $\underset{\sim}{\partial}_i$, though it is a vector field on U. The reason is that we are looking at it not as a vector-valued map $U \to TU$ but as a function-valued map on functions, taking f to $\partial_i f$; thus some inconsistency in notation is inevitable. This is at least consistent with writing dx^i not $\underset{\sim}{d}x^i$ in 4.01, which follows from the use of df not $\underset{\sim}{d}f$ in 3.02.)

As in 4.01, for "named" rather than "numbered" coordinates we shall write, for instance, ∂_x, ∂_y, ∂_z or ∂_r, ∂_θ. (Or $\frac{\partial}{\partial x}$, $\frac{\partial}{\partial y}$, $\frac{\partial}{\partial z}$, etc.)

We have encountered here another of the right ways to construct TM: Each "$\partial_{\underset{\sim}{t}}$" is linear and has the property that

$$\partial_{\underset{\sim}{t}}(fg) \quad = \quad (\partial_{\underset{\sim}{t}} f)g + f(\partial_{\underset{\sim}{t}} g)$$

(Ex. 4). Also, any map δ from the space of smooth real-valued functions on M to itself that is linear and has

$$(\delta(fg))(x) \quad = \quad \delta f(x)g(x) + f(x)\ \delta g(x)$$

for each point $x \in M$, called a <u>derivation</u> on M, turns out - though we shall not prove this - to have

$$(\delta f)x \quad = \quad \partial_{\underset{\sim}{t}(x)} f$$

for some unique vector field $\underset{\sim}{t}$. Given an object corresponding, linearly and naturally, to the collection of vector fields, it is clearly possible to reconstruct the tangent bundle. (This particular construction only works properly for M strictly smooth, i.e. C^{∞} not just C^k for some large k. The difficulty is that differentiating C^{∞} functions gives C^{∞} functions, but C^k only C^{k-1}. This is why we omit the technical details of this approach, and avoid proofs based on it.)

4.03 Tensors of higher degree

The basis for $(T^k_h M)_u$ induced by a chart (U,ϕ), where $\phi(u) = (x^1(u),\ldots,x^n(u))$, is exactly the basis constructed from $\partial_1,\ldots,\partial_n$ and its dual as in V.1.12: Thus the basis for, say, $(T^3_2 M)_u$ is the set of all n^5 tensors at u of the form

$$\partial_i \otimes \partial_j \otimes \partial_k \otimes dx^{\ell} \otimes dx^m$$

where $\{i,j,k,\ell,m\} \subseteq \{1,\ldots,n\}$. Doing this for each u, we have n^5 fields.

We follow convention in abbreviating a tensor field, given as a linear combination

$$\underset{\sim}{w} = w^{ijk}_{\ell m}(\partial_i \otimes \partial_j \otimes \partial_k \otimes dx^{\ell} \otimes dx^m)$$

of these basis tensor fields, to $w^{ijk}_{\ell m}$. For "named" rather than "numbered" coordinates, (x,y,z) not (x^1,x^2,x^3), write w^{xyz}_{yx} for w^{123}_{21} (and do not apply the summation convention).

4.04 Transformation formulae

Recall that a tangent vector $\underset{\sim}{t} \in T_u M$ was formally constructed (Ex. 2.3) as an equivalence class of vectors representing it via charts. It follows that the components t^i of $\underset{\sim}{t}$ in the basis $\partial_1,\ldots,\partial_n$ induced by the chart (U,ϕ), with $\phi(u) = (x^1,\ldots,x^n)$, are exactly those of its representative $\underset{\sim}{D}_u\phi(\underset{\sim}{t}) \in T_{\phi(u)}R^n$ in the standard basis ε for $T_{\phi(u)}R^n \cong R^n$. This is because $\partial_1,\dots,\partial_n$ is exactly $(\underset{\sim}{D}_u\phi)^{\leftarrow}(\mathcal{E})$. (cf. Ex. 2.6b)

If therefore (U',ϕ') is another chart, with $\phi'(u) = (x'^1,\dots,x'^n)$, and t'^i are the components of $\underset{\sim}{t}$ with respect to the basis $\partial'_1,\dots,\partial'_n$ induced by (U',ϕ'), we know by Ex. 2.4 that at any $u \in U \cap U'$,

$$t'^i = t^j \frac{\partial x'^i}{\partial x^j}. \tag{1}$$

This is the <u>transformation formula</u> (or <u>rule</u>, or <u>law</u>) for contravariant vectors and vector fields.

For covariant vector fields, represented in the dual bases, we therefore know the formula by III.1.07 once we know the inverse of the matrix $\left[\frac{\partial x'^i}{\partial x^j}\right]$. But by 1.03 this matrix is just the Jacobian of $\phi'\circ\phi^{\leftarrow}$, which is just the matrix of $\underset{\sim}{D}_{\phi(u)}(\phi'\circ\phi^{\leftarrow})$. Therefore its inverse is the Jacobian of $(\phi'\circ\phi^{\leftarrow})^{\leftarrow} = \phi\circ\phi'^{\leftarrow}$, which is $\left[\frac{\partial x^j}{\partial x'^i}\right]$. Hence the formula we want is

$$v'_i = v_j \frac{\partial x^j}{\partial x'^i}. \tag{2}$$

(One of the more baffling things last century, when people tried to look at $\frac{\partial x^j}{\partial x'^i}$ as the ratio of two "infinitesimals" ∂x^j and $\partial x'^i$, was the way $\frac{\partial x'^i}{\partial x^j}$ is <u>not</u> one over $\frac{\partial x^j}{\partial x'^i}$; it is the Jacobians as whole matrices that are inverse to each other. This means that for instance the chain rule (Ex. 1.6),

$$\frac{\partial x''^i}{\partial x^k} = \frac{\partial x''^i}{\partial x'^j}\,\frac{\partial x'^j}{\partial x^k} \quad \text{in components,}$$

is not the simple cancellation it formally resembles if you do not realise the summation it involves. The room for confusion here is immense - and was fully taken up; it is greatly reduced by starting from the coordinate-free view point and finding components as needed.

Notice that in a "change of variables" from (x^i)'s to (x'^i)'s you are given the (x'^i)'s in terms of the (x^i)'s. This means that for each $(x'^1,\dots,x'^n)$ you are told the corresponding $(x^1,\dots,x^n)$. That is, you have the formula for $\phi\circ\phi'^{\leftarrow}$, not $\phi'\circ\phi^{\leftarrow}$. Differentiating it gives directly what is needed for formula (2), while to apply formula (1) you need to invert the Jacobian at each point. This is even messier than with just one matrix, at one point, to invert, and it was in this context that the words "covariant" and "contravariant" were chosen the way they were, (cf. III.1.07).

Combining (1) and (2) with the discussion of V.1.12, we have immediately the transformation formulae for tensors of all types. For example if $\underset{\sim}{w}$ is a tensor field of type $\binom{3}{2}$,

$$w'^{i'j'k'}_{\ell'm'} = w^{ijk}_{\ell m} \frac{\partial x'^{i'}}{\partial x^i} \frac{\partial x'^{j'}}{\partial x^j} \frac{\partial x'^{k'}}{\partial x^k} \frac{\partial x^\ell}{\partial x'^{\ell'}} \frac{\partial x^m}{\partial x'^{m'}},$$

$$= w^{ijk}_{\ell m}\, \partial_i(x'^{i'})\partial_j(x'^{j'})\partial_k(x'^{k'})\partial'_{\ell'}(x^\ell)\partial'_{m'}(x^m)$$

and so forth for other types (Ex. 5): the old definition of $\binom{3}{2}$-tensor fields.

Sometimes $w'^{i'j'k'}_{\ell'm'}$ is written as $w^{i'j'k'}_{\ell'm'}$, but this is as logically peculiar as writing $[a^j_i]$ for the inverse of $[a^i_j]$ (I.2.08) and for the same reason: what is the difference between " $w^{i'j'k'}_{\ell'm'}$ with $i' = 1$, $j' = 2$, $k' = 1$, $\ell' = 3$, $m' = 2$ " and " $w^{ijk}_{\ell m}$ with $i = 1$, $j = 2$, $k = 1$, $\ell = 3$, $m = 2$ "?

4.05 Raising and lowering indices

If M has a metric tensor field $\underset{\sim}{G}$, then this defines $(\underset{\sim}{G}_x)_\downarrow : T_xM \to T^*_xM$ and its inverse $(\underset{\sim}{G}_x)_\uparrow$ for each x, and maps to "raise and lower indices" (V.1.13) for tensors of higher order at x. These glue together to give maps that alter the variance of tensor fields (Ex. 6).

The results are altered even more drastically by a change of metric tensor field $\underset{\sim}{G}$ than a change of metric tensor alters things in the linear case. For example, two different Riemannian metrics can take the same $\binom{0}{1}$-tensor field to contravariant fields for which the flows (§6) are crucially different. So be wary above all of using one metric tensor to raise indices, and another to lower them (or vice versa). Chalk might become cheese.

Exercises VII.4

1) Let M be a smooth manifold and (ϕ,U) a chart on M with $\phi(u) = (x^1(u),\ldots,x^n(u))$.

a) Show that each dx^i is a smooth field.

b) Prove from this that a covariant vector field $\underset{\sim}{v}$, where $\underset{\sim}{v} = v_i dx^i$ on U, is C^k on U if and only if each $v_i : U \to R$ is C^k on U.

2) If a cotangent vector $\underset{\sim}{v}_u$ at a point $u \in M$ has $\underset{\sim}{v}_u = v_i dx^i(u)$ (notice that the v_i are in this case just numbers, not maps) show that $\underset{\sim}{v}_u$ is the derivative at u of the real-valued function

$$v_i x^i : M \to R : u \leadsto v_1(x^1(u)) + \ldots + v_n(x^n(u)).$$

(This does <u>not</u> imply that a vector <u>field</u> $\underset{\sim}{v}$ with $\underset{\sim}{v}(u) = \underset{\sim}{v}_u$ need be the <u>differential</u> of the function $v_i x^i$, or of any other.)

3) Show that if $\underset{\sim}{t} \in T_uM$, $dx^i(\underset{\sim}{t}) = \partial_{\underset{\sim}{t}}(x^i)$ in the notation of 4.02. Deduce that, writing $\underset{\sim}{t} = t^j\partial_j$

$$dx^i(t^j\partial_j) = t^i$$

and in particular

$$dx^i(\partial_j) = \delta^i_j$$

so that $dx^1,\ldots,dx^n$ and $\partial_1,\ldots,\partial_n$ are dual bases to each other.

4) Deduce from Ex. 1.5b that for any $\underset{\sim}{t} \in T_uM$, and real-valued functions f,g on M with their product fg, defined as usual by $(fg)(u) = f(u)g(u)$ for each u,

$$(d(fg))\underset{\sim}{t} = (df(\underset{\sim}{t}))g(u) + f(u)(dg(\underset{\sim}{t})).$$

Deduce that for $\underset{\sim}{t}$ a vector <u>field</u>, we have the <u>Leibniz Rule</u>

$$(d(fg))\underset{\sim}{t} = (df(\underset{\sim}{t}))g + f(dg(\underset{\sim}{t})).$$

or in the alternative notation of 4.02

$$\underset{\sim}{t}(fg) = \underset{\sim}{t}(f)g + f\underset{\sim}{t}(g).$$

5) Write down the transformation formulae for tensors and tensor fields of types $\binom{1}{2}$, $\binom{0}{2}$ (notice that these latter include the metric tensor fields), and $\binom{3\ 1}{1\ 2}$.

6) Define $\underset{\sim}{G}_{\downarrow} : TM \to T^*M$ (equivalently, $\underset{\sim}{G}_{\downarrow} : M \to T^1_1M$) and $\underset{\sim}{G}_{\uparrow}: T^*M \to TM$ by $\underset{\sim}{G}_{\downarrow}|T_xM = (\underset{\sim}{G}_x)_{\downarrow}$, $\underset{\sim}{G}_{\uparrow}|T^*_xM = (\underset{\sim}{G}_x)_{\uparrow}$ (so that $(\underset{\sim}{G}_{\downarrow})_x = (\underset{\sim}{G}_x)_{\downarrow}$, etc.)

a) Prove that $\underset{\sim}{G}_{\downarrow}$, $\underset{\sim}{G}_{\uparrow}$ are diffeomorphisms.

b) Write down the coordinate formulae for raising and lowering various indices (take your pick, but specify your choice) in tensor fields of type $\binom{3\ 1}{1\ 2}$, using the metric $\underset{\sim}{G}$.

7) Show that if $\phi : U \to R^n$ is a chart, the map

$$T^k_h U \to R^{n+n^{(k+h)}},$$

taking a tensor at x to the n coordinates of x and its own $n^{(k+h)}$ components, is exactly the chart $\underset{\sim}{D}^k_h \phi$ constructed in Ex. 3.4.

5. CURVES

5.01 Definition

A curve or path in a manifold or affine space M is a (smooth unless otherwise stated) map $c : J \to M$, where J is an interval in the real line. The interval may be open or closed, finite or infinite, at either end. If J is $[a,b]$, for some $a<b \in R$, c is a curve from $c(a)$ to $c(b)$.

If for all choices of $x, y \in M$ there is a curve from x to y, M is path-connected. ($R\setminus\{0\}$ for instance is not path-connected as by the Intermediate Value Theorem there is no path from -1 to +1.) We shall include "path-connected", like "smooth", in our concept of a manifold, unless otherwise stated.

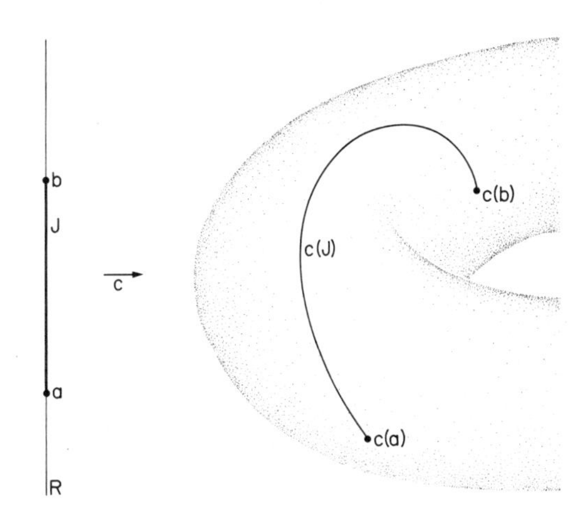

Fig. 5.1

Notice that 5.01 is not the notion of "curve" used in elementary geometry; that refers rather to a set in M, such as the parabola $\{(x,y) | x^2 = y\} \subseteq R^2$. The two curves $f : R \to R^2 : t \leadsto (t,t^2)$ and $g : t \leadsto (2t,4t^2)$ both have this set as image, but are different as maps and therefore as curves, in our sense. In this instance, however, we can "give one in terms of the other" : if $h : R \to R : t \leadsto 2t$, $g = f \circ h$. This leads to

5.02 Definition

If for two curves $f : J \to M$, $g : J' \to M$ there is a continuous [respectively smooth, affine] bijection $J \to J'$ such that $f = g \circ h$, then f is a continuous [respectively smooth, affine] reparametrisation of g. In the special affine case where $h(t) = t+m$, $m \in R$, we shall call f a constant reparametrisation of g.

Two curves need not, however, be reparametrisations of each other even if both injective and with the same image set. For example, consider $f,g : [0,1[\to R^2$ with $f(t) = (\sin 2\pi t, \cos 2\pi t)$, $g(t) = (\cos 2\pi t, \sin 2\pi t)$.

"Curve" does not imply "not straight", even when "straight" is defined in M (which it is not, for a general manifold): an affine map $R \to X$ for X an affine space, for instance, satisfies Definition 5.01. Remember that a mathematical term for which a definition is given means exactly, and only, what it is defined to mean, independently of ordinary language.

We shall generally, as above, use t as the "parameter" (name for a typical point in the domain) of a curve. This is suggested by the notion of a curve c as specifying a motion through the manifold, with position $c(t)$ at time t. Sometimes we want to avoid this suggestion of time involvement; when convenient for this or other reasons we generally replace t by s.

The discussion of maps from T to an affine space in 1.03 was purely local, and hence applies equally to curves in M. If we think of t as "time", the vector $f^*(t) = \underset{\sim}{D}_t f(1)$ introduced there emerges naturally as a "velocity vector". (If this is not transparent, think about writing in coordinates the velocity of a particle moving in R^n.) In general we shall call it <u>the tangent vector to the curve</u> f <u>at</u> t : not "at f(t)", as we might have f(t) = f(t'), but $f^*(t) \neq f^*(t')$ if f crosses itself (Fig. 5.2). This would give <u>two</u> tangent vectors "at f(t)".

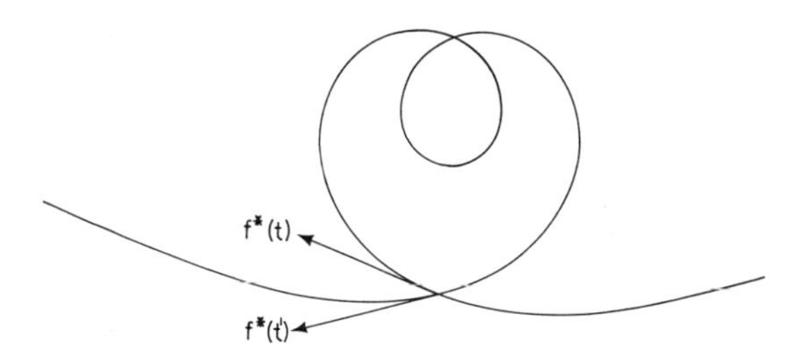

Fig. 5.2

Thus far we have a "velocity" but not a "speed": a non-zero vector in a general vector space V has no "size" except in comparison to others in the same direction, unless V has a metric tensor. Such a tensor for each tangent space, in which the tangent vectors $f^*(t)$ are located, is given by a metric tensor field $\underset{\sim}{G}$, say, on M.

If $\underset{\sim}{G}$ is positive definite (so that M with $\underset{\sim}{G}$ is a Riemannian manifold), $f^*(t).f^*(t)$ is naturally to be thought of as (length of $f^*(t))^2$. This leads us to the idea of

$$\int_a^b \sqrt{f^*(t).f^*(t)}\ dt = \ell, \quad \text{say,}$$

as the length of the whole curve $f : [a,b] \to M$. (In Euclidean space, where one has already a notion of "length" for <u>straight</u> curves, one can show that this integral coincides with the limit obtained by approximating f ever more finely by polygonal curves; cf also Ex. 5) If we define

$$s(k) = \int_a^k \sqrt{f^*(t).f^*(t)}\ dt$$

then s(k) is the length of $f|[a,k]$ and $s : [a,b] \to [0,\ell]$ is a smooth surjective map. If s has a smooth inverse h (which always holds when $f^*(t)$ is not a zero vector for any t, by Thm 1.04) then $g = f \circ h : [0,\ell] \to M$ is a smooth reparametrisation of f, with

$$(\text{length of } g|[0,k]) = k.$$

Such a curve g is <u>parametrized</u> <u>by</u> <u>arc</u> <u>length</u>.

The length of f is infinite in the "negative direction" if the lengths of $f|[a,k]$ increase without bound as a takes lower values in J. (Note that by Ex. 5c an open interval of finite length, like]-1,1[, can have a continuous image of infinite length.) Then we have to choose some $x \in J$ as $\overleftarrow{s}(0)$ and allow negative k. We call such a curve parametrised by arc length if the length of $g|[t,t']$ is $t'-t$ for any t,t' in its domain. For any curve of finite length, then, we have a unique reparametrisation by arc length, while that for a negatively infinite one is unique only up to a constant reparametrisation.

These concerns explain the classical notation used to specify a metric tensor field in coordinates: The length of an arbitrary curve is denoted by s, as above. The curve is written as $(x^1,x^2,\ldots,x^n)$, with these x^i being functions of a suppressed argument t, and $\frac{dx^1}{dt}, \ldots, \frac{dx^n}{dt}$ are as in 1.03. Then

$$\left(\frac{ds}{dt}\right)^2 = g_{ij}\frac{dx^i}{dt}\frac{dx^j}{dt} \qquad \bigstar$$

with the argument of $g_{ij}(x^1(t),\ldots,x^n(t))$ also suppressed. A typical example would be written out as

$$ds^2 = (dx^1)^2 + \tfrac{1}{2}dx^1dx^2 + (dx^2)^2 + ((x^1)^2+1)(dx^3)^2, \qquad **$$

giving the g_{ij} explicitly and "multiplying out" the dt's. The ds on the left was interpreted as the length of an infinitesimal piece of the curve, and called a "line element"; the dx^i were "infinitesimal displacements" in the "infinitesimal time interval" dt. Of course ds/dt was not <u>defined</u> as a ratio of infinitesimals, but as a limit, and until recently infinitesimals were not objects you could safely do algebra with. (For if there is just one "infinity", ∞, $a/ds = \infty = b/dt$ for any a,b non-zero real numbers, so by the ordinary rules of algebra $ds/dt = a/b$ for any a,b. It is <u>not</u> trivial to erect a consistent theory of infinitesimals.) This is a good instance of physics' usage holding onto a highly formal and manipulative - and thus abstract - approach, long after mathematics had developed a language that was geometric, visualisable and essentially concrete.

Sectarian jibes apart, though, it is clear that ** above is sufficient to specify★, and gives $\underset{\sim}{G}_{f(t)}(f^*(t),f^*(t))$ for any f and t. Since any tangent vector can arise as an $f^*(t)$ (a point we elaborate in VIII. §1) we have $\underset{\sim}{G}_x(\underset{\sim}{v},\underset{\sim}{v})$ for any $x \in M$, $\underset{\sim}{v} \in T_xM$, and hence $\underset{\sim}{G}_x(\underset{\sim}{u},\underset{\sim}{v})$ for $\underset{\sim}{u},\underset{\sim}{v} \in T_xM$ by the polarisation identity (Ex. IV.1.7d). Thus specifying the "line element" ds^2 gives the metric tensor, in a single equation rather than a separate formula for each g_{ij}. If the chart used makes the ∂_i everywhere orthogonal (though they cannot in general be ortho<u>normal</u>, as we see in Ch. X) it is much the most succinct way to write down a particular metric tensor in coordinates, and we shall use it freely.

If $\underset{\sim}{G}$ is not positive definite, the "length" $\sqrt{\underset{\sim}{G}_x(\underset{\sim}{v},\underset{\sim}{v})}$ for a vector at x is not a very practical quantity (cf. IV.1.04) and we do better to use $\| \ \|_{\underset{\sim}{G}_x}$ (IV.1.06). Even with this we do well to restrict the kinds of curve we examine.

<u>5.03 Definition</u>

A curve $f : J \to M$ in a pseudo-Riemannian manifold is

(i) <u>timelike</u> if $\underset{\sim}{G}(f^*(t),f^*(t)) > 0, \quad \forall t \in J$

(ii) <u>null</u>, or <u>lightlike</u> if $\underset{\sim}{G}(f^*(t),f^*(t)) = 0, \quad \forall t \in J$

(iii) spacelike if $\underset{\sim}{G}(f^*(t), f^*(t)) < 0, \quad \forall t \in J$

(iv) like if timelike, spacelike, or null.

We shall generally be interested in the length only of like curves.

5.04 Definition

The length of a curve $f : J \to M$ is

$$L(f) = \int_J \|f^*(t)\|_{\underset{\sim}{G}_{f(t)}} dt$$

which may - though need not - be infinite if J is not compact (Ex. 5c.).

Since the definition of a like curve requires that $f^*(t)$ be never zero, we may extend the discussion above to get reparametrisations of such curves by arc length. The length of a null curve, however, is automatically zero; since $[0,0]$ is just a point, we cannot therefore parametrise a null curve by arc length. In Chapters XI and XII we shall use σ to denote arc length for timelike curves.

5.05 Note

We have defined differentiation rigorously, but not integration. We cannot treat integration in general without introducing differential forms, which we do not cover in this volume, but for our uses of it in one dimension the following will suffice ("integral as anti-differential").

An indefinite integral of a function $f : J \to R$ is a differentiable function $g : J \to R$ such that

$$\frac{dg}{ds}(t) = f(t), \quad \forall t \in J.$$

If f has such an indefinite integral, the definite integral

$$\int_a^b f(t)dt$$

of f from a to b ($a,b \in R$) is defined as $(g(b)-g(a))$. (cf. Ex. 2) If J is a closed interval $[a,b]$ we write also

$$\int_J f(t)dt = \int_a^b f(t)dt.$$

If J is a non-closed interval (say, R or $[0,1[$), we choose a decreasing sequence a_n and an increasing sequence b_n such that every $x \in J$ is in $[a_n, b_n]$ for some n (say, $a_n = -n$, $b_n = +n$ for R, or $a_n = 0$, $b_n = 1-1/n$ for $[0,1[$). Then the definite integral of f over J is defined as $\lim_{n\to\infty} (g(b_n) - g(a_n))$ if this limit exists and is the same for all choices of sequences a_n, b_n (cf. Ex. 3). Otherwise, the integral is non-existent, or <u>divergent</u>. If for any $x \in R$, however large, there is a subinterval J' of J such that for any closed subinterval $[a,b]$ containing J' we have

$$\int_a^b f(t)dt > x$$

then the integral is <u>positively infinite</u> and similarly for <u>negatively infinite</u>.

If f has an indefinite integral we say f is <u>integrable</u>.

The proof of the existence by these definitions of the integral we have called "length" can be direct only when the image f(J) of a curve is a subset of an affine straight line (Ex. 5). The reason is the essential triviality of <u>defining</u> integration as the reverse of differentiation. Apart from applying only to continuous functions, it does not say anything about the way an integral is a glorified sum. (The very symbol $\int$ is just an olde Engliſhe "s" for "sum"). The integral of a function is more fundamentally the "area under the curve" (Fig. 5.3a)

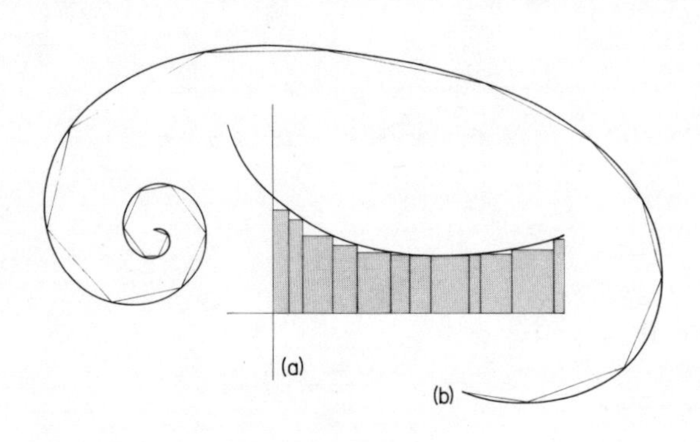

Fig. 5.3

defined as a limit of approximating sums of rectangular areas. The length of a curve is the limit of the sums of the lengths of the straight bits in a polygonal approximation (Fig. 5.3b), and so on. Then the fact that the integral exists, and is an "anti-differential" when f is continuous, says something significant, and requires work to prove. Any Maths student reading this has a proof among his first Analysis lecture notes. Physics students with the mathematician's hunger for a proof are referred to any reliable introductory Real Analysis text, for example [Moss and Roberts].

Exercises VII.5

1a) If $f : [a,b] \rightarrow R$ is C^1 at $t \in]a,b[$, and $f(s) \leq f(t)$ for all $s \in [a,b]$, show that $f^*(t) = \underset{\sim}{0}$. (If not, $\underset{\sim}{D}_t f$ is injective. Apply 1.05 to contradict the assumed maximum for f at t.)

b) Deduce that f^* similarly is $\underset{\sim}{0}$ at a differentiable minimum, strict (so $f(s) > f(t)\ \forall s \neq t$), or otherwise.

c) If $f : [a,b] \rightarrow R$ is C^1 and $f(a) = f(b) = k$ show that f has a maximum or a minimum at some $t \in]a,b[$. (If $f([a,b]) = \{k\}$, set $t = \frac{1}{2}(a+b)$; if not, use VI.4.11.)

d) Deduce from a)-c) that if $f : [a,b] \rightarrow R$ is C then there exists $t \in]a,b[$ with $\underset{\sim}{D}_t f = \underset{\sim}{0}$.

e) [Not used in the book.] Extend d) to the case that f is differentiable but not C^1. (Replace Theorem 1.05 which fails, by a proof that an injective derivative at t for $f : R \rightarrow R$ implies that f takes values on both sides of $f(t)$.)

2a) Deduce from 1c) that if $f : [a,b] \to R$ is C^1 then $f(a)\neq f(b) \Rightarrow f^*(t)\neq \underset{\sim}{0}$ for some $t \in]a,b[$.

b) Deduce that if $f^*(t) = \underset{\sim}{0}$ for all $t \in]a,b[$, then f is constant.

c) Deduce that if f,g are two indefinite integrals for a continuous function $h : J \to R$, J any interval, then $g(t) = h(t)+m \quad \forall t$, where m is some real constant.

d) Deduce that any indefinite integral for h gives the same definite integral from any a to any b, if h is defined everywhere in [a,b].

e) If $h(t) = -1/t^2$, $g(t) = 1/t$, $f(t) = 1/t + t/|t|$, then

$$\frac{df}{dt} = \frac{dg}{dt} = h$$

wherever f, g and h are defined. (Thus two "anti-differentials" need <u>not</u> in general differ everywhere by the same constant. The rule "one integration, one constant" is valid only if the domain of definition of the functions involved is path-connected, cf. 5.01.)

3) Assuming that sin is an indefinite integral for cos on R, show that cos has no integral over all R by producing sequences $a_n \to -\infty$, $b_n \to +\infty$ such that

(i) $(\sin(b_n) - \sin(a_n)) = k \ \forall n$, for any given constant $k \in [-2,2]$, or

(ii) $\lim_{n\to\infty} (\sin(b_n) - \sin(a_n))$ does not exist.

4) If $f : J \to R$ is integrable and $a,b,k \in J$, show that

$$\int_a^b f(t)dt + \int_b^k f(t)dt = \int_a^k f(t)dt.$$

5a) Show that if $f : J \to R$ is C^1 and injective, the length of f using Definition 5.04 with the usual Riemannian metric on R is exactly the length of the set

$f(J)$, defined in the usual way. (Just combine definitions.)

b) Deduce that if f is a smooth injective curve in Euclidean space whose image is a subset of a straight line, then even if f is not affine the length defined in 5.04 coincides with the usual length of the set $f(J)$.

c) The curve $f :]-1,1[\to R : t \rightsquigarrow t/(1-t^2)$ has infinite length, and $g : R \to R : t \rightsquigarrow t^2/(1+t^2)$ has length 2. (Since g is not injective, apply Ex. 4)

6. VECTOR FIELDS AND FLOWS

6.01 Examples

Given a tangent vector at each point in a region U, it is natural to try to "join the arrows up" : fill U with curves, so that at every point x the curve through x is in the direction pointed by the vector at x. Familiar examples are the "lines of force" defined by a magnetic field (the vector field being defined at each point by the effect on a hypothetical "free north pole"), and the "stream lines" defined by the velocity vector field of a moving fluid. In steady flow, the stream lines are realised physically as the paths followed by particles in the fluid. Moreover if we express such a movement by a curve c in U with $c(t)$ = position at time t, the velocity at the point $c(t)$ is exactly the tangent vector $c^*(t)$, not merely in the same direction. Can we produce such a set of curves for an arbitrary vector field?

First, let us consider some examples.

On R, if we have a vector field

$$\underset{\sim}{v}(x) = \left(x^2 + \frac{2x^4}{1+2x^2-\sqrt{1+4x^2}}\right) \underset{\sim}{e}_1(x) = v(x)\underset{\sim}{e}_1(x) \quad \text{for short,}$$

the only curve c with $c^*(t) = \underset{\sim}{v}(c(t))$ is

$$c :]-1,1[\rightarrow R : t \rightsquigarrow t/(1-t^2),$$

up to a constant change in parameter or restriction to a small domain (Ex. 1a). There is no way to extend c to a continuous map with domain all of R.

So in general we cannot expect to do better than find a curve $c :]-\varepsilon,\varepsilon[\rightarrow U$ with $c(0)$ a given $x \in M$, $c^*(t) = \underset{\sim}{v}(c(t))$ for all $t \in]-\varepsilon,\varepsilon[$, for some ε.

Moreover, we cannot expect to use the same ε for the curves through all the different points in M. Let M be R^2 and put (cf. 4.01, 4.02 for notation)

$$\underset{\sim}{w}(x,y) = v((1+y^2)x)\partial_x + 0\partial_y$$

with the function v as before. Then the unique curve c through $(0,y_0)$ with $c^*(t) = \underset{\sim}{v}(c(t))$, (again up to a constant or restriction) is, by Ex. 1b,

$$c :]-\varepsilon,\varepsilon[\rightarrow R^2 : t \rightsquigarrow \left(\frac{\varepsilon t}{\varepsilon^2-t^2}, y_0\right), \quad \text{where } \varepsilon = 1/(1+y_0^2).$$

Thus no one $\varepsilon > 0$ will do for all points, since for any given choice a large enough y_0 requires a smaller one. The best we can expect in general is a result local both in R (curves with limited domain) and in M (the choice of ε depending on where in M we are).

The reader should have noticed by now that we are talking about solutions of <u>differential equations</u>. With the field $\underset{\sim}{v}$ on R^2 above, for example, the equation $c^*(t) = \underset{\sim}{v}(c(t))$ is equivalent to

$$\frac{dc^1}{dt} = v, \quad \frac{dc^2}{dt} = 0$$

in the notation of 1.03. Hence we use the language:-

6.02 Definition

A solution curve, or integral curve of a vector field $\underset{\sim}{v}$ on a manifold M or on a region U in M is a curve $c : J \to M$ such that $c^*(t) = \underset{\sim}{v}(c(t))$, $\forall t \in J$.

A first order differential equation on M is a vector field on M.

Thus differential equations are as essentially geometric as linear algebra, though from many treatments you would guess it for neither. An excellent, highly pictorial (and cheap) introduction to the geometric point of view on differential equations is [Schwarzenberger (1)], based on a first-year undergraduate course.

"Solving a differential equation for given initial conditions $x^i = x_0^i$ at time $t = 0$" now translates exactly into finding a solution curve c of a vector field $\underset{\sim}{v}$, with $c(0)$ the point x_0 labelled $(x_0^1,\ldots,x_0^n)$ by the chart being used. In any real calculation we do not know x_0 exactly, so exact solutions for time t are worthless unless $c(t)$ depends continuously on the x_0 through which the curve c is required to pass (compare VI §1.) Very conveniently, if $\underset{\sim}{v}$ is continuously differentiable, it always does. First we define a flow.

6.03 Definition

A C^k local flow for a vector field $\underset{\sim}{v}$ on M is a C^k map

$$\phi : U \times]-\varepsilon,\varepsilon[\to M$$

where U is an open set in M and ε a positive real number, such that

(i) $\phi(y,0) = y$, $\forall y \in U$.

(ii) For any $y \in U$, if we set $c(t) = \phi(y,t)$ for $t \in]-\varepsilon,\varepsilon[$ then $c :]-\varepsilon,\varepsilon[\to M$ is a solution curve of $\underset{\sim}{v}$.

The local flow is on U, and is around any $x \in U$. We now have the language to state

6.04 Theorem

If $\underset{\sim}{v}$ is a C^k vector field on a manifold M, there is such a C^k local flow for $\underset{\sim}{v}$ around every $x \in M$, which is unique in the following sense:

(i) If $\phi' : U' \times]-\varepsilon',\varepsilon'[\to M$ is another local flow for $\underset{\sim}{v}$, then setting $\varepsilon'' = \min(\varepsilon,\varepsilon')$ and $U'' = U \cap U'$ we have

$$\phi|U'' \times]-\varepsilon'',\varepsilon''[= \phi'|U'' \times]-\varepsilon'',\varepsilon''[.$$

So ϕ and ϕ' agree where they are both defined.

(ii) If f is a solution curve with $f(t) = x$, then $f(s) = \phi(x,s-t)$ whenever both sides are defined. (Thus there is essentially just <u>one</u> solution curve through x, up to constant reparametrisation. This need not be true if $\underset{\sim}{v}$ is merely continuous : cf. Ex. 2.)

We give a recent, simple geometric proof of this theorem in the Appendix. Our use of the result depends only on what it says, not how it is proved, so the reader will miss nothing essential to the rest of the book by taking the theorem on trust or from a proof already encountered - though he <u>will</u> miss a nice proof. ■

6.05 Corollary

Let $\phi : U \times J \to M$ be a C^k local flow for $\underset{\sim}{v}$. If for $t \in J$ we define

$$\phi_t : U \to M : x \rightsquigarrow \phi(x,t)$$

then $\phi_{t+s} = \phi_t \circ \phi_s$ whenever t, s and t+s are all in J.

Proof

Let $f(t) = \phi_t(\phi_s(x)) = \phi(\phi_s(x),t)$. Then f is a solution curve of $\underset{\sim}{v}$ and hence by the theorem a constant reparametrisation of the solution curve g defined by $g(r) = \phi(x,r)$. Since $f(0) = g(s)$, we must therefore have

$$f(t) = g(t+s), \text{ and hence}$$

$$\begin{aligned}(\phi_t \circ \phi_s)(x) &= f(t) \\ &= g(t+s) \\ &= \phi_{t+s}(x).\end{aligned}$$

6.06 Corollary

Each set $\phi_t(U)$ is an open set in M, and, giving U and $\phi_t(U)$ the differential structure restricted from M, each map

$$\phi_t : U \to \phi_t(U)$$

is a diffeomorphism.

Proof

By Ex. 3.2b each map $i_t : U \to U \times J : x \mapsto (x,t)$ is smooth, so the composites $\phi_t = \phi \circ i_t$ are C^k.

By 6.05 $\phi_t \circ \phi_{-t}(x) = \phi_0(x) = x$, so each ϕ_t has the C^k inverse ϕ_{-t} (modulo minor technicalities about domains of definition). The result follows. ■

We shall confine our attention largely to local flows where $\underset{\sim}{v}$ is non-zero on U, as we shall not be needing results on the behaviour of flows around zeros of $\underset{\sim}{v}$.

(Some samples of the latter are shown in Fig. 6.1; the study of these is a large

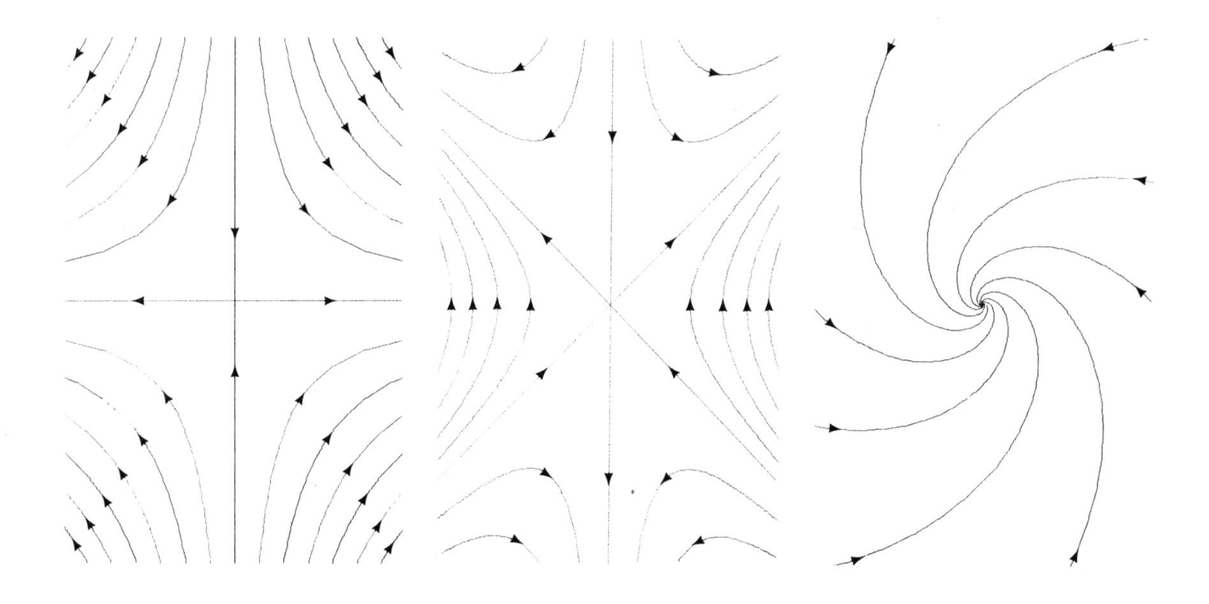

Fig. 6.1

part of the theory of dynamical systems.) For such non-zero fields we have the following "straightening out" locally for a flow.

6.07 Lemma

Let M be a manifold modelled on an affine space X with vector space T.

If $\underset{\sim}{v}$ is a smooth vector field on M and $x \in M$ has $\underset{\sim}{v}(x) \neq \underset{\sim}{0}$, then there is a local flow $\phi : U \times J \to M$ for $\underset{\sim}{v}$ around x, a chart $\psi : U \to X$, and a vector $\underset{\sim}{w} \in T$, such that

$$\psi(\phi(y,t)) \quad = \quad \psi(\phi(y,0)) + t\underset{\sim}{w}$$

whenever both sides are defined. (Thus the flow looks locally like a family of translations in the direction of $\underset{\sim}{w}$, by this chart.)

Proof

By continuity, x has a neighbourhood V_1 on which $\underset{\sim}{v}$ is non-zero; by 6.04 $\underset{\sim}{v}$ has a local flow $\theta : V_2 \times J$ around x; by the definition of a manifold there is a chart $\alpha : V_3 \to X$. Let $\alpha(x) = y$, $\underset{\sim}{D}\alpha(\underset{\sim}{v}(x)) = \underset{\sim}{w}$. Choose a linear functional $\underset{\sim}{f} : T \to R$ with $\underset{\sim}{f}(\underset{\sim}{d}_{\alpha(x)}(\underset{\sim}{w})) > 0$; by continuity x has a neighbourhood $V_4 \subseteq V_3$ with $\underset{\sim}{f}(\underset{\sim}{d}_{\alpha(y)}(\underset{\sim}{D}\alpha(\underset{\sim}{v}(x))))$ for all $z \in V_4$. Let $V_1 \cap V_2 \cap V_4 = V$, and denote the restrictions of θ to $V \times J$ and α to V likewise by θ and α. We now have the situation of Fig. 6.2, where all vectors $\underset{\sim}{v}(z) \in T_zM$ with $\alpha(z)$ in the affine hyperplane $K = y + \ker\underset{\sim}{f}$ are carried to vectors $\underset{\sim}{D}\alpha(\underset{\sim}{v}(z))$ pointing across K in the

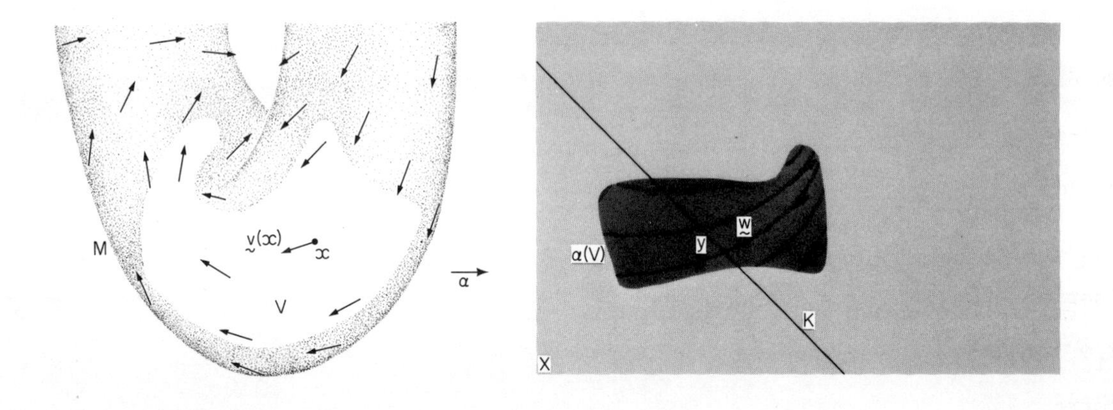

Fig. 6.2

same way. Thus no solution curve in V crosses $\overleftarrow{\alpha}(K)$ more than once. So if we set $U = V \cap \phi(\overleftarrow{\alpha}(K) \times J)$ and

$$\eta : (\alpha(U) \cap K) \times J \to M : (k,t) \rightsquigarrow \phi_t(\overleftarrow{\alpha}(k))$$

we have a local inverse $\gamma : U \to K \times J$ (that is, $\gamma(\eta(k,t)) = (k,t)$ when defined) with $\gamma|\overleftarrow{\alpha}(K) = \alpha|\overleftarrow{\alpha}(K)$. Since η is evidently smooth and $D\eta$ always invertible, γ is also smooth. Define $\delta(k,t) = k+t\underset{\sim}{w}$ and $\psi = \delta\circ\gamma$. Then $\phi|U \times J$, ψ and $\underset{\sim}{w}$ satisfy the conditions above. ∎

Exercises VII.6

1a) Show that if $c(t) = t/(1-t^2)$, then $c^*(t) = \underset{\sim}{v}(c(t))$, where $\underset{\sim}{v}$ is the vector field on R introduced in 6.01.

b) Show that the curves in R^2 given in 6.01 are solutions of $c^*(t) = \underset{\sim}{w}(c(t))$.

2) If the vector field $\underset{\sim}{v}$ on R has $\underset{\sim}{v}(x) = x^{\frac{2}{3}}\underset{\sim}{e}_1(x)$, show that $\underset{\sim}{v}$ is not differentiable and that $c(t) = \frac{1}{3}t^3$ and $c(t) = 0$ are both solution curves for $\underset{\sim}{v}$ through 0.

7. LIE BRACKETS

By 6.07 we can make one vector field look locally like translation, where it is non-zero. It will be important later to know when we can do it for several vector fields at once. Evidently this need not be true in general, since translations always commute $((x+\underset{\sim}{t})+\underset{\sim}{s} = (x+\underset{\sim}{s})+\underset{\sim}{t})$, while there is no reason for flows to. (For instance if M is R^2, $\phi((x,y),t) = (x+ty,y)$ and $\psi((x,y),t) = (x,y+t)$ we have $\phi_1\psi_1(0,0) = (0,1)$ while $\psi_1\phi_1(0,0) = (1,1)$. To what vector fields do ϕ and ψ correspond?) It turns out that there is a purely local condition on the vector fields which decides the question for the flows.

We have mentioned, in 4.02, the view of a vector field $\underset{\sim}{v}$ as a derivation on functions : $(\underset{\sim}{v}(f))(x) = df(\underset{\sim}{v}(x))$. If we have two vector fields $\underset{\sim}{v}$, $\underset{\sim}{w}$ we may or may not have $\underset{\sim}{v}(\underset{\sim}{w}(f)) = \underset{\sim}{w}(\underset{\sim}{v}(f))$ for all functions f. (Consider the vector fields of the flows ϕ,ψ above, and the function $f(x,y) = x+y$.) It turns out that we do have this property exactly when the corresponding flows commute.

7.01 Definition

The Lie bracket or commutator of two vector fields $\underset{\sim}{v},\underset{\sim}{w}$ on a manifold M is the unique vector field, denoted $[\underset{\sim}{v},\underset{\sim}{w}]$, such that

$$[\underset{\sim}{v},\underset{\sim}{w}](f) = \underset{\sim}{v}(\underset{\sim}{w}(f)) - \underset{\sim}{w}(\underset{\sim}{v}(f)), \text{ for all smooth } f : M \to R.$$

If we had shown that every derivation corresponds to a unique vector field, this would guarantee the existence and uniqueness of $[\underset{\sim}{v},\underset{\sim}{w}]$, since

$$\begin{aligned}\underset{\sim}{v}(\underset{\sim}{w}(fg)) - \underset{\sim}{w}(\underset{\sim}{v}(fg)) &= \underset{\sim}{v}(g\underset{\sim}{w}(f) + f\underset{\sim}{w}(g)) - \underset{\sim}{w}(g\underset{\sim}{v}(f) + f\underset{\sim}{v}(g)) \\ &= \underset{\sim}{v}(g)\underset{\sim}{w}(f) + g\underset{\sim}{v}(\underset{\sim}{w}(f)) + \underset{\sim}{v}(f)\underset{\sim}{w}(g) + f\underset{\sim}{v}(\underset{\sim}{w}(g)) \\ &\quad - \underset{\sim}{w}(g)\underset{\sim}{v}(f) - g\underset{\sim}{w}(\underset{\sim}{v}(f)) - \underset{\sim}{w}(f)\underset{\sim}{v}(g) - f\underset{\sim}{w}(\underset{\sim}{v}(g)) \\ &= g(\underset{\sim}{v}(\underset{\sim}{w}(f)) - \underset{\sim}{w}(\underset{\sim}{v}(f))) + f(\underset{\sim}{v}(\underset{\sim}{w}(g)) - \underset{\sim}{w}(\underset{\sim}{v}(g)))\end{aligned}$$

so we have a new derivation. As it is, the most direct method is to use coordinates (Ex. 2). (Note that $f \leadsto \underset{\sim}{v}(\underset{\sim}{w}(f))$ does not generally give a derivation - consider, say, the examples above on R^2 - and so cannot correspond to a vector field. It is very special that $f \leadsto \underset{\sim}{v}(\underset{\sim}{w}(f)) - \underset{\sim}{w}(\underset{\sim}{v}(f))$ does.)

7.02 Theorem

Let $\underset{\sim}{v},\underset{\sim}{w}$ be vector fields defined in a region U of a manifold M, with local flows ϕ, ψ on U for $\underset{\sim}{v}$ and $\underset{\sim}{w}$. Then ϕ and ψ satisfy

$$\phi_t \circ \psi_s = \psi_s \circ \phi_t$$

wherever defined if and only if

$$[\underset{\sim}{v},\underset{\sim}{w}]_x = \underset{\sim}{0}$$

for all $x \in U$.

Proof

If $\phi_t \circ \phi_s = \psi_s \circ \phi_t$ everywhere, for any function f on U then

$$\underset{\sim}{v}(\underset{\sim}{w}(f))-\underset{\sim}{w}(\underset{\sim}{v}(f)) = \underset{\sim}{v}\left(\lim_{s\to 0}\left(\frac{f\circ\psi_s - f}{s}\right) - \underset{\sim}{w}\lim_{t\to 0}\left(\frac{f\circ\phi_t - f}{t}\right)\right) \qquad \text{(Ex. 3)}$$

$$= \lim_{t\to 0}\left(\frac{\left(\lim_{s\to 0}\left(\frac{f\circ\psi_s - f}{s}\right)\right)\circ\phi_t - \lim_{s\to 0}\left(\frac{f\circ\psi_s - f}{s}\right)}{t}\right)$$

$$- \lim_{s\to 0}\left(\frac{\left(\lim_{t\to 0}\left(\frac{f\circ\phi_t - f}{t}\right)\right)\circ\psi_s - \lim_{t\to 0}\left(\frac{f\circ\phi_t - f}{t}\right)}{s}\right)$$

$$= \lim_{(s,t)\to(0,0)}\left(\frac{f\circ\psi_s\circ\phi_t - f\circ\phi_t - f\circ\psi_s + f - f\circ\phi_t\circ\psi_s + f\circ\psi_s + f\circ\phi_t - f}{st}\right)$$

$$= 0.$$

Conversely, if at $x \in M$ both $\underset{\sim}{v}$ and $\underset{\sim}{w}$ are zero, then $\phi_t(x) = \psi_s(x) = x,\ \forall s,t,$ so the result if trivial. If, say, $\underset{\sim}{v}(x) \neq \underset{\sim}{0}$ we have by continuity a neighbourhood of x in which $\underset{\sim}{v}$ is non-zero and, applying 6.07 in coordinate form, a chart θ in which $\underset{\sim}{v} = \partial_1$, $\phi_t(x^1,\ldots,x^n) = (x^1+t,\ldots,x^n)$. If $[\underset{\sim}{v},\underset{\sim}{w}] = \underset{\sim}{0}$, then

$$0 = v^i\partial_i w^j - w^i\partial_i v^j, \quad \forall_j, \quad \text{by Ex. 2a.}$$

That is, $0 = \partial_1 w^j, \quad \forall j,$

Fig. 7.1

so the w^j are constant in the x^1-direction. Hence for any solution curve $c = (c^1,\ldots,c^n)$ for $\underset{\sim}{w}$, $\tilde{c} = (c^1+t,c^2,\ldots,c^n)$ is also a solution curve where it lies in the range of the chart θ. Hence $\psi_s(x+t\underset{\sim}{e}_1) = \psi_s(x)+t\underset{\sim}{e}_1$, i.e. $\psi_s\circ\phi_t = \phi_t\circ\psi_s$.

■

7.03 Language

The equation $[\underset{\sim}{v},\underset{\sim}{w}] = \underset{\sim}{0}$ is thus the "infinitesimal version" of $\phi_t\circ\psi_s = \psi_s\circ\phi_t$. We say that $\underset{\sim}{v}$ and $\underset{\sim}{w}$ commute in a region U when their Lie bracket vanishes there. A larger set of vector fields is said to commute if any two of them do.

$[\underset{\sim}{v},\underset{\sim}{w}]$ can in fact be defined as the infinitesimal failure of ϕ and ψ to commute, just as df/dt is the infinitesimal failure of f to be constant. In an affine space this takes the form

$$[\underset{\sim}{v},\underset{\sim}{w}]_x = \overleftarrow{\underset{\sim}{d}}_x\left(\lim_{h\to 0}\frac{\underset{\sim}{d}(x,\ \psi_{-s}\phi_{-t}\psi_s\phi_t(x))}{h^2}\right)$$

which clearly vanishes if $\phi_t\circ\psi_s = \psi_s\circ\phi_t$ always, since this implies $\psi_{-s}\phi_{-t}\psi_s\phi_t = \psi_{-s}\psi_s\phi_{-t}\phi_t = I$ where defined. In a general manifold the equivalent definition is a little more complicated.

We shall omit the proof that this definition is equivalent to 7.01, as we shall not need to use it (indeed, we have yet to see a use for it except as motivation.)

7.02 gives us a result that will be crucial when we come to decide which spaces are intrinsically "curved" in Chapter X.

7.04 Theorem

Suppose around a point x in an n-manifold M we have n vector fields $\underset{\sim}{v}_1,\ldots,\underset{\sim}{v}_n$ such that for all y in a neighbourhood U of x we have $\underset{\sim}{v}_1(y),\ldots,\underset{\sim}{v}_n(y)$ linearly independent and $[\underset{\sim}{v}_i,\underset{\sim}{v}_j]_y = \underset{\sim}{0}\ \ \forall\, i,j = 1,\ldots,n$.

Then there is a chart $\psi : U \to R^n$ around x with respect to which $\underset{\sim}{v}_i = \partial_i$, $i = 1,\ldots,n$ and the corresponding flows $\phi^1,\ldots,\phi^n$ have

$$\phi^i_t(x^1,\ldots,x^n) = (x^1,\ldots,x^i+t,\ldots,x^n) \quad \forall i,t.$$

Proof

If $\phi^1,\ldots,\phi^n$ are defined on $V_i \times J_i$, $i = 1,\ldots,n$ let

$$\theta(t^1,\ldots,t^n) = \phi^1_{t^1}\,\phi^2_{t^2}\cdots\phi^n_{t^n}(x)$$

where defined, and denote its domain of definition by $J \subseteq J_1 \times \ldots \times J_n \subseteq R^n$. Evidently J is open, and θ is smooth by the smoothness of the ϕ^i. Now $\underset{\sim}{D}\theta$ takes the vector $\underset{\sim}{e}_i(t^1,\ldots,t^n)$ to $c^*(0)$ at $\theta(t^1,\ldots,t^n)$, where we set

$$\begin{aligned} c(s) &= \phi^1_{t^1}\cdots\phi^i_{t^i+s}\cdots\phi^n_{t^n}(x) \\ &= \phi^i_s(\phi^1_{t^1}\cdots\phi^i_{t^i}\cdots\phi^n_{t^n})(x) \qquad \text{by 6.05 and 7.02} \\ &= \phi^i_s(\phi(t^1,\ldots,t^n)) \end{aligned}$$

so c is a solution curve for $\underset{\sim}{v}_i$ through $\theta(t^1,\ldots,t^n)$, and hence $c^*(0) = \underset{\sim}{v}_i(\theta(c(0)))$.

In particular, $\underset{\sim}{D}\theta$ takes the standard basis $\underset{\sim}{e}_1,\ldots,\underset{\sim}{e}_n$ for $T_{(0,\ldots,0)}R^n$ to $((\underset{\sim}{v}_1)_x,\ldots,(\underset{\sim}{v}_n)_x)$. That is a linearly independent subset of the n-dimensional space $T_x M$, by assumption, so $\underset{\sim}{D}\theta_{(0,\ldots,0)}$ is an isomorphism. Hence by the Inverse Function Theorem (1.04) there is a neighbourhood U of x and a local diffeomorphism $\psi : U \to R^n$ with $\psi \circ \theta = I_{\psi(U)}$, which can be used as a chart. With respect to this chart, $\underset{\sim}{v}_i = \partial_i$, $i = 1,\ldots,n$ as required.

Notice that by Ex. 2c this theorem gives a necessary as well as sufficient condition for $\underset{\sim}{v}_1,\ldots,\underset{\sim}{v}_n$ to have a realisation as $\partial_1,\ldots,\partial_n$ in some chart. ∎

Exercises VII.7

1a) Show that if for $f : R^n \to R$ the partial derivatives $\partial_i f$, $\partial_j f$ and $\partial_i(\partial_j f)$ exist and are continuous, then so does $\partial_j(\partial_i f)$ and it is equal to $\partial_i(\partial_j f)$. Hint : show that both are equal to

$$\lim_{(h,k)\to(0,0)} \frac{\begin{pmatrix} f(x^1,\ldots,x^i+h,\ldots,x^j+k,\ldots,x^n) - f(x^1,\ldots,x^i+h,\ldots,x^j,\ldots,x^n \\ -f(x^1,\ldots,x^i,\ldots,x^j+k,\ldots,x^n) + f(x^1,\ldots,x^i,\ldots,x^j,\ldots,x^n) \end{pmatrix}}{hk}$$

This is known as the equality of second mixed partials (cf. also Ex. X.2.1).

b) Show that the continuity conditions above cannot be dropped, by proving that if $f(x,y) = xy(x^2-y^2)/(x^2+y^2)$ then $\partial_1 f$, $\partial_2 f$, $\partial_1\partial_2 f$ and $\partial_2\partial_1 f$ all exist, but that $(\partial_1\partial_2 f)(0,0) \neq (\partial_2\partial_1 f)(0,0)$.

c) Show that existence and continuity of the $\partial_i f^j$ for $f : R^n \to R^m$ imply existence and continuity of $\underset{\sim}{D}f$. (Show that the linear map defined by the Jacobian matrix in 1.03 satisfies the definition of $\underset{\sim}{D}f$, if the $\partial_i f^j$ are continuous.)

2a) Use Ex. 1a to show that if two vector fields $\underset{\sim}{v}, \underset{\sim}{w}$ on M have $\underset{\sim}{v} = v^i\partial_i$, $\underset{\sim}{w} = w^i\partial_i$ with respect to some chart $U \to R^n$ then the vector field $\underset{\sim}{u} = (v^i\partial_i w^j - w^i\partial_i v^j)\partial_j$ has $\underset{\sim}{u}(f) = \underset{\sim}{v}(\underset{\sim}{w}(f)) - \underset{\sim}{w}(\underset{\sim}{v}(f))$ for all smooth $f : U \to R$, and that it is the only vector field with this property. (Hint : the coordinate functions $x^i : U \to R$ are smooth.)

b) Deduce that $\underset{\sim}{u}$ does not depend on the chart used to define it.

c) Deduce from 1a that if $\partial_1, \ldots, \partial_n$ are the basis vector fields given by any chart of an at least C^2 manifold, then $[\partial_i, \partial_j] = \underset{\sim}{0}$ for $i,j = 1,\ldots,n$.

3) If ϕ is a local flow for $\underset{\sim}{v}$ around x and $f : M \to R$ is a smooth function, then

$$(\underset{\sim}{v}(f))(x) = \lim_{t\to 0} \left(\frac{f(\phi_t(x)) - f(x)}{t} \right).$$

4) If ϕ is a local flow for $\underset{\sim}{u}$ around x and $\underset{\sim}{v}$ is another vector field, then

$$[\underset{\sim}{u},\underset{\sim}{v}]_x = \lim_{h\to 0} \left(\frac{\overleftarrow{(\underset{\sim}{D}_x\phi_h)}\,\underset{\sim}{v}_{\phi_h(x)} - \underset{\sim}{v}_x}{h} \right).$$

5) If $\underset{\sim}{u},\underset{\sim}{v}$ are vector fields and $f : M \to R$, show by comparing effects on a typical $g : M \to R$ that

$$[\underset{\sim}{u}, f\underset{\sim}{v}] = \underset{\sim}{u}(f)\underset{\sim}{v} + [\underset{\sim}{u},\underset{\sim}{v}].$$

6) For any vector fields $\underset{\sim}{u}$, $\underset{\sim}{v}$, $\underset{\sim}{w}$ on M, prove similarly the <u>Jacobi identity</u> :

$$[\underset{\sim}{u},[\underset{\sim}{v},\underset{\sim}{w}]] + [\underset{\sim}{v},[\underset{\sim}{w},\underset{\sim}{u}]] + [\underset{\sim}{w},[\underset{\sim}{u},\underset{\sim}{v}]] = \underset{\sim}{0}.$$

VIII Connections and Covariant Differentiation

"Whither the spirit was to go, they went;
and they turned not as they went."
Ezekiel 1.12

1. CURVES AND TANGENT VECTORS

We have remarked (VII.5.02) that any vector in TM can arise as a tangent vector to a curve. It can moreover be defined in this way; Exercises 1-3 outline this construction of the tangent bundle. This way of looking at tangent vectors is central to the notation and thinking of this chapter, so if you do not do these exercises in full, at least be sure you are clear what is asserted in them. The tangent bundle is like compactness: not to be grokked in fullness from any one point of view.

Exercises VIII.1

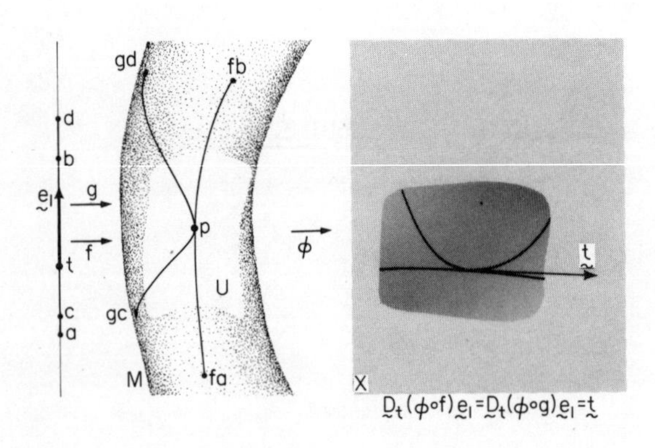

Fig. 1.1

Suppose we have two curves $f : [a,b] \to M$, $g : [c,d] \to M$ in a manifold M modelled on an affine space X, with $t \in [a,b] \cap [c,d]$, $f(t)=g(t)=p$, say, and p in the domain U of a chart $\phi : U \to X$ on M. We define f and g to be <u>tangent at</u> t and p if and only if $\underset{\sim}{D}_t(\phi\circ f) = \underset{\sim}{D}_t(\phi\circ g)$ as linear maps $T_tR \to T_{\phi(p)}X$.

1a) Prove that this definition is independent of the chart used, and that tangency at t is an equivalence relation on the set of paths taking t to p.

Thus we have a rigorous definition of tangency, intuitively amounting to f and g going in the same direction through p, at the same speed. (Notice that this is stronger than just requiring the curves tangent as sets, in the elementary sense; the parametrisations are involved.) We can use this to define the collection of speeds-with-direction - that is tangent vectors - at p, as follows.

b) Define the sum of two paths h_1, h_2 in X with $h_1(t) = h_2(t) = x$ by $(h_1+h_2)(s) = x+\underset{\sim}{d}(h_1(s),x)+\underset{\sim}{d}(h_2(s),x)$. Using this definition, show that if f, g are tangent to f', g' respectively at t and p, then for any chart ϕ around p

$$\underset{\sim}{D}_t(\phi\circ f + \phi\circ g) = \underset{\sim}{D}_t(\phi\circ f' + \phi\circ g')$$

so that $\overset{\leftarrow}{\phi}(\phi\circ f + \phi\circ g)$ is tangent to $\overset{\leftarrow}{\phi}(\phi\circ f' + \phi\circ g')$ at t and p.

Use this to define and prove well-defined, an addition on the set T_pM of tangency classes of curves at 0 and p. Define a scalar multiplication similarly, and show that this gives a vector space canonically isomorphic to the tangent space T_pM as defined in the last chapter. (cf. Ex. VII.2.5)

c) Show that if X is R^n, $\phi(p) = (p^1,\dots,p^n)$ and a curve c_i is given by $c_i(s) = \overset{\leftarrow}{\phi}(p^1,\dots,p^i+s,\dots,p^n)$, then c_i is a member of the tangency class corresponding to the vector $(\partial_i)_p \in TM$ defined in VII.4.02.

d) Define a topology and differential structure, on the set $\bigcup_{p\in M}\{T_pM\}$ of all tangency classes of curves in M tangent at 0 and any $p \in M$, such that they coincide with those of the last chapter.

2) If $\underset{\sim}{t} \in T_pM$ is represented by a curve f in M, and g is a smooth function $M \to R$

show that

$$\underset{\sim}{t}(g) = dg(\underset{\sim}{t}) = \frac{d(g \circ f)}{ds}(0) \quad \text{in the notation of VII.1.03.}$$

Thus $dg(\underset{\sim}{t})$ is "the rate at which g is changing for an observer passing through p with velocity $\underset{\sim}{t}$", and we have come very close to the earliest idea of directional derivative; "the ratio of the infinitesimal change in f to the standard infinitesimal time dt in which an infinitesimal displacement given by $\underset{\sim}{t}$ is made" or similar formulations.

3) Let $f : M \to N$ be a smooth map, and define $\underset{\sim}{D}_x f$ on any vector $\underset{\sim}{t} \in T_x M$ by choosing a representative curve c in the tangency class $\underset{\sim}{t}$, and setting $\underset{\sim}{D}_x f(\underset{\sim}{t})$ to be the class of $f \circ c$ at $f(x)$.

a) Prove that $\underset{\sim}{D}_x f(\underset{\sim}{t})$ is well defined (does not depend on the choice of representing curve).

b) Prove that $\underset{\sim}{D}_x f$ is linear.

c) Prove that this definition of $\underset{\sim}{D}_x f$ coincides with the previous one (VII.2.03) via the isomorphism of 1b.

4a) In Ex. VII.2.8b a manifold structure is defined on a set $\overset{\leftarrow}{f}(p) = P$, say. For each $x \in \overset{\leftarrow}{f}(p)$, $T_x P$ corresponds exactly to the kernel in $T_x M$ of $\underset{\sim}{D}_x f$.

b) Give a metric vector space $(X, \underset{\sim}{G})$ the canonical affine structure and the constant metric tensor field obtained from $\underset{\sim}{G}$. By Ex. VII.2.1b and Ex. VII.2.8c, $\{\underset{\sim}{x} \in X \mid \underset{\sim}{x} \cdot \underset{\sim}{x} = 1\}$ is a manifold. From a) above its tangent space at a point $\underset{\sim}{x}$ is just the set of vectors in $T_{\underset{\sim}{x}} X$ that are orthogonal to $\overset{\leftarrow}{\underset{\sim}{d}}_{\underset{\sim}{x}}(\underset{\sim}{x})$.

c) Deduce using IV.2.06 that a metric tensor field is induced on $\{\underset{\sim}{x} \mid \underset{\sim}{x} \cdot \underset{\sim}{x} = 1\}$. In particular, the metric tensor field induced on

$$\{\underset{\sim}{A} \in L(R^2; R^2) \mid \det \underset{\sim}{A} = 1\}$$

by the determinant metric tensor on R^4 (cf. IV.1.03 and Ex. IV.3.6) is indefinite, giving a pseudo-Riemannian manifold. What is its signature?

2. ROLLING WITHOUT TURNING

The differential df of a function f on an n-manifold M is a covariant vector field on M, as we have seen. That is, at each point we have a linear function that for each tangent vector tells us how fast the value of the function will change initially, if we whizz off in that direction and at that speed. It is a very useful object: for example, if the function f is thought of as a potential, df is its gradient. Obviously, we would like to generalise this powerful operation that gets df from f to tensor fields of higher order than $\binom{0}{0}$. So, what is the change in the value of a tensor field $\underset{\sim}{w}$ at p, if we move p a bit?

Out of this world.

Out, that is, of our universe of discourse up to now: that of tensors on M. If for instance $\underset{\sim}{w}$ is a contravariant vector field, thought of by the embedded picture as shown in Fig. 2.1, it is clear that as we move along the curve c towards p the tips of vectors at successive points are moving <u>at right angles</u> to the tangent plane at p. Thus the direction and rate of change at q of $\underset{\sim}{w}$ along c is not, itself, a tangent vector to M. Nor is it any sort of tensor on M. The path of the ends of the attached vectors is a

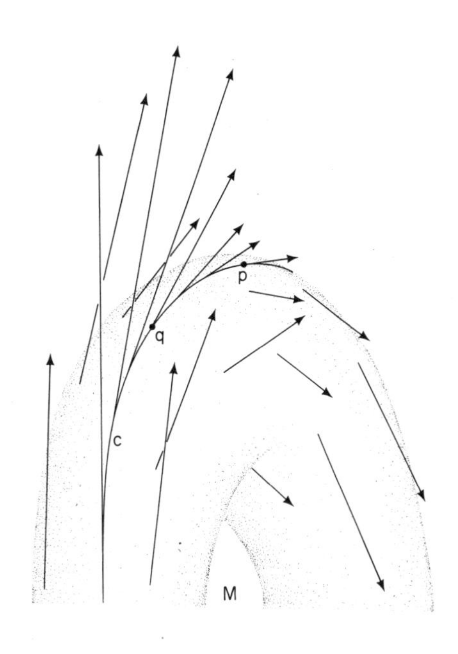

Fig. 2.1

curve not in M but in the affine space X, in which M is embedded, representing a vector. But the vector is not tangent to M, nor even located at q: it is at the tip of the vector $\underset{\sim}{w}(q)$, which is not even a point in M. So it is useless to look for it in some $(T^k_h M)_p$; it is a vector, but in the wrong place.

If, however, we

(i) replace this vector, tangent to a point in X, by the corresponding free vector (II.1.02) in T, which we shall call $\underset{\sim}{t}$,

(ii) replace $T_q M$ by the subspace of free vectors $\underset{\sim}{d}_q(T_q M) \subseteq T$,

(iii) Use a metric tensor on T to take the "tangential component" of $\underset{\sim}{t}$, that is project it orthogonally into $\underset{\sim}{d}_q(T_q M)$, and finally

(iv) Move the result back to $T_q M$ by applying $\underset{\sim}{d}_q$,

we get a vector in $T_q M$ which will serve as a derivative of $\underset{\sim}{w}$ at q in the direction represented by c.

Another way of obtaining this derivative is to roll the affine subspace of X tangent to M along the curve c without turning or slipping. Technically, this can be taken as giving a family of affine maps $\{f_t : E^n \to X \mid t \in [a,b]\}$ (where [a,b] is the domain of c) such that :

(a) Each f_t is a "rigid position" for the Euclidean space E^n, that is it preserves lengths and angles (its linear part $\underset{\sim}{f}_t$ must have $\underset{\sim}{f}_t(\underset{\sim}{v}) \cdot \underset{\sim}{f}_t(\underset{\sim}{w}) = \underset{\sim}{v} \cdot \underset{\sim}{w}$).

(b) For each x in [a,b], $f_t(E^n)$ is the affine subspace of X tangent to M at c(t). (This is the "rolling" condition.)

(c) For any point x in E^n, the smooth curve

$$c_x : [a,b] \to X : t \rightsquigarrow f_t(x)$$

traced out by x as we roll E^n has no component of velocity tangent to $f_t(E^n)$;

$c^*(t).\underset{\sim}{v} = 0$ for any $\underset{\sim}{v} \in \underset{\sim}{D}_x f_t(T_x E^n)$. (This is the "without slipping or turning" condition: a sliding of the subspace, for instance, would break it.)

(We defer to 6.07 the proof that if we fix a rigid position f_0 tangent to M at c(0) there exists a unique such family, and that the derivative we get is independent of the choice we make of f_0.)

Now, as the subspace rolls, the vector field $\underset{\sim}{w}$ on M specifies a vector in it tangent to M at c(t), for each successive position f_t. The result is a curve $\tilde{c}$ in E^n, with $\tilde{c}(t) = \overleftarrow{f_t}(c(t)+\underset{\sim}{w}(c(t)))$. We can differentiate this to give $\tilde{c}^*(t)$, translate the result to a vector, $\underset{\sim}{v}$ say, at $\overleftarrow{f_t}(c(t))$, and get the same vector $\underset{\sim}{D}f_t(\underset{\sim}{v})$ as a derivative of $\underset{\sim}{w}$ by $c^*(t)$ as by the previous procedure. We leave the formalities to Ex. 1b since as usual we shall do our more detailed work with the bundle picture.

What we have achieved is a way of "connecting" the successive tangent spaces along a curve. A "direction" tangent to M is assigned to a change from a vector in one tangent space to one in another (hence the name "connection" for the formulation we shall introduce shortly). It has a slightly curious feature. If we connect the tangent space at p to that at q by rolling along a curve between them, we get a map

$$T_pM \to T_qM : \underset{\sim}{v} \rightsquigarrow \text{(vector } \underset{\sim}{v} \text{ is rolled to)}$$

which is plainly affine, but need not be linear. Rolling the tangent line at

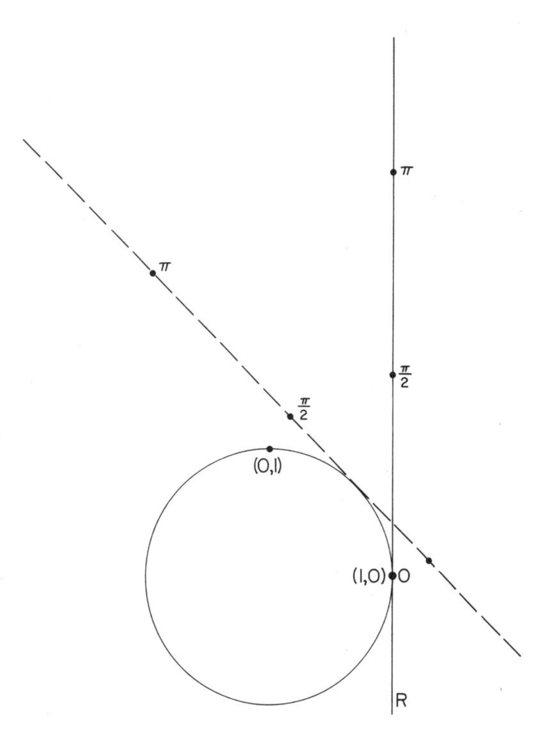

Fig. 2.2

(0,1) once clockwise around the unit circle in R^2 (Fig. 2.2) carries its origin to the old position of the point -2π. This comes of rolling entirely without slipping: in the previous description, it comes of looking only at changes in the tips of vectors along c, ignoring movement of the roots. Not quite so natural at first glance, but often more useful, is a version which in our first description translates the tangent vectors at the points c(t) to some standard point before differentiating the movement of the ends, and in our second keeps sliding the tangent space to keep the origin always at the point of contact as the space rolls, still without turning. This makes the vector field on R^2 sketched in Fig. 2.3b, rather than a, the constant one by the test of the vector tip at one point being carried to that at another: to roll the tangent plane to a flat plane in R^3 along a curve in it, without sliding, is to keep it fixed. Since this involves linear maps between the tangent spaces, while the first involved affine ones, we have correspondingly linear and affine connections. Since linear connections are much more widely useful than affine ones, they are often called simply connections (and in a few books miscalled affine connections. Be warned.)

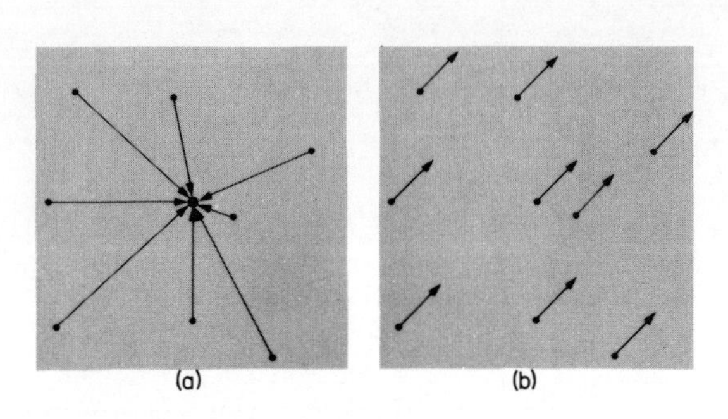

Fig. 2.3

Notice that the effect of rolling a vector from p to q depends on the route. Fig. 2.4 illustrates this (for the slide-back-to-the-origin, linearised way of rolling) for a vector at a point A on the equator rolled to another point B,

a) along the equator b) via the North Pole.

The other crucial dependence is on the metric on X. We can get a different derivative by using a different metric (most dramatically, if we switch from a

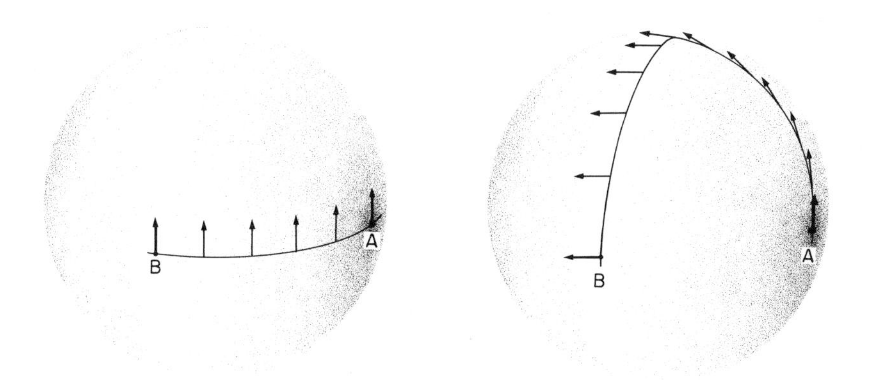

Fig. 2.4

definite to an indefinite metric. In this case we get very different isometries.) However it turns out to be a dependence only on the metric as restricted to the tangent spaces: any embedding of M in an affine space with a constant metric tensor that induces the same metric tensor on M gives the same connection. This is a remarkable fact, best proved by showing the derivative above to be the same as that defined in the next section, which only uses the metric on the tangent spaces, a proof outlined in Ex. 6.3. We leave it as an exercise, because it is the intrinsic, not the embedded description which is important for spacetime in current theories. (Science fiction has embedded the universe in all kinds of things, including abstract mathematics as a concrete object [Kagan], but physics to date has kept to the thing in itself.) We shall just use embeddings for illustrations.

Exercises VIII.2

1) This exercise is a formalisation of the above, to allow a later proof that it gives exactly the Levi-Cività connection on M (defined in 6.05 below), so that we can illustrate in the embedded picture such things as parallel transport (defined in §4 below). If you are happy with the pictures, and

prepared to believe that they correspond to the Levi-Cività connection, you can ignore it.

Let M be embedded in an affine space X with vector space T, freeing maps $\underset{\sim}{d}_x$ and a constant metric tensor $\underset{\sim}{G}$, in such a way that $\underset{\sim}{G}$ restricts to a metric on each tangent space (cf. VII.3.05). Assume the existence for any curve $c : [a,b] \to M$ with $c(0) = p$, say, of a family of linear maps $\underset{\sim}{A}_t : R^n \to T$ (not affine to X, now) for each $t \in [a,b]$, with R^n having the standard inner product or one of the standard indefinite metrics, such that (cf. 6.07 below)

(i) $\underset{\sim}{G}(\underset{\sim}{A}_t\underset{\sim}{v}, \underset{\sim}{A}_t\underset{\sim}{w}) = \underset{\sim}{v}.\underset{\sim}{w}$ for all $\underset{\sim}{v}, \underset{\sim}{w} \in R^n$, $t \in [a,b]$.

(ii) $\overset{\leftarrow}{\underset{\sim}{d}}_{c(t)}(\underset{\sim}{A}_t(R^n)) = T_{c(t)}M$, as sets, for all t.

(iii) For any point $\underset{\approx}{x} \in R^n$, the curve $c_{\underset{\approx}{x}} : [a,b] \to T : t \rightsquigarrow \underset{\sim}{A}_t(x)$ has $c^*_{\underset{\approx}{x}}(t).\underset{\sim}{v} = 0$ whenever $\underset{\sim}{v}$ is a vector at $c_{\underset{\approx}{x}}(t)$ tangent to $\underset{\sim}{A}_t(R^n)$. We think of $\underset{\sim}{A}_t$ as a "position" of R^n in T; the change of $\underset{\sim}{A}_t$ with t copies at $\underset{\sim}{0} \in T$ the rolling around of the tangent spaces at c(t) as t changes.

(This is the formalisation appropriate for the idea of rolling <u>with</u> slipping, to keep the origin as the point of tangency, but without turning.)
For any vector field $\underset{\sim}{w}$ on M we define a curve $w : [a,b] \to R^n$ by $w(t) = \underset{\sim}{A}_T \overset{\leftarrow}{\underset{\sim}{d}}_{c(t)} \underset{\sim}{w}(c(t))$, and set

$$\nabla_{\underset{\approx}{t}}\underset{\sim}{w} = \overset{\leftarrow}{\underset{\sim}{d}}_p \underset{\sim}{A}_t(w^*(0)) \text{ where } \underset{\sim}{t} = c^*(0).$$

Then

a) Prove that this coincides with the result of setting

$$\hat{\underset{\sim}{w}}(t) = \underset{\sim}{d}_{c(t)}\underset{\sim}{w}(c(t)), \qquad \nabla_{\underset{\sim}{t}}\underset{\sim}{w} = \overleftarrow{\underset{\sim}{d}}_{p}\underset{\sim}{P}_t(\hat{\underset{\sim}{w}}^*(t))$$

where $\underset{\sim}{P}_t$ is orthogonal projection $T \to \underset{\sim}{d}_{c(t)}(T_{c(t)}M)$.

(This is the linear version of the affine construction first discussed in §2.)

b) Show that the rolling without slipping discussed in the text gives the same derivative as the one obtained by differentiating the path defined by the tips of the tangent vectors (ignoring the changes in their roots), translating what you get to the point of interest, and taking the component tangential to M of the result. Can you prove from this definition of $\nabla_{\underset{\sim}{t}}\underset{\sim}{w}$ that it is independent of the choice of path c representing $\underset{\sim}{t}$? (This result follows from the intrinsic approach without a separate proof.)

3. DIFFERENTIATING SECTIONS

We turn now to considering a vector field as a section of the tangent bundle, without involving embeddings. This is logically tidier, since the relationship between a manifold and its tangent bundle is fixed, while the embedded picture requires a choice of embedding. In coordinates it emerges as far more convenient, using the charts on TM we constructed from those on M in VII.3.01.

Now the vector field $\underset{\sim}{w}$ is just a smooth map $M \to TM$, and the results of "changing p a bit" in various directions are summed up by its differential (cf. Ex. 1.3,VII.3.02). Now the differential of a map between any two manifolds goes from the tangent bundle of one to the tangent bundle of the other, so we have

$$\underset{\sim\sim}{Dw} \; : \; TM \to T(TM)$$

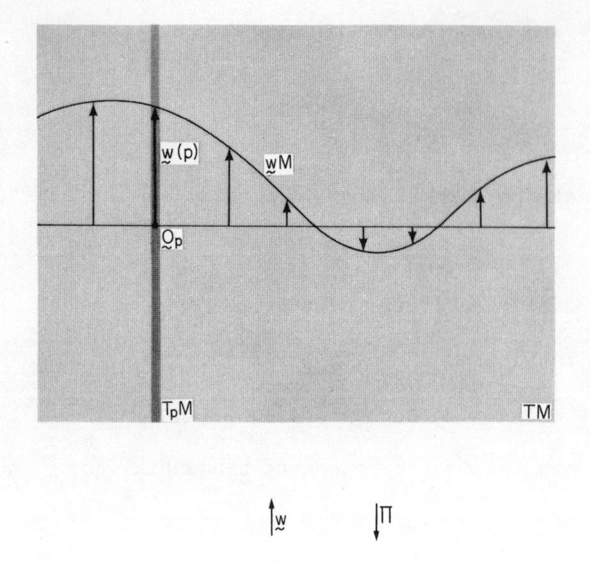

Fig. 3.1

The domain of this is as we want it; we are looking for a derivative corresponding to each vector tangent to M; but the image is in the wrong place. We get vectors tangent to TM, not to M. (In classical coordinate notation - with most of the functions and arguments involved suppressed - it shows up at once that differentiating a vector field on M gives no sort of tensor field on M directly. But what it does give is less than clear. Historically this made it much harder to find a way of "correcting" the result to give something more manageable.)

In Fig. 3.2 we draw M and its tangent bundle slightly bent, to let us use the embedded picture to represent the tangent spaces at p and $\underset{\sim}{w}(p)$ to M and TM. For a vector $\underset{\sim}{t}$ tangent to M at p we have $(\underset{\sim}{D}_p\underset{\sim}{w})\underset{\sim}{t}$ at $\underset{\sim}{w}(p)$ tangent to TM; a vector, but in the wrong place. We want, as in the previous section, to swap it for a vector tangent at p to M.

In this picture, the interesting part of $\underset{\sim}{D}_p\underset{\sim}{w}(\underset{\sim}{t})$ is obviously its "vertical component"; how $\underset{\sim}{w}$ is changing at p for us as we go through p at velocity $\underset{\sim}{t}$. This unfortunately is less easy to discover than its "horizontal component". For we have a natural projection $T_{\underset{\sim}{w}(p)}(TM) \to TM$ in the form of $\underset{\sim}{D}_{\underset{\sim}{w}(p)}\Pi$. This just expresses the fact that we are going at velocity $\underset{\sim}{t}$:

$$\underset{\sim}{D}_{\underset{\sim}{w}(p)}\Pi(\underset{\sim}{D}_p\underset{\sim}{w}(\underset{\sim}{t})) = \underset{\sim}{D}_p(\Pi\circ\underset{\sim}{w})(\underset{\sim}{t}) = \underset{\sim}{D}_p(\underset{\sim}{I}_M)(\underset{\sim}{t}) = \underset{\sim}{I}_{T_pM}(\underset{\sim}{t}) = \underset{\sim}{t}$$

by the chain rule and the definition of a vector field. What we want is a projection of $T_{\underset{\sim}{w}(p)}(TM)$ onto the subspace of the "vertical" vectors tangent to TM at $\underset{\sim}{w}(p)$, that is those tangent to the fibre $T_pM \subseteq TM$. Once we have taken $\underset{\sim}{D}_{\underset{\sim}{p}}\underset{\sim}{w}(\underset{\sim}{t})$ to its component tangent to T_pM, we can use T_pM's nice flat vector space structure to look at this as a vector <u>in</u> T_pM in an unambiguous way because T_pM is an affine space with itself as vector space. Then we have - it turns out - a derivative, with nice properties.

Now, taking the component of a vector in a subspace S is exactly applying orthogonal projection onto S (IV.2.01), which depends on a metric and is different for different metrics. So we must now work with M a Riemannian or pseudo-Riemannian manifold with metric tensor field $\underset{\sim}{G}$. Regrettably, this does not solve the problem at once, since it gives a metric tensor on each T_pM, not on each $T_{\underset{\sim}{w}}(TM)$. $\underset{\sim}{G}$ on M does in fact produce a canonical metric on TM (Ex. 6.6), but a direct definition of it from $\underset{\sim}{G}$ would not be geometrically intuitive. We shall therefore concentrate on the orthogonal projection, which by Ex. 1 is logically equivalent to the metric. Let us look at the consequences of having such a projection $P_{\underset{\sim}{v}}$ at each point $\underset{\sim}{v} \in TM$. This is the most geometric definition of a connection (Ex. 1), and we use it to motivate the most formally powerful.

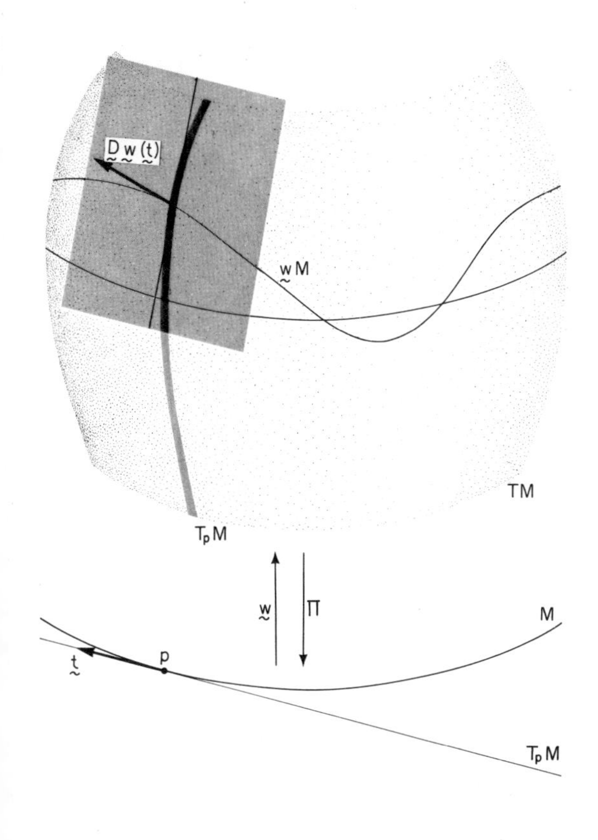

Fig. 3.2

An orthogonal projection $\underset{\sim}{P}$ in a space X gives a decomposition of a space into the direct sum of its image $\underset{\sim}{P}(X)$ and its kernel ker $\underset{\sim}{P} = (\underset{\sim}{P}(X))^{\perp}$ (Ex. VII.3.1d).

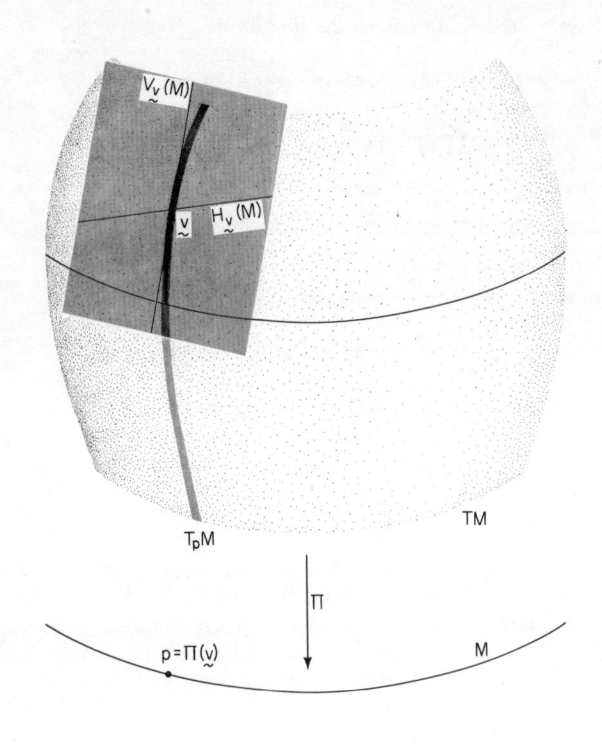

Fig. 3.3

In this instance we shall call the image $\underset{\sim}{P}_{\underset{\sim}{v}}(T_{\underset{\sim}{v}}(TM))$, which can be identified naturally with $T_{\underset{\sim}{v}}(T_{\Pi(\underset{\sim}{v})}M)$, the space of <u>vertical</u> vectors $V_{\underset{\sim}{v}}(M)$ at $\underset{\sim}{v}$, and its orthogonal complement the space of <u>horizontal</u> vectors $H_{\underset{\sim}{v}}(M)$ at $\underset{\sim}{v}$. (Fig. 3.3 is drawn as it is to emphasise that the idea of "orthogonal" varies with the metric. "Horizontal" means, by definition, "in ker $\underset{\sim}{P}_{\underset{\sim}{v}}$", not "looks level in the picture". Remember the variety of orthogonal projections in Fig. IV.2.2.)

For any vector field $\underset{\sim}{w}$ on M and vector $\underset{\sim}{t} \in T_pM$, we can use these projections to define exactly a "directional derivative" of $\underset{\sim}{w}$ by $\underset{\sim}{t}$ at p, in T_pM where we want it:

$$\nabla_{\underset{\sim}{t}}\underset{\sim}{w} = \underset{\sim}{d}_{\underset{\sim}{w}(p)}(\underset{\sim}{P}_{\underset{\sim}{w}(p)}(\underset{\sim}{D}_p\underset{\sim}{w}(\underset{\sim}{t})))$$

Here $\underset{\sim}{D}_p\underset{\sim}{w}$ is the derivative at p of $\underset{\sim}{w}$ as a map from M to TM, $\underset{\sim}{P}_{\underset{\sim}{w}(p)}$ is the projection $T_{\underset{\sim}{w}}(TM) \to V_{\underset{\sim}{w}}(M)$ we are assuming we have, and $\underset{\sim}{d}_{\underset{\sim}{w}(p)}$ is the freeing map taking vectors in $V_{\underset{\sim}{w}}(M) = T_{\underset{\sim}{w}}(T_pM)$ to vectors in T_pM itself, using the vector/affine space structure of T_pM. (∇ is generally pronounced "del", or sometimes "nabla", after an ancient Hebrew instrument of the same shape.)

Clearly, $\nabla_{\underset{\sim}{t}}\underset{\sim}{w}$ will depend linearly on $\underset{\sim}{t}$, for $\underset{\sim}{d}_{\underset{\sim}{w}(p)}$, $\underset{\sim}{P}_{\underset{\sim}{w}(p)}$ and $\underset{\sim}{D}_p\underset{\sim}{w}$ are all linear.

How would it behave for different $\underset{\sim}{w}$? Since we have not formulated the conditions that the $P_{\underset{\sim}{v}}$ must satisfy as we vary $\underset{\sim}{v}$, we cannot deduce this behaviour from the foregoing: we are free to decide what properties would be nice to have[†].

First, we obviously want linearity. If $\underset{\sim}{u},\underset{\sim}{w}$ are two vector fields and λ a real number, we want

$$\nabla_{\underset{\sim}{t}}(\underset{\sim}{u} + \underset{\sim}{w}) = \nabla_{\underset{\sim}{t}}\underset{\sim}{u} + \nabla_{\underset{\sim}{t}}\underset{\sim}{w}, \qquad \nabla_{\underset{\sim}{t}}(\lambda\underset{\sim}{w}) = \lambda\nabla_{\underset{\sim}{t}}\underset{\sim}{w}.$$

For the whole idea of the differential calculus is to make everything linear whenever possible; the differential of a map is just its replacement by a linear approximation at each point. However, we need rather more: $\underset{\sim}{w}$ being a vector <u>field</u> ($D_{\underset{\sim}{p}}\underset{\sim}{w}$, and hence $\nabla_{\underset{\sim}{t}}\underset{\sim}{w}$, are not defined if $\underset{\sim}{w}$ is just a vector at p) we may want to multiply it not just by a single constant everywhere, but by a function f on M. We cannot expect simply $\nabla_{\underset{\sim}{t}}(f\underset{\sim}{w}) = f(p)\nabla_{\underset{\sim}{t}}\underset{\sim}{w}$. For instance suppose $\underset{\sim}{w}$ and f on M = R are as illustrated in Fig. 3.4, so that $f\underset{\sim}{w}$ must be as shown in (c). Now, $\nabla_{\underset{\sim}{t}}(f\underset{\sim}{w})$ is supposed to measure the rate of change in $\underset{\sim}{w}$ at p in a way related to the usual metric on R. Therefore $\nabla_{\underset{\sim}{t}}\underset{\sim}{w}$ and hence $f(p)\nabla_{\underset{\sim}{t}}\underset{\sim}{w}$ should plainly be positive, as $\underset{\sim}{w}$ is "increasing" to the right. But equally plainly $\nabla_{\underset{\sim}{t}}(f\underset{\sim}{w})$ should be negative. The next simplest formula is the analogue of that for differentiating products of functions (Ex. VII. 1.5b), the Leibniz rule

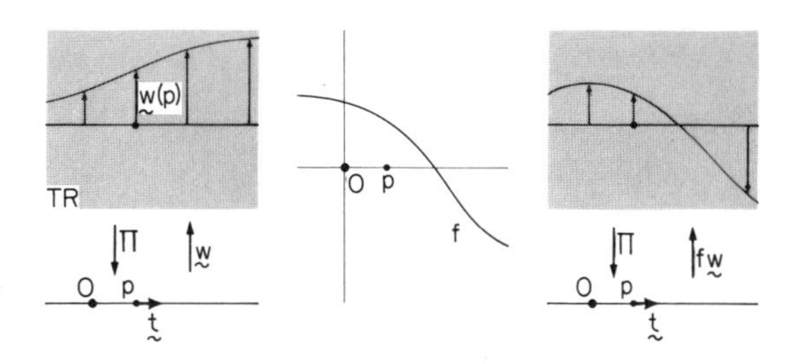

Fig. 3.4

[†] Subject of course to arriving at the usual answer, which this book is supposed to communicate. It all rather suggests a theological scheme in which you have free will in the matter of deciding <u>why</u> it is a good idea to do what you are predestined to. The elaboration of unusual answers is called research.

$$\underset{\sim}{t}(fg) = (\underset{\sim}{t}(f))g(p) + f(p)(\underset{\sim}{t}(g)).$$

This suggests

$$\nabla_{\underset{\sim}{t}}(f\underset{\sim}{w}) = (\underset{\sim}{t}(f))\underset{\sim}{w}(p) + f(p)\nabla_{\underset{\sim}{t}}w.$$

For we certainly want the effect of $\nabla_{\underset{\sim}{t}}$ to generalise the effect of $\underset{\sim}{t}$ on functions (alias $\binom{0}{0}$-tensor fields): when we define $\nabla_{\underset{\sim}{t}}$ for general tensor fields we want it to coincide with what we already have for $\binom{0}{0}$-tensors (cf. also Ex. 2).

Finally, we want everything to stay smooth. If $\underset{\sim}{w}$ is a smooth vector field, and instead of differentiating just at one point, with respect to a single vector, we take another smooth vector field $\underset{\sim}{t}$ and find $\nabla_{\underset{\sim}{t}(p)}\underset{\sim}{w}$ at each point p, we get a new vector field. If this is not smooth, our projections $P_{\underset{\sim}{v}}$ were not smoothly chosen and could not have come from a smooth Riemannian metric on TM.

We can summarise these requirements as follows:

3.01 Definition

A connection on a manifold M is a function ∇ which assigns to every tangent vector $\underset{\sim}{t}$ and C^{∞} vector field $\underset{\sim}{w}$ on M a vector $\nabla_{\underset{\sim}{t}}\underset{\sim}{w}$ in T_pM (where $\underset{\sim}{t}$ is at $p \in M$), such that

Ci) $\nabla_{(\underset{\sim}{s}+\underset{\sim}{t})}\underset{\sim}{w} = \nabla_{\underset{\sim}{s}}\underset{\sim}{w} + \nabla_{\underset{\sim}{t}}\underset{\sim}{w}$ for any $\underset{\sim}{s},\underset{\sim}{t}$ in the same tangent space, and vector field $\underset{\sim}{w}$.

Cii) $\nabla_{\underset{\sim}{t}}(\underset{\sim}{u}+\underset{\sim}{w}) = \nabla_{\underset{\sim}{t}}\underset{\sim}{u} + \nabla_{\underset{\sim}{t}}\underset{\sim}{w}$ for any $\underset{\sim}{t} \in TM$, $\underset{\sim}{u}$, $\underset{\sim}{w}$ vector fields on M.

Ciii) $\nabla_{\lambda\underset{\sim}{t}}\underset{\sim}{w} = \lambda\nabla_{\underset{\sim}{t}}\underset{\sim}{w}$ for any $\underset{\sim}{t} \in TM$, $\underset{\sim}{w}$ a vector field on M.

Civ) $\nabla_{\underset{\sim}{t}}(f\underset{\sim}{w}) = (\underset{\sim}{t}(f))\underset{\sim}{w}(p)+f(p)\nabla_{\underset{\sim}{t}}\underset{\sim}{w}$ for any $\underset{\sim}{t} \in T_pM$, $\underset{\sim}{w}$ a vector field on M, and C^{∞} $f : M \to R$.

Cv) If $\underset{\sim}{t}$, $\underset{\sim}{w}$ are C^{∞} vector fields, so is

$$\nabla_{\underset{\sim}{t}}\underset{\sim}{w} : p \rightsquigarrow \nabla_{\underset{\sim}{t}(p)}\underset{\sim}{w}.$$

It turns out that many of the important tools in differential geometry can be derived from a connection, to the point that we could almost forget about metrics. We shall not so forget, but we shall investigate the geometry of a "manifold with connection" as a thing in itself for a while before coming back to relate connections to metrics.

3.02 Coordinates

First, let us see what a connection looks like in coordinates, as some proofs will be easiest that way. We use a chart $\phi : U \to R^n : p \rightsquigarrow (x^1(p),\ldots,x^n(p))$ and the basis vector fields $\partial_1,\ldots,\partial_n$ set up in VII.4.02. Writing $\underset{\sim}{t} = t^i\partial_i$, $\underset{\sim}{w} = w^j\partial_j$ we have

$$\begin{aligned}
\nabla_{\underset{\sim}{t}(p)}\underset{\sim}{w} &= \nabla_{t^i(p)\partial_i(p)}(w^j\partial_j) \\
&= t^i(p)\nabla_{\partial_i(p)}(w^j\partial_j) && \text{by Ci)} \\
&= t^i(p)\left[\partial_i(w^j)\partial_j + w^j(p)\nabla_{\partial_i(p)}(\partial_j)\right] && \text{by Civ)} \\
&= t^i\partial_i(w^j)\partial_j + t^i w^j(\nabla_{\partial_i}\partial_j),
\end{aligned}$$

suppressing reference to p. Now the first term in this sum is already "in coordinates". For $t^i\partial_i(w^j)\partial_j$ means exactly

$$\left(t^1\frac{\partial w^1}{\partial x^1} + \ldots + t^n\frac{\partial w^1}{\partial x^n},\ t^1\frac{\partial w^2}{\partial x^1} + \ldots + t^n\frac{\partial w^2}{\partial x^n},\ \ldots,\ t^1\frac{\partial w^n}{\partial x^1} + \ldots + t^n\frac{\partial w^n}{\partial x^n}\right),$$

disentangling summations (cf. VII.4.02); but $\nabla_{\partial_i}\partial_j$ is just some vector in T_pM determined by ∇, ∂_i and ∂_j. We shall represent this vector in components by

$$\nabla_{\partial_i}\partial_j = \Gamma^k_{ij}\partial_k .$$

The n^3 functions Γ^k_{ij} for $i,j,k = 1,\ldots,n$ so defined are called the <u>Christoffel symbols</u> of the connection ∇ with respect to this chart and ∇ thus has the coordinate form

$$\nabla_{\underset{\sim}{t}}\underset{\sim}{w} = \left(t^i \frac{\partial w^k}{\partial x^i} + t^i w^j \Gamma^k_{ij}\right)\frac{\partial}{\partial x^k} ,$$

changing one dummy index.

3.03 Transformation formula

If we change to another chart $\tilde{\phi} : U \to R^n : p \rightsquigarrow (\tilde{x}^1(p),\ldots,\tilde{x}^n(p))$ we get a new basis $\tilde{\partial}_1,\ldots,\tilde{\partial}_n$ for T_pM and a new lot $\tilde{\Gamma}^\gamma_{\alpha\beta}$ of Christoffel symbols. The two are related by the formula

$$\bigstar \quad \tilde{\Gamma}^\gamma_{\alpha\beta} = \Gamma^k_{ij}\tilde{\partial}_\alpha(x^i)\tilde{\partial}_\beta(x^j)\partial_k(\tilde{x}^\gamma) + (\tilde{\partial}_\alpha(\tilde{\partial}_\beta(x^k)))(\partial_k(\tilde{x}^\gamma))$$

(Ex. 3a) or equivalently

$$\tilde{\Gamma}^\gamma_{\alpha\beta}\,\tilde{\partial}_\gamma(x^\ell) = \Gamma^\ell_{ij}\tilde{\partial}_\alpha(x^i)\tilde{\partial}_\beta(x^j) + \tilde{\partial}_\alpha(\tilde{\partial}_\beta(x^\ell)).$$

(Recall that $\partial_k(\tilde{x}^\gamma)\tilde{\partial}_\gamma(x^\ell) = \delta^\ell_k$, since $\partial_k(\tilde{x}^\gamma)$ and $\tilde{\partial}_\gamma(x^\ell)$ are components of the two change-of-basis matrices for T_pM.) This constitutes the classical definition of a connection ("a set of numbers that transform according to $\bigstar$"). The essential equivalence of this to 3.01 follows from Ex. 3b.

It is clear that ★ is not the transformation law for the components of any sort of tensor (the first term is just the formula for $\binom{2}{1}$-tensors, but the other involves a second differential). This is reasonable, as the Γ^k_{ij} are a kind of correction term to bring erring derivatives back into the tensor fold. Roughly, "if $t^i \frac{\partial w^j}{\partial x^i} \frac{\partial}{\partial x^j}$ differed only by a tensor from being a tensor it would be a tensor anyway". (The more classical texts derive ★ from the requirement that the expression $(u^i \partial_i (v^k) + u^i v^j \Gamma^k_{ij})\partial_k$ - usually omitting the basis vectors ∂_k - should transform as a vector, define "connection" from ★, and proceed from there.) The Γ^k_{ij} are not, then, the components of a tensor; to anticipate the language of 3.07, $\Gamma^k_{ij}\partial_k$ is the vertical part of the derivative of ∂_j in the direction ∂_i (the significant change in ∂_j as we move in the x^i direction) as measured by this connection. Γ^k_{ij} is its k-th component.

We shall make some use of 3.02, but none of ★, since the coordinate-free characterisation Ci),...,Cv) of connections is far more convenient; so ★ is there as part of our general programme of relating "numbers that transform right" to geometrically defined objects.

Returning to the geometry of a manifold with connection: let us recover the decomposition $V_{\underset{\sim}{v}}(M) \oplus H_{\underset{\sim}{v}}(M)$ via which we motivated 3.01, from a connection satisfying Ci),...,Cv). First we need :

3.04 Definition

For a curve $c : [a,b] \to M$, a C^∞ vector field along c is a C^∞ function giving a vector tangent to M at c(t) for each $t \in [a,b]$; that is a map $\underset{\sim}{v} : [a,b] \to TM$ such that $\Pi \circ \underset{\sim}{v} = c$.

Important examples of this are the tangent vector field c* of c (Fig. 3.5a), and the restriction to c of a vector field $\underset{\sim}{w}$ on M (Fig. 3.5b) which assigns the

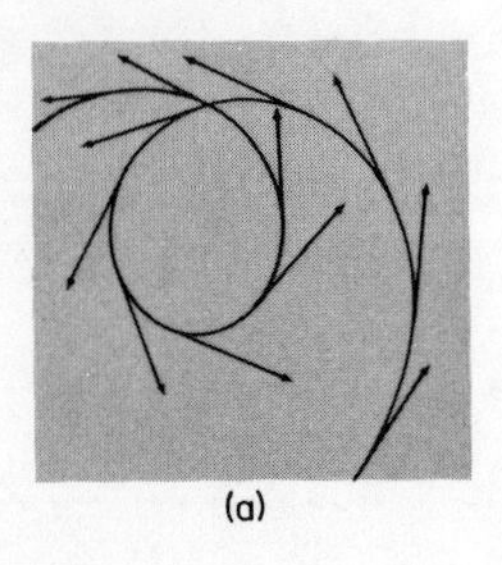
(a)

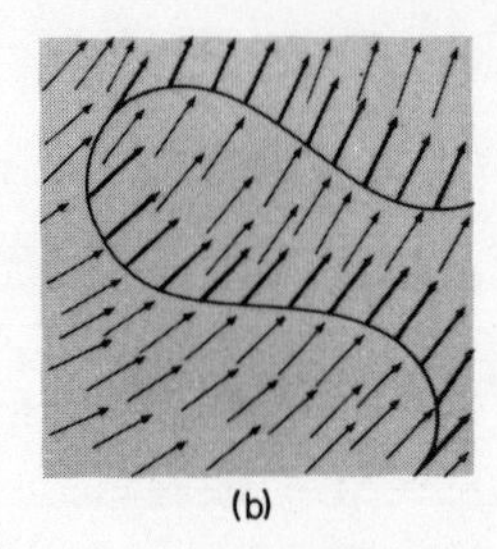
(b)

Fig. 3.5

vector $\underset{\sim}{w}(c(t))$ to the point $t \in [a,b]$, precisely $\underset{\sim}{w} \circ c$.

3.05 Differentiating, along curves, fields along curves

Our first approach, in §2, to finding a candidate for $\nabla_{\underset{\sim}{t}}\underset{\sim}{w}$ involved only the values of $\underset{\sim}{w}$ at points $c(t)$ in M, and so extends at once to vector fields $\underset{\sim}{w}$ along c as well as on M. We have chosen 3.01 as our formal starting point, however. We must therefore show that $\nabla_{\underset{\sim}{t}}\underset{\sim}{w}$ for a connection satisfying Ci),...,Cv) depends only on the restriction of $\underset{\sim}{w}$ to a typical representative c with $\underset{\sim}{t} = c^*(t)$ of the tangency class $\underset{\sim}{t}$, and that we can extend this differentiation, of restrictions to c of vector fields on M, to a differentiation of general vector fields along c. This necessary check of a credible fact is left as Ex. 4.

We denote the resulting linear map, taking vector fields along c (not vector fields on M) to vector fields along c, by $\mathbb{V}_{c*}$ not ∇_{c*} (although $\mathbb{V}_{c*}\hat{\underset{\sim}{w}} = \nabla_{c*}\underset{\sim}{w}$ whenever $\hat{\underset{\sim}{w}}$ is the restriction of $\underset{\sim}{w}$ to c) to emphasise the difference in their domains.

Now we are equipped to decompose $T_{\underset{\sim}{v}}(TM)$.

3.06 Definition

A vector $\underset{\sim}{v} \in T_{\underset{\sim}{w}}(TM)$, where $\underset{\sim}{w} \in T_pM$, is vertical if $\underset{\sim}{D}_{\underset{\sim}{w}}\Pi : T_{\underset{\sim}{w}}(TM) \to T_pM$ has $\underset{\sim}{D}_{\underset{\sim}{w}}\Pi(\underset{\sim}{v}) = \underset{\sim}{0}$. We denote the space $\ker(\underset{\sim}{D}_{\underset{\sim}{w}}\Pi)$ of such vectors by $V_{\underset{\sim}{w}}(M)$, as in our less formal discussion at the beginning of this section.

If $\underset{\sim}{v}$ is not vertical we can find a path $\tilde{c} : [a,b] \to TM$ to represent it and get a curve $c = \Pi \circ \tilde{c}$ in M with

$$c(0) = p$$

$$c^*(0) = D_{\underset{\sim}{w}}\Pi(\tilde{c}^*(0)) = D_{\underset{\sim}{w}}\Pi(\underset{\sim}{t}) \neq \underset{\sim}{0}$$

using §1. For each $t \in [a,b]$, $\tilde{c}(t)$ is a vector at $c(t)$; $\tilde{c}$ exactly gives a vector field $\underset{\sim}{c}$ along c. In this notation, we make

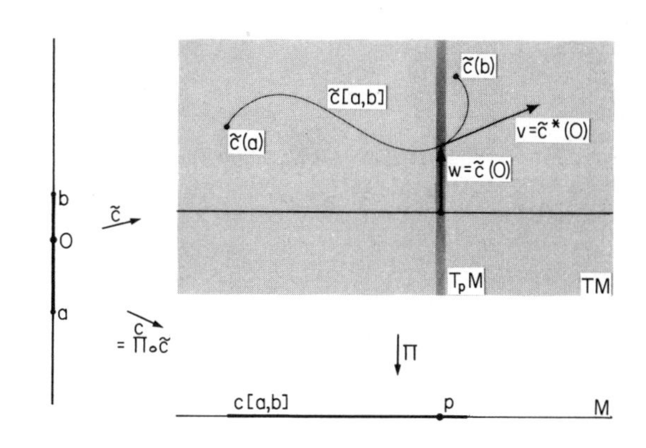

Fig. 3.6

3.07 Definition

The vertical part of $\underset{\sim}{v}$ with respect to ∇ is $\underset{\sim}{v}$ itself if $D_{\underset{\sim}{w}}\Pi(\underset{\sim}{v}) = \underset{\sim}{0}$, otherwise it is

$$\overleftarrow{d}_{\underset{\sim}{w}}(\nabla_{c^*}\underset{\sim}{c}(0)) \in T_{\underset{\sim}{w}}(T_pM) \subseteq T_{\underset{\sim}{w}}(TM)$$

where $\underset{\sim}{c}$ is as above. (That this is well defined, depending on only $\underset{\sim}{v}$ and $\underset{\sim}{w}$, is Ex. 5a.)

Defining the projection (cf. Ex. 5b)

$$P_{\underset{\sim}{w}} : T_{\underset{\sim}{w}}(TM) \to V_{\underset{\sim}{w}}(M) : \underset{\sim}{v} \leadsto (\text{vertical part of } \underset{\sim}{v}),$$

we say $\underset{\sim}{v}$ is horizontal if its vertical part $P_{\underset{\sim}{w}}(\underset{\sim}{v})$ is $\underset{\sim}{0}$, and set $H_{\underset{\sim}{w}}(M) = \ker P_{\underset{\sim}{w}}$. Exs. 5c, d show that these $P_{\underset{\sim}{w}}$ are exactly those that give ∇ as discussed at the beginning of §3, so that the $P_{\underset{\sim}{w}}$ and ∇ are equivalent structures containing the same information.

The horizontal part of $\underset{\sim}{t} \in T_{\underset{\sim}{w}}(TM)$ is $\underset{\sim}{t} - P_{\underset{\sim}{w}}\underset{\sim}{t}$.

3.08 Language

A connection defined as in 3.01 is called a Koszul connection; the corresponding splitting of the $T_{\underset{\sim}{w}}(TM)$ into horizontal and vertical parts is an Ehresmann connection. The two equivalent conceptions illuminate each other and the coordinate definition as the various constructions of tangent spaces do. Pioneering work on the geometrical role of connections in spacetime was done by Hermann Weyl about sixty years ago, using the coordinate description.

Exercises VIII.3

1) If $\underset{\sim}{G}$ is a metric tensor field on M, the metric tensor $\underset{\sim}{G}_p$ on T_pM gives a corresponding constant metric tensor $\hat{\underset{\sim}{G}}$ on T_pM. Identifying $T_{\underset{\sim}{v}}(T_pM)$ with $V_{\underset{\sim}{v}}(M) = \ker(\underset{\sim}{D}_{\underset{\sim}{v}}\Pi)$ in the natural way, show that for any idempotent operator $\underset{\sim}{P}_{\underset{\sim}{v}}$ on $T_{\underset{\sim}{v}}(TM)$ with image $V_{\underset{\sim}{v}}(M)$,

   a) $\underset{\sim}{x}\cdot\underset{\sim}{y} = \underset{\sim}{G}_p((\underset{\sim}{D}_{\underset{\sim}{v}}\Pi)\underset{\sim}{x},(\underset{\sim}{D}_{\underset{\sim}{v}}\Pi)\underset{\sim}{y}) + \hat{\underset{\sim}{G}}_{\underset{\sim}{v}}(\underset{\sim}{P}_{\underset{\sim}{v}}\underset{\sim}{x},\underset{\sim}{P}_{\underset{\sim}{v}}\underset{\sim}{y})$ defines a metric tensor on $T_{\underset{\sim}{v}}(TM)$, and

   b) $\underset{\sim}{P}_{\underset{\sim}{v}}$ is orthogonal projection onto $V_{\underset{\sim}{v}}(M)$ with respect to this metric tensor.

2) Define a projection $\underset{\sim}{P}_x : T_x(T^0_0M) \to T_x((T^0_0M)_p)$ where x is a $\binom{0}{0}$-tensor at p, using the natural identification of T^0_0M with M × R, such that for $\underset{\sim}{t} \in T_pM$ the map $\nabla_{\underset{\sim}{t}}$ defined on $\binom{0}{0}$-tensor fields by

$$\nabla_{\underset{\sim}{t}}(f) = \underset{\sim}{d}_{f(p)}\underset{\sim}{P}_{f(p)}\underset{\sim}{D}_p f(\underset{\sim}{t})$$

coincides with $\underset{\sim}{t} : f \rightsquigarrow df(\underset{\sim}{t})$, so that Civ) becomes

$$\nabla_{\underset{\sim}{t}}(f\underset{\sim}{w}) = (\nabla_{\underset{\sim}{t}}f)\underset{\sim}{w}(p) + f(p)\nabla_{\underset{\sim}{t}}\underset{\sim}{w}.$$

3a) Establish equation ★ of 3.03 by using Ci),...,Cv), VII.4.04 and the fact that $\nabla_{\partial_i}\partial_j = \Gamma^k_{ij}\partial_k$ is the same vector however it is labelled.

b) Suppose that we have a rule that, given a chart $U \to R^n$, produces n^3 functions $\Gamma^k_{ij} : U \to R$ in such a way that where two charts overlap the results of applying the rule are related by ★. Show that the formula

$$\nabla_{\underset{\sim}{u}}\underset{\sim}{v} = u^i\partial_i(v^k)\partial_k + u^i v^j \Gamma^k_{ij}\partial_k$$

defines the same vector field around any point whatever chart is used, and that ∇ so defined satisfies 3.01.

4a) Show that if $c^*(t) \neq \underset{\sim}{0}$ for $c : [a,b] \to M$, some $t \in [a,b]$, any vector field $\underset{\sim}{v}$ along c is locally a restriction of some vector field $\hat{\underset{\sim}{v}}$, on a neighbourhood W of $c(t)$ in M, to $c|J$ where J is a neighbourhood of t. (By Ex. VII.2.7a there is a choice of coordinates that makes this very easy.)

b) Define

$$\nabla_{c^*(t)}\underset{\sim}{v} = \nabla_{c^*}\underset{\sim}{v}(t) = \begin{cases} \nabla_{c^*(t)}\hat{\underset{\sim}{v}}, & \hat{\underset{\sim}{v}} \text{ being as in a, if } c^*(t) \neq \underset{\sim}{0} \\[2ex] \underset{\sim}{0} & \text{if } c^*(t) = \underset{\sim}{0} \end{cases}$$

and show that it depends only on $\underset{\sim}{v}$, not on the choice $\hat{\underset{\sim}{v}}$ of extension. (Work in coordinates to get

$$\nabla_{c^*}\underset{\sim}{v} = \left(\frac{d\tilde{v}^k}{ds} + (\Gamma^k_{ij}\circ c)(\tilde{v}^j\circ c)\frac{dc^i}{dt}\right)\partial_k$$

and deduce the result from this.)

c) Show that $\nabla_{c^*}(\underset{\sim}{v} + \underset{\sim}{v}') = \nabla_{c^*}\underset{\sim}{v} + \nabla_{c^*}\underset{\sim}{v}'$, that $\nabla_{c^*}(\lambda\underset{\sim}{v}) = \lambda\nabla_{c^*}\underset{\sim}{v}$ for $\lambda \in R$,

that $\nabla_{c^*}(f\underset{\sim}{v}) = \frac{df}{dt}\underset{\sim}{v} + f\nabla_{c^*}\underset{\sim}{v}$, for $f : [a,b] \to R$, and that if $\underset{\sim}{v}$ is a smooth field along c, then so is $\nabla_{c^*}\underset{\sim}{v}$. (In particular, check at points where c*(t) becomes zero and the definition changes from "$\nabla_{c^*(t)}$ of a local extension of $\underset{\sim}{v}$ to M" to simply "$\underset{\sim}{0}$".)

5a) In the notation of 3.07, show that $\nabla_{c^*}\underset{\sim}{c}$ is independent of the choice of path $\tilde{c}$ representing $\underset{\sim}{t}$. (Either use geometrical devices with paths, or bash it with the coordinate equation of Ex. 4b.)

b) Show that $\underset{\sim}{P}_{\underset{\sim}{w}}$ of 3.07 is linear, despite the different definitions on different subsets of $T_{\underset{\sim}{w}}(TM)$.

c) Show that

$$\nabla_{c^*(t)}\underset{\sim}{c} = \underset{\sim}{d}_{\underset{\sim}{c}(t)}(\underset{\sim}{P}_{\underset{\sim}{c}(t)}(\underset{\sim}{D}_t\underset{\sim}{c}(\underset{\sim}{e}_t)))$$

where $\underset{\sim}{e}_t \in T_tR$ is the standard unit basis vector.

d) Deduce that for $\underset{\sim}{w}$ a vector field on M, $\underset{\sim}{t} \in T_pM$.

$$\nabla_{\underset{\sim}{t}}\underset{\sim}{w} = \underset{\sim}{d}_{\underset{\sim}{w}(p)}(\underset{\sim}{P}_{\underset{\sim}{w}(p)}(\underset{\sim}{D}_p\underset{\sim}{w}(\underset{\sim}{t}))).$$

e) Using the coordinates on TM induced by a chart $U \to R^n$ on M (cf. Ex. VII.4.7) and the corresponding basis for $T_{\underset{\sim}{v}(p)}(TM)$ (if the coordinates of p are $(x^1,\ldots,x^n)$, and those of $\underset{\sim}{v}$ are $(x^1,\ldots,x^n,v^1,\ldots,v^n)$, the basis is $\frac{\partial}{\partial x^1}, \ldots, \frac{\partial}{\partial x^n}, \frac{\partial}{\partial v^i}, \ldots, \frac{\partial}{\partial v^n}$) find the components of $\underset{\sim}{P}_{\underset{\sim}{v}(p)}$.

6a) Show that if a linear map $\underset{\sim}{A} : X \to Y$ is surjective, and $X \cong (\ker \underset{\sim}{A}) \oplus B$ for some subspace $B \subseteq X$, then $\underset{\sim}{A}|B$ is an isomorphism.

b) Deduce that there is exactly one horizontal vector $\hat{\underset{\sim}{t}}$ at any $\underset{\sim}{v} \in T_pM$ such that $\underset{\sim}{D}_{\underset{\sim}{v}}\Pi(\hat{\underset{\sim}{t}})$ is a given vector $\underset{\sim}{t} \in T_pM$.

4. PARALLEL TRANSPORT

4.01 Definition

A vector field $\underset{\sim}{v}$ along a curve c in a manifold M (with a connection ∇) is parallel if

$$\nabla_{c_*}\underset{\sim}{v} = \underset{\sim}{0}$$

or equivalently if $\underset{\sim}{v}$, considered as a curve in TM, has $v^*(t)$ horizontal for all t.

A vector field $\underset{\sim}{w}$ on M is parallel along c if $\underset{\sim}{w}\circ c$ is parallel, and $\underset{\sim}{w}$ is parallel if it is parallel along all curves. (Most such M have no parallel vector fields on them, as we shall see in Chapter X.)

Fig. 4.1 illustrates parallel fields along curves in S^2 and R^2, with their usual connections (associated with their usual metrics in a way we discuss in §6).

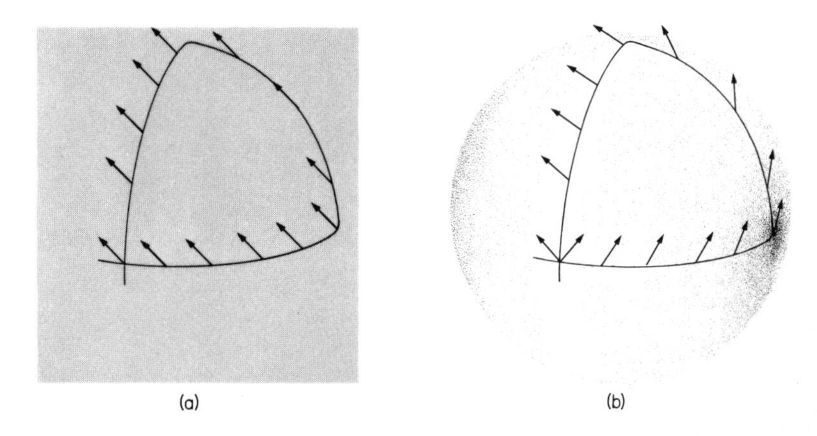

Fig. 4.1

In "rolling" terms, we shall see (6.07) that a parallel vector field along a curve c is one for which any vector $\underset{\sim}{v}(t)$ is carried to $\underset{\sim}{v}(t')$, for any t, t', by rolling without turning of the tangent spaces along c from c(t) to c(t'). This suggests that each vector in $T_{c(t)}M$ should be part of one and only one parallel vector field along c. This is indeed true:

4.02 Theorem

If $c : J \to M$ is a curve in a manifold with connection, and $t_0 \in J$, then (putting $c(t_0) = x$) for each $\underset{\sim}{v} \in T_xM$ there is exactly one parallel vector field $\underset{\sim}{w}_{\underset{\sim}{v}}$ along c with $\underset{\sim}{w}(t_0) = \underset{\sim}{v}$. Moreover the map

$$\underset{\sim}{\tau}_{t-t_0} : T_xM \to T_{c(t)}M : \underset{\sim}{v} \rightsquigarrow \underset{\sim}{w}_{\underset{\sim}{v}}(t)$$

is linear and an isomorphism.

Proof

Without loss of generality (why?), suppose $J = \mathbb{R}$, $t = 0$.

We need a differentiable manifold structure on the set $N = \Pi^{\leftarrow}(c(J))$ for this proof. Since this is tricky if c or $\underset{\sim}{D}c$ is not injective, we replace M by $M \times \mathbb{R}$, c by $\tilde{c} : t \rightsquigarrow (c(t),t)$, and the connection ∇ by the product connection on $M \times \mathbb{R}$ (Ex. 1); by Ex. 2 we get an equivalent problem. Ex. 3 reduces the question to the solution of a differential equation, so that we can apply VII.6.04 to get

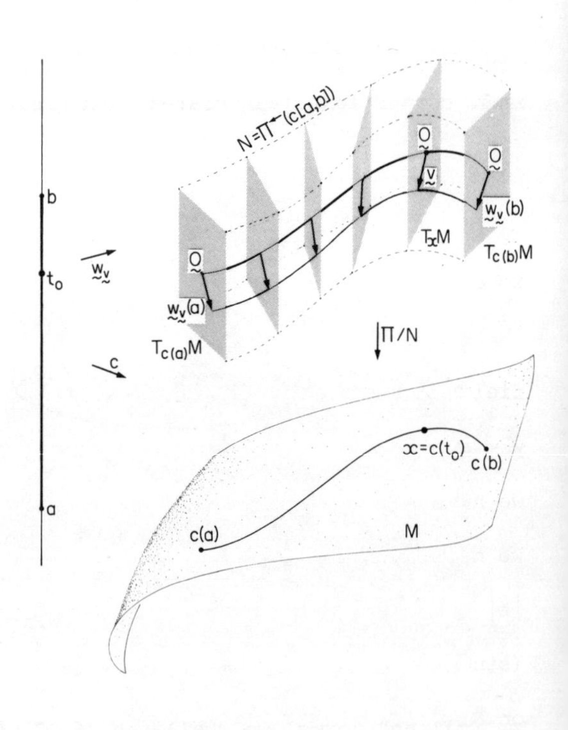

Fig. 4.2

* For any $\underset{\sim}{v} \in T_x M$ there is $\varepsilon \in R$ and a unique parallel vector field $\underset{\sim}{w}_{\underset{\sim}{v}}$ along $c|]-\varepsilon,\varepsilon[$ with $\underset{\sim}{w}(0) = \underset{\sim}{v}$.

Before replacing $]-\varepsilon,\varepsilon[$ by J we prove linearity. By Ex. 3.4c if $\underset{\sim}{w}$ is parallel along $c|]-\varepsilon,\varepsilon[$, $\underset{\sim}{w}'$ along $c|]-\varepsilon',\varepsilon'[$ with $\underset{\sim}{w}(0) = \underset{\sim}{v}$, $\underset{\sim}{w}'(0) = \underset{\sim}{v}'$, then

$$\nabla_{c*}(\lambda \underset{\sim}{w}) = \lambda \nabla_{c*}\underset{\sim}{w} = \underset{\sim}{0}, \quad \text{and} \quad \lambda\underset{\sim}{w}(0) = \lambda\underset{\sim}{v}, \quad \text{for any } \lambda \in R$$

$$\nabla_{c*}(\underset{\sim}{w}+\underset{\sim}{w}') = \nabla_{c*}\underset{\sim}{w} + \nabla_{c*}\underset{\sim}{w}' = \underset{\sim}{0}, \quad (\underset{\sim}{w}+\underset{\sim}{w}')(0) = \underset{\sim}{v}+\underset{\sim}{v}'$$

where defined. Accordingly, $\lambda\underset{\sim}{w}$ and $\underset{\sim}{w}+\underset{\sim}{w}'$ are the unique solution curves through $\lambda\underset{\sim}{v}$ and $\underset{\sim}{v}+\underset{\sim}{v}'$ where defined. Hence if $|t| < \min\{\varepsilon,\varepsilon'\}$

$$\underset{\sim}{\tau}_t(\lambda\underset{\sim}{v}) = \lambda\underset{\sim}{w}(t) = \lambda\underset{\sim}{\tau}_t(\underset{\sim}{v})$$

$$\underset{\sim}{\tau}_t(\underset{\sim}{v}+\underset{\sim}{v}') = (\underset{\sim}{w}+\underset{\sim}{w}')(t) = \underset{\sim}{w}(t)+\underset{\sim}{w}'(t) = \underset{\sim}{\tau}_t(\underset{\sim}{v})+\underset{\sim}{\tau}_t(\underset{\sim}{v}')$$

so each $\underset{\sim}{\tau}_t$ is linear.

Hence if $\underset{\sim}{v}_1,\ldots,\underset{\sim}{v}_n$ form a basis for $T_x M$ and $\underset{\sim}{w}_i :]-\varepsilon_i,\varepsilon_i[\to TM$ are the parallel fields along c with $\underset{\sim}{w}_i(0) = \underset{\sim}{v}_i$ given by *, and $\varepsilon = \min\{\varepsilon_1,\ldots,\varepsilon_n\}$, we have $\underset{\sim}{w} = a^i\underset{\sim}{w}_i$ defined on $]-\varepsilon,\varepsilon[$ with $\underset{\sim}{w}(0) = \underset{\sim}{v}$ for any $\underset{\sim}{v} \in T_x M$, where $\underset{\sim}{v} = a^i\underset{\sim}{v}_i$. (Thus we have <u>one</u> ε that works for all $\underset{\sim}{v} \in T_x M$: contrast the field $\underset{\sim}{w}$ in VII 6.01, where we needed smaller and smaller ε as we got further from the x-axis.) So if $|t| < \varepsilon$, then $\underset{\sim}{\tau}_t : T_x M \to T_{c(t)} M$ is everywhere defined, and bijective by VII.6.06 (since it is just $\phi_t|T_x M$ where ϕ is a local flow for the horizontal vector field on N). Hence it is an isomorphism.

Now if $\underset{\sim}{w}$ cannot be extended to a parallel field along c with domain all of R, there is some E in R that we cannot reach. Let

$$S = \{\varepsilon \mid \exists \text{ parallel } \underset{\sim}{w} :]-\varepsilon,\varepsilon[\to TM, \text{ with } \underset{\sim}{w}(0) = \underset{\sim}{v}, \text{ along } c|]-\varepsilon,\varepsilon[\} \subseteq [-E,E].$$

By Ex. VI.4.6b there is a real number $e = \sup S$. It is clear that we can define $\underset{\sim}{w}$ on $]-e,e[$. But we can use * on $c' : t \leadsto c(t-e)$, $c'' : t \leadsto c(t+e)$ to get local flows that extend $\underset{\sim}{w}$ forwards past e, backwards past $-e$. (The $\underset{\sim}{\tau}_t$ are isomorphisms, so that some $\underset{\sim}{u} \in T_{c(e)}M$, for instance, is mapped back to $\underset{\sim}{w}$ of some s in both $]-e,e[$ and the interval $]e-\varepsilon,e+\varepsilon[$ that we have around e by *). So e cannot be sup S after all, so S has no supremum and $\underset{\sim}{w}$ can be defined for all $\mathbb{R}$. ■

4.03 Definition

The map $\underset{\sim}{\tau}_t : T_xM \to T_{c(t)}M$ introduced in 4.02 is called parallel transport along c from $x = c(0)$ to $c(t)$, with respect to the connection on M. We shall relate this to the effect of "rolling without turning" along c in 6.07.

Notice again that in general which vector in T_yM is parallel to a given vector in T_xM depends on our choice of curve from x to y. On the sphere, for instance, any two unit vectors, anywhere, are "parallel" by transport along a suitably chosen curve (Fig. 2.4). The study of which vectors can be parallel in a general manifold with connection is the theory of holonomy groups, outside our present scope; see [Kobayashi and Nomizu].

Evidently, a vector field $\underset{\sim}{v}$ on M is parallel if and only if $\underset{\sim}{\tau}_t(\underset{\sim}{v}_p) = \underset{\sim}{v}_q$ along all paths from p to q, for all $p,q \in M$. We shall examine parallel vector fields in more detail in Ch. X.

Intuitively, the effect of rolling a tangent space along a curve from p to q should be independent of whether we go by a slow roll or a fast: the parametrisation should be irrelevant. (For instance, if we stop for a while to admire the view we shall not alter the final result.) We prove this now for our more precise and intrinsic formulation of parallel transport:

4.04 Lemma

If $f = c \circ h$ is a smooth reparametrisation of c, and $\overleftarrow{h}(a) = \alpha$, $\overleftarrow{h}(b) = \gamma$, parallel transport along f from $f(\alpha)$ to $f(\gamma)$ is the same as parallel transport along c from c(a) to c(b).

Proof

Let $\underset{\sim}{v}$ be any parallel vector field along c. Then $\underset{\sim}{v}' = \underset{\sim}{v} \circ h$ is a vector field along f, and if $p = f(t)$, $s = h(t)$ we have

$$\nabla_{c_*}\underset{\sim}{v}'(t) = \underset{\sim}{d}_{\underset{\sim}{v}'(t)}(\underset{\sim}{P}_{\underset{\sim}{v}'(t)}(\underset{\sim}{D}_t\underset{\sim}{v}'(\underset{\sim}{e}_t))) \quad \text{by Ex. 3.5c}$$

$$= \underset{\sim}{d}_{\underset{\sim}{v}(s)}(\underset{\sim}{P}_{\underset{\sim}{v}(s)}(\underset{\sim}{D}_s\underset{\sim}{v} \circ \underset{\sim}{D}_t h(\underset{\sim}{e}_t))) \quad \text{by the chain rule}$$

$$= \left(\frac{dh}{dr}(t)\right) \underset{\sim}{d}_{\underset{\sim}{v}(s)}(\underset{\sim}{P}_{\underset{\sim}{v}(s)}(\underset{\sim}{D}_s\underset{\sim}{v}(\underset{\sim}{e}_s))) \quad \text{by linearity,}$$

since $\underset{\sim}{D}_t h$ is just scalar multiplication by $\frac{dh}{dr}(t)$, relative to the bases $\{\underset{\sim}{e}_t\}$ and $\{\underset{\sim}{e}_s\}$

$$= \left(\frac{dh}{dr}(t)\right)\nabla_{c_*}\underset{\sim}{v}(s)$$

$$= \underset{\sim}{0} \quad \text{since } \underset{\sim}{v} \text{ is parallel.}$$

Thus $\underset{\sim}{v}'$ is also parallel. The result follows by uniqueness. ∎

We first used the $\underset{\sim}{\tau}_t$ (in their avatar of "rolling without turning") in §2 to produce a ∇; we have now obtained them as "solutions" of a given connection ∇. To show that they constitute yet another equivalent form of the connection, we obtain from them the projections $\underset{\sim}{P}_{\underset{\sim}{v}}$ and, as a corollary, the given ∇ :-

4.05 Theorem

Let $\underset{\sim}{v} \in T_pM$, and $c : J \to TM$ represent $\underset{\sim}{w} \in T_{\underset{\sim}{v}}(TM)$. Then if we define

$$\underset{\sim}{Q}_{\underset{\sim}{v}}(\underset{\sim}{w}) = \bar{c}^*(0) \in T_{\underset{\sim}{v}}(T_pM), \text{ where } \bar{c}(t) = \overleftarrow{\underset{\sim}{\tau}}_t(c(t)) \text{ along } \Pi\circ c,$$

$\underset{\sim}{Q}_{\underset{\sim}{v}}$ coincides with the projection $\underset{\sim}{P}_{\underset{\sim}{v}} : T_{\underset{\sim}{v}}(TM) \to T_{\underset{\sim}{v}}(T_pM)$ of 3.07.

<u>Proof</u>

A) If $\underset{\sim}{w} \in T_{\underset{\sim}{v}}(T_pM)$ already, $\bar{c}$ must be tangent at 0 and $\underset{\sim}{v}$ to c, by the smoothness of parallel transport, so that

$$\underset{\sim}{Q}_{\underset{\sim}{v}}(\underset{\sim}{w}) = \bar{c}^*(0) = c^*(0) = \underset{\sim}{w}.$$

B) If $\underset{\sim}{Q}_{\underset{\sim}{v}}(\underset{\sim}{w}) = \underset{\sim}{0}$, c must be tangent to the horizontal curve through $\underset{\sim}{v}$ along $\Pi\circ c$, that is, $\underset{\sim}{w}$ has vertical part $P_{\underset{\sim}{v}}(\underset{\sim}{w}) = \underset{\sim}{0}$, and conversely.

C) Restrict attention to the subspace

$$S = (\underset{\sim}{D}\pi)^{\leftarrow}\{\lambda\underset{\sim}{t} \mid \lambda \in R\} \subseteq T_{\underset{\sim}{v}}(TM)$$

for some non-zero $\underset{\sim}{t} \in T_pM$ represented by injective $f : J \to M$. Define a chart on $N = \Pi^{\leftarrow}(f(J))$ with image in the affine space $T_pM \times R$ by

$$\phi(\underset{\sim}{n}) = (\overleftarrow{\underset{\sim}{\tau}}_{t(\underset{\sim}{n})}(\underset{\sim}{n}), t(\underset{\sim}{n})) \text{ where } \underset{\sim}{n} \in N,\ t(\underset{\sim}{n}) = \overleftarrow{f}\,\Pi(\underset{\sim}{n}).$$

Then if $P : T_pM \times R \to T_pM : (x,t) \rightsquigarrow x$ and $P' = P\circ\phi : N \to T_pM$, $\underset{\sim}{Q}_{\underset{\sim}{v}}|S$ is exactly $\underset{\sim}{D}_{\underset{\sim}{v}}P'$ by 4.04. For, any $\underset{\sim}{w} \in S$ can be represented by a curve in TM which is a vector field along some parametrisation of f (unless $\underset{\sim}{D}\Pi(\underset{\sim}{w}) = \underset{\sim}{0}$, in which case we are covered by A).

So $\underset{\sim}{Q}_{\underset{\sim}{v}}|S$ is linear, being a derivative, and by A and B is idempotent with the same image and kernel as $\underset{\sim}{P}_{\underset{\sim}{v}}|S$. Hence by simple linear algebra $\underset{\sim}{Q}_{\underset{\sim}{v}}|S = \underset{\sim}{P}_{\underset{\sim}{v}}|S$.

D) Since any $\underset{\sim}{w} \in T_{\underset{\sim}{v}}(TM)$ is in some such S, it follows that $\underset{\sim}{Q}_{\underset{\sim}{v}} = \underset{\sim}{P}_{\underset{\sim}{v}}$.

■

4.06 Corollary

If $\underset{\sim}{w}$ is a vector field on M, and $f : J \to M$ represents $\underset{\sim}{t} \in T_pM$,

$$\nabla_{\underset{\sim}{t}}\underset{\sim}{w} = \lim_{h\to 0}\left(\frac{\overleftarrow{\underset{\sim}{\tau}_h}\,\underset{\sim}{w}_{f(h)} - \underset{\sim}{w}_p}{h}\right).$$

Proof

If $c = \underset{\sim}{w}\circ f$, in the notation of 4.05 we have

$$\begin{aligned}\lim_{h\to 0}\left(\frac{\overleftarrow{\underset{\sim}{\tau}_h}\,\underset{\sim}{w}_{f(h)} - \underset{\sim}{w}_p}{h}\right) &= \lim_{h\to 0}\frac{\bar{c}(h)-\bar{c}(0)}{h} \\ &= \underset{\sim}{d}_{\underset{\sim}{v}}(\bar{c}^*(0)) \in T_pM \text{ putting } \underset{\sim}{w}_p = \underset{\sim}{v} \\ &= \underset{\sim}{d}_{\underset{\sim}{v}}(\underset{\sim}{P}_{\underset{\sim}{v}}(c^*(0))) \quad \text{by 4.05} \\ &= \nabla_{f^*}\underset{\sim}{w}(0) \\ &= \nabla_{\underset{\sim}{t}}\underset{\sim}{w}.\end{aligned}$$

Thus the $\underset{\sim}{\tau}_h$ serve to "connect" vectors in nearby tangent spaces so as to let us differentiate vector fields in the ordinary way, because they give us $\nabla_{\underset{\sim}{t}}\underset{\sim}{w}$ as the ordinary tangent vector (freed) to the curve $h \rightsquigarrow (\overleftarrow{\underset{\sim}{\tau}_h}\underset{\sim}{w}_{f(h)} - \underset{\sim}{w}_p)$. So the equivalence between the $\underset{\sim}{\tau}_h$ and ∇ may be thought of as one being an "integrated" version of the other, the other a "differential", local, tangent-vectorial version of the one. Hence one old name for the Γ^k_{ij} of "infinitesimal connection". In exactly the same way a vector field's property of being the differential of the transformations ϕ_t obtained by integrating it (VII §6) explains the old term "infinitesimal transformation" for a vector field. Similarly, "infinitesimal displacement" for a tangent vector at a point.

4.07 Corollary

$$\nabla_{c_*}\underset{\sim}{w}(t) = \lim_{h\to 0}\left(\frac{\overleftarrow{\underset{\sim}{\tau}}_h^w \underset{\sim}{c}(t+h) - \underset{\sim}{w}c(t)}{h}\right).$$

Exercises VIII.4

1a) Suppose that manifolds M, N have connections ∇^M, ∇^N respectively. Using the decomposition in Ex. VII. 3.2c of any vector $\underset{\sim}{v}$ in $T_{(p,q)}M \times N$ into $\underset{\sim}{v}^M + \underset{\sim}{v}^N$, where $\underset{\sim}{v}^M \in T_pM$, $\underset{\sim}{v}^N \in T_qN$, show that

$$\nabla_{\underset{\sim}{t}}\underset{\sim}{v} = \nabla^M_{\underset{\sim}{t}^M}\underset{\sim}{v}^M + \nabla^N_{\underset{\sim}{t}^N}\underset{\sim}{v}^N$$

defines a connection on M × N, the <u>product</u> connection $\nabla^{M\times N}$ of ∇^M and ∇^N.

b) Show that a vector field $\underset{\sim}{w}$ along a curve $c : J \to M \times N$ is parallel with respect to $\nabla^{M\times N}$ if and only if both $\underset{\sim}{w}^M$ and $\underset{\sim}{w}^N$ are parallel with respect to ∇^M, ∇^N along c^M, c^N respectively, where $c(t) = (c^M(t), c^N(t)) \in M \times N$.

c) Deduce that a vector field $\underset{\sim}{w}$ along a curve c in M is parallel if and only if $\hat{\underset{\sim}{w}}$, defined by

$$\hat{\underset{\sim}{w}}(t) = (\underset{\sim}{w}(t)+\underset{\sim}{k}(t)) \in T_{c(t)}M \oplus T_{\bar{c}(t)}R \stackrel{\sim}{=} T_{(c(t),\bar{c}(t))}M \times R,$$

is parallel along $\tilde{c} : t \rightsquigarrow (c(t),\bar{c}(t)) \in M \times R$, where $\underset{\sim}{k}$ is a given vector field along a curve $\bar{c}$ in R, parallel with respect to the connection used on R.

2a) Show that if R has the connection given by

$$\nabla_{\underset{\sim}{t}} w = \overset{\leftarrow}{d}_{\underset{\sim}{x}}(w^*(0))$$

where t is a curve representing $\underset{\sim}{t}$, $x = t(0)$, and $w(s) = \underset{\sim}{d}_{t(s)}(\underset{\sim}{w}(t(s)))$, then a vector field along a curve c in R is parallel if and only if it is the restriction to c of a constant vector field on R (in the sense of VII.3.05).

b) Deduce that if $c : R \to R$ is the identity, then c^* is a parallel vector field along c with respect to this connection.

3a) If $c : J \to M$ is a curve in M, define $\tilde{c} : J \to M \times R : t \leadsto (c(t),t)$ and define a manifold structure on the set $N = \overset{\leftarrow}{\Pi}(\tilde{c}(J))$ of tangent vectors to $M \times R$ at points $\tilde{c}(t)$, $t \in J$.

b) Show that if $\underset{\sim}{v} \in T(T(M \times R))$ has $\underset{\sim}{D\Pi}(\underset{\sim}{v}) = \tilde{c}^*(t)$ for some t, then $\underset{\sim}{v}$ can be represented by a smooth curve in N and hence can be considered as a tangent vector to N.

c) Deduce that the map taking $\underset{\sim}{v} \in T_{\tilde{c}(t)}(M \times R)$ to $\underset{\sim}{t}_v = \tilde{c}^*(t) \in T_{\underset{\sim}{v}}(T_{\tilde{c}(t)}M \times R)$, in the notation of Ex. 3.6b, defines a smooth vector field $\underset{\sim}{t}$ on N.

This is called the <u>horizontal vector field</u> on N; any solution curve of $\underset{\sim}{t}$, considered as a curve in $T(M \times R)$ is a <u>horizontal curve</u> of the connection. (Evidently a curve c in $T(M \times R)$ is horizontal if and only if, considered as a vector field along $\Pi \circ c$, it is parallel.)

d) Show that any solution curve $w :]-\varepsilon,\varepsilon[\to N$ with $\Pi(w(0)) = \tilde{c}(0)$ has $\Pi(w(t)) = \tilde{c}(t)$ for all $t \in]-\varepsilon,\varepsilon[$, so that w may be thought of as a vector field $\underset{\sim}{w}$ along $\tilde{c}|]-\varepsilon,\varepsilon[$.

e) Deduce via 1c and 2b that $\underset{\sim}{w}^M$ is parallel along $c|]-\varepsilon,\varepsilon[$.

4) If $M = \{(r,\theta) | r = 1\}$, the circle in polar coordinates, with respect to any chart $U : M \to R : (r,\theta) \leadsto \theta$ let ∇ on M be given by $\Gamma^1_{11} = 1$. Drawing TM as a cylinder, sketch the horizontal curves in TM and show that although by 4.02 there is a parallel vector field through any vector along any curve, there is

no non-zero parallel vector field on M with respect to ∇.

5. TORSION AND SYMMETRY

If, as in 3.01.Cv), we have two vector fields $\underset{\sim}{t}$ and $\underset{\sim}{w}$ we can use a connection to differentiate either with respect to the other, getting either $\nabla_{\underset{\sim}{t}}\underset{\sim}{w}$ or $\nabla_{\underset{\sim}{w}}\underset{\sim}{t}$. By 4.06 $\nabla_{\underset{\sim}{t}(p)}\underset{\sim}{w}$ is "how $\underset{\sim}{w}$ varies as we flow along $\underset{\sim}{t}$ through p" and vice versa for $\nabla_{\underset{\sim}{w}(p)}\underset{\sim}{t}$. It would be nice if the results were necessarily the same, but in general things cannot be quite so neat. In Fig. 5.1 $\underset{\sim}{u}$ is "constant" along the solution curves of $\underset{\sim}{v}$ (parallel along them with the usual connection), so $\nabla_{\underset{\sim}{v}}\underset{\sim}{u} = \underset{\sim}{0}$ everywhere, while clearly $\nabla_{\underset{\sim}{u}}\underset{\sim}{v} \neq \underset{\sim}{0}$. So before deciding whether ∇ fails to be symmetrical in its effects on $\underset{\sim}{u}$ and $\underset{\sim}{v}$, we must correct for any lack of symmetry between flowing along $\underset{\sim}{u}$ and along $\underset{\sim}{v}$, themselves. As this should suggest, the appropriate fudge factor is the Lie bracket discussed in VII §7:

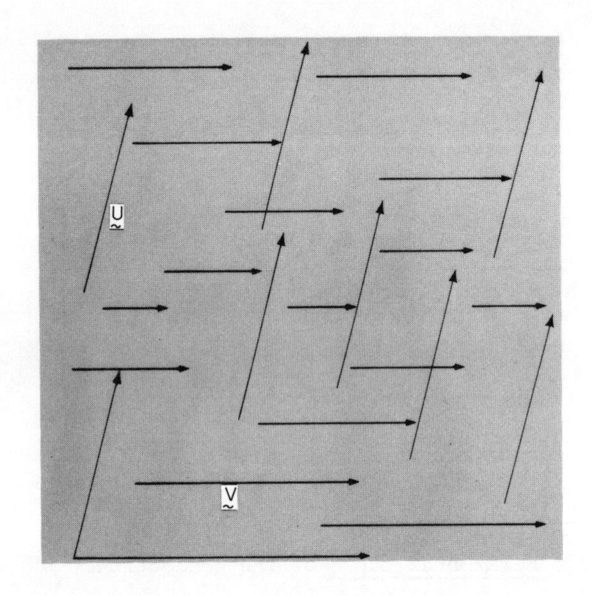

Fig. 5.1

5.01 Definition

The torsion of a connection ∇ on M is the map

$$\underset{\sim}{T} : T^1M \times T^1M \to T^1M : (\underset{\sim}{u},\underset{\sim}{v}) \rightsquigarrow \nabla_{\underset{\sim}{u}}\underset{\sim}{v} - \nabla_{\underset{\sim}{v}}\underset{\sim}{u} - [\underset{\sim}{u},\underset{\sim}{v}]$$

(Recall - VII.3.04 - that T^1M is the space of contravariant vector fields on M.)

If $\underset{\sim}{T}$ is identically zero, ∇ is symmetric.

5.02 Lemma

∇ is symmetric if and only if, whenever $\underset{\sim}{u}$, $\underset{\sim}{v}$ commute,

$$\nabla_{\underset{\sim}{u}}\underset{\sim}{v} = \nabla_{\underset{\sim}{v}}\underset{\sim}{u}.$$

Proof

i) If ∇ is symmetric, then for any $\underset{\sim}{u}$, $\underset{\sim}{v}$ with $[\underset{\sim}{u},\underset{\sim}{v}] = \underset{\sim}{0}$

$$\underset{\sim}{0} = \underset{\sim}{T}(\underset{\sim}{u},\underset{\sim}{v}) = \nabla_{\underset{\sim}{u}}\underset{\sim}{v} - \nabla_{\underset{\sim}{v}}\underset{\sim}{u} - [\underset{\sim}{u},\underset{\sim}{v}] = \nabla_{\underset{\sim}{u}}\underset{\sim}{v} - \nabla_{\underset{\sim}{v}}\underset{\sim}{u}.$$

So $$\nabla_{\underset{\sim}{u}}\underset{\sim}{v} = \nabla_{\underset{\sim}{v}}\underset{\sim}{u}.$$

ii) If $\nabla_{\underset{\sim}{u}}\underset{\sim}{v} = \nabla_{\underset{\sim}{v}}\underset{\sim}{u}$ for all commuting $\underset{\sim}{u}$, $\underset{\sim}{v}$, then for non-commuting $\underset{\sim}{u}$, $\underset{\sim}{v}$ we work locally. In a chart we have

$$\underset{\sim}{u} = u^i\partial_i, \quad \underset{\sim}{v} = v^j\partial_j$$

(VII.4.02) and the ∂_i all commute (Ex. VII.7.2c). Hence in the domain of the chart the vector field $\underset{\sim}{T}(\underset{\sim}{u},\underset{\sim}{v})$ has

$$\underset{\sim}{T}(\underset{\sim}{u},\underset{\sim}{v}) = \nabla_{u^i\partial_i}(v^j\partial_j) - \nabla_{v^j\partial_j}(u^i\partial_i) - [u^i\partial_i, v^j\partial_j]$$

$$= u^i(\partial_i v^j)\partial_j + u^i v^j(\nabla_{\partial_i}\partial_j) - v^j(\partial_j u^i)\partial_i - v^j u^i(\nabla_{\partial_j}\partial_i) - (u^i(\partial_i v^j)\partial_j - v^j(\partial_j u^i)\partial_i)$$

by 3.02, and Ex. VII.7.2a with a switch of dummy indices

$$* \qquad = u^i v^j (\nabla_{\partial_i} \partial_j - \nabla_{\partial_j} \partial_i)$$

$= \underset{\sim}{0}$ by hypothesis, since ∂_i, ∂_j commute. ■

5.03 Corollary

If around every $x \in M$ there is some chart giving Christoffel symbols for ∇ such that $\Gamma^k_{ij} = \Gamma^k_{ji}$, ∇ is symmetric. Conversely, if ∇ is symmetric all charts give Christoffel symbols with $\Gamma^k_{ij} = \Gamma^k_{ji}$.

Proof

The above proof showed that symmetry in the domain of a chart is equivalent to

$$\nabla_{\partial_i} \partial_j = \nabla_{\partial_j} \partial_i \qquad \forall i,j,$$

and to
$$\Gamma^k_{ij}\partial_k = \Gamma^k_{ji}\partial_k \qquad \forall i,j \text{ by } 3.02,$$

and therefore to
$$\Gamma^k_{ij} = \Gamma^k_{ji} \qquad \forall i,j,k \text{ since the } \partial_k \text{ are a basis.}$$
■

Fairly plainly, $(\nabla_{\underset{\sim}{u}}\underset{\sim}{v})_p$, $(\nabla_{\underset{\sim}{v}}\underset{\sim}{u})_p$ and $[\underset{\sim}{u},\underset{\sim}{v}]_p$ can be changed by substituting new fields $\underset{\sim}{v}'$ and $\underset{\sim}{u}'$ with $\underset{\sim}{v}'_p$ and $\underset{\sim}{v}_p$ and $\underset{\sim}{u}'_p = \underset{\sim}{u}_p$ but passing differently <u>through</u> these values (because we can always make $[\underset{\sim}{u}',\underset{\sim}{v}']_p = \underset{\sim}{0}$ for any given $\underset{\sim}{u}'_p$, $\underset{\sim}{v}'_p$). In contrast, for their combination into $\underset{\sim}{T}$ we have

5.04 Lemma

The vector $(\underset{\sim}{T}(\underset{\sim}{u},\underset{\sim}{v}))_p \in T_pM$ depends only on $\underset{\sim}{u}_p$ and $\underset{\sim}{v}_p$, and depends on them bilinearly, given ∇.

Proof

Putting the Γ^k_{ij} into * of 5.02, in a chart

$$\underset{\sim}{T}(\underset{\sim}{u},\underset{\sim}{v}) = u^i v^j (\Gamma^k_{ij} - \Gamma^k_{ji})\partial_k,$$

so at a point, $(\underset{\sim}{T}(\underset{\sim}{u},\underset{\sim}{v}))_p = u^i(p)v^j(p)(\Gamma^k_{ij}(p) - \Gamma^k_{ji}(p))\partial_k(p)$.

The result follows. ■

5.05 The torsion tensor

$\underset{\sim}{T}$ thus specifies, and is specified by, a bilinear map

$$T_pM \times T_pM \to T_pM$$

for each p, taking $(\underset{\sim}{u}_p,\underset{\sim}{v}_p)$ to $(\nabla_{\underset{\sim}{u}(p)}\underset{\sim}{v} - \nabla_{\underset{\sim}{v}(p)}\underset{\sim}{u} - [\underset{\sim}{u},\underset{\sim}{v}]_p)$ where $\underset{\sim}{u}$, $\underset{\sim}{v}$ are arbitrary vector fields with $\underset{\sim}{u}(p) = \underset{\sim}{u}_p$, $\underset{\sim}{v}(p) = \underset{\sim}{v}_p$. This corresponds to specifying an element $\underset{\sim}{T}_p$ of

$$\overset{*}{T}_pM \otimes \overset{*}{T}_pM \otimes T_pM$$

for each p, by a proof like that of V.1.08; in coordinates,

$$\underset{\sim}{T}_p = (\Gamma^k_{ij} - \Gamma^k_{ji})dx^i \otimes dx^j \otimes \partial_k.$$

From this we can recover the map $\underset{\sim}{T} : T^1M \times T^1M \to T^1M$ as a double contraction.

$$[(\Gamma^k_{ij}-\Gamma^k_{ji})dx^i \otimes dx^j \otimes \partial_k] \otimes u^\alpha\partial_\alpha \otimes v^\beta\partial_\beta \rightsquigarrow (\Gamma^k_{ij}-\Gamma^k_{ji})dx^i(u^\alpha\partial_\alpha)dx^j(v^\beta\partial_\beta)\partial_k.$$

That is, $\underset{\sim}{T} \otimes (\underset{\sim}{u} \otimes \underset{\sim}{v}) \rightsquigarrow u^i v^j (\Gamma^k_{ij}-\Gamma^k_{ji})\partial_k$, locally.

$\underset{\sim}{T}$, then, is essentially a $\binom{1}{2}$-tensor. We defer the discussion of its geometric nature to a future volume where space will permit consideration of significant examples with $\underset{\sim}{T} \neq 0$. For instance, $\underset{\sim}{T}$ can describe a "crystal dislocation density" in a continuum model for matter. For the present we are concerned only with the geometric implications of its vanishing.

Given a connection ∇ for which $\underset{\sim}{T} \equiv \underset{\sim}{0}$, we have

$$[\underset{\sim}{u},\underset{\sim}{v}]_p = \nabla_{\underset{\sim}{u}_p}\underset{\sim}{v} - \nabla_{\underset{\sim}{v}_p}\underset{\sim}{u}$$

which is sometimes used to <u>define</u> the Lie bracket. However this obscures the latter's independence of any connection or metric tensor, and conceals its closer relation to <u>flowing</u> along solution curves than <u>rolling</u> along them with the $\underset{\sim}{I}_t$ (compare Ex. VII.7.4 with 4.06). Notice in particular that for a flow ϕ, $\underset{\sim}{D}_x\phi_t$ is always an isomorphism but not usually an isometry: for instance, on R^2, if $\underset{\sim}{v}(x,y) = x\underset{\sim}{e}_1 + y\underset{\sim}{e}_2$, what is $\underset{\sim}{D}_{(0,0)}\phi_1$ for the corresponding flow? (Ex. 1)

<u>Exercises VIII.5</u>

1a) Show that if $\underset{\sim}{v}(x,y) = x\underset{\sim}{e}_1 + y\underset{\sim}{e}_2$, the corresponding flow has $\phi_t(x,y) = (e^t x, e^t y)$.

b) Deduce that $\underset{\sim}{D}_{(0,0)}\phi_1 : T_{(0,0)}R^2 \to T_{(0,0)}R^2$ is $e\underset{\sim}{I}$.

2a) Prove (similarly to Lemma 5.04) that if $\nabla, \tilde{\nabla}$ are connections on M, their difference

$$\underset{\sim}{S}(\underset{\sim}{u},\underset{\sim}{v}) = \tilde{\nabla}_{\underset{\sim}{u}}\underset{\sim}{v} - \nabla_{\underset{\sim}{u}}\underset{\sim}{v}$$

is essentially a $\binom{1}{2}$-tensor field.

b) Show that for any connection ∇ and any $\binom{1}{2}$-tensor field S on M, the formula

$$\tilde{\nabla}_{\underset{\sim}{u}}\underset{\sim}{v} = \nabla_{\underset{\sim}{u}}\underset{\sim}{v} + \underset{\sim}{S}(\underset{\sim}{u},\underset{\sim}{v})$$

defines another connection $\tilde{\nabla}$ (that is, $\tilde{\nabla}$ satisfies Def^n 3.01).

6. METRIC TENSORS AND CONNECTIONS

6.01 Definition

A connection ∇ on a manifold M with a metric tensor field $\underset{\sim}{G}$ is compatible with $\underset{\sim}{G}$ if all the parallel transports $\underset{\sim}{\tau}_t$ it defines are isometries: we require

$$\underset{\sim}{G}_q(\underset{\sim}{\tau}_t\underset{\sim}{u}_p,\underset{\sim}{\tau}_t\underset{\sim}{v}_p) = \underset{\sim}{G}_p(\underset{\sim}{u}_p,\underset{\sim}{v}_p)$$

for all $p,q \in M$, $\underset{\sim}{u}_p$, $\underset{\sim}{v}_p \in T_pM$, along all curves.

That is, parallel transport must preserve lengths and angles: an obvious condition if it is to correspond to the rollings around of §2. A slightly less obvious one relates to the torsion tensor of ∇.

On R^2, for instance, we can define a connection in the usual coordinates with $\Gamma^1_{12} = 1$ and the other $\Gamma^k_{ij} = 0$. The corresponding parallel transport, along any curve from (x,y) to (x',y'), amounts to "rotation through $\overleftarrow{\sin}(x-x')$" relative to the usual idea of parallelism (Ex. 1). Since rotations are isometries, this connection is thus compatible with the metric, but clearly it is not the one we usually want.

In general, the torsion tensor relates to this twisting of parallel transport away from the kind we want, which corresponds to rolling without turning. Hence we shall generally require connections to have torsion zero : non-zero torsion represents "extra structure" in a sense we outline in 6.09.

These two conditions, compatibility with $\underset{\sim}{G}$ and symmetry, suffice to determine the connection appropriate to M with $\underset{\sim}{G}$. In this volume our only further straying from zero torsion is in Ex. IX.1.2, whose only purpose is to highlight the odd geometry of the above example.

We confine ourselves henceforward to a specific M and $\underset{\sim}{G}$, and abbreviate $\underset{\sim}{G}(\underset{\sim}{u},\underset{\sim}{v})$ to $\underset{\sim}{u}.\underset{\sim}{v}$. First, let us reduce 6.01 to a local condition:

6.02 Lemma

∇ is compatible with $\underset{\sim}{G}$ if and only if, for any vector fields $\underset{\sim}{u}$, $\underset{\sim}{v}$ along any curve $c : J \to M$ and $t \in J$,

$$* \qquad \frac{d(\underset{\sim}{u}.\underset{\sim}{v})}{ds}(t) = (\nabla_{c^*}\underset{\sim}{u}(t)).\underset{\sim}{v}(t) + \underset{\sim}{u}(t).\nabla_{c^*}\underset{\sim}{v}(t)$$

along c, where ∇_{c^*} is obtained from ∇ as in 3.05.

Proof

First suppose * holds everywhere. Then for any parallel vector field $\underset{\sim}{w}$ along c we have

$$\frac{d}{ds}(\underset{\sim}{w}.\underset{\sim}{w}) = \nabla_{c^*}\underset{\sim}{w}.\underset{\sim}{w} + \underset{\sim}{w}.\nabla_{c^*}\underset{\sim}{w} = \underset{\sim}{0}.\underset{\sim}{w} + \underset{\sim}{w}.\underset{\sim}{0} = 0.$$

Thus $\underset{\sim}{w}.\underset{\sim}{w}$ is constant along c, so that if $c(0) = p$

$$** \qquad \underset{\sim}{\tau}_t\underset{\sim}{w}_p.\underset{\sim}{\tau}_t\underset{\sim}{w}_p = \underset{\sim}{w}_{c(t)}.\underset{\sim}{w}_{c(t)} = \underset{\sim}{w}_p.\underset{\sim}{w}_p \quad \text{along } c$$

for any $\underset{\sim}{w}_p \in T_pM$. Consequently

$$\tfrac{1}{4}(\underset{\sim}{\tau}_t(\underset{\sim}{u}_p+\underset{\sim}{v}_p).\underset{\sim}{\tau}_t(\underset{\sim}{u}_p+\underset{\sim}{v}_p)-\underset{\sim}{\tau}_t(\underset{\sim}{u}_p-\underset{\sim}{v}_p).\underset{\sim}{\tau}_t(\underset{\sim}{u}_p-\underset{\sim}{v}_p)) = \tfrac{1}{4}((\underset{\sim}{u}_p+\underset{\sim}{v}_p).(\underset{\sim}{u}_p+\underset{\sim}{v}_p)-(\underset{\sim}{u}_p-\underset{\sim}{v}_p).(\underset{\sim}{u}_p-\underset{\sim}{v}_p))$$

applying ** with $\underset{\sim}{u}_p+\underset{\sim}{v}_p$ and $\underset{\sim}{u}_p-\underset{\sim}{v}_p$ for $\underset{\sim}{w}_p$. Expanding and cancelling,

$$\underset{\sim}{\tau}_t\underset{\sim}{u}_p\cdot\underset{\sim}{\tau}_t\underset{\sim}{v}_p = \underset{\sim}{u}_p\cdot\underset{\sim}{v}_p.$$

Conversely, if ∇ is compatible with $\underset{\sim}{G}$, choose an orthonormal basis (IV.3.05) $\underset{\sim}{b}_1,\ldots,\underset{\sim}{b}_n$ for T_pM. Then the parallel fields $\underset{\sim}{\beta}_i(t) = \tau_t(\underset{\sim}{b}_i)$ along any $c : J \to M$ with $c(0) = p$ give an orthonormal basis for $T_{c(t)}M$, $\forall t \in J$, since $\underset{\sim}{\tau}_t$ is an isometry. Hence any $\underset{\sim}{u}$, $\underset{\sim}{v}$ along c can be written as $\underset{\sim}{u} = u^i\underset{\sim}{\beta}_i$, $\underset{\sim}{v} = v^j\underset{\sim}{\beta}_j$, and

$$\begin{aligned}
\frac{d}{ds}(\underset{\sim}{u}\cdot\underset{\sim}{v}) &= \frac{d}{ds}((u^i\underset{\sim}{\beta}_i)\cdot(v^j\underset{\sim}{\beta}_j)) \\
&= \frac{d}{ds}(u^1v^1 + \ldots + u^nv^n) \\
&= (\frac{du^1}{ds}v^1 + \ldots + \frac{du^n}{ds}v^n) + (u^1\frac{dv^1}{ds} + \ldots + u^n\frac{dv^n}{ds}) \\
&= (\frac{du^i}{ds}\underset{\sim}{\beta}_i)\cdot\underset{\sim}{v}+\underset{\sim}{u}\cdot(\frac{dv^i}{ds}\underset{\sim}{\beta}_i).
\end{aligned}$$

Now $\nabla_{c*}(f\,\underset{\sim}{\beta}_i) = \frac{df}{ds}\underset{\sim}{\beta}_i + f\,\nabla_{c*}\underset{\sim}{\beta}_i$, by Ex. VIII.3.4b for any f, $\underset{\sim}{\beta}_i$;

$= \frac{df}{ds}\underset{\sim}{\beta}_i$ since $\underset{\sim}{\beta}_i$ is parallel by construction.

$$\begin{aligned}
\text{Thus } \frac{d}{ds}(\underset{\sim}{u}\cdot\underset{\sim}{v}) &= \left(\nabla_{c*}u^i\underset{\sim}{\beta}_i\right)\cdot\underset{\sim}{v}+\underset{\sim}{u}\cdot\left(\nabla_{c*}v^i\underset{\sim}{\beta}_i\right) \\
&= \nabla_{c*}\underset{\sim}{u}\cdot\underset{\sim}{v} + \underset{\sim}{u}\cdot\nabla_{c*}\underset{\sim}{v}, \text{ as required.}
\end{aligned}$$

■

<u>6.03 Corollary</u>

∇ is compatible with G if and only if

$$\underset{\sim}{w}(\underset{\sim}{u}\cdot\underset{\sim}{v}) = (\nabla_{\underset{\sim}{w}}\underset{\sim}{u})\cdot\underset{\sim}{v} + \underset{\sim}{u}\cdot(\nabla_{\underset{\sim}{w}}\underset{\sim}{v})$$

for all $\underset{\sim}{w} \in T_pM$, $\underset{\sim}{u}, \underset{\sim}{v} \in T^1M$.

Proof

Apply 6.02 to a curve representing $\underset{\sim}{w}$. ■

6.04 Theorem

There exists exactly one symmetric connection ∇ compatible with any given metric tensor field $\underset{\sim}{G}$.

Proof

Suppose ∇ exists: we must express it in terms of $\underset{\sim}{G}$.

Rather than look for $\nabla_{\underset{\sim}{u}}\underset{\sim}{v}$ directly, given $\underset{\sim}{u}$ and $\underset{\sim}{v}$, the trick is to find first $\underset{\sim}{G}_{\downarrow}(\nabla_{\underset{\sim}{u}}\underset{\sim}{v})$ (cf. VII.4.05), that is, the covariant vector field $\underset{\sim}{w} \rightsquigarrow (\nabla_{\underset{\sim}{u}}\underset{\sim}{v}).\underset{\sim}{w}$.

By 6.03 we have

$$* \quad (\nabla_{\underset{\sim}{u}}\underset{\sim}{v}).\underset{\sim}{w} = \underset{\sim}{u}(\underset{\sim}{v}.\underset{\sim}{w}) - \underset{\sim}{v}.(\nabla_{\underset{\sim}{u}}\underset{\sim}{w})$$

and by 5.01 and ∇'s supposed symmetry,

$$** \quad \nabla_{\underset{\sim}{u}}\underset{\sim}{w} = \nabla_{\underset{\sim}{w}}\underset{\sim}{u} + [\underset{\sim}{u},\underset{\sim}{w}].$$

Hence

$$\begin{aligned}
(\nabla_{\underset{\sim}{u}}\underset{\sim}{v}).\underset{\sim}{w} &= \underset{\sim}{u}(\underset{\sim}{v}.\underset{\sim}{w})-\underset{\sim}{v}.(\nabla_{\underset{\sim}{w}}\underset{\sim}{u}+[\underset{\sim}{u},\underset{\sim}{w}]) \\
&= \underset{\sim}{u}(\underset{\sim}{v}.\underset{\sim}{w})-(\nabla_{\underset{\sim}{w}}\underset{\sim}{u}).\underset{\sim}{v}-\underset{\sim}{v}.[\underset{\sim}{u}.\underset{\sim}{w}] \qquad (\underset{\sim}{G} \text{ symmetric}) \\
&= \underset{\sim}{u}(\underset{\sim}{v}.\underset{\sim}{w})-(\underset{\sim}{w}(\underset{\sim}{u}.\underset{\sim}{v})-\underset{\sim}{u}.(\nabla_{\underset{\sim}{w}}\underset{\sim}{v}))-\underset{\sim}{v}.[\underset{\sim}{u}.\underset{\sim}{w}] \qquad \text{by } * \\
&= \underset{\sim}{u}(\underset{\sim}{v}.\underset{\sim}{w})-\underset{\sim}{w}(\underset{\sim}{u}.\underset{\sim}{v})+\underset{\sim}{u}.(\nabla_{\underset{\sim}{v}}\underset{\sim}{w}+[\underset{\sim}{w},\underset{\sim}{v}])-\underset{\sim}{v}.[\underset{\sim}{u},\underset{\sim}{w}] \qquad \text{by } ** \\
&= \underset{\sim}{u}(\underset{\sim}{v}.\underset{\sim}{w})-\underset{\sim}{w}(\underset{\sim}{u}.\underset{\sim}{v})+(\nabla_{\underset{\sim}{v}}\underset{\sim}{w}).\underset{\sim}{u}+\underset{\sim}{u}.[\underset{\sim}{w},\underset{\sim}{v}]-\underset{\sim}{v}.[\underset{\sim}{u},\underset{\sim}{w}]
\end{aligned}$$

$$= \ u(v.w)-w(u.v)+(v(w.u)-w(\nabla_v u))+u[w,v]-v.[u,w] \quad \text{by } *$$

$$= \ u(v.w)-w(u.v)+v(w.u)-w.(\nabla_u v+[v,u])+u.[w,v]-v.[u,w] \quad \text{by } **$$

$$= \ u(v.w)-w(u.v)+v(w.u)-(\nabla_u v).w-w.[v,u]+u.[w,v]-v.[u,w]$$

$$2(\nabla_u v).w \ = \ u(v.w)-w(u.v)+v(w.u)-w.[v,u]+u.[w,v]-v.[u,w]. \qquad ***$$

Thus if ∇ exists it satisfies ***, which fixes $G_{\downarrow}(\nabla_u v)$, and therefore $\nabla_u v$ uniquely. It remains to prove existence.

The value at $p \in M$ of the expression on the right of *** - call it $z(u,v,w)$ for short - depends for fixed vector fields u and v only, and linearly, on the value w_p of w at p. (Ex. 2a-c) Hence since all vectors in T_pM can occur as w_p, we have a well-defined linear functional

$$(Y(u,v))_p : T_pM \to R : w_p \rightsquigarrow z(u,v,w)$$

where w is any vector field with $w(p) = w_p$. If we now <u>define</u>

$$\nabla_u v \ = \ G_{\uparrow}(Y(u,v))$$

we have a contravariant vector field on M which satisfies ***. A precisely similar proof to Ex. 2 shows that $z(u,v,w)$, and hence $\nabla_u v$, depends only, and linearly, on u_p if we fix vector fields v and w, so we have a well defined $\nabla_{u_p} v$ with Ci) and Ciii) of 3.01 satisfied. Cii) follows from the immediate fact that $z(u,v+v',w) = z(u,v,w) + z(u,v',w)$, and Civ) by expanding $z(u,fv,w)$ to get (Ex. 2d)

$$z(u,fv,w) \ = \ fz(u,v,w) + u(f)(v.w)$$

so that $$\nabla_{u_p}(fv).w_p \ = \ (f(p)\nabla_{u_p} v).w_p + (u_p(f)v_p).w_p \quad \forall p \quad \text{and hence}$$

$$\nabla_{u_p}(fv) \ = \ f(p)\nabla_{u_p} v + u_p(f)v_p \quad \text{as required.}$$

Cv) follows from the fact that $\underset{\sim}{G}_{\uparrow}$ and the operations by which $\underset{\sim}{z}(\underset{\sim}{u},\underset{\sim}{v},\underset{\sim}{w})$ is defined all give C^{∞} results from C^{∞} data.

Hence ∇ is indeed a connection. Similarly to the proof of Civ), compatibility with $\underset{\sim}{G}$ follows directly from

$$\underset{\sim}{z}(\underset{\sim}{u},\underset{\sim}{v},\underset{\sim}{w}) + \underset{\sim}{z}(\underset{\sim}{u},\underset{\sim}{w},\underset{\sim}{v}) = \underset{\sim}{u}(\underset{\sim}{v}\cdot\underset{\sim}{w})$$

and 6.03, and symmetry from

$$\underset{\sim}{z}(\underset{\sim}{u},\underset{\sim}{v},\underset{\sim}{w}) - \underset{\sim}{z}(\underset{\sim}{v},\underset{\sim}{u},\underset{\sim}{w}) = [\underset{\sim}{u},\underset{\sim}{v}]\cdot\underset{\sim}{w} .$$

These equations result from simply writing out their left-hand sides in full and collecting terms, using the symmetry of $\underset{\sim}{G}$ and the skew-symmetry of [,]. ■

6.05 Definition

The unique symmetric connection compatible with $\underset{\sim}{G}$ is called the Levi-Cività connection for $\underset{\sim}{G}$. From now on, ∇ (on a manifold for which we have a metric tensor field) will always refer to this connection unless we explicitly state otherwise. (cf. Ex. 2.1 above.)

6.06 Components

For the $\partial_1,\ldots,\partial_n$ given by a chart, the Lie brackets in *** of the proof of 6.04 vanish, so

$$(\nabla_{\partial_i}\partial_j)\cdot\partial_k = \tfrac{1}{2}(\partial_i(\partial_j\cdot\partial_k) - \partial_k(\partial_i\cdot\partial_j) + \partial_j(\partial_k\cdot\partial_i))$$

$$= \tfrac{1}{2}(\partial_i(g_{jk}) - \partial_k(g_{ij}) + \partial_j(g_{ki}))$$

by the definition (IV.3.01) of the components of $\underset{\sim}{G}$. The function $(\nabla_{\partial_i}\partial_j).\partial_k$, giving (for $\underset{\sim}{G}$ Riemannian) the component, by orthogonal projection at each point, of $\nabla_{\partial_i}\partial_j$ in the ∂_k direction, is called for historical reasons a <u>Christoffel symbol of the first kind</u> and denoted by Γ_{ijk}. The full name of the Γ^k_{ij} we have already met is "Christoffel symbols of the <u>second</u> kind." We have

$$\begin{aligned} \tfrac{1}{2}(\partial_i g_{j\ell} - \partial_\ell g_{ij} + \partial_j g_{\ell i}) &= (\Gamma^m_{ij}\partial_m).\partial_\ell \\ &= \Gamma^m_{ij}\, g_{m\ell} \end{aligned}$$

so

$$\begin{aligned} \tfrac{1}{2}g^{k\ell}(\partial_i g_{j\ell} - \partial_\ell g_{ij} + \partial_j g_{\ell i}) &= \Gamma^m_{ij}\, g_{m\ell}g^{k\ell} \\ &= \Gamma^m_{ij}\, \delta^k_m \\ &= \Gamma^k_{ij} \end{aligned}$$

by the usual formulae. So we can apply the usual formula for raising indices, setting

$$\Gamma^k_{ij} = g^{k\ell}\Gamma_{ij\ell},$$

although this is <u>not</u> simply an application of $\underset{\sim}{G}_{\downarrow}$ since the $\Gamma_{ij\ell}$ do not constitute a tensor.

An occasionally useful fact is that

$$\begin{aligned} \Gamma_{ijk} + \Gamma_{jki} &= \tfrac{1}{2}(\partial_i g_{jk} - \partial_k g_{ij} + \partial_j g_{ki}) + \tfrac{1}{2}(\partial_j g_{ki} - \partial_i g_{jk} + \partial_k g_{ij}) \\ &= \partial_k g_{ij} \quad \text{for all } i,j,k. \end{aligned}$$

6.07 Rolling

Neither Theorem 6.04 nor the formulae above are outstandingly geometrical, so it is worth rigorously tying the Levi-Civitȁ connection to the "rolling" approach we first discussed.

If M is embedded in an affine space X with vector space T and a constant metric inducing the metric tensor field $\underset{\sim}{G}$ on M, and $c : J \to M$ has $c(0) = p$, let $\underset{\sim}{A}$ be any isometry from R^n (with a metric of the appropriate signature) to T_pM. Then if we define

$$\underset{\sim}{A}_t = \underset{\sim}{d}_{c(t)} \circ \underset{\sim}{\tau}_t \circ \underset{\sim}{A} : R^n \to X,$$

where $\underset{\sim}{\tau}_t$ is parallel transport along c from T_pM to $T_{c(t)}M \subseteq T_{c(t)}X$ with respect to the Levi-Civitȁ connection for $\underset{\sim}{G}$, we have the unique family $\underset{\sim}{A}_t$ satisfying (i), (ii), (iii) of Ex. 2.1. For condition (i) follows from the requirement that ∇ be compatible with $\underset{\sim}{G}$, and (ii) holds by construction. It is clear that the curve $c_{\underset{\sim}{x}}$ of Ex. 2.1 is exactly the curve $t \rightsquigarrow \underset{\sim}{d}_{c(t)}\hat{\underset{\sim}{x}}(t)$, where $\hat{\underset{\sim}{x}}$ is the parallel vector field along c through $\underset{\sim}{A}(\underset{\sim}{x})$. It remains only to show that (iii) follows from the facts that $\hat{\underset{\sim}{x}}$ <u>is</u> parallel, for all $\underset{\sim}{x} \in R^n$, and that ∇ is symmetric: this we leave as an exercise (Ex. 3) in components. Notice that this proves the existence of the $\underset{\sim}{A}_t$, and that uniqueness follows similarly, since essentially what is involved is the equivalence of the system of differential equations defining parallelism to condition (iii), for ∇ symmetric. Hence the differentiation of §2 is indeed that given by the Levi-Civitȁ connection.

Thus we have rigorously established the "rolling" picture, which is useful for testing our intuition about the Levi-Civitȁ connection: Fig. 4.1 for instance had the connections for the normal Riemannian metrics on the plane and sphere in mind.

On a point of language: notice that with the normal connection on the plane the field shown in Fig. 6.1 is <u>not</u> parallel in the sense of 4.01, since parallel transport preserves lengths as well as angles, though in the elementary sense of "having the same direction" the vectors are parallel. The alternatives are to redefine "parallel", to use some other word like "constant" which would involve worse confusion, or invent something like "translationally congruent". The universal choice is the first.

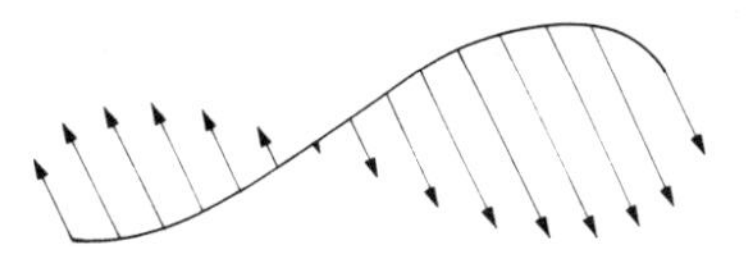

Fig. 6.1

Rolling vectors around on a manifold with an indefinite metric takes a little more imagination: notice in particular that since it preserves the lengths of vectors it takes timelike/null/spacelike vectors to others of the same kind. There can be nothing like the way any direction on the sphere (with the usual connection) can be rolled/parallel transported to any other.

6.08 Signs

In terms of rolling, it is clear that replacing $\underset{\sim}{G}$ by $-\underset{\sim}{G}$ should not change the resulting parallel transport (nor, hence, the connection) since the same maps are isometries. In components it can be seen that replacing g_{ij} by $-g_{ij}$ leaves $\Gamma^k_{ij} = \frac{1}{2}g^{k\ell}(\partial_i g_{j\ell} - \partial_\ell g_{ij} + \partial_j g_{\ell i})$, (by 6.06), unchanged.

Thus neither the sign of ∇ nor that of the Riemann tensor we define from it in Chapter X is altered if we change from a (+---) metric tensor on spacetime to the equally popular (-+++)

6.09 Naturality

We have seen the way that a finite-dimensional vector space V is isomorphic to its dual V*, but not "naturally" so. To choose a particular isomorphism $\underset{\sim}{A} : V \to V^*$ is equivalent to choosing a non-degenerate bilinear form $\underset{\sim}{B}$ such that $\underset{\sim}{B}(\underset{\sim}{u},\underset{\sim}{v}) = \underset{\sim}{A}\underset{\sim}{u}(\underset{\sim}{v})$ on V; for example, a metric tensor. Such an extra structure contains all the geometric possibilities of Ch. IV, and more besides. Similarly, a metric tensor field is considerable extra structure for a manifold. Given such a field, how much extra does a connection represent?

We have seen that a metric $\underset{\sim}{G}$ determines a particular ∇ via the conditions of compatibility and symmetry. It is a recent result in [Stredder] that $\underset{\sim}{G}$ determines this ∇ via merely the condition that ∇ should <u>not</u> represent a further choice of structure, in a sense roughly as follows.

If N' is an open subset of a manifold N, it is also a manifold in an obvious way. It is clear how to "restrict" a connection ∇_N or metric tensor $\underset{\sim}{G}_N$ on N to $\nabla_N|N'$ or $\underset{\sim}{G}_N|N'$ on N'. Suppose we have some rule that assigns to every manifold-with-metric-tensor $(M,\underset{\sim}{G}_M)$ a connection $\nabla^{\underset{\sim}{G}_M}$ on M. Suppose also that whenever N' is an open subset of $(N,\underset{\sim}{G}_n)$, the connection $\nabla^{\underset{\sim}{G}_N|N'}$ we get by applying the rule to $(N',\underset{\sim}{G}_N|N')$ coincides with the connection $\nabla^{\underset{\sim}{G}_N}|N'$ we get by applying the rule to $(N,\underset{\sim}{G}_n)$ and restricting the result to N'. Then the rule can <u>only</u> be "Choose the Levi-Cività connection"! Both compatibility with the metric and symmetry turn out to be consequences of "naturality with respect to restrictions".

So given $(M,\underset{\sim}{G})$ we get the Levi-Cività connection ∇ free in the same package: any other connection $\tilde{\nabla}$ we pay for with a special choice, representing extra structure. (In physics this would generally mean another force or form of matter added to the theory.) By Ex. 5.2 we see that the special choice involved is exactly that of the $\binom{1}{2}$-tensor field describing the difference between $\tilde{\nabla}$ and ∇.

<u>Exercises VIII.6</u>

1) Prove that parallel transport on R^2 by the connection given in the usual coordinates by $\Gamma^1_{12} = 1$, the other $\Gamma^k_{ij} = 0$, is as described in the discussion after 6.01. (Either translate the description into a formula for a general horizontal curve and prove that it solves the differential equation, or apply Th^m. 4.06).

2) In the proof of 6.04:

a) Show that $\underset{\sim}{z}(\underset{\sim}{u},\underset{\sim}{v},\underset{\sim}{w}+\underset{\sim}{w}') = \underset{\sim}{z}(\underset{\sim}{u},\underset{\sim}{v},\underset{\sim}{w}) + \underset{\sim}{z}(\underset{\sim}{u},\underset{\sim}{v},\underset{\sim}{w}')$

b) Show that $\underset{\sim}{u}(\underset{\sim}{v}.f\underset{\sim}{w}) = \underset{\sim}{u}(f)(\underset{\sim}{v}.\underset{\sim}{w})+f(\underset{\sim}{u}(\underset{\sim}{v}.\underset{\sim}{w}))$. Use this and Ex. VII.7.5 to show that $\underset{\sim}{z}(\underset{\sim}{u},\underset{\sim}{v},f\underset{\sim}{w}) = f(\underset{\sim}{z}(\underset{\sim}{u},\underset{\sim}{v},\underset{\sim}{w}))$.

c) Show that for any linear $F : T^1_0M \to T^0_0M$ with $F(f\underset{\sim}{w}) = f(F(\underset{\sim}{w}))$ $\forall \underset{\sim}{w}$, f, $(F(\underset{\sim}{w}))_p$ depends only on $\underset{\sim}{w}_p$. (Show that the question is local by taking f zero outside the domain of a chart p. In this chart write $\underset{\sim}{w} = w^i\partial_i$ and consider $\underset{\sim}{w}$ and $\underset{\sim}{w}'$ with $\underset{\sim}{w}'(p) = \underset{\sim}{w}(p)$.) Deduce that $\underset{\sim}{Y}(\underset{\sim}{u},\underset{\sim}{v})$ is a covariant vector field.

(Note the similarity between this proof and the proof (5.04) that the torsion of a connection is a tensor field.)

d) Prove that $\underset{\sim}{z}(\underset{\sim}{u},f\underset{\sim}{v},\underset{\sim}{w}) = f\underset{\sim}{z}(\underset{\sim}{u},\underset{\sim}{v},\underset{\sim}{w}) + 2\underset{\sim}{u}(f)(\underset{\sim}{v}.\underset{\sim}{w})$.

3a) In the situation of 6.07, take coordinates $(x^1,\dots,x^N)$ on X corresponding to some choice of an origin for X and orthonormal basis for T, take a chart giving coordinates $(q^1,\dots,q^n)$ for points q in a neighbourhood of $p \in M$, and define $x^i(q)$ = i-th coordinate of q as a point in X, i = 1,...,N. Write out everything in these coordinates, and prove condition (iii).

b) Deduce via Th^m. 4.04 that the definition of ∇ in Ex. 2.1 gives exactly the Levi-Cività connection for the metric induced by the embedding.

4) The Levi-Cività connection for any constant metric tensor on an affine space has parallel transport

$$\underset{\sim}{\tau} = \overleftarrow{\underset{\sim}{d}_q}\underset{\sim}{d}_p : T_pX \to T_qX$$

independently of the curve.

5a) Using the charts on the sphere S^2 set up in Ex. VII.2.1, find the components g_{ij} of the metric induced by the standard embedding in R^3 with the standard metric. (These are <u>not</u> just δ_{ij}, since the ∂_i's produced are not orthonormal except at special points.)

b) Find the Γ^k_{ij} for the corresponding Levi-Cività connection.

c) Show that around any $p \in S^2$ there is a chart making the Γ^k_{ij} all zero at p (though not around p).

d) Repeat for the general sphere $S^n \subseteq R^{n+1}$.

6) Use Ex. 3.1, Ex. 3.4v and 6.06 to give in components a metric tensor on TM such that the resulting orthogonal projections $\underset{\sim}{P}_{\underset{\sim}{v}}$ onto vertical subspaces give the Levi-Cività connection by

$$\nabla_{\underset{\sim}{u}_p} \underset{\sim}{v} = \underset{\sim}{d}_{\underset{\sim}{v}_p}(\underset{\sim}{P}_{\underset{\sim}{v}_p}(\underset{\sim}{D}_p \underset{\sim}{v}(\underset{\sim}{u}_p))).$$

7) Show that the connection in Ex. 4.4 is incompatible with any metric.

8) If M, N and M × N have metrics $\underset{\sim}{G}^M$, $\underset{\sim}{G}^N$, and $\underset{\sim}{G}^{M \times N}$ (Ex. VII.3.2v), show that the Levi-Cività connection on M × N is the product (Ex. 4.1) of those on M, N, and that a vector field along $c : t \mapsto (c^M(t), c^N(t)) \in M \times N$ is parallel if and only if its M, N components are parallel along c^M, c^N in M, N respectively.

7. COVARIANT DIFFERENTIATION OF TENSORS

In sections 2 to 6 of this chapter we have established how to differentiate $\binom{1}{0}$-tensor fields when we have a metric. We already knew how to differentiate $\binom{0}{0}$-tensor fields: what about other types? Fortunately, we do not have to invent new machinery for each - what we already have will rapidly set up differentiation for them all.

7.01 Transporting tensors

If $\underset{\sim}{\tau}_t : T_pM \to T_{c(t)}M$ is parallel transport of vectors along $c : J \to M$ from p to $c(t) = q$, define parallel transport of $\binom{k}{h}$-tensors along c from p to q by

$$(\underset{\sim}{\tau}{}^k_h)_t : T_pM \otimes \ldots \otimes T_pM \otimes T^*_pM \otimes \ldots \otimes T^*_pM \to T_qM \otimes \ldots \otimes T_qM \otimes T^*_qM \otimes \ldots \otimes T^*_qM$$

$$\underset{\sim}{v}_1 \otimes \ldots \otimes \underset{\sim}{v}_k \otimes \underset{\sim}{f}_1 \otimes \ldots \otimes \underset{\sim}{f}_k \rightsquigarrow \underset{\sim}{\tau}_t\underset{\sim}{v}_1 \otimes \ldots \otimes \underset{\sim}{\tau}_t\underset{\sim}{v}_k \otimes (\underset{\sim}{\tau}^*_t)^{\leftarrow}\underset{\sim}{f}_1 \otimes \ldots \otimes (\underset{\sim}{\tau}^*_t)^{\leftarrow}\underset{\sim}{f}_h$$

on simple tensors. (Recall that $(\underset{\sim}{\tau}^*_t)^{\leftarrow}\underset{\sim}{f}$ just means the functional whose value on a vector $\underset{\sim}{v}$ at q is obtained by transporting $\underset{\sim}{v}$ back to p and evaluating $\underset{\sim}{f}$ on the result (cf. III.1.03): we have $\underset{\sim}{\tau}^*_t\underset{\sim}{f} = \underset{\sim}{f}\circ\underset{\sim}{\tau}_t$, $(\tau^*_t)^{\leftarrow}\underset{\sim}{f} = \underset{\sim}{f}\circ\underset{\sim}{\tau}^{\leftarrow}_t$). Evidently this reduces back to $\underset{\sim}{\tau}$ on $\binom{1}{0}$-tensors, and for $\binom{0}{0}$-tensors it is just the identity $R \to R$, the usual tensor product of zero copies of a map.

7.02 Definition

If $\underset{\sim}{v}$ is a $\binom{k}{h}$-tensor field on M, $\underset{\sim}{u} \in T_pM$, and $c : J \to M$ any representative of $\underset{\sim}{u}$, we define the <u>covariant directional derivative</u> of $\underset{\sim}{v}$ with respect to $\underset{\sim}{u}$ as

$$\nabla_{\underset{\sim}{u}}\underset{\sim}{v} = \lim_{t\to 0}\left(\frac{(\underset{\sim}{\tau}{}^k_h)^{\leftarrow}_t\,\underset{\sim}{v}_{c(t)} - \underset{\sim}{v}_p}{t}\right)$$

That this is independent of the choice of representing curve c follows from:

7.03 Theorem

The derivative defined in 7.02 has the following properties.

A) If $f \in T^0_0M$, $\nabla_{\underset{\sim}{u}}f = \underset{\sim}{u}(f)$.

B) If $\underset{\sim}{v} \in T^1_0M$, $\nabla_{\underset{\sim}{u}}\underset{\sim}{v}$ is the vector given by the connection we used to define $\underset{\sim}{\tau}$.

C) $\nabla_{\underset{\sim}{u}}(\underset{\sim}{v}+\underset{\sim}{w}) = \nabla_{\underset{\sim}{u}}\underset{\sim}{v} + \nabla_{\underset{\sim}{u}}\underset{\sim}{w}$, for $\underset{\sim}{v}, \underset{\sim}{w} \in T^k_hM$.

D) $\nabla_{\underset{\sim}{u}}(a\underset{\sim}{v}) = a\nabla_{\underset{\sim}{u}}\underset{\sim}{v}$, for $\underset{\sim}{v} \in T^k_hM$, $a \in R$.

E) $\nabla_{\underset{\sim}{u}}(\underset{\sim}{v}\otimes\underset{\sim}{w}) = (\nabla_{\underset{\sim}{u}}\underset{\sim}{v})\otimes\underset{\sim}{w} + v\otimes(\nabla_{\underset{\sim}{u}}\underset{\sim}{w}) \in T^{k+\ell}_{h+m}M$, for $\underset{\sim}{v} \in T^k_hM$, $\underset{\sim}{w} \in T^\ell_mM$.

F) If $\underset{\sim}{C} : T^k_h M \to T^{k-1}_{h-1} M$ is a contraction map (VII.3.04),

$\nabla_{\underset{\sim}{u}} (\underset{\sim}{C} \circ \underset{\sim}{v}) = \underset{\sim}{C}(\nabla_{\underset{\sim}{u}} \underset{\sim}{v})$ for $\underset{\sim}{v} \in T^k_h M$.

G) For $\underset{\sim}{u}, \underset{\sim}{u}' \in T_p M$, $a \in R$,

$$\nabla_{\underset{\sim}{u}+\underset{\sim}{u}'} \underset{\sim}{v} = \nabla_{\underset{\sim}{u}} \underset{\sim}{v} + \nabla_{\underset{\sim}{u}'} \underset{\sim}{v} , \quad \nabla_{a\underset{\sim}{u}} \underset{\sim}{v} = a \nabla_{\underset{\sim}{u}} \underset{\sim}{v} .$$

Proof

A) is essentially just (one) definition of $\underset{\sim}{u}(f)$ (Ex. 1.2).

B) is Theorem 4.06.

C) and D) follow from the linearity of $\underset{\sim}{\tau}_t$, which implies that of $(\underset{\sim}{\tau}^k_h)_t$.

E) is straightforward and left to the reader (Ex. 1), with a full outline.

Notice that if either $\underset{\sim}{v}$ or $\underset{\sim}{w}$ is just a function, $\underset{\sim}{v} \otimes \underset{\sim}{w}$ reduces to a pointwise scalar multiple. Thus E is a generalised Leibniz rule, reducing to the usual one when both $\underset{\sim}{v}$ and $\underset{\sim}{w}$ are functions (Ex. VII.4.4) and to 3.01 Civ) when $\underset{\sim}{v}$ is a function and $\underset{\sim}{w}$ a contravariant vector field.

By E it clearly suffices to prove F for $\binom{1}{1}$-tensor fields, since we can arrange any contraction as

$$T^k_h M \cong T^1_1 M \otimes T^{k-1}_{h-1} M \xrightarrow[\underset{\sim}{C} \circ \underset{\sim}{I}]{} T^0_0 M \otimes T^{k-1}_{h-1} M \cong T^{k-1}_{h-1} M.$$

This simplifies notation in the proof

$$\begin{aligned} \underset{\sim}{C}((\tau^1_1)_t (\underset{\sim}{v}_p \otimes \underset{\sim}{f}_p)) &= \underset{\sim}{C}(\underset{\sim}{\tau}_t \underset{\sim}{v}_p \otimes (\underset{\sim}{f}_p \circ \overleftarrow{\underset{\sim}{\tau}_t})), \text{ by 7.01,} \\ &= \underset{\sim}{f}_p(\overleftarrow{\underset{\sim}{\tau}_t}(\underset{\sim}{\tau}_t \underset{\sim}{v}_p)) \end{aligned}$$

$$= \underset{\sim}{f}_p(\underset{\sim}{v}_p)$$

$$= \underset{\sim}{C}(\underset{\sim}{v}_p \otimes \underset{\sim}{f}_p)$$

$$= (\underset{\sim}{\tau}_0^0)_t(\underset{\sim}{C}(\underset{\sim}{v}_p \otimes \underset{\sim}{f}_p)), \text{ since } (\underset{\sim}{\tau}_0^0)_t = \underset{\sim}{I}_R ,$$

that parallel transport commutes with contraction.

F) follows immediately, since $\underset{\sim}{C}$ is continuous and so commutes with limits.

G) is left as an easy exercise (1b - d). ■

7.04 Corollary

$\nabla_{\underset{\approx}{u}}\underset{\sim}{v}$ as defined above is independent of the choice of representing curve c.

Proof

By 7.03E $\nabla_{\underset{\approx}{u}}$ is determined by its values on vector fields.

By Ex. 1b $\nabla_{\underset{\approx}{u}}$ is determined on covariant vector fields by its values on contravariant ones.

By 7.03B $\nabla_{\underset{\approx}{u}}$ coincides on T^1M with the original connection, whose values do not depend on c. ■

7.05 Definition

In VII.1.02 we met a directional derivative with respect to $\underset{\approx}{u}$ as the image of $\underset{\approx}{u}$ under the derivative of a map. Similarly we define the covariant derivative at p of an $\binom{h}{k}$-tensor field $\underset{\sim}{v}$ to be the map

$$\nabla_p\underset{\sim}{v} : T_pM \to (T^h_kM)_p : \underset{\approx}{u} \mapsto \nabla_{\underset{\approx}{u}}\underset{\sim}{v} .$$

Linear by 7.03G, this corresponds canonically by V.1.08 to a vector in the space

$$(T_pM)^* \otimes (\underbrace{T_pM \otimes \ldots \otimes T_pM}_{h \text{ times}} \otimes \underbrace{T_p^*M \otimes \ldots \otimes T_p^*M}_{k \text{ times}}).$$

Switching the dual space on the left round to the right (Ex. V.8b) for convenience, we get an element of $T^k_{h+1}M$, which we shall also denote by $\nabla_p \underset{\sim}{v}$. (Applying this $\nabla_p \underset{\sim}{v} \in T^k_{h+1}M$, as our initial linear map, to $u \in T_pM$ just means taking the tensor product $(\nabla_p \underset{\sim}{v}) \otimes \underset{\sim}{u}$ and contracting over the last two places. In coordinates the isomorphism $L(T_pM; (T^h_kM)_p) \cong (T^h_{k+1}M)_p$ vanishes into invisibility, since the same "sets of numbers" serve as components on both sides.)

Evidently $\nabla_p \underset{\sim}{v}$ depends smoothly on p by 3.01 Cv) and Theorem 7.03. So we have a new smooth tensor field $\nabla\underset{\sim}{v}$ on M, the <u>covariant differential</u> of $\underset{\sim}{v}$, of the same contravariant order as $\underset{\sim}{v}$ and covariant order one higher. (Hence, it is sometimes asserted, we use the term "covariant". But at the time it was christened, "covariant" was also used to mean "independent of the choice of coordinates". It seems more plausible that the name is just due to this property, which took a lot of work to reach when working entirely in components (Ex. 2b). Over to the historians.)

7.06 Ricci's Lemma

If ∇ is the Levi-Cività connection for $\underset{\sim}{G}$, $\underset{\sim}{G}$ has the covariant differential $\nabla\underset{\sim}{G} = \underset{\sim}{0}$.

Proof

By definitions 6.05 and 6.01, all the $\underset{\sim}{\tau}_t$ for ∇ are isometries. But this means exactly that $(\underset{\sim}{\tau}^0_2)_t(\underset{\sim}{G}_p) = \underset{\sim}{G}_q$, as expansion of the definitions will show. Hence

$$(\underset{\sim}{\tau}^0_2)_t(\underset{\sim}{G}_q) - \underset{\sim}{G}_p = \underset{\sim}{0} \quad \text{always}$$

Application of definitions 7.02 and 7.05 gives the result. ■

7.07 Corollary

Covariant differentiation commutes with applications of $\underset{\sim}{G}_{\downarrow}$ and $\underset{\sim}{G}_{\uparrow}$ to "raise and lower indices".

Proof

By 7.03 it suffices to work with vector fields.

Lowering indices; for $\underset{\sim}{v} \in T^1M$,

$$\begin{aligned}
\nabla(\underset{\sim}{G}_{\downarrow}\underset{\sim}{v}) &= \nabla(\underset{\sim}{C}(\underset{\sim}{G} \otimes \underset{\sim}{v})) \\
&= \underset{\sim}{C}(\nabla(\underset{\sim}{G} \otimes \underset{\sim}{v})) \\
&= \underset{\sim}{C}(\nabla\underset{\sim}{G} \otimes \underset{\sim}{v} + \underset{\sim}{G} \otimes \nabla\underset{\sim}{v}) \\
&= \underset{\sim}{C}(\underset{\sim}{G} \otimes \nabla\underset{\sim}{v}) \quad \text{since} \quad \nabla\underset{\sim}{G} = \underset{\sim}{0} \\
&= (\underset{\sim}{G}_{\downarrow} \otimes \underset{\sim}{I}_{T^*M})\nabla\underset{\sim}{v}
\end{aligned}$$

that is, $$\nabla\circ\underset{\sim}{G}_{\downarrow} = (\underset{\sim}{G}_{\downarrow} \otimes \underset{\sim}{I}_{T^*M})\circ\nabla.$$

Raising indices; from the above,

$$\begin{aligned}
(\underset{\sim}{G}_{\uparrow} \otimes \underset{\sim}{I}_{T^*M})\circ(\nabla\circ\underset{\sim}{G}_{\downarrow})\circ\underset{\sim}{G}_{\uparrow} &= (\underset{\sim}{G}_{\uparrow} \otimes \underset{\sim}{I}_{T^*M})\circ((\underset{\sim}{G}_{\downarrow} \otimes I_{T^*M})\circ\nabla)\circ\underset{\sim}{G}_{\uparrow} \\
((\underset{\sim}{G}_{\uparrow} \otimes \underset{\sim}{I}_{T^*M})\circ\nabla)\circ\underset{\sim}{G}_{\downarrow}\circ\underset{\sim}{G}_{\uparrow} &= (\underset{\sim}{G}_{\uparrow} \otimes \underset{\sim}{I}_{T^*M})\circ(\underset{\sim}{G}_{\downarrow} \otimes I_{T^*M})\circ(\nabla\circ\underset{\sim}{G}_{\uparrow}) \\
(\underset{\sim}{G}_{\uparrow} \otimes I_{T^*M})\circ\nabla &= \nabla\circ\underset{\sim}{G}_{\uparrow}.
\end{aligned}$$

■

7.08 Components

If in coordinates $\underset{\sim}{u} = u^{\eta}\partial_{\eta}$ and $\underset{\sim}{w} \in T^k_hM$ has components $w^{i_1\ldots i_k}_{j_1\ldots j_k}$, then we define the components of $\nabla\underset{\sim}{w}$ by the equation

$$\nabla_{\underset{\sim}{u}} \underset{\sim}{w} = w^{i_1 \dots i_k}_{j_1 \dots j_n ; \eta} u^{\eta} (\partial_{i_1} \otimes \dots \otimes \partial_{i_k} \otimes dx^{j_1} \otimes \dots \otimes dx^{j_h}).$$

We leave it to the reader (Ex. 2a) to prove from Theorem 7.03 and the formulae of 3.02 that for these components

$$★ \quad w^{i_1 \dots i_k}_{j_1 \dots j_h ; \eta} =$$

$$\partial_{\eta} (w^{i_1 \dots i_k}_{j_1 \dots j_h}) + \sum_{\ell=1}^{k} w^{i_1 \dots i_{\ell-1} t i_{\ell+1} \dots i_k}_{j_1 \dots j_n} \Gamma^{i_\ell}_{\eta t} \quad \sum_{\ell=1}^{h} w^{i_1 \dots i_k}_{j_1 \dots j_{\ell-} \, t \, j_{\ell+1} \dots j_h} \Gamma^{t}_{\eta j_\ell} .$$

(The trick is first to extend 3.02, to get $\nabla_{\partial_i} dx^j$, by Ex. 1b. Then extend to general $\underset{\sim}{w}$ by 7.03E.)

Notice that, if for some η we have $\underset{\sim}{u} = \partial_{\eta}$, which has components δ^{i}_{η} (since $\delta^{i}_{\eta} \partial_{i} = \partial_{\eta}$).

$$\nabla_{\partial_\eta} \underset{\sim}{w} = \nabla_{\underset{\sim}{u}} \underset{\sim}{w} = w^{i_1 \dots i_k}_{j_1 \dots j_n ; \ell} \delta^{\ell}_{\eta} (\partial_{i_1} \otimes \dots \otimes dx^{j_h}) \quad \text{by } ★$$

$$= w^{i_1 \dots i_k}_{j_1 \dots j_h ; \eta} (\partial_{i_1} \otimes \dots \otimes dx^{j_h}),$$

giving an alternative definition of $w^{i_1 \dots i_k}_{j_1 \dots j_h ; \eta}$.

The generalised Leibniz rule 7.03E becomes in coordinates

$$\left(v^{i_1 \dots i_k}_{j_1 \dots j_h} w^{a_1 \dots a_\ell}_{b_1 \dots b_m} \right)_{; \eta} = v^{i_1 \dots i_k}_{j_1 \dots j_n ; \eta} w^{a_1 \dots a_\ell}_{b_1 \dots b_m} + v^{i_1 \dots i_k}_{j_1 \dots j_h} w^{a_1 \dots a_\ell}_{b_1 \dots b_m ; \eta}$$

by plugging V.1.12 into the definition.

Note that some books use the notation $\nabla_{\eta} w^{i_1 \dots i_k}_{j_1 \dots j_h}$ for our $w^{i_1 \dots i_k}_{j_1 \dots j_h ; \eta}$.

We introduce here and use subsequently an abbreviation common in the literature:

denote $\partial_\eta(w^{i_1\ldots i_k}_{j_1\ldots j_h})$ by $w^{i_1\ldots i_k}_{j_1\ldots j_h,\eta}$. (Notice the vital difference between comma and semi-colon: one means a derivative of a component, the other a component of a derivative.) The computation rule ★ above then becomes

$$w^{i_1\ldots i_k}_{j_1\ldots j_h;\eta} = w^{i_1\ldots i_k}_{j_1\ldots j_h,\eta} + \text{junk as before.}$$

We also abbreviate $(w^{i_1\ldots i_k}_{j_1\ldots j_h;\eta})_{;\mu}$ and $(w^{i_1\ldots i_k}_{j_1\ldots j_h,\eta})_{,\mu}$ to $w^{i_1\ldots i_k}_{j_1\ldots j_h;\eta\mu}$ and $w^{i_1\ldots i_k}_{j_1\ldots j_h,\eta\mu}$, respectively.

Notice that if $\underset{\sim}{w}$ is just a function $w : M \to R$,

$$w_{;\eta} = w_{,\eta} = \partial_\eta(w).$$

In components, 7.07 takes the form, for instance,

$$g^{ij}w^k_{j;\eta} = w^{ik}_{;\eta}, \quad g_{ij}w^{im}_{st;\eta} = w^m_{jst;\eta}.$$

Hence we can ignore the presence of semi-colons (but not commas) when raising and lowering indices. Equally usefully;

<u>7.09 Lemma</u>

The covariant differential of the constant $\binom{1}{1}$-tensor field $\underset{\sim}{I}$ on M (cf. VII.3.05iii) is $\underset{\sim}{0}$.

<u>Proof</u>

Clearly parallel transport takes the identity on any tangent space to that on any other (just apply the definitions) and the result follows. ■

The utility of 7.09 is its coordinate form:

$$\delta^i_{j;\eta} = 0 \quad \forall i, j, \eta.$$

This is often expressed by the statement that δ^i_j, like g_{ij} and g^{ij} is "a constant with respect to covariant differentiation". It has the consequence, frequently invaluable in manipulations, that "change of indices", such as using $\delta^i_j v^j = v^i$, commutes like raising and lowering indices with covariant differentiation. For example,

$$v^i_{;\eta} = (\delta^i_j v^j)_{;\eta} = \delta^i_j v^j_{;\eta} \quad \text{(using the Leibniz rule)},$$

which can also be seen by considering the component functions directly. Notice that change of indices is essentially contraction of $\underset{\sim}{I} \otimes \underset{\sim}{v}$.

7.10 Definition

A tensor field $\underset{\sim}{t}$ on M is constant relative to a connection ∇ or metric tensor field $\underset{\sim}{G}$ if the corresponding parallel transport along any curve takes its value at any point to its value at any other: equivalently, if its covariant differential is $\underset{\sim}{0}$ identically.

Lemmas 7.06, 7.09 state that a metric is constant relative to itself and that the identity is constant relative to any connection. The other constant fields we have encountered so far are constant functions and parallel vector fields. Evidently, any multiple of a constant field by a scalar constant, or more generally a tensor product of constant fields, is again constant by 7.03E.

Notice that the constant fields on affine spaces of VII.3.05 need not necessarily be constant relative to metrics that are not constant in the sense used there.

Exercises VIII.7

1a) From 7.03C, show that to prove E it suffices to consider $\underset{\sim}{v} = f\underset{\sim}{v}'$, $\underset{\sim}{w} = g\underset{\sim}{w}'$ with each of $\underset{\sim}{v}'$, $\underset{\sim}{w}'$ a parallel field along c, and $f,g : J \to R$.

(Hint: pick bases and parallel-transport them.)

Establish the equation

$$\frac{d(fg)}{ds}(0)\underset{\sim}{v}_p \otimes \underset{\sim}{w}_p = \left(\frac{df}{ds}(0)\underset{\sim}{v}_p\right) \otimes (g(0)\underset{\sim}{w}_p) + (f(0)\underset{\sim}{v}_p) \otimes \left(\frac{dg}{ds}(0)\underset{\sim}{w}_p\right)$$

by bilinearity and Ex. VII.4.4, and deduce E.

b) Prove from A, F, E that if $\underset{\sim}{f} \in T^0_1M$, $\underset{\sim}{w} \in T^1_0M$ then

$$(\nabla_{\underset{\sim}{u}}\underset{\sim}{f})\underset{\sim}{w} = \underset{\sim}{u}(\underset{\sim}{f}(\underset{\sim}{w})) + \underset{\sim}{f}(\nabla_{\underset{\sim}{u}}\underset{\sim}{w}).$$

c) Deduce G when $\underset{\sim}{v}$ is the covariant vector field $\underset{\sim}{f}$.

d) Use E to prove G for tensors of the form $\underset{\approx}{x}_1 \otimes \ldots \otimes \underset{\approx}{x}_s$, where each $\underset{\approx}{x}_i$ is in T^1_0M or T^0_1M, and C to extend it to general tensors.

2a) Prove the equation ★ of 7.08.

b) Use 7.08 and 3.03 to show that the n^{h+k+1} functions $w^{i_1\ldots i_k}_{j_1\ldots j_k;\eta}$ obey the transformation rule (cf. VII.4.04) for a $\binom{k}{h+1}$-tensor field.

(The work involved will show you why "covariance" with respect to these rules was such a triumph in the original approach.)

c) Use 7.08 and 6.06 to prove Ricci's Lemma (7.06) in its coordinate form

$$g_{ij;\eta} = 0, \quad g^{ij}_{;\eta} = 0$$

and derive its corollary, again using 7.08. (There are some places where coordinates give the quickest and most convenient proof. On the other hand, there are places where geometry not only gives more insight but is <u>much</u> quicker.)

IX Geodesics

"The voice of him that crieth in the wilderness,
Prepare ye the way of the Lord,
make straight in the desert a highway for our God."
Isaiah 40,3.

The ancient custom in the Eastern Mediterranean of the straight, royal road for the exclusive use of the semi-divine ruler (cf. Aristotle telling Alexander there was no royal road to geometry - he had to go the same way as everyone else) involved a clear, if unformulated, idea of "straight". With the rigid formalisation of geometry into the Euclidean system, "straight" became a more restricted notion which clearly would <u>not</u> fit a road that bent over the horizon, as a long enough road must. Hence a new word was needed. Earth had been considered a perfect sphere since early Greek times, and on such if you keep "straight on", deviating neither to the left nor to the right, for long enough you return to your starting point and your starting direction. Your path, then, unambiguously divides the earth into two parts, to its left and to its right: hence the chosen word for such a path was "geodesic" or "divides the earth". This name has become fixed for an undeviating path, though only on a <u>perfect</u> sphere does such a path always have this dividing property (and the earth is no such thing).

1. LOCAL CHARACTERISATION

When is a curve "undeviating"? Its direction at $c(t)$ is given by $c_*(t) \in T_{c(t)}M$, so not deviating must mean that $c_*(t)$ does not vary with t, in some sense. Since we move it from one tangent space to another, it cannot be "constant" in the strict sense of that word. The previous chapter, though, was almost entirely devoted to the study of what "rate of change along a curve" ought to mean for vectors. For a manifold M with a metric tensor $\underset{\sim}{G}$, Levi-Cività connection ∇ and associated differentiation ∇_{c*} along c (VIII.3.05), then, the natural definition is

1.01 Definition

A curve c in M is a <u>geodesic</u> if $\nabla_{c*}c^*(t) = \underset{\sim}{0}$ $\forall t$; that is, if its tangent vector field (VIII.3.04) is parallel.

If c is thought of as describing the motion of a particle, $c^*(t)$ becomes "velocity at time t", and $\nabla_{c*}c^*$ becomes "rate of change of velocity" or "acceleration". So the geodesic is the path of a particle "subject to no forces", constrained only by the geometry of the manifold. (We give another interpretation in §3.)

It is clear that this definition depends on $\underset{\sim}{G}$, since ∇ does. This is entirely reasonable: Fig. 1.1 illustrates a diffeomorphism between two manifolds (which are thus "the same" topologically and differentially) carrying an intuitively "undeviating" curve to an obviously "bent" one.

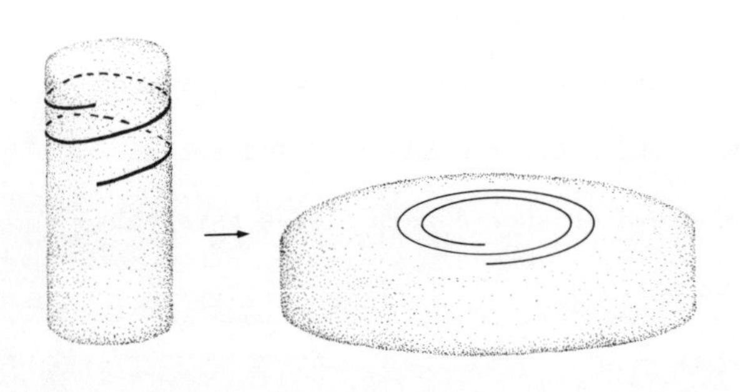

Fig. 1.1

Since parallel transport is an isometry, parallelism is a somewhat stronger condition than simply that $c^*(t)$ not be "turning" with respect to the connection: it must stay the same size. This is the most convenient formulation, as if we allowed the size to change, $c^*(t)$ could go to zero unless we added a separate condition to forbid it - and when you have stopped, you no longer have a "direction you are going in" to preserve. One consequence of this is

1.02 Lemma

A geodesic is always a like curve (VII.5.03). ■

In spacetime, null and timelike geodesics are often called world-lines though usually this term is allowed to include other timelike or null curves (cf. XI.1.02).

We already see that geodesics have a more elaborate geometry if M has an indefinite metric than in the Riemannian case. Various facts of strictly Riemannian geometry fail for indefinite metrics. For example, if M is Riemannian, connected and "geodesically complete" (defined in 2.03), any two points in M can be joined by a geodesic, while in the example of §6 this fails even for points connected by a timelike curve. We leave such "strictly Riemannian" results to the "strictly pure" mathematics texts. (Even if spacetime does have a geodesic between any two points this would be without physical significance - pending the discovery of tachyons - since if x,y can be joined only by a spacelike geodesic, events at either cannot affect events at the other.)

1.03 Closed geodesics

In Euclidean geometry (or, for that matter Lobachevskian : Ex. 2.3) no straight line "meets itself" as a great circle does. But on the sphere geodesics are "closed curves" in the obvious sense. (Ex. 1.) This suggests that we define a closed geodesic to be a smooth curve $c : R \to M$ with $c*$ parallel and some $0 \neq k \in R$ such that $c(t+k) = c(t)$, $\forall t \in R$. We can also define a crossed geodesic to be a geodesic c with domain either $J \subseteq R$, or S^1, and $c(x) = c(y)$ for some $x \neq y$. On a compact Riemannian 2-manifold, unless there is a good deal of symmetry, a typical geodesic will not be closed but will cross itself infinitely many times. In particular on the earth the bulges away from sphericity of the "geoid", which is geodesy's name for whatever shape the ideal "sea-level" surface currently has, mean that geodesics which really divide the earth in two are highly unusual. (A theorem of global analysis [Liusternik and Schnirelman] says that on any manifold homeomorphic to S^2, given any metric tensor, there must be at least three closed geodesics. But it gives no indication how to find them. For generalisations of this to n-manifolds, see [Klingenberg].)

We do not draw pictures to illustrate how irregularities in the geoid cause geodesics to deviate, cross, etc; by §3, you can get a clearer notion than from any figure by pulling string tight around a potato.

1.04 Components

Relative to a chart, a curve c takes the form $t \rightsquigarrow (c^1(t),\ldots,c^n(t))$. In the corresponding basis for $T_{c(t)}X$, $c^*(t)$ has components $\left(\frac{dc^1}{ds}(t),\ldots,\frac{dc^n}{ds}(t)\right)$. The geodesic equation,

$$\nabla_{c^*(t)}c^* = \underset{\sim}{0},$$

thus takes the form (using VIII.3.02)

$$\left(\frac{d^2c^k}{ds^2} + \Gamma^k_{ij}\frac{dc^j}{ds}\frac{dc^i}{ds}\right)\partial_k = \underset{\sim}{0},$$

or

$$\frac{d^2c^k}{ds^2} + \Gamma^k_{ij}\frac{dc^j}{ds}\frac{dc^i}{ds} = 0, \quad \forall k.$$

Exercises IX.1

1) Use Ex. VIII.6.5 to show that the geodesics on a sphere S^n of any dimension, with the usual metric tensor, are the great circles.

2) Compute, and draw, the geodesics given by the asymmetric connection on R of Ex. VIII 6.1. (Notice that they are of two kinds - what goes up need not come down.)

3a) We can define a manifold RP^2, the real projective plane, as the set of unordered pairs $\{x,-x\}$ where $x \in S^2$ (so that $\{x,-x\}$ is the same as $\{-x,x\}$) with charts defined from those of Ex. VII.2.1 by

$$U_i = \left\{\{x,-x\} \,\middle|\, x \in U_{i+}\right\}$$

$\phi_i(\{x,-x\}) = \phi_{i+}$(whichever of $x,-x$ is in U_{i+}).

(This manifold, "S^2 with opposite points identified", sits in R^3 even less comfortably than the Klein bottle, but embeds in R^4.)

b) Define a Riemannian metric on RP^2 such that, with the standard metric on S^2, the derivative of $Q : S^2 \to RP^2 : x \rightsquigarrow \{x,-x\}$ at any point is an isometry.

c) Show - easier without coordinates - that the geodesics on RP^2 are the images by Q of the great circles. Deduce that any two distinct geodesics meet at exactly one point.

(Taking "straight line" to mean "geodesic in RP^2" this contradicts Euclid's parallel postulate, which asserts the existence of straight lines that never meet. But since all his other axioms are true for such "lines", if the others implied the parallel postulate that also would hold for them. Two millenia of attempts to prove the parallel axiom from the others hit this and similar rocks last century.

RP^2 is the standard elliptic non-Euclidean geometry (cf. also Ex. 2.3).)

4) Show that in the situation of Ex. VIII.6.8, a curve

$$c : J \to M \times N : t \rightsquigarrow (c^M(t),\ c^N(t))$$

is a geodesic if and only if c^M, c^N are geodesics in M, N.

2. GEODESICS FROM A POINT

2.01 The horizontal field

From any point $p \in M$, we would expect to be able to go off with any given "starting velocity" vector $\underset{\sim}{v}_p \in T_pM$, and by "not deviating" get a well defined geodesic through p, with tangent vector $\underset{\sim}{v}_p$ at 0. The proof of this is a question in differential equations, as follows.

At each point $\underset{\sim}{v} \in TM$ there is a unique horizontal vector $\tilde{\underset{\sim}{v}} \in T_{\underset{\sim}{v}}(TM)$ with $D_{\underset{\sim}{v}}\Pi(\tilde{\underset{\sim}{v}}) = \underset{\sim}{v}$, by Ex. VIII 3.6b. (In Fig. 2.1, T_pM and $\underset{\sim}{v}$ are drawn twice, using the embedded picture and the bundle picture.) This gives a (clearly smooth) vector field $\underset{\sim}{x}$ on TM, with $\underset{\sim}{x}(\underset{\sim}{v}) = \tilde{\underset{\sim}{v}}$, with the properties:

(i) $D_{\underset{\sim}{w}}\Pi(\underset{\sim}{x}_{\underset{\sim}{w}}) = \underset{\sim}{v} \iff \underset{\sim}{w} = \underset{\sim}{v}$.

(ii) Each $\underset{\sim}{x}_{\underset{\sim}{v}}$ is a horizontal vector.

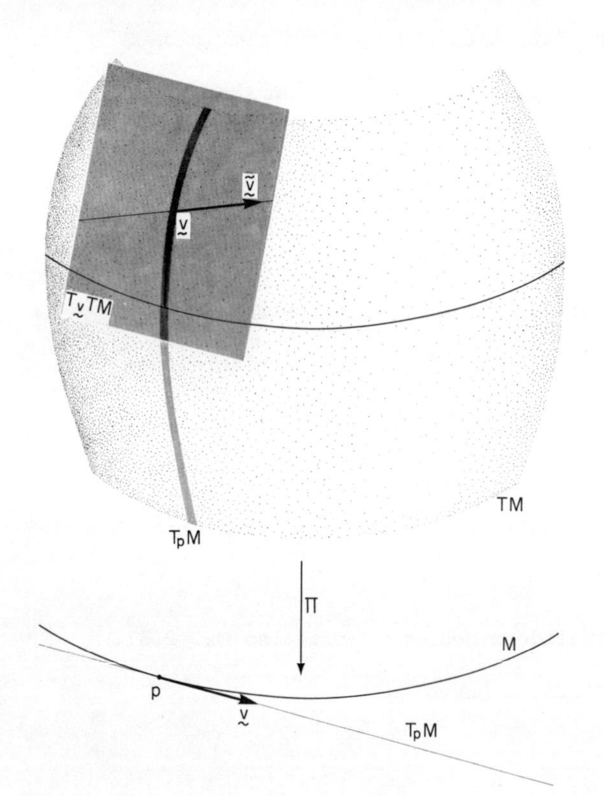

Fig. 2.1

If $\tilde{c}$ is a solution curve in TM of $\underset{\sim}{x}$ (cf. VII.6.01), and $c = \Pi \circ \tilde{c}$ is its projection down to M, then $\tilde{c}$ becomes a vector field along c with $D_{\tilde{c}(t)}\Pi(\tilde{c}^*(t)) = c^*(t)$, as in VIII.3.06. Since by assumption $\tilde{c}^*(t) = \underset{\sim}{x}(\tilde{c}(t))$, by (i) we have $\tilde{c}(t) = c^*(t)$; $\tilde{c}$ is exactly the tangent vector field c^* along c. But $\tilde{c}(t)$ is horizontal for all t, and examining Definition VIII.3.07 this means exactly that

$$\nabla_{c^*} c^* = \underset{\sim}{0}\ , \quad \text{so } c \text{ is a geodesic.}$$

So, appealing to Theorem VII.6.04 for the existence and uniqueness of such a $\tilde{c}$ through any $\underset{\sim}{v} \in TM$, we have

2.02 Theorem

For any given $p \in M$, $\underset{\sim}{v} \in T_pM$, there is a geodesic $c : J \to M$ with $c(0) = p$, $c^*(0) = \underset{\sim}{v}$, unique in the sense that any other such geodesic $f : K \to M$ has $f|K \cap J = c|K \cap J$. ■

(Notice that the geodesic equation 1.01, 1.04 is a second order ordinary differential equation, and that geometrically this means a vector field on TM. Similarly a 3rd order equation is a vector field on T(TM), and so forth.)

2.03 Geodesic completeness

Theorem 2.02 is a local fact. Unlike VIII.4.02, which says we have parallel transport as far as we like along a given path, a geodesic through p cannot necessarily be extended to a geodesic with domain all of R. (For example if $M = R$ with the metric $g_{11}(x) = (\sqrt{1+x^2}-1)/x^2\sqrt{1+x^2})$ the only geodesic with $c(0) = 0$, $c^*(0) = 2\underset{\sim}{e}$ is $t \rightsquigarrow 2t/1-t^2$, by Ex. 4.) If all geodesics on M can be extended in this way, M is called geodesically complete. Quite mild conditions (a strong one is compactness, but others work) guarantee that a Riemannian manifold is geodesically complete (cf. 3.10 and see [Kobayashi and Nomizu]), but recent work (see [Hawking and Ellis]) has shown that physically reasonable assumptions make it impossible for a spacetime to be complete with reference to the interesting geodesics: the timelike and null ones. (In fact the situation is worse. For there are spacetimes that are geodesically complete but which have incomplete timelike curves of bounded acceleration, so particles or people in them may vanish. In consequence, a new idea of completeness has been devised in terms of a certain bundle over the spacetime.) An obstruction to extending geodesics forward in time may be a black hole or collapse (local or global); a barrier backwards, a bang or (in theories in which all geodesics extended backwards meet the same singularity) a Big Bang, (cf. XII. 2.04).

Locally however we do have geodesics in all directions from a point, and with them we construct a special map that has various useful features. The idea is to carry a tangent vector $\underset{\sim}{v} \in T_pM$ to the effect of "travelling unit time" by the geodesic with initial vector $\underset{\sim}{v}$. (Thus $\underset{\sim}{0}$ will go to p, $2\underset{\sim}{v}$ will go "twice as far" as $\underset{\sim}{v}$, etc.) If M is not geodesically complete, the geodesic through $\underset{\sim}{v}$ may not be extendable to a domain that contains 1, but a small enough $\underset{\sim}{v}$ starts us travelling sufficiently slowly to meet no obstruction before unit time is up. (How small is "small enough" will generally vary from one point in M to another.)

2.04 Definition

The exponential map from a subset $E \subseteq T_pM$ to M is defined as follows.

$$E = \{\underset{\sim}{v} \mid \exists \text{ a geodesic } c_{\underset{\sim}{v}} \text{ s.t. } c_{\underset{\sim}{v}}(0) = p,\ c^*_{\underset{\sim}{v}}(0) = \underset{\sim}{v},\ \&\ c_{\underset{\sim}{v}}(1) \text{ is defined}\}$$

$$\exp_p : E \to M : \underset{\sim}{w} \mapsto c_{\underset{\sim}{w}}(1).$$

By Ex. 1, E contains an open neighbourhood of $\underset{\sim}{0} \in T_pM$. (In fact E is itself open and so a neighbourhood of $\underset{\sim}{0}$, but the proof is somewhat technical and we shall not need it.)

The map $\exp_p$ is well defined on E by the uniqueness property in 2.02. (The name "exponential" is due to a special case. Using the usual metric on S^1, considered as the set $\{z \mid |z| = 1\}$ of complex numbers, and the obvious parameter on the tangent space at 1, $\exp_1$ is given exactly by $x \mapsto e^{ix}$, (Fig. 2.2). A more elaborate example, not involving complex notation, is discussed in §6.)

If M is geodesically complete, of course, $\exp_p$ is defined on all of T_pM; if any $p,q \in M$ can be joined by a geodesic $\exp_p$ is surjective. Thus it is not in general injective, as a typical manifold is not smoothly bijective with any vector space. Fig. 2.3 illustrates $\exp_{(N.Pole)}$ on S^2 with its usual metric, for those vectors in $T_{(N.Pole)}S^2$ of length $\leq \pi$ (what happens to the longer ones?) The images of the straight lines shown in the tangent space are the curves shown on S^2: notice that though by Ex.1f the straight lines through $\underset{\sim}{0}$ are carried to geodesics, no other straight lines in $T_{(N.Pole)}S^2$ are. Indeed, if all others were then the sphere would be "flat" in the sense we discuss next chapter.

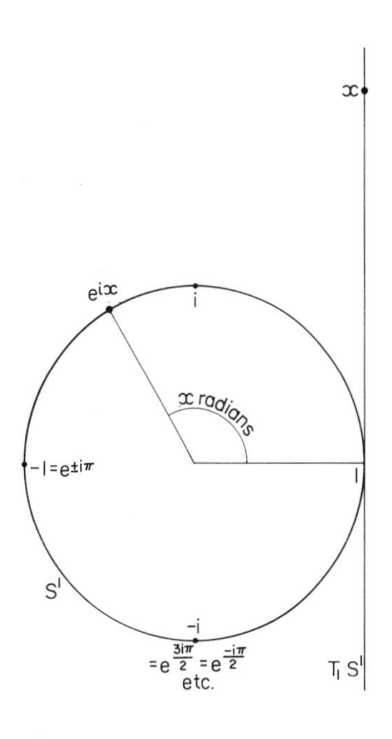

Fig. 2.2

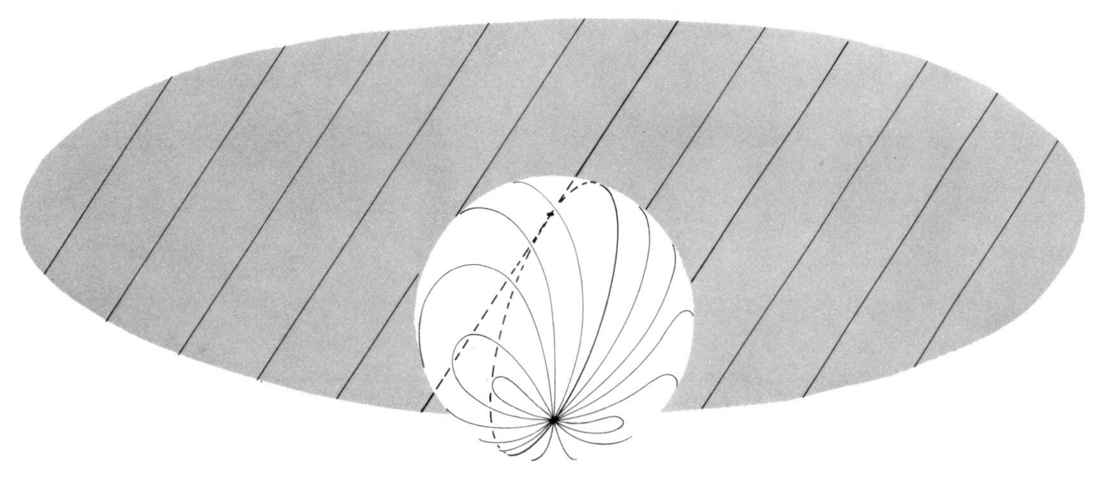

Fig. 2.3

The exponential map is not, then, a diffeomorphism. (The study of its singularities, the places where not even $\underset{\sim}{D}_{\underset{\sim}{x}}(\exp_p)$ is injective, is an active topic in differential geometry.) But it is always smooth (Ex. 2), and near $\underset{\sim}{0}$ even better:

2.05 Lemma

$\underset{\sim}{0} \in T_pM$ has a neighbourhood $U \subseteq T_pM$ such that $\exp_p|U$ is a diffeomorphism.

Proof

By Ex. 1f, $c : t \mapsto \exp_p(t\underset{\sim}{v})$ has $c(0) = p$, $c^*(0) = \underset{\sim}{v}$. But evidently $\tilde{c} : t \mapsto t\underset{\sim}{v}$ represents $\underset{\sim}{v}$ (considered as a tangent vector to T_pM at $\underset{\sim}{0}$ in the natural way), so $c = \exp\circ\tilde{c}$ represents $\underset{\sim}{D}_{\underset{\sim}{0}}\exp_p(\underset{\sim}{v})$. Thus $\underset{\sim}{v} = c^*(0)$ is in the image of $\underset{\sim}{D}_{\underset{\sim}{0}}\exp_p$. Since this holds for any $\underset{\sim}{v}$, $\underset{\sim}{D}_{\underset{\sim}{0}}\exp_p$ is a surjective linear map and so, since $\dim(T_{\underset{\sim}{0}}(T_pM)) = \dim T_pM$, an isomorphism by I.2.13.

The result follows by VII.1.04. ■

2.06 Normal coordinates

By 2.05 we have an open set $V = \exp_p(U) \subseteq M$ and a C^∞ map $\overleftarrow{\exp_p} : V \to T_pM$. The pair $(V, \overleftarrow{\exp_p})$ constitutes an admissible chart (VII.2.01), using the canonical affine structure of T_pM. This chart played a crucial role in the early development of differential geometry, because of the simplicity it can bring to a wilderness of coordinates. Choosing a basis $\beta = \{\underset{\sim}{b}_1,\dots,\underset{\sim}{b}_n\}$ for T_pM, orthonormal with respect to $\underset{\sim}{G}_p$, we get an isomorphism $B : T_pM \to R^n$, and a chart $B\circ\overleftarrow{\exp}$ with domain V and range R^n. Such a chart is called a <u>system of normal coordinates about</u> p, with respect to $\underset{\sim}{G}$. (Notice that we have as many such as there are choices of orthonormal basis for T_pM.) It is very convenient in computations - if used with care - because:

2.07 Lemma

With respect to a system of normal coordinates about $p \in M$,

A) $g_{ij}(p) = \pm\delta_{ij}$, $\forall i,j$

B) $\Gamma^k_{ij}(p) = 0$, $\forall i,j,k$

C) $\partial_k g_{ij}(p) = 0 \quad , \quad \forall i,j,k$

Proof

A) $g_{ij}(p) = \partial_i(p).\partial_j(p)$ by definition

$= c_i^*(0).c_j^*(0)$, where $c_i(t) = \exp_p(\overleftarrow{B}(0,\ldots,0,t,0,\ldots,0))$, by Ex. VIII.1.1a (t in the i-th place)

$= \underset{\sim}{b}_i.\underset{\sim}{b}_j$ since $c_i(t) = \exp(t\underset{\sim}{b}_i)$ so that $c_i^*(0) = \underset{\sim}{b}_i$ by Ex. 1f

$= \pm\delta_{ij}$ since β was chosen orthonormal.

B) $\nabla_{\partial_i}\underset{\sim}{v} = \mathbb{V}_{c_i^*}\underset{\sim}{v}$, for any field $\underset{\sim}{v}$, since $c_i^*(t) = (\partial_i)_{c_i(t)}$.

Hence $\nabla_{\partial_i}\partial_i(t) = \underset{\sim}{0}$ $\forall i,t$ since c_i is a geodesic.

Similarly,

$\nabla_{(\partial_i+\partial_j)}(\partial_i+\partial_j) = \mathbb{V}_{c_{i+j}^*}(\partial_i+\partial_j)$, along $c_{i+j} : t \rightsquigarrow \exp_p(\overleftarrow{B}(0,\ldots,t,\ldots,t,\ldots,0)$ (t in the i-th and j-th places)

$= \underset{\sim}{0}$

Hence by Ci), Cii) of VIII.3.01,

$$\begin{aligned}
\underset{\sim}{0} &= \nabla_{(\partial_i+\partial_j)_p}(\partial_i+\partial_j) \\
&= \nabla_{\partial_i(p)}\partial_i + \nabla_{\partial_j(p)}\partial_i + \nabla_{\partial_i(p)}\partial_j + \nabla_{\partial_j(p)}\partial_j \\
&= \underset{\sim}{0} + \nabla_{\partial_j(p)}\partial_i + \nabla_{\partial_i(p)}\partial_j + \underset{\sim}{0} \\
&= 2\nabla_{\partial_i(p)}\partial_j \text{ by symmetry of } \nabla, \text{ since } [\partial_i,\partial_j] = \underset{\sim}{0}.
\end{aligned}$$

Thus all the vectors $\nabla_{\partial_i(p)}\partial_j$ at p, and hence their components the Γ_{ij}^k, vanish.

C) is an immediate consequence of B) and the last equation of VII.6.06, since all the Γ_{ijk} vanish when all the Γ^k_{ij} do. ■

Notice that the above simplifications occur only at p; in general g_{ij} and Γ^k_{ij} will be more awkward at all other points. (The Γ^k_{ij} just pass through 0 at p, for this chart.) The proof in part B that $\Gamma^k_{ii} = 0$ is valid only along the x^i-axis, since lines in R^n parallel to the x^i-axis need not correspond to geodesics (as we noticed for S^2). Hence the proof that $\Gamma^k_{ij} = 0$, which needs both $\Gamma^k_{ii} = 0$ and $\Gamma^k_{jj} = 0$, applies only to points on both the x^i and the x^j-axis: that is only at p. We study the question of when the g_{ij} can be made constant (or the Γ^k_{ij} zero) in a whole chart, next chapter.

It is not unknown for a course in an otherwise respectable physics department to "prove" things on the assumption that the Γ^k_{ij} can be vanished, by a suitable choice of coordinates, over a small region at a time. This error rests above all on the habit of leaving off the argument of a function, so that distinction is lost between the numbers $\Gamma^k_{ij}(p)$, which we can make zero by a suitable choice of chart, and the functions Γ^k_{ij} which we usually can't. (The distinction is often lost even between the words "number" and "function", as when a "vector" is introduced as a set of "numbers" that transform nicely when the equations given concern vector fields and their coordinate functions. Choice of normal coordinates is one context where this sin against the light leads not just to confusion but to serious error.)

Exercises IX.2

1a) Use Theorem VII.6.04 to show that there is a neighbourhood U of $\underset{\sim}{0}_p \in TM$ and a local flow $U \times]-\varepsilon,\varepsilon[\to TM$ for the vector field $\underset{\sim}{x}$ of 2.01.

b) Deduce that for the neighbourhood $V = U \cap T_pM$ of the zero of T_pM there is a smooth map $\phi : V \times]-\varepsilon,\varepsilon[\to M$ such that for $\underset{\sim}{v} \in V$, $\phi_{\underset{\sim}{v}} : t \mapsto \phi(\underset{\sim}{v},t)$ is a geodesic with $\phi^*_{\underset{\sim}{v}}(0) = \underset{\sim}{v}$.

c) Show that if $c :]-\alpha,\alpha[\to M$ is a geodesic and $a \in R$, then the curve $c_a :]-\alpha/a,\alpha/a[\to M : t \rightsquigarrow c(at)$ is also a geodesic, with $c_a^*(0) = a(c^*(0))$.

d) Deduce that if $W = \{\frac{1}{2}\varepsilon \underset{\sim}{v} \mid \underset{\sim}{v} \in V\}$, then W is an open neighbourhood of $\underset{\sim}{0}_p$ in T_pM with a smooth map $\psi : W \times]-2,2[\to M$ such that for $\underset{\sim}{w} \in W$, $\psi_{\underset{\sim}{w}}$ is a geodesic with $\psi^*_{\underset{\sim}{w}}(0) = \underset{\sim}{w}$.

e) Deduce that the map $\exp_p : \underset{\sim}{w} \rightsquigarrow \psi_{\underset{\sim}{w}}(1)$ is well defined on W.

f) Show that the map $t \rightsquigarrow \exp_p(t\underset{\sim}{w})$ is exactly $\psi_{\underset{\sim}{w}}$.

2) Deduce from the smoothness (by VII.6.04) of the geodesic flow that whenever $\exp_p$ is defined in an open neighbourhood U of any tangent vector, it is C^∞ in U.

3) Let $P^{\frac{1}{2}}$ be the manifold $\{(x,y) \mid y > 0\} \subseteq R^2$ with the obvious chart, and the metric tensor field

$$\underset{\sim}{G}_{(x,y)}\left(u\frac{\partial}{\partial x}, v\frac{\partial}{\partial y}\right) = \frac{u^2+v^2}{y^2} .$$

(Our notation $P^{\frac{1}{2}}$ is taken from the standard name, <u>Poincaré upper half-plane</u>, for this manifold with this metric.)

a) Show that

$$\Gamma^2_{22} = \Gamma^1_{12} = \Gamma^1_{21} = -\frac{1}{y} , \quad \Gamma^2_{11} = \frac{1}{y} , \quad \Gamma^1_{11} = \Gamma^2_{12} = \Gamma^2_{21} = \Gamma^1_{22} = 0.$$

b) Show that the image of any geodesic is confined to some set $\{(x,y) \mid (x-a)^2+y^2=r^2\}$, $a,r \in R$, or $\{(x,y) \mid x=a\}$, and that $P^{\frac{1}{2}}$ is geodesically complete (Fig. 2.4).

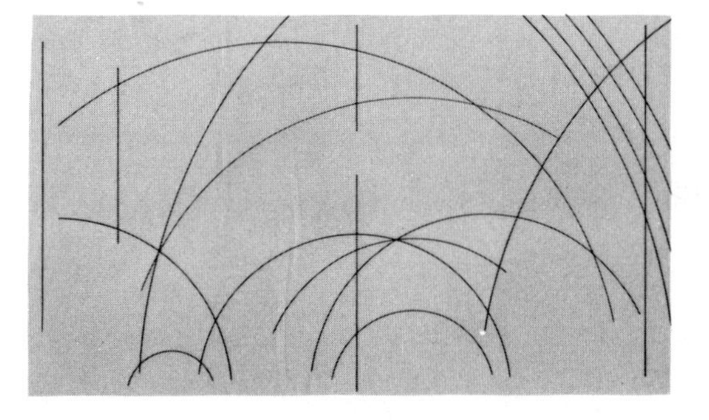

Fig. 2.4

c) Prove that for any point p and geodesic c not through p, with domain R, there are infinitely many geodesics with domain R that fail to meet c.

(This breaks Euclid's parallel postulate in the opposite way to that in Ex. 1.3: instead of having no "lines" through p not meeting c, or exactly one as Euclid proposed, there are more than one. $P^{\frac{1}{2}}$ is an example of a hyperbolic or Lobachevskian non-Euclidean geometry.)

4) Show that the example given in 2.03 is indeed a geodesic, and that any other with $c(0) = 0$, $c^*(0) = 2\underset{\sim}{e}$ is a restriction of it to a subinterval of $]-1,1[$.

3. GLOBAL CHARACTERISATION

"Straight" above was interpreted as the opposite of "bent": c is a geodesic if at each t its direction $c^*(t)$ is unchanging, by the measure for rate of change along curves developed in the last chapter. Now for a long road on the earth this local point of view is natural, but for the top of a wall we have another test - we compare it to a stretched string.

Why is a stretched string straight? It has some give - perfectly inelastic strings occur only in Applied Maths exams - so there are other positions it could occupy. Disturbing it into them, however, takes effort and, disturbing force removed, it will relax back into straightness because (neglecting gravity) this is its position of least energy. Now being of least energy is a property of the position as a whole, a global property, in contrast to the local condition 1.01, and here it is clearly the physically decisive property. We can generalise this global approach to straightness from the Euclidean to the Riemannian and pseudo-Riemannian contexts, and show the result equivalent to the local one.

Start by fixing $p, q \in M$ and considering paths $c : [a,b] \to M$ with $c(a) = p$, $c(b) = q$. If M is embedded in a Euclidean space, and each c represents a possible way to lie in M for a piece of elastic of length b-a, we have a clear intuitive idea of "position of least energy". Notice that this includes being evenly stretched: push the midpoint of the elastic along the set of points occupied and it will slide back fast, when released, to its previous position, even though the elastic has been moved to a position of the same length.

We make one simplification: real elastic, if p and q are too close, has many positions of zero energy and will rest in any of them. So a section δ long when relaxed, short enough that we may make the linear approximation of supposing it evenly stretched in Euclidean space to a length v, will have tension proportional to $(v-\delta)/\delta$ and energy to $(v-\delta)^2/2\delta$ when $v \geq \delta$, 0 when $v < \delta$. This is simplified if we imagine "ideal elastic" for which unstretched length is always negligible compared to stretched length, and so suppose the energy of the piece is $\frac{1}{2}(v^2/\delta)$. Completing the process of linear approximation around a point by taking the limit as $\delta \to 0$ (going to the derivative), this suggests that we adopt $c^*(t)$ as the "tension" vector and $\frac{1}{2}(\text{length of } c^*(t))^2$ as the "energy per unit unstretched length" at c(t). The $(\text{length})^2$ of a vector $\underset{\sim}{v}$ is given by $\underset{\sim}{v}.\underset{\sim}{v}$, so we are led to make

<u>3.01 Definition</u>

The <u>energy</u> of a smooth curve $c : [a,b] \to M$ when M has metric tensor field $\underset{\sim}{G}$ is the quantity

$$E(c) = \tfrac{1}{2}\int_a^b \underset{\sim}{G}_{c(s)}(c^*(s),c^*(s))ds, \quad \text{or} \quad \tfrac{1}{2}\int_a^b c^*(s).c^*(s)ds \quad \text{for short.}$$

Similar remarks to those about the length integral (VII.5.05; cf. also Ex. 1) apply to the existence and meaning of E(c): our motivation above of the definition again appeals to the idea of an integral as a kind of total, rather than an anti-derivative.

Warning: only in the Riemannian situation, as above, does E bear any relation whatever to anything else called "energy" in physics, and our main interest is not in this case. But the discussion above explains the standard use of "energy" as its name, and the equally standard factor $\frac{1}{2}$ which has historically stuck to it despite being quite irrelevant in the analysis of geodesics.

Let us then start looking for c with E(c) minimal. (Not exactly what we'll <u>find</u> but, as the elastic example suggests, this is a good place to start.) Now, in the case of functions $f : R \to R$, the first move in finding minima is to find those x where $\frac{df}{dt}(x) = 0$, since as we vary t through a minimum f can be neither increasing or decreasing, so $\frac{df}{dt}(x)$ can be neither strictly positive nor negative (Ex. VII.5.1b). Essentially the same idea applies here: we look for the curves c such that varying through them, in whatever way, involves at c a zero rate of change of E. Of course, there are infinitely many independent ways to vary c, but fortunately we do not here need the theory of infinite-dimensional manifolds of maps, and of the derivatives of functions on them. We just consider the "directional derivatives" of E at c. That is, we require the vanishing of the rate of change of E at c, as we change path smoothly through c in any particular way. To do this formally we first need:

3.02 Definition

A <u>smooth variation</u> of a curve $c : [a,b] \to M$ from p to q is a smooth map $V :]-\varepsilon,\varepsilon[\times [a,b] \to M$ with the properties

i) $V(t,a) = p$, $V(t,b) = q$, $\forall t \in]-\varepsilon,\varepsilon[$

ii) $V(0,s) = c(s)$, $\forall s \in [a,b]$.

We can think of this as a family of paths $V_t : [a,b] \to M : s \rightsquigarrow V(t,s)$ for $t \in]-\varepsilon,\varepsilon[$ with each V_t having the same end-points as c, and V_0 actually coinciding with c. The formulation in terms of one map V makes the idea of smooth family of curves more precise: it is possible for each V_t to be smooth, and each $V_s : t \rightsquigarrow V(t,s)$ across c to be smooth, without V as a whole being smooth (Ex. 2.).

We want to examine the behaviour of $E(V_t)$ as t varies through 0. Since this is just a real-valued function of t, we know just what we mean by a derivative of it. We are looking for curves c such that for any variation of c, this derivative is zero. To carry out the necessary calculations neatly we need a bit more language. We have moved from curves, with domain an interval in R, to maps with their domains in R^2. We give these a special name and make an analogy with definition VIII.3.04 (though "along" looks a little odd in this context):

3.03 Definition

A parametrised surface in a manifold M is a smooth map S from a product of intervals (open or closed) $I \times J \subseteq R^2$ to M. (So that a smooth variation of a curve is a highly special parametrised surface).

Just as a smooth curve S is allowed to cross itself, have zero derivative etc., so a parametrized surface need not sit very neatly in M.

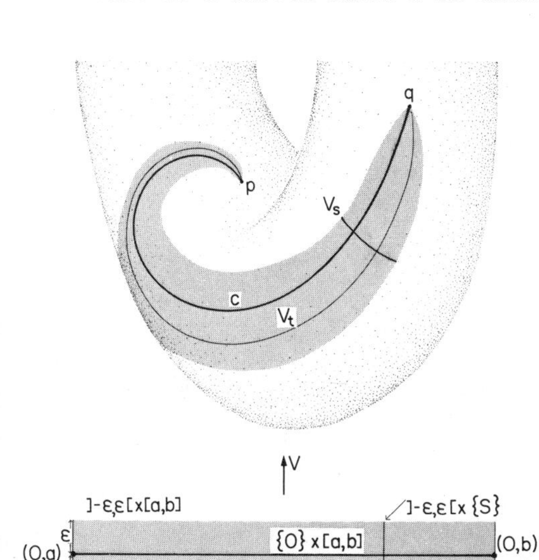

Fig. 3.1

A vector field along S is a smooth map $\underset{\sim}{v} : I \times J \to TM$ such that $\Pi \circ \underset{\sim}{v} = S$.

Analogously to the tangent vector field along a curve, if for a point $(x,y) \in I \times J$ we define

$c_1(t) = (x+t,y)$, $c_2(s) = (x,y+s)$ we can set

$$S_1^*(x,y) = (S \circ c_1)^*(0), \quad S_2^*(x,y) = (S \circ c_2)^*(0).$$

Doing this for each $(x,y) \in I \times J$ gives us vector fields S_1^*, S_2^* along S.

As with fields along curves, we can use the connection on M to define covariant differentiation "along the surface" of a vector field $\underset{\sim}{u}$ along S that need not be the restriction $\underset{\sim}{w} \circ S$ of a vector field $\underset{\sim}{w}$ on M. We need in particular two partial differentials of $\underset{\sim}{u}$ along S:

$$\Delta_x \underset{\sim}{u}, \text{ where } \Delta_x \underset{\sim}{u}(x,y) = \nabla_{S_1^*}(\underset{\sim}{u} \circ c_1)(0), \text{ along } S \circ c_1 : t \rightsquigarrow S(x+t,y), \text{ and}$$

$$\Delta_y \underset{\sim}{u}, \text{ where } \Delta_y \underset{\sim}{u}(x,y) = \nabla_{S_2^*}(\underset{\sim}{u} \circ c_2)(0), \text{ along } S \circ c_2.$$

(When we are using other labels such as (t,s) or (x^1,x^2) for points in $I \times J$, we rename these fields accordingly.)

For the particular case of S being the variation V, we see $V_2^*(t,s)$ in the "ideal elastic" conception as the tension at $V(t,s)$ of a piece P of elastic whose position is given by the curve V_t. Thinking of V_t as "position of P at time t" and so of $V(t,s)$ as "position of a point s of P at time t" leads to seeing $V_1^*(t,s)$ as "velocity of the point s at time t". The vector field $s \rightsquigarrow V_1^*(0,s)$ along c is called a <u>variation vector field</u> along c (cf Ex. 6a). Also, we shall need the vector field $\Delta_s V_2^*$ along V, whose value at (t,s) gives "the variation (considering size <u>and</u> direction), along V_t, through $V_t(s)$, of tension in elastic with the position V_t". It is the limit as $\delta \to 0$ of the difference in tension forces at the front and back ends for a piece of elastic, δ long and centred on s, per unit length. Hence, this field may be thought of as giving the instantaneous force on each point s of P, at each time t.

(For curves in spacetime, of course, the above motivations do not hold, but the geometry does. We have already encountered the "elastic force" vectors as "acceleration" vectors when s is thought of as time, not position on the elastic: a curve is a more general object than anything it represents)

For manipulative purposes we need

3.04 Lemma

If the connection on M is symmetric, then

$$\Delta_s S_2^*(t,s) \;=\; \Delta_t S_1^*(t,s), \quad \forall (t,s)$$

along any parametrised surface $S : I \times J \to M$.

Proof

If we had set up the language of connections along general maps (Ex. 3) this would be a corollary of the fact that one symmetric connection induces another. As it is, the quickest proof is to write the statement down in coordinates - since it is purely local we lose nothing by working in a chart - and compute (Ex. 4a).
(N.B. This calculation is more than a check on the Lemma's assertion: it is a good check on your grasp of the various objects involved, and well worth doing. Better still, do Ex. 4b.) ■

3.05 Definition

Writing the energy of the curve $V_t : s \rightsquigarrow V(t,s)$ as $E_V(t)$, the <u>first variation</u> of the energy function at a curve c with respect to a variation V of c is

$$\left(\frac{d}{dt} E_V\right)(0)$$

or, expanding from 3.01 and 3.03

$$\tfrac{1}{2}\left(\frac{d}{dt} \int_a^b (V^*(\ ,s) . V^*(\ ,s))ds\right)(0).$$

(cf. VII.1.03 on notation.)

3.06 Definition

If c has least energy among "nearby" curves from a to b, E_V must have a minimum at O for any variation V. Since it is a smooth function $R \to R$, we must therefore have

$$\left(\frac{d}{dt} E_V\right)(0) = 0$$

for any variation V through c: O is a critical point for every E_V. Thus we shall look for curves that are energy-critical: that is, those whose first variations vanish with respect to all V. Not all such curves are minima, but it turns out that "critical" is more important than "minimal" anyway. Our main tool in the search is the following

3.07 Theorem (First Variation Formula)

Using the Levi-Civitȧ connection on M,

$$\left(\frac{d}{dt} E_V\right)(0) = -\int_a^b \Delta_s V_2^*(0,s).V_1^*(0,s)ds.$$

In words, for "ideal elastic", this says that the total rate of change of elastic energy at time O is the integral over $s \in [a,b]$ of the dot product [-(net force on point s at time O).(velocity of s at time O)] which would be trivial if we were summing over a finite set of s's. (Their "kinetic energy" increases at the expense of the "elastic energy" producing the force - hence the minus sign.)

Proof

For any $s \in [a,b]$, applying VIII.6.02, 6.05 to the curve $V_s : t \rightsquigarrow V(t,s)$,

$$* \qquad \frac{\partial}{\partial t}(V_2^*.V_2^*) = \Delta_t V_2^*.V_2^* + V_2^*.\Delta_t V_2^* = 2V_2^*.\Delta_t V_2^*$$

as functions $]-\varepsilon,\varepsilon[\to R$. Hence

$$\left(\frac{d}{dt} E_V\right)(0) = \tfrac{1}{2}\left(\frac{d}{dt}\int_a^b V_2^*(\ ,s).V_2^*(\ ,s)ds\right)(0)$$

$$= \tfrac{1}{2}\int_a^b\left(\frac{d}{dt}(V_2^*(\ ,s).V_2^*(\ ,s))(0)\right)ds, \quad \text{by Ex. 5}$$

$$= \tfrac{1}{2}\int_a^b \frac{\partial}{\partial t}(V_2^*.V_2^*)(0)ds$$

$$= \int_a^b V_2^*(0,s).\Delta_t V_2^*(0,s)ds \qquad \text{by } *$$

$$** \qquad = \int_a^b V_2^*(0,s).\Delta_s V_1^*(0,s)ds, \quad \text{by 3.04.}$$

By VIII.6.02 (along V_t this time) we have for $t \in]-\varepsilon,\varepsilon[$

$$\frac{d}{ds}(V_2^*(t,\ ).V_1^*(t,\ )) = \Delta_s V_2^*(t,\ ).V_1^*(t,\ ) + V_2^*(t,\ ).\Delta_s V_1^*(t,\ )$$

as functions $[a,b] \to R$. That is, $s \rightsquigarrow V_2^*(t,s).V_1^*(t,s)$ is an indefinite integral for the function on the right, so, setting $t = 0$

$$\int_a^b\left(\Delta_s V_2^*(0,s).V_1^*(0,s) + V_2^*(0,s).\Delta_s V_1^*(0,s)\right)ds = V_2^*(0,b).V_1^*(0,b) - V_2^*(0,a).V_1^*(0,a).$$

But $V_1^*(t,a)$ and $V_1^*(t,b)$ are zero for all t, since the variation keeps the end points of c fixed. Therefore $V_1^*(0,b) = \underset{\sim}{0}$, $V_1^*(0,a) = \underset{\sim}{0}$, and so

$$\int_a^b (V_2^*(0,s).\Delta_s V_1^*(0,s))ds = -\int_a^b (\Delta_s V_2^*(0,s).V_1^*(0,s))ds$$

which combines with ** to prove the theorem. ■

Since $V_2(0,s)$ is exactly $c(s)$, of course, 3.07 can also be expressed as

$$\left(\frac{d}{dt} E_V\right)(0) = -\int_a^b \left(\nabla_{c^*} c^*(s) . \underset{\sim}{w}(s)\right) ds$$

where $\underset{\sim}{w}$ is the variation vector field corresponding to V. Intuitively for elastic a position c will be an equilibrium exactly when the "force" vector field $\nabla_{c^*} c^*$ vanishes. In general,

3.08 Corollary

A curve c is energy-critical if and only if it is a geodesic.

Proof

If c is a geodesic, for any variation V we have

$$\left(\frac{d}{dt} E_V\right)(0) = -\int_a^b \underset{\sim}{0} . V_1^*(0,s) ds = \int_a^b 0\, ds = 0$$

so c is energy-critical. Conversely, if c is energy-critical,

$$\int_a^b \left(\nabla_{c^*} c^*(s) . \underset{\sim}{u}\right) ds = 0$$

whenever $\underset{\sim}{u}$ is a variation vector field; hence by Ex. 6a, whenever $\underset{\sim}{u}$ is a vector field along c with $\underset{\sim}{u}(a) = \underset{\sim}{0}_p$, $\underset{\sim}{u}(b) = \underset{\sim}{0}_q$. But this implies (Ex. 6b-e) that

$$\nabla_{c^*} c^* = \underset{\sim}{0}$$

identically, so c is a geodesic. ■

3.09 Length

If we had set out to generalise the idea of "shortest distance" between two points we would have used instead of E the integral

$$L(c) = \int_a^b \sqrt{|c^*(s).c^*(s)|}\,ds = \int_a^b \|c^*(s)\|\,ds$$

for length introduced in VII.5.04, and proceeded similarly. This would have had two disadvantages: first $\sqrt{|\ |}$ is not differentiable at 0, which means harder technicalities to handle; second, there are more length-critical curves than energy-critical, in an unhelpful way. Consider elastic stretched in Euclidean space: there is one position only which is minimal (or even critical) for energy, but we need only pull the middle along to find infinitely many other positions achieving the same minimal length. (It is also much clearer, for our motivation, why elastic should "want" to minimise energy, as distinct from length.) However, a non-null curve of critical length once found, it is always possible to rearrange it "evenly" along its image and get a curve of critical energy. More precisely, we state without proof:

3.10 Fact

A non-null length-critical curve is always a reparametrisation of an energy-critical one: that is, of a geodesic. (On null curves, cf. 4.02).

Thus we would find no more interesting curves in M by considering length than by using energy, while working harder to find them. Moreover we would have missed the unique canonical parametrisation of those we did find, which comes almost free with the energy approach (Ex. 7). The interested reader is referred to [Spivak (2)], Vol I, for a proof, though he will have to extend the arguments slightly for the non-Riemannian case. In the Riemannian case infimum of arc length between two points in a connected manifold actually gives a (nontensorial;

VI.1.02) metric, and the metric space is complete (in the sense given in Appendix, 1.02) if and only if the manifold is geodesically complete: see [Kobayashi and Nomizu].

Exercises IX.3

1) Show from Def^{n}s VII.5.04, 5.05 and Def^n 3.01 that the energy of an affine curve, in an affine space with a constant metric tensor, is proportional to the square of its length.

2a) Show that the function f of Ex. VII.1.2 has all functions $f_x : R \to R : t \leadsto f(x,t)$ and $f_y : t \leadsto f(t,y)$ smooth, though it has no derivative at (0,0).

b) Deduce that $V = f|]-1,1[\times [-2,2]$ does not constitute a smooth variation of the curve $[-2,2] \to R : t \leadsto 0$, though each V_t and V_s is smooth.

3) The extension in 3.03 of Definition VIII.3.04 should have stimulated the reader's generalisation reflexes:

What is the appropriate definition for a vector field along any smooth map $f : M \to N$? (It should reduce to an earlier definition when f is the identity $M \to M$.)

Define a connection along $f : M \to N$. (Δ_t gives a connection along a curve, for instance, and Δ_x, Δ_y give a connection along S, but in a coordinate-dependent way.)

If ∇ is a connection on N, define the induced connection along $f : M \to N$.

4a) Prove Lemma 3.04 in components. (You will need the fact that $\frac{\partial}{\partial t}\frac{\partial}{\partial s} = \frac{\partial}{\partial s}\frac{\partial}{\partial t}$ (Ex. VII.7.1), which is why we do not allow V of Ex. 2b as a variation or parametrised surface.)

b) Or, follow the direction pointed by Ex. 3 far enough to get Lemma 3.04 as part of a more general theory.

5) Prove from VII.5.05 and Ex. VII.7.1c that if $f : R^2 \to R$ is C^1 then

$$\frac{d}{dt}\int_a^b f(\ ,s)ds = \int_a^b \frac{\partial f}{\partial t}(\ ,s)ds.$$

6a) If $\underset{\sim}{u}$ is any vector field along $c : [a,b] \to M$ from p to q with $\underset{\sim}{u}(a) = \underset{\sim}{0}_p$, $\underset{\sim}{u}(b) = \underset{\sim}{0}_q$, construct a variation V through c such that $V_1(0,s) = u(s)$. (One method: use $\exp_{c(s)}$ to define $V(t,s)$.)

b) If $\underset{\sim}{w}$ along c is not identically $\underset{\sim}{0}$, use continuity to show there is $x \in [a,b]$, $\delta \in R$ with $B(x,\delta) \subseteq [a,b]$, $y \in B(x,\delta) \Rightarrow \underset{\sim}{w}(y) \neq \underset{\sim}{0}$.

c) Use the smoothness and non-degeneracy of $\underset{\sim}{G}$ to construct a vector field $\underset{\sim}{t}$ along c with $\underset{\sim}{t}(t).\underset{\sim}{w}(t) > 0$, $t \in B(x,\delta)$. (If M is Riemannian, $\underset{\sim}{t} = \underset{\sim}{w}$ will do.)

d) Show that the function

$$f : R \to R : s \rightsquigarrow \begin{cases} e^{\left(\frac{1}{s^2-1}\right)} & \text{if } |s| < 1 \\ 0 & \text{otherwise} \end{cases}$$

is smooth and deduce that

$$\underset{\sim}{u} : [a,b] \to TM : t \rightsquigarrow f\left(\frac{t-x}{\delta}\right)\underset{\sim}{t}(t)$$

is a smooth vector field along c with

$$\underset{\sim}{w}(x).\underset{\sim}{u}(x) > 0$$

$$\underset{\sim}{w}(t).\underset{\sim}{u}(t) \geq 0, \quad \forall t \in [a,b]$$

$$\underset{\sim}{u}(a) = \underset{\sim}{0}_p,\ \underset{\sim}{u}(b) = \underset{\sim}{0}_q.$$

e) Deduce that if

$$\int_a^b \underset{\sim}{w}(s).\underset{\sim}{u}(s) \quad = \quad 0$$

for all variation vector fields $\underset{\sim}{u}$, then $\underset{\sim}{w}$ is identically zero.

7a) Show that for a geodesic $c : [a,b] \to M$ and diffeomorphism $f : [c,d] \to [a,b]$, the reparametrised curve $c \circ f$ is a geodesic if and only if f is affine. Deduce that no reparametrisation of c by a map $g : [a,b] \to [a,b]$ is a geodesic unless f is the identity. (Thus we have a <u>unique canonical parametrisation</u> of c with domain $[a,b]$.)

b) If $c : [a,b] \to M$ is an arbitrary reparametrisation of a non-null geodesic $\tilde{c}$, show that the reparametrisation $\bar{c}$ of c by arc length is a geodesic. Deduce that $\bar{c}$ is the unique affine reparametrisation of $\tilde{c}$ with domain $[0,L(c)]$.

c) Why does b) fail for c null? And why does the condition that parallel transport $\underset{\sim}{\tau}$ be an isometry guarantee that a null vector $\underset{\sim}{w}$ is carried to a specific $\underset{\sim}{\tau}(\underset{\sim}{w})$ (rather than, say, $2\underset{\sim}{\tau}(\underset{\sim}{w})$ which has the same size), and hence that a) does not fail for c null?

4. MAXIMA, MINIMA, UNIQUENESS

Elastic will sit, stably, only in a position that is a minimum for energy at least locally (that is, among nearby paths). Energy and length are both critical, but not minimal, for the great circle route from Greenwich (England) to Tema (on the coast of Ghana) via both Poles. By suitable small changes of the curve we can either diminish or increase its energy, so that even locally this geodesic is neither a maximum nor a minimum for energy among curves from Greenwich to Tema. It is called a <u>saddle point</u> for energy, by analogy with the picture of the simplest situation where this sort of non-extremal criticality can arise. Fig. 4.1 shows the graph in R^3 of the function $R^2 \to R : (x,y) \rightsquigarrow x^2-y^2$, which has neither a maximum nor a minimum at $(0,0)$ but has zero derivative there.

We could investigate systematically whether an energy-critical curve was minimal, saddle-type or what, by looking at the second variation, which corresponds to differentiating a function $R \to R$ a second time. However unless $E(c)$ means physical energy it is usually important that a curve be critical while wholly irrelevant physically whether it be minimal; likewise when the interesting integral is action, or time. Why nature should behave so was a great mystery in classical mechanics. Even a "least whatever" principle seemed a bit mystical and mediaeval in flavour, when a particle had no way of comparing the integral along its actual history with other possibilities. The "critical whatever" conditions actually apparent in, for example, Fermat's Principle (often misstated as "a light ray follows the path of least time": easy critical but not locally minimal examples are given in [Poston and Stewart]) could not even be seen as a kind of Divine economy drive. In quantum theory, however, variation principles are entirely reasonable: "the particle goes all possible ways and probably arrives by a route that delivers it in phase with the result of nearby routes", and this turns out to involve the criticality condition directly, with no reference to minima. This rationale then motivates variational techniques in classical mechanics, considered as an approximation to quantum descriptions. This is not the book in which to go further into this point, however, particularly as it is so lucidly discussed in [Feynman] - a work which the reader should in any case read, mark, learn and inwardly digest.

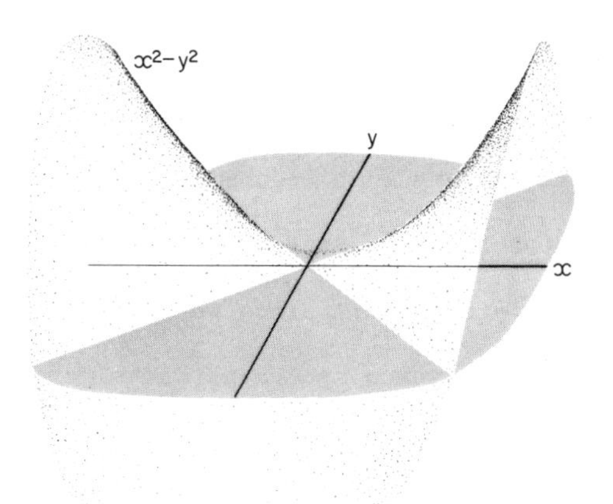

Fig. 4.1

We shall not, then, set up the machinery of the second variation. But it is worth looking at some particular facts which help geometrical insight into geodesics.

First, we have seen that a Riemannian geodesic need not be minimal. If it is minimal, it need not be the unique such : consider the circle's worth of geodesics from N. Pole to S. Pole on a sphere, all of the least possible length and energy.

Next - since all along we have assumed in illustration that an affine curve in Euclidean space is the unique minimum energy curve with that domain and end points - let us prove it.

4.01 Lemma

Given points p, q in an affine space X with vector space T and any constant Riemannian metric $\underset{\sim}{G}$, the unique affine map $f : [a,b] \to X$ with $f(a) = p$, $f(b) = q$ has less energy than any other curve $c : [a,b] \to X$ from p to q.

Proof

It follows at once from the definitions (Ex. 2) and 3.08 that no other curve can be critical or, therefore, minimal, but we have still to show that every other curve has more energy. (These are not equivalent: we can find differentiable curves $f : [-1,1] \to R$ from -1 to 1 with

$$Q(f) = \int_{-1}^{1} (s^2-1)^2((f(s))^2-1)^2 \left(\left(\frac{df}{dt}(s)\right)^2 + 1\right)ds$$

arbitrarily small, but not zero. [Why? Interpret Q(f) as the energy of the curve $t \rightsquigarrow (t,f(t))$ with respect to a sometimes vanishing "metric tensor" on R^2.] So although there is a Q-critical function, there is no f with $Q(f) \le Q(g)$, $\forall g$. Energy is better-behaved if M is complete and Riemannian (cf. 3.10) - there is then always at least one path of least energy between any two points - but we have not proved this.)

Define $\underset{\sim}{c}_1, \underset{\sim}{c}_2 : [a,b] \to T$, by setting $\underset{\sim}{d}(p,q) = \underset{\sim}{v}$ and

$$\underset{\sim}{c}_1(t) = \left(\frac{\underset{\sim}{d}(p,c(t)).\underset{\sim}{v}}{\underset{\sim}{v}.\underset{\sim}{v}}\right)\underset{\sim}{v}, \qquad \underset{\sim}{c}_2(t) = \underset{\sim}{d}(p,c(t)) - \underset{\sim}{c}_1(t).$$

The curve $\underset{\sim}{c}_1$ is "the part along $\underset{\sim}{v}$ of c", with $\underset{\sim}{c}_1(a) = \underset{\sim}{0}$, $\underset{\sim}{c}_1(b) = \underset{\sim}{v}$, and $\underset{\sim}{c}_2$ "the part orthogonal to $\underset{\sim}{v}$" (Fig. 4.2).

It is clear that (using the corresponding Riemannian metric on T)

$$E(c) = E(\underset{\sim}{c}_1)+E(\underset{\sim}{c}_2)$$

since, at each point in the integration,

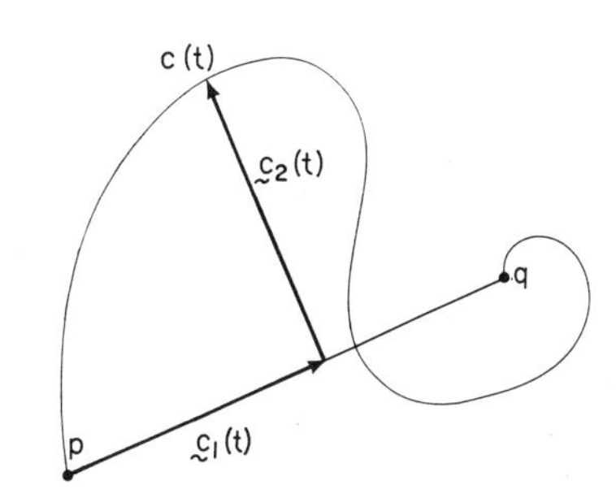

Fig. 4.2

$$c^*(s).c^*(s) = (\underset{\sim}{c}_1+\underset{\sim}{c}_2)^*(s).(\underset{\sim}{c}_1+\underset{\sim}{c}_2)^*(s)$$

$= (\underset{\sim}{c}_1^*(s)+\underset{\sim}{c}_2^*(s)).(\underset{\sim}{c}_1^*(s)+\underset{\sim}{c}_2^*(s))$, freeing the tangent vectors $\underset{\sim}{c}_i^*$ to T

$= \underset{\sim}{c}_1^*(s).\underset{\sim}{c}_1^*(s)+\underset{\sim}{c}_2^*(s).\underset{\sim}{c}_2^*(s)$, as $\underset{\sim}{c}_1^*$, $\underset{\sim}{c}_2^*$ are tangent to orthogonal subspaces and so themselves orthogonal.

Hence, since in the Riemannian situation E is always positive and only vanishes on constant curves,

$$E(c) \geq E(\underset{\sim}{c}_1)$$

with equality only when $\underset{\sim}{c}_2$ is identically $\underset{\sim}{0}$. So if the image of c is not confined to the affine hull V of {p,q} (cf. II.1.03), then we can reduce its energy by orthogonal projection onto V. It therefore suffices to consider c of the form

$c(t) = p+\tilde{c}(t)\underset{\sim}{v}$, where $\tilde{c} : [a,b] \to R$ has $\tilde{c}(a) = 0$, $\tilde{c}(b) = 1$. We then have

$$c^*(t) = \frac{d\tilde{c}}{ds}(t)\underset{\sim}{v}, \text{ while } f^*(t) = \frac{\underset{\sim}{v}}{a-b}$$

(freeing vectors where convenient), and

$$* \quad E(f) = \tfrac{1}{2}\int_a^b \frac{\underset{\sim}{v}}{a-b} \cdot \frac{\underset{\sim}{v}}{a-b}\, ds = \tfrac{1}{2}\frac{\underset{\sim}{v}\cdot\underset{\sim}{v}}{(a-b)^2}\int_a^b ds = \tfrac{1}{2}\frac{\underset{\sim}{v}\cdot\underset{\sim}{v}}{a-b}.$$

So if $\underset{\sim}{c}(t) = \underset{\sim}{d}(f(t),c(t))$, we have

$$E(\underset{\sim}{c}) = \tfrac{1}{2}\int_a^b (c^*(s)-f^*(s)).(c^*(s)-f^*(s))ds \quad \text{(freeing the tangent vectors)}$$

$$= \tfrac{1}{2}\int_a^b c^*(s).c^*(s) - \int_a^b\left(\frac{d\tilde{c}}{dt}(s)\underset{\sim}{v}\right)\cdot\frac{\underset{\sim}{v}}{b-a}\,ds + \tfrac{1}{2}\int_a^b f^*(s).f^*(s)ds$$

$$= E(c) - \frac{\underset{\sim}{v}\cdot\underset{\sim}{v}}{b-a}\int_a^b\left(\frac{d\tilde{c}}{dt}(s)\right)ds + E(f)$$

$$= E(c) - \frac{\underset{\sim}{v}\cdot\underset{\sim}{v}}{b-a}(\tilde{c}(b)-\tilde{c}(a)) + E(f)$$

$$= E(c) - E(f), \quad \text{by } *.$$

So $E(c) \geq E(f)$,

with equality only when $E(\underset{\sim}{c}) = 0$, that is when $\underset{\sim}{c}$ is constant: precisely when $c = f$, since $c(a) = f(a)$.

4.02 Other cases

A similar proof to 4.01 (Ex. 3) shows the affine path from p to q, in a Riemannian X to have the shortest possible <u>length</u>, sharing this length only with its reparametrisations. A minor adaptation (Ex. 4) of the same technique shows that if exactly one vector in an orthonormal basis for T is timelike, a timelike geodesic has <u>maximum</u> length among timelike curves from p to q measured in the corresponding constant metric, though not maximum energy. If there is more than one spacelike dimension, spacelike geodesics are neither maximal nor minimal. Fig. 4.3 shows such a geodesic f from p to q, together with spacelike curves c_1, c_2 (with length and energy closer to 0 than those of f, and further, respectively).

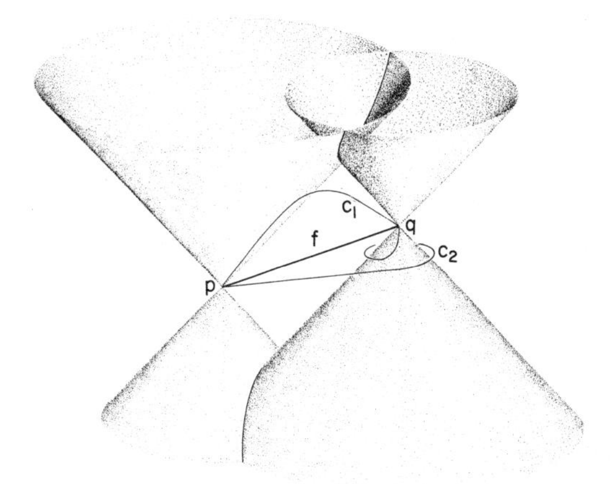

Fig. 4.3

Defining the length L(c) of c in terms of $\| \ \|$ rather than $| \ |$ makes it automatically real and non-negative, so that any null curve is trivially minimal, geodesic or not. With respect to energy, a null geodesic f is a saddle point. For we can take a variation V of f that changes only the parametrisation. Thus each V_t is a null curve, so E_V is identically zero. But since only one parametrisation with the given domain is energy-critical, this means that arbitrarily near f there are non-critical curves <u>with the same</u> (zero) <u>energy</u>. But this is only possible in the saddle situation (Fig. 4.4; there must be points arbitrarily near x

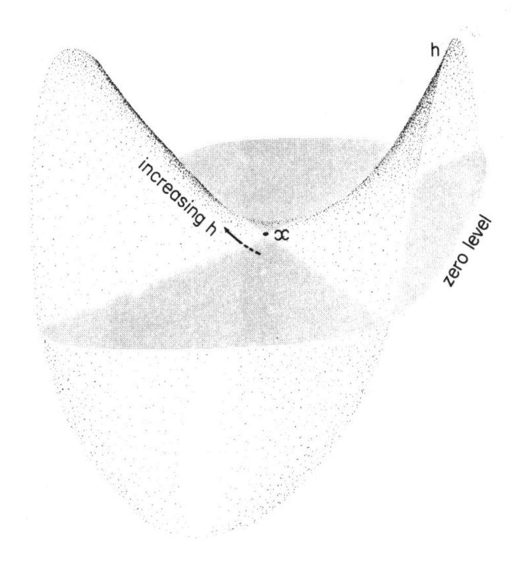

Fig. 4.4

where h takes values either above or below $h(x)$.)

4.03 Riemannian geodesics

We have seen that a geodesic on the sphere need not be - even among nearby geodesics - a minimal curve between its ends. Intuitively the failure of the Polar route from Greenwich to Tema to be a minimum stems from taking too large a piece of a great circle. In fact, any sufficiently small piece of a geodesic in a Riemannian manifold will be a minimum (Ex. 5).

One way to develop intuition about Riemannian geodesics, clearly, is to tighten elastic strings in various ways around potatoes, husbands, and any other objects with interesting surfaces. Another, allowing 3-dimensional manifolds not just surfaces, is to consider systems of lenses and refractive materials generally.

To keep things C^∞, assume that the refractive index $k(x)$ at a point x in the interesting space X gives a smooth function $X \to R$. (Assuming that a lens "fades out" through a thin boundary layer violates reality no worse than supposing that k has a discontinuity through a C^∞ surface defining a boundary of zero thickness, and in this context saves technicalities.) Now, in classical optics, ignoring polarisation, light travels through X with speed $1/k(x)$, choosing units as in Ch. 0 §3 to make its speed 1 in vacuum. If however we define a new Riemannian metric from the usual constant one $\underset{\sim}{G}$ on X (supposed affine) by

$$\underset{\sim}{\tilde{G}}(x) = k^2(x)\underset{\sim}{G}(x)$$

according to $\underset{\sim}{\tilde{G}}$ a path f of a light ray has constant speed

$$[\underset{\sim}{\tilde{G}}_{f(t)}(f^*(t),f^*(t))]^{\frac{1}{2}} = k(x)[\text{usual speed}] = 1,$$

independently of t. Moreover Fermat's principle (light rays take paths of critical time) becomes the statement that f has critical length as measured by $\tilde{G}$: time taken is given exactly by L(f). So f is a parametrisation of a geodesic, and in fact is a geodesic since it has constant speed. So by altering the geometry, we have rescued the principle that light travels in (now generalised) straight lines even when not in a medium of constant refractive index, and incorporated the physics into the geometry (cf also Ex. 6).

This is closely analogous to general relativity's changing Newton's "particles follow straight lines at constant speed in the absence of gravitational forces" to "particles follow geodesics in spacetime", except that it is avoidable. For gravitation, only geometry seems to work.

4.04 Pseudo-Riemannian geodesics

In the pseudo-Riemannian case, as in the Riemannian, there can be more than one geodesic between two points. Indeed, let $M = R \times S^2$ with the metric tensor induced by the obvious inclusion $R \times S^2 \hookrightarrow R \times R^3 = R^4$ into R^4 with the Minkowski metric. (So that each $\{t\} \times S^2$ has all vectors tangent to it spacelike). Ex. 7 illustrates on M some of the things that are possible; more appear in §5 and §6.

Sufficiently small bits of a timelike geodesic, in a manifold with one timelike dimension, are maximal for length (Ex. 8) though not for energy, since that is false even in the affine case (4.02, Ex. 4b). As with minimality in the Riemannian case, this may fail for larger pieces (cf. for example, Ex. 7b). Null and spacelike geodesics are always saddle-type criticalities, even in small pieces, by the same arguments as before.

4.05 Twins

The Twins "paradox" (Ch. 0 §3) is not a logical problem. It is an experimental fact about measurements of time which is neatly modelled (hence "explained") by pseudo-Riemannian geometry. But since from time to time any physicist is trapped by some Philosopher of Science who is proud of not understanding equations, he should be equipped with some arguments simple enough for the Philosopher to understand.

The quantity ($\text{time}^2 - \text{distance}^2$), separating two events, is known experimentally to be independent of who measures the separate times and distances and how fast he is going, up to differences in velocity very close to the speed of light. (This is contrary to Newton's theory, since there distance between non-simultaneous events depends on choice of "rest velocity" while time does not.) A new physical theory might alter the philosophy of this fact as profoundly as relativity alters Newtonian gravitation, but would involve only very small numerical changes if it still agreed with experiment. (Just as Newtonian mechanics remains accurate enough for ICBMs.) So the consequences of this experimental fact will not alter much numerically for situations already studied.

One of these consequences is that along an affine world-line (supposing for the present that spacetime is affine), the length given by the Lorentz metric for $\underset{\sim}{d}(f(b),f(a))$ is the appropriate perceived time for an observer following f from f(a) to f(b), since for him $(\text{distance})^2$ between f(a) and f(b) is zero. It is then less a postulate of relativity, special or general, than a basic notion of the calculus that $\sqrt{f^*(s).f^*(s)}$ is the <u>rate</u> of change of perceived or <u>proper</u>[†] time for an observer whose motion is described by any timelike curve f - not just an affine curve. Differentiation is linear or affine approximation, so if the calculus is applicable we can make an arbitrarily good affine description by taking a small enough bit. In an affine motion, the arguments for $\sqrt{f^*(s).f^*(s)}$ being the (constant) rate of change of proper time with s are overwhelming.

Equally basic is the notion that, for a quantity changing with s at a rate

---

[†] "Proper" here does not mean "right": it is older English usage for "tied to the particular person or thing", as in "property".

depending on s, you get the total change between s = a and s = b by integrating the rate. Hence the appropriate "elapsed proper time" along a timelike curve is the integral we have called length. This will remain an exceedingly good approximation to observed or predicted lapse of proper time, even if relativity is replaced by something else and the same happens to the calculus: both fit the facts too well to fail to approximate anything that fits better.

Suppose spacetime is isomorphic to Minkowski space (such an isomorphism, preserving the metric, is called an inertial frame and only exists - even locally - if spacetime is flat: cf. Chapter X). We see then by 4.02 that along any timelike curve c from p to q, that is not a reparametrisation of the affine one f parametrised by arc length, proper time measured along c is less than that along f. This conclusion does not depend on c being an "inertial movement", it depends only on the existence of an inertial frame. The Philosopher who says "But if one observer is accelerating, his observations are no longer referred to an inertial frame and special relativity is inapplicable" either has never seen the right pictures drawn to explain the calculus, or has not been told that special relativity assumes only that spacetime has the geometry of Minkowski space (a statement without "observers") and that the calculus is applicable.

General relativity is diferent in two respects. First, it ceases to suppose that spacetime has an affine structure, and makes the weaker assumption (justified by local experiments) that it is a Lorentz manifold. Second, it relates the metric to the distribution of matter. We shall discuss the second point in Chapter XII; for the present discussion, all we need is the first. The above reasoning still justifies considering the length integral as elapsed proper time (recall that T_pM is exactly the flat approximation to the manifold M at p, just as $\underset{\sim}{D}_p f : T_pM \to T_{f(p)}N$ is the linear approximation at p to the map $f : M \to N$) and the observations of 4.04 apply.

In this situation two timelike geodesics, not just two curves, from the same point in spacetime can meet again, and observers travelling them can compare watches. This confuses even some of the Philosophers who have got used to the special theory. Before, they could see the difference between the geodesic curve and the other, but now with both observers "inertial" shouldn't symmetry guarantee equal elapsed times?

Not if the matter distribution influencing the metric has any asymmetry of its own. Exactly analogously, why should the two geodesics in Ex. 6 have the same length? Even if the spacetime is highly symmetric and the end points symmetrically placed, the geodesics need not be symmetrically related (Ex. 7c).

Enough, or even too much, on points that should be obvious. One final remark: certain journals are given to carrying acrimonious disputes about this "paradox", in coordinates yet. To print material at that level is a waste of precious trees.

Exercises IX.4

1) Find all three minima, and both maxima, of the function $f : R \to R : x \rightsquigarrow |x^3-6x^2+11x-6|$. Comment on the relationship of smoothness to the rule "differentiate to find the minimum", and on the relative difficulty of the length and energy variational problems.

2) A geodesic curve in an affine space with a constant metric tensor field is necessarily affine (argue geometrically, or use 1.04 and VIII.6.06 for components.), and vice versa.

3a) Reduce the minimum length problem from p to q, in an affine space X with a constant Riemannian metric, to one dimension in the manner of 4.01.

b) Any non-injective curve in R with the usual metric has greater length than an injective one with the same end points.

c) Deduce that any path with a length no greater than $\|\underset{\sim}{d}(p,q)\|$ is a reparametrisation of the affine curve from p to q with domain [0,1].

4a) If the affine space X has a constant indefinite metric of signature (2-dimX), and $\underset{\sim}{d}(p,q)$ is timelike, let f be an affine curve from p to q. Show that any timelike curve from p to q that is not a reparametrisation of f has greater length than f.

b) Find a variation of f along which E_V attains a maximum at 0 (vary the route), and one along which E_V attains a minimum at 0 (vary the parametrisation.)

Thus f is a local maximum for L, a saddle point for E.

5a) Let M be Riemannian, and $U \subseteq T_xM$, $V \subseteq M$ have $\exp_x|U$ a diffeomorphism $U \to V$. Find a ball $\bar{B}_\delta = \{\underset{\sim}{t} \mid \|\underset{\sim}{t}\| \leq \delta\} \subseteq U$, $\underset{\sim}{v} \in T_xM$ with $\|\underset{\sim}{v}\| = \delta$, and let $y = \exp_x(\underset{\sim}{v})$, $\underset{\sim}{f}(t) : [0,1] \to T_xM : t \leadsto t\underset{\sim}{v}$. Decompose an arbitrary curve $\underset{\sim}{c} : [0,1] \to T_xM$ from $\underset{\sim}{0}$ to $\underset{\sim}{v}$ into a "radial part" $c_1 : t \leadsto \|\underset{\sim}{c}(t)\|$ and a "spherical part" $\underset{\sim}{c}_2 : t \leadsto \underset{\sim}{c}(t)/c_1(t)$, to show that if the image C of $\underset{\sim}{c}$ is contained in U, but not in $\bar{B}_\delta$ then

$$E(c) > \tfrac{1}{2}\delta^2 = E(f), \quad L(c) > \delta = L(f)$$

where $c = \exp_x \circ \underset{\sim}{c}$, $f = \exp_x \circ \underset{\sim}{f}$.

Deduce that this is true even if $C \not\subseteq U$, and show that for any $g : [0,1] \to M$ from x to y,

$$E(g) \geq E(f) \quad \text{with equality only if} \quad g = f$$

$$L(g) \geq L(f) \quad \text{with equality only if} \quad g = f\circ h,\ h : [0,1] \to [0,1].$$

b) Deduce that for a geodesic $c : [a,b] \to M$ and any point $t \in [a,b]$ there is an ε such that whenever $s \in]t,t+\varepsilon[$, $c|[t,s]$ is

(i) the curve of least energy from c(t) to c(s) with domain [t,s]

(ii) shorter than any other curve from c(t) to c(s) that is not a reparametrisation of it.

6) A mirage is due to air near the ground becoming hotter than that above, and consequently having a lower density and refractive index. Model this mathematically as in 4.03 (using accurate numbers, or making them up) to get the two geodesic light rays shown in Fig. 4.5. Which of these (if either) is minimal for length and/or energy among nearby curves? (See [Poston and Stewart] for more of the geometry of mirages.)

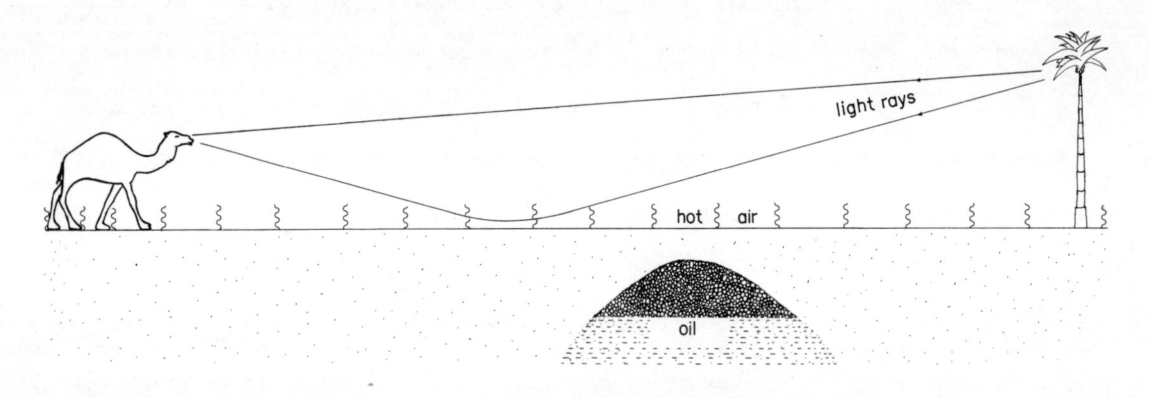

Fig. 4.5

7) Let M be $S^2 \times R$ with the indefinite metric described in 4.04.

a) Show that any geodesic in M has the form $f(s) = (f^1(s), f^2(s)) \in S^2 \times R$, where f^1 is a geodesic in S^2 with its usual Riemannian metric and f^2 is an affine map $J \to R$, and that any curve of this form is a geodesic. (Hint: Ex. 1.4)

b) Show that there are infinitely many distinct spacelike geodesics parametrised by arc length between any two points.

c) Show that unless (p^1,p^2), $(q^1,q^2) \in S^2 \times R$ have p^1 equal to or diametrically opposite to q^1, there are only finitely many timelike geodesics parametrised by arc length (if any) between them. If there are several, are they of equal lengths?

d) If p^1, q^1 <u>are</u> equal or opposite, then to each of the infinite families of equal length geodesics from p^1 to q^1 in S^2 there corresponds a family of geodesics from p to q in M; only finitely many of these families are timelike.

8) Let M be an (n+1)-manifold, with metric $\underset{\sim}{G}$ of signature (1-n), and $p \in M$. Choose normal coordinates $(x^0,\ldots,x^n)$ around p (x^0 being timelike) and if $y = \exp_p(\underset{\sim}{v})$ for $\underset{\sim}{v}$ timelike has coordinates $(v^0,\ldots,v^n)$, define

$$\phi(y) = (y^0,\ldots,y^n) = (t(\underset{\sim}{v}), \frac{v^1}{t(\underset{\sim}{v})},\ldots,\frac{v^n}{t(\underset{\sim}{v})}) \text{ where}$$

$t(\underset{\sim}{v}) = \underset{\sim}{G}(\underset{\sim}{v},\underset{\sim}{v})$.

a) Show that this defines an admissible chart (whose domain U does not include p, though its closure does) and that for any vector $\underset{\sim}{u} = u^i\partial_i$ at a point in U,

$$\underset{\sim}{G}(\underset{\sim}{u},\underset{\sim}{u}) = (u^0)^2 - \gamma_{ij}u^iu^j$$

where the γ_{ij} are the components of a negative definite bilinear form.

b) Deduce that if $q \in U$, there is a unique geodesic $c : [0,1] \to M$ from p to q with image confined to $U \cup \{p\}$ and that any timelike curve from p to q with image in $U \cup \{p\}$ is either shorter than, or a reparametrisation of, c.

c) Compare and contrast this result with Ex. 5.

9a) Suppose spacetime X is isomorphic to Minkowski space, and that someone has invented a "hyperdrive" by which we can "travel at twice the speed of light": that is, suppose that curves $f = (f^0,f^1,f^2,f^3)$ in R^4 are descriptions of possible motions provided that

$$* \quad \left[\left(\frac{df^1}{dt}\right)^2 + \left(\frac{df^2}{dt}\right)^2 + \left(\frac{df^3}{dt}\right)^2\right]^{\frac{1}{2}} \leq 2\,\frac{df^0}{dt} \quad \text{always.}$$

If any unit timelike vector tangent to X may be chosen as "rest velocity" in setting up affine coordinates before applying *, show that for <u>any</u> two points p,q with $\underset{\sim}{d}(p,q)$ spacelike the geodesic between them satisfies * for a suitable choice of rest velocity.

b) Suppose that the choice of "rest velocity" is automatically the velocity of your spacecraft at the moment of pressing the "hyperdrive" button B. If p is "here and now", and q is "Alpha Centauri one year ago", to get from p to q how fast and in what direction would you initially leave the Solar System before pressing B? (Assume Alpha Centauri is 4 light years off, for simplicity, and ignore acceleration time.)

Give your answer in "sun is at rest" terms, as we have given q.

c) Suggest a way to control your "spacelike direction" after pressing B.

d) Comment on the prevalence of science-fiction stories in which "hyperdrive", but not time travel, is involved. How many include an idea which prevents uses like the above for the "hyperdrive"?

5. GEODESICS IN EMBEDDED MANIFOLDS

Clearly, a geodesic in a manifold M embedded in an affine space X (like a great circle in $S^2 \subseteq R^3$) is not in general a geodesic in X; M forces it to bend. But it must bend only as M forces it to:

5.01 Theorem

If M is embedded by the inclusion $i : M \hookrightarrow X$ in an affine space with constant metric tensor $\underset{\sim}{G}$, denote covariant differentiation along curves in X by $\tilde{\nabla}_{\tilde{c}*}$, and along curves in M (with respect to the Levi-Cività connection of the metric induced on M) by ∇_{c*}. Then for $c : J \to M$ we have, with $i \circ c = \tilde{c} : J \to X$,

$$\mathbb{\nabla}_{c^*}c^*(t) = \underset{\sim}{0} \text{ if and only if } \underset{\sim}{G}\left(\tilde{\mathbb{\nabla}}_{\tilde{c}^*}\tilde{c}^*(t), \underset{\sim}{v}\right) = 0, \quad \forall \underset{\sim}{v} \in T_{c(t)}M.$$

Thus, c is a geodesic in M if and only if the "net elastic force" on each point (in the "elastic" conception) or the "acceleration vector" at each moment (in the "motion of a particle") is always orthogonal to M.

Proof

Apply Ex. VIII.2.1a and Ex. VIII.6.3b and the relation VIII.3.05 between ∇ and $\mathbb{\nabla}$. (The theorem is true for X a general manifold and $\underset{\sim}{G}$ arbitrary, as long as we use the metric induced by $\underset{\sim}{G}$ on M, but we shall not need this.) ■

Among the "motion of a particle" examples are the spherical pendulum (M is an S^2 in R^3) with no external forces, or if c(t) is a position in R^3 for a classical pair of point masses joined by a light rigid rod of length ℓ it corresponds to a point in the 5-dimensional manifold (cf. Ex. VII.2.1b, Ex. VII.2.8c)

$$M = \{(x^1,x^2,x^3,y^1,y^2,y^3) \mid (x^1-y^1)^2+(x^2-y^2)^2+(x^3-y^3)^2 = \ell^2\} \subseteq R^3 \times R^3 = R^6,$$

and the condition that "no external forces act" is exactly that the derivative of c* be always normal to M, in the usual Riemannian metric on R^6.

5.01 is the differential justification for the "string-stretching" idea of geodesics we have been using, in embedded Riemannian manifolds (the local minimisation of length, subject to the constraint of lying in M, being an integral one). It is useful equally as a way to build intuition in the pseudo-Riemannian case, to which string will not stretch.

5.02 An indefinite metric on part of the sphere

Let $M = \{(x,y,z) \in R^3 \mid x^2+y^2+z^2 = 1,\ z^2 < \frac{1}{2}\}$, the part of the 2-sphere lying strictly between the 45°S and 45°N parallels of latitude. Give R^3 the constant metric $\underset{\sim}{G}$

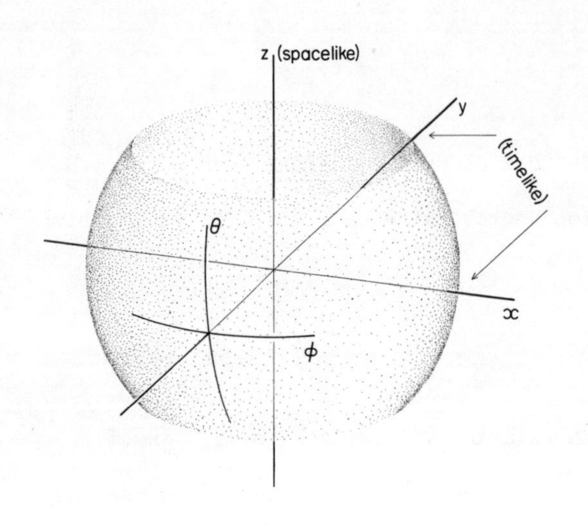

Fig. 5.1

$$(ds)^2 = (dx)^2+(dy)^2-(dz)^2$$

in "line element" notation, and call R^3 with this metric X.

Then define a chart $\psi : U \to R^2$ on M by $\psi(p) = (\theta(p),\phi(p))$ where $\theta(p)$ means "latitude Northwards" and $\phi(p)$ "longitude Eastward" of p (Fig. 5.1). The metric $\underset{\sim}{G}$ induced on M is, by Ex. 1c,

$$(ds)^2 = \cos^2\theta(d\phi)^2 - \cos 2\theta(d\theta)^2.$$

Thus the ϕ direction is timelike and θ spacelike. (Since $\cos 2\theta = 0$ for $\theta = \pm 45^\circ$, $\underset{\sim}{G}$ would become degenerate if we included more of the sphere than M. Geometrically, this is because the tangent plane to the sphere at any point with $z^2 = \frac{1}{2}$ is degenerate (recall Fig. VII.3.7), and the metric induced on the top and bottom caps ($z^2 > \frac{1}{2}$) is Riemannian.)

What curves in M are geodesic? Clearly by symmetry we have the meridians and the equator (suitably parametrised). But the other pieces of great circles are not geodesics in this metric.

Consider a general curve $c : t \rightsquigarrow (c^\theta(t),c^\phi(t))$ in M, with $c^\theta(0) > 0$ and $c^*(0) \neq \underset{\sim}{0}$. Without loss of generality suppose $c^\phi(0) = \pi/2$, so that $p = c(0)$ lies in the x-z plane of X, which we may call Q.

Now from Fig. 5.2a it is clear that the "acceleration vector" $\underset{\sim}{a} = \nabla_{c^*}c^*(0)$ (differentiating in X) must be non-zero and point to the left of the plane P in X geometrically tangent to M at p, since c "bounces off" P on that side. If c is geodesic, $\underset{\sim}{a}$ must lie in the one-dimensional subspace $V = (T_pM)^\perp \subseteq T_pX$. By symmetry V is tangent to Q (otherwise reflection in Q would give us another V), so

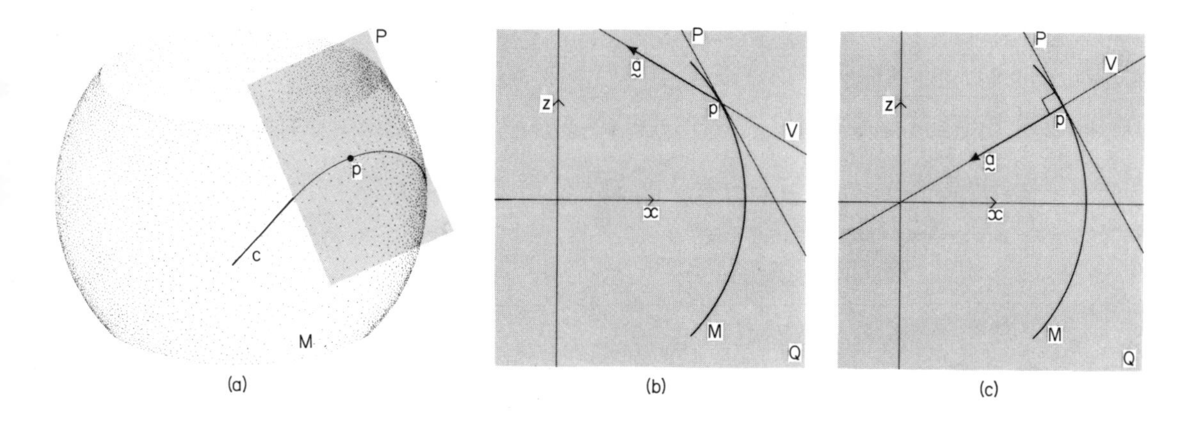

Fig. 5.2

as Q has the indefinite metric

$$(ds)^2 = (dx)^2 - (dz)^2$$

we see (recalling IV.1.08) that V is as shown in Fig. 5.2b, in contrast with the Riemannian picture 5.2c.

Thus $\underset{\sim}{a}$ is non-zero, points left and is in V. In terms of the x-z plane Q and of R^3, then, c has at O an "acceleration" with its z-component upwards. If $c^\theta(0)$ is negative, of course, a similar analysis shows that $\underset{\sim}{a}$ points downwards. Thus the geodesics in M are qualitatively as illustrated in Fig. 5.3. (Null vectors, and hence the spacelike ones squeezed between them, tend to multiples of $d\theta$ as $\theta \to \pm 45°$. Thus null and spacelike geodesics tend to tangency with meridians

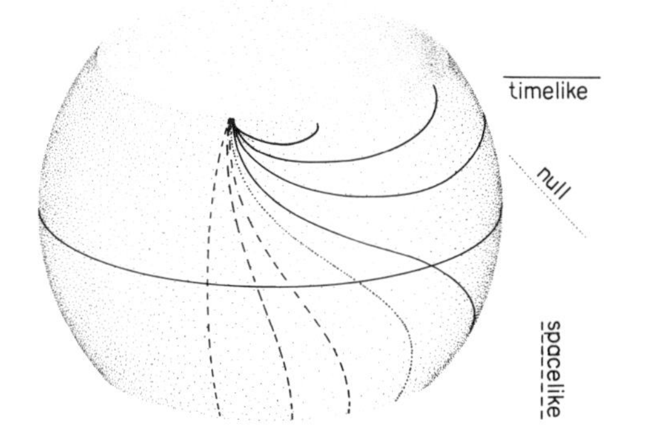

Fig. 5.3

as $c^\theta \to \pm 45^\circ$. The fact that timelike geodesics do likewise can be seen by examining the way that "upward acceleration" goes to infinity.)

Theorem 5.01 thus allows us qualitatively to analyse geodesics without hard work on the equations

$$\frac{d^2c^k}{ds^2} + \Gamma^k_{ij}\frac{dc^j}{ds}\frac{dc^i}{ds} = 0,$$

when we can obtain the metric from a convenient embedding. (In the words of Dirac "I feel I understand a differential equation when I can see what the solutions look like without actually solving it".)

5.03 An example on the plane minus a point

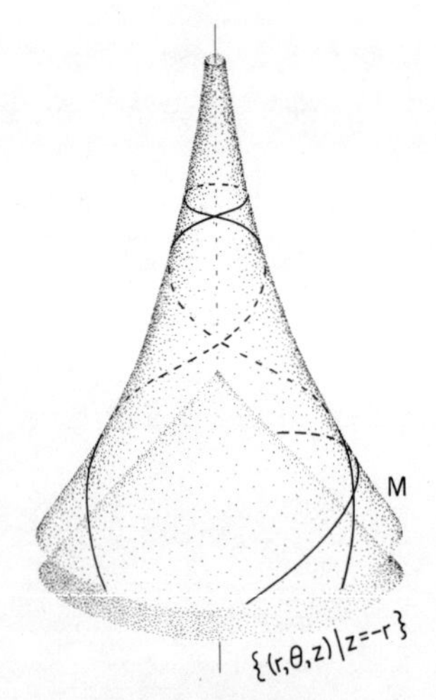

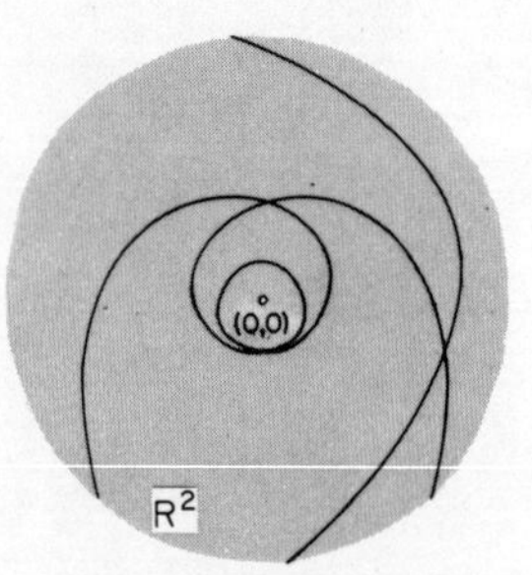

Fig. 5.4

Consider the submanifold

$$M = \{(r,\theta,z) \in R^3 \mid z = \frac{1}{r} - r\},$$

using cylindrical coordinates on R^3. Evidently we may use (r,θ) as coordinates on M, so that M is an embedding of $R^2\backslash\{(0,0)\}$ by the map

$$\psi : (r,\theta) \rightsquigarrow (r,\theta,\frac{1}{r} - r).$$

If R^3 has its usual Riemannian metric, string-stretching shows that the geodesics in M are as shown in Fig. 5.4a, or, in $R^2\backslash\{(0,0)\}$ with the metric induced by ψ, as in Fig. 5.4b. Evidently a "free end" of any such geodesic is asymptotic to a straight line in R^2 with its usual metric, and M is geodesically complete

(Ex. 2), cf. 2.03.

If M has the indefinite metric induced by the metric on R^3 of 5.02, a curve tangent to one with c^r constant (that is a "static" timelike one) has an acceleration $\underset{\sim}{a}$ in R^3 that points "upwards"; Fig. 5.5 is analogous to Fig. 5.2a,b. Such tangency is equivalent to $\frac{dc^r}{ds}(t) = 0$, so we see that no geodesic has a point where c^r is minimal, as with the previous metric: we only have maxima. Thus a non-radial geodesic must spiral out from the origin and escape (Fig. 5.6a) spiral in from r infinite (b), or (c) spiral out, reach a maximum value of c^r and spiral back in again (Ex. 3). At such a maximum, c* is (trivially) timelike, so only timelike geodesics can look like (c).

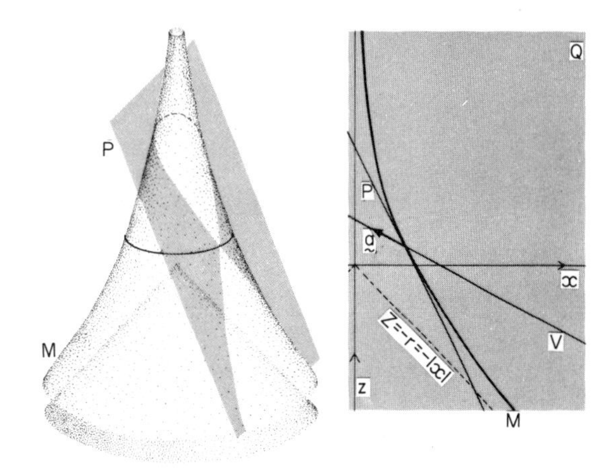

Fig. 5.5

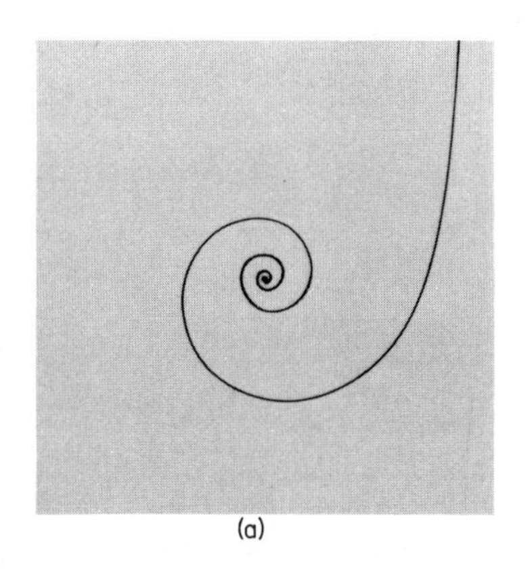

(a)

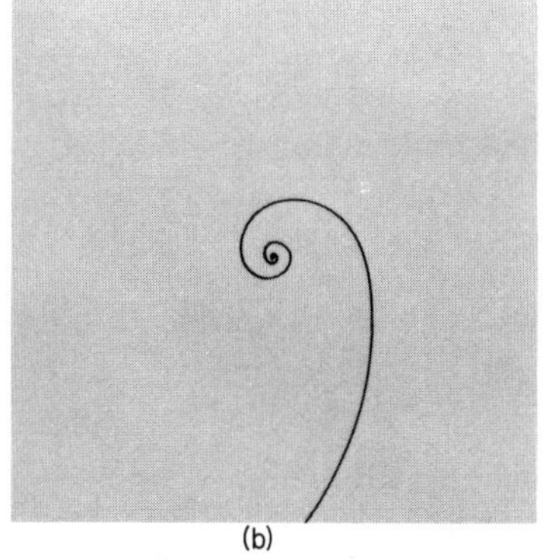

(b)

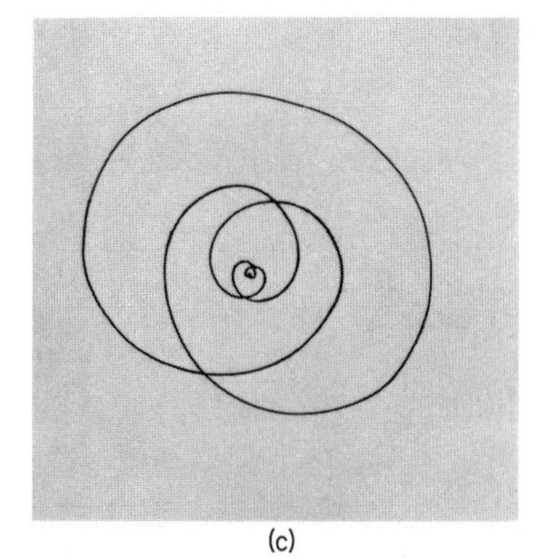

(c)

Fig. 5.6

Exercises IX.5

1a) Let $c : J \to R^3$ have $c(J) \subseteq S^2 \subseteq R^3$, and set $c(t) = (c^x(t), c^y(t), c^z(t)) \in R^3$, $c(t) = (c^\theta(t), c^\phi(t)) \in S^2$ using the labels discussed in 5.02 for as much of S^2 as possible. Show that

$$\frac{dc^\phi}{ds} = \cos(c^\theta)(\cos(c^\phi)\frac{dc^x}{ds} + \sin(c^\phi)\frac{dc^y}{ds}$$

$$\frac{dc^\theta}{ds} = \sin(c^\theta)\left(-\sin(c^\phi)\frac{dc^x}{ds} + \cos(c^\phi)\frac{dc^y}{ds}\right) + \cos(c^\theta)\frac{dc^z}{ds}$$

as functions $J \to R$.

b) Deduce that using the indefinite metric $\tilde{G}$ given on R^3 in 5.02

$$c^*.c^* = \cos^2(c^\theta)\left(\frac{dc^\phi}{ds}\right)^2 - \cos(2c^\theta)\left(\frac{dc^\theta}{ds}\right)^2.$$

c) Deduce that in "line element" notation the induced tensor field $G = \tilde{G}|T_0^2(S^2)$ on S is given by

$$(ds)^2 = \cos^2\theta(d\phi)^2 - \cos 2\theta(d\theta)^2$$

and show this is not a metric tensor on $T_{(\theta,\phi)}S^2$ if $\theta = \pm 45°$.

d) Describe the geodesics on the upper and lower caps ($z^2 > \frac{1}{2}$) with this induced metric tensor. (Note that $(\theta - \frac{\pi}{2}, \phi)$ on a cap essentially gives polar coordinates (r,ϕ), so if you wish to work in coordinates you can use the metric as $(ds)^2 = \cos 2r(dr)^2 + \sin^2 r(d\phi)^2$.)

2a) Show that M as in 5.03, with the Riemannian metric given, is geodesically complete. (Hint: suppose a geodesic $c :]a,b[\to M$ cannot be extended past b. Use the fact that a geodesic in M from p to q is longer than that in R^3 to show by compactness that $\lim_{t\to b} c(t)$ exists in R^3. Deduce that $\lim_{t\to b} (c(t))$

exists in M, and obtain a contradiction.)

b) In Fig. 5.4b the "ordinary distance" $c^r(t)$ of the point $c(t)$ from the origin has a minimum for each geodesic shown. Can c^r have a maximum for any geodesic?

c) Do all geodesics except the "radial" ones (c^θ = constant) have minima for c^r? If there are others with no minimum, do they go finitely or infinitely many times around the origin?

d) Make precise and prove carefully the statement that the free end(s) of a geodesic in $R^2\backslash\{(0,0)\}$ with this metric are asymptotic to affine path(s) $f : R \to R^2$.

3) In this exercise M is as in 5.03, with the indefinite metric.

a) Using the coordinates (r,θ) for a point $(r,\theta,z) \in M$, show that the metric is given by $(ds)^2 = r^2(d\theta)^2 - \frac{1}{r^2}\left(\frac{1}{r^2}+2\right)(dr)^2$

b) Find the coordinate equation for a geodesic $c : t \rightsquigarrow (c^r(t), c^\theta(t))$ in M.

c) Show, with or without (b), that null geodesics spiral infinitely many times around the origin (what does a null geodesic look like in the embedded picture?) Deduce that the same is true for timelike geodesics. What about spacelike ones? (cf. Ex. 2c)

d) Find a reparametrisation of $]0,\infty[\to M : t \rightsquigarrow (t,0,t^{-1}-t)$ that makes it a geodesic, and deduce that M is not geodesically complete.

e) Can a timelike geodesic have a free end, as in Fig. 5.6a,b, or must it always be "trapped" as in c?

6. AN EXAMPLE OF LIE GROUP GEOMETRY

We have not set up any of the general machinery of Lie group theory. But this example does not require that; we can analyse it explicitly with what we have.

We include it mainly as an illuminating example in pure pseudo-Riemannian geometry. The fact that it brings out some geometric matter often neglected in courses on Lie groups is a bonus for readers already studying them. Others can enjoy the example by thinking of a Lie group as a group that is also a smooth manifold with smooth composition. (Bring together Ex. I.1.10, VII.2.01 and 2.02)

Consider the set of unimodular operators on R^2, that is those with determinant 1. We have established (IV.1.03, Ex. IV.3.6, Ex. VII.2.8c, Ex. VIII.1.4c) that this has the structure of a 3-manifold with an indefinite metric tensor field. Via the polarisation identity (Ex. IV.1.7d), everything involved is defined starting with the function det, which is a "natural" object definable without coordinates (Ex. V.1.11). In the same sense, then, we have found a "natural" structure on the set of unimodular operators $R^2 \to R^2$, usually called SL(2;R). (This stands for "the Special Linear group on 2 Real variables". The corresponding "General" group, GL(2;R) mentioned in I.3.01, consists of all operators with non-zero determinant, that is all invertible ones.) Now, as a group, SL(2;R) has a natural multiplicative structure - we can compose two unimodular operators and (by I.3.08) get another. To investigate this, let us use the standard basis for R^2 and the corresponding matrix labels for operators. The metric tensor we are using on $L(R^2;R^2)$ can be written

$$\begin{bmatrix} a & b \\ c & d \end{bmatrix} \cdot \begin{bmatrix} p & q \\ r & s \end{bmatrix} = \tfrac{1}{2}(as+dp) - \tfrac{1}{2}(br+cq).$$

6.01 Tangent vectors at the identity

By Ex. VIII.1.4 the tangent vectors to SL(2;R) at any point (operator) $\underset{\sim}{A}$ are exactly those orthogonal to $\underset{\sim}{A}$ in the determinant metric tensor, transferred to the space of vectors tangent to $L(R^2;R^2)$ at $\underset{\sim}{A}$. Using "matrix" coordinates on $T_{\underset{\sim}{A}}(L(R^2;R^2))$

corresponding to those on $L(R^2;R^2)$ itself, this means in particular that for $\mathbf{B} \in T_{\mathbf{I}}(L(R^2;R^2)$ with matrix $\begin{bmatrix} a & b \\ c & d \end{bmatrix}$, we have

$$\mathbf{B} \in T_{\mathbf{I}}(SL(2;R)) \iff \begin{bmatrix} a & b \\ c & d \end{bmatrix} \cdot \begin{bmatrix} 1 & 0 \\ 0 & 1 \end{bmatrix} = 0$$

$$\iff \tfrac{1}{2}(a+d) = 0$$

$$\iff \operatorname{tr} \mathbf{B} = 0.$$

Thus the tangent space at $\mathbf{I}$ to $SL(2;R)$ consists of the operators with zero trace. Such "traceless" operators on R^2 have a curious property: using matrix multiplication (not the metric tensor),

$$[\mathbf{B}]^2 = \begin{bmatrix} a & b \\ c & -a \end{bmatrix}\begin{bmatrix} a & b \\ c & -a \end{bmatrix} = \begin{bmatrix} a^2+bc & ab-ba \\ ca-ac & cb+a^2 \end{bmatrix} = (a^2+bc)\begin{bmatrix} 1 & 0 \\ 0 & 1 \end{bmatrix} = (-\det \mathbf{B})[\mathbf{I}].$$

In consequence, for any integer k,

$$\mathbf{B}^{2k} = (-\det \mathbf{B})^k \mathbf{I}, \qquad \mathbf{B}^{2k+1} = (-\det \mathbf{B})^k \mathbf{B}.$$

6.02 Power series of operators

The usual series defining $\exp(x)$, alias e^x, for a real number x is

$$1 + x + \frac{x}{2!} + \frac{x}{3!} + \frac{x}{4!} + \ldots \quad .$$

There is an obvious analogue for an operator $\mathbf{B}$. Using a capital E to distinguish it from the maps defined in 2.04, we set

$$\operatorname{Exp}(\mathbf{B}) = \mathbf{I} + \mathbf{B} + \frac{1}{2!}\mathbf{B}^2 + \frac{1}{3!}\mathbf{B}^3 + \frac{1}{4!}\mathbf{B}^4 + \ldots$$

The usual proofs (for which see any elementary analysis text) that the series for e^x converges absolutely for all x transfer at once to $\text{Exp}(\underset{\sim}{B})$. Recall that a series $\sum_{i=1}^{\infty} x_i$ converges, by definition, if its finite partial sums $s_j = \sum_{i=1}^{j} x_i$ converge as a sequence. The partial sums are all in the finite dimensional vector space $L(R^2;R^2)$, so we can use the usual topology on this and define convergence as in VI.2.01.

Now if $\underset{\sim}{B}$ is in $T_{\underset{\sim}{I}}(SL(2;R))$, by 6.01 we have

$$\text{Exp}(\underset{\sim}{B}) = \underset{\sim}{I} + \underset{\sim}{B} + \frac{1}{2!}(-\det \underset{\sim}{B})\underset{\sim}{I} + \frac{1}{3!}(-\det \underset{\sim}{B})\underset{\sim}{B} + \frac{1}{4!}(-\det \underset{\sim}{B})^2\underset{\sim}{I} + \ldots$$

Now, the convergence is absolute so we can rearrange the series and leave the sum unaltered:

$$\text{Exp}(\underset{\sim}{B}) = \left(1 - \frac{1}{2!}\det \underset{\sim}{B} + \frac{1}{4!}(\det \underset{\sim}{B})^2 - \ldots\right)\underset{\sim}{I} + \left(1 - \frac{1}{3!}\det \underset{\sim}{B} + \frac{1}{5!}(\det \underset{\sim}{B})^2 - \ldots\right)\underset{\sim}{B}.$$

The coefficients of $\underset{\sim}{I}$ and $\underset{\sim}{B}$ are now power series in the ordinary real number $\det \underset{\sim}{B}$, and can be made to look very familiar. Set $d = \sqrt{|\det \underset{\sim}{B}|} = \sqrt{|\underset{\sim}{B}.\underset{\sim}{B}|}$, the "size" $\|\underset{\sim}{B}\|$ of $\underset{\sim}{B}$ (IV.1.07) by the metric tensor we are using. (This is <u>not</u> the "norm of an operator" used in IV.4.01.) Then if $\det \underset{\sim}{B}$ is positive it is d^2, and

$$\begin{aligned}\text{Exp}(\underset{\sim}{B}) &= \left(1 - \frac{d^2}{2!} + \frac{d^4}{4!} - \ldots\right)\underset{\sim}{I} + \frac{1}{d}\left(d - \frac{d^3}{3!} + \frac{d^5}{5!} - \ldots\right)\underset{\sim}{B} \\ &= \cos(d)\underset{\sim}{I} + \sin(d)\left(\frac{1}{d}\underset{\sim}{B}\right).\end{aligned}$$

If $\det \underset{\sim}{B} = 0$, then we have

$$\text{Exp}(\underset{\sim}{B}) = \underset{\sim}{I} + \underset{\sim}{B}\ .$$

If det $\underset{\sim}{B}$ is negative it is $(-d^2)$, so that

$$\begin{aligned}\operatorname{Exp}(\underset{\sim}{B}) &= (1 + \frac{d^2}{2!} + \frac{d^4}{4!} + \ldots)\underset{\sim}{I} + \frac{1}{d}(d + \frac{d^3}{3!} + \frac{d^5}{5!} + \ldots)\underset{\sim}{B} \\ &= \cosh(d)\underset{\sim}{I} + \sinh(d)\,(\frac{1}{d}\underset{\sim}{B}).\end{aligned}$$

The operator $\frac{1}{d}\underset{\sim}{B}$ that appears here is just the normalisation (IV.1.06) of $\underset{\sim}{B}$, the unit vector in the same direction. When $\underset{\sim}{B}$ is taken, as in 6.01, as a tangent vector at $\underset{\sim}{I}$ we shall denote by $\bar{\underset{\sim}{B}}$ the result of normalising $\underset{\sim}{B}$ if it is non-null and freeing it (shifting it to the origin of $L(R^2;R^2)$.) If det $\underset{\sim}{B} = 0$ we define $\bar{\underset{\sim}{B}}$ by freeing $\underset{\sim}{B}$ without - since we can't - normalising it. Then as a map $T_{\underset{\sim}{I}}(SL(2;R)) \to L(R^2;R^2)$, Exp takes the form

$$\operatorname{Exp}(\underset{\sim}{B}) = \begin{cases}\cos(d)\underset{\sim}{I} + \sin(d)\bar{\underset{\sim}{B}}, & \det \underset{\sim}{B} > 0 \\ \underset{\sim}{I} + \underset{\sim}{B}, & \det \underset{\sim}{B} = 0 \\ \cosh(d)\underset{\sim}{I} + \sinh(d)\bar{\underset{\sim}{B}}, & \det \underset{\sim}{B} < 0.\end{cases}$$

6.03 The geometry of Exp

The image $\operatorname{Exp}(T_{\underset{\sim}{I}}(SL(2;R)))$ lies <u>in</u> $SL(2;R)$, by Ex. 1. What does the mapping Exp look like? It is convenient to use the orthonormal basis

$$\underset{\sim}{b}_1 = \begin{bmatrix}1 & 0\\ 0 & 1\end{bmatrix},\quad \underset{\sim}{b}_2 = \begin{bmatrix}0 & 1\\ -1 & 0\end{bmatrix},\quad \underset{\sim}{b}_3 = \begin{bmatrix}1 & 0\\ 0 & -1\end{bmatrix},\quad \underset{\sim}{b}_4 = \begin{bmatrix}0 & 1\\ 1 & 0\end{bmatrix}$$

found in Ex. IV.3.6, giving coordinates (a^1,a^2,a^3,a^4) say to an operator $\underset{\sim}{A}$. In these coordinates $SL(2;R)$ is the set of $\underset{\sim}{A}$ satisfying

$$(a^1)^2 + (a^2)^2 - (a^3)^2 - (a^4)^2 = 1,$$

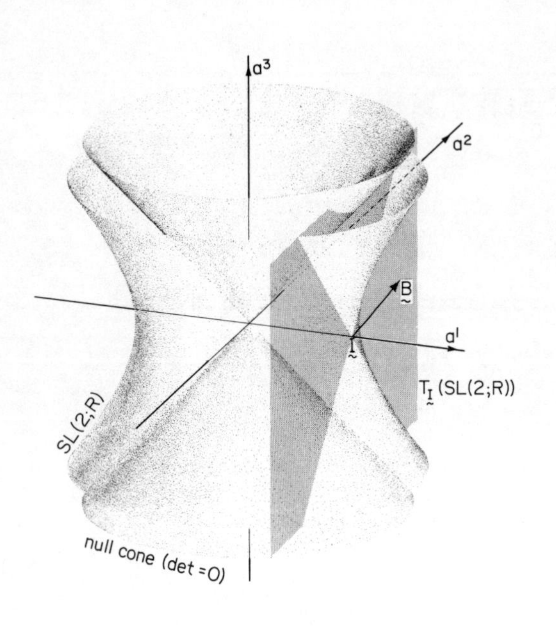

Fig. 6.1

a sort of 3-dimensional "hyperboloid" in R^4. We can only draw slices of this; for instance, fixing $a^4 = 0$ gives Fig. 6.1. This picture turns out to be fairly adequate for our present purposes.

Suppose $\underset{\sim}{B}$ is a tangent vector at $\underset{\sim}{I}$, not just to SL(2;R) but to the $a^4 = 0$ slice of it. Then by 6.02 $\mathrm{Exp}(\underset{\sim}{B})$ is a linear combination of $\underset{\sim}{B}$ itself, shifted to the origin, and $\underset{\sim}{I}$. Hence it lies in the same slice.

Consider $\underset{\sim}{B}$ with $\det \underset{\sim}{B} = 1$, a unit timelike vector in $T_{\underset{\sim}{I}}(SL(2;R))$. Then for any $t\underset{\sim}{B} \neq \underset{\sim}{0}$, the normalised vector $\overline{t\underset{\sim}{B}}$ is either $+\bar{\underset{\sim}{B}}$ or $-\bar{\underset{\sim}{B}}$, according to the sign of t. So for such a $\underset{\sim}{B}$,

$$\mathrm{Exp}(t\underset{\sim}{B}) = \cos(t)\underset{\sim}{I} + \sin(t)\bar{\underset{\sim}{B}}.$$

The line $\{t\underset{\sim}{B} | t \in R\}$ is thus wrapped around an ellipse as in Fig. 6.2a. Since every timelike vector is $t\underset{\sim}{B}$ for some unit timelike $\underset{\sim}{B}$, this describes Exp on such vectors completely.

When $\det \underset{\sim}{B} = 0$, $\underset{\sim}{B}$ lies in the intersection of SL(2:R) and its geometric tangent space. In the slice shown in Figs. 6.1 , 6.2 it is a pair of lines. (Recall that such a hyperboloid contains two straight lines through each point - a fact useful in making string models of it but, contrary to many schoolbooks, nothing to do with cooling tower design.) For SL(2;R) proper the intersection is a cone (a

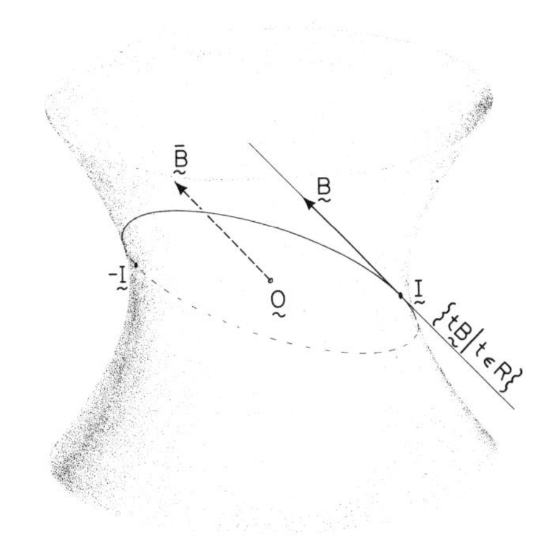

(a)

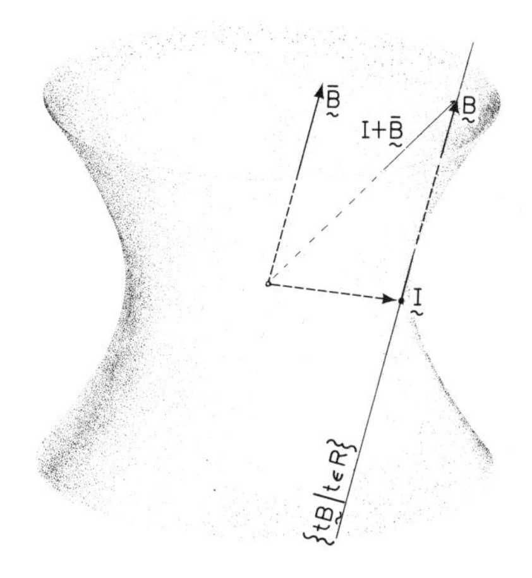

(b)

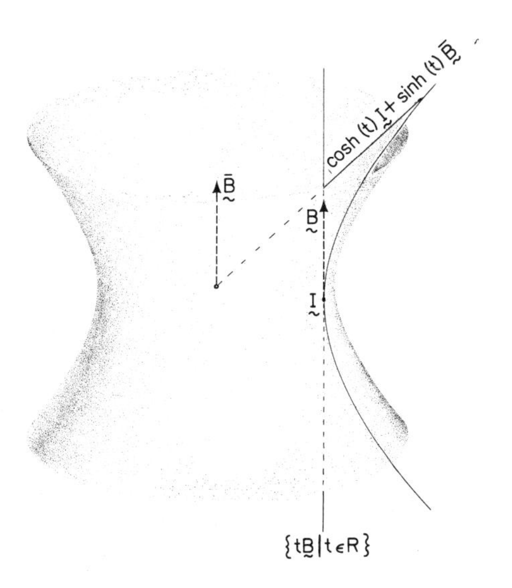

(c) Fig. 6.2

union of straight lines) - the null cone of the tangent space, symmetrical around the vector B of Fig. 6.1. (The tangent space has orthonormal basis the bound vectors $\overleftarrow{d}_I(b_2)$, $\overleftarrow{d}_I(b_3)$, $\overleftarrow{d}_I(b_4)$ orthogonal to $\overleftarrow{d}_I(I) = \overleftarrow{d}_I(b_1)$, of which only the first is timelike. So with this basis the "timelike axis", vertical in Fig.IV.1.5 for H^3, is parallel to the a_2 axis.) The effect of freeing B from I to give $\bar{B}$, then adding that to I, is to leave the vector tip where it started (Fig. 6.2b). The line $\{tB | t \in R\}$, thought of as part of the geometric tangent space, is mapped to SL(2;R) by simple inclusion.

Finally, suppose det $B = -1$. Then

$$\mathrm{Exp}(t\underset{\sim}{B}) = \cosh(t)\underset{\sim}{I} + \sinh(t)\underset{\sim}{B},$$

so $\{t\underset{\sim}{B} \mid t \in R\}$ is mapped to a hyperbola as in Fig. 6.2c.

6.04 Relation to $\exp_{\underset{\sim}{I}}$

The three different formulae we have used in studying Exp do fit neatly together. To demonstrate this we could appeal to more theorems on convergent power series (again generalised from real numbers to operators) or examine carefully the behaviour of $\mathrm{Exp}(\underset{\sim}{B})$ as $\det\underset{\sim}{B}$ tends to 0. But we need not. This map coincides exactly with the differential-geometric map $\exp_{\underset{\sim}{I}}$ from a tangent space to a manifold defined in 2.04 (using the metric tensor we have chosen), so that smoothness follows by Ex. 2.2. To establish this agreement, we need only prove it on rays $\{t\underset{\sim}{B} \mid t \in R\}$, since every point is on some ray. Thus it suffices to prove that the curves

$$E_{\underset{\sim}{B}} : R \to SL(2;R) : t \rightsquigarrow \mathrm{Exp}(t\underset{\sim}{B})$$

studied in 6.03 are geodesics, and that the map

$$\underset{\sim}{B} \rightsquigarrow E^{*}_{\underset{\sim}{B}}(0)$$

given by differentiating them at 0 is the identity. (We have affine coordinates on $L(R^2;R^2)$, so the differentiation of curves in it is simple.)

For the case $\det \underset{\sim}{B} = 0$, both statements are trivial.

In the case $\det \underset{\sim}{B} = d^2$, $\quad E_{\underset{\sim}{B}}(t) = \cos(td)\underset{\sim}{I} + \sin(td)\,\frac{1}{d}\,\underset{\sim}{B}$

$$E^{*}_{\underset{\sim}{B}}(t) = -d\,\sin(td)\underset{\sim}{I} + \cos(td)\underset{\sim}{B}$$

$$E^*_{\underset{\sim}{B}}(0) = \underset{\sim}{B}.$$

In the case $\det \underset{\sim}{B} = -d^2$, $\quad E_{\underset{\sim}{B}}(t) = \cosh(td) + \sinh(td)\,\frac{1}{d}\,\underset{\sim}{B}$

$$E^*_{\underset{\sim}{B}}(t) = d\sinh(td) + \cosh(td)\,\underset{\sim}{B}$$

$$E^*_{\underset{\sim}{B}}(0) = \underset{\sim}{B}.$$

It remains to check that each $E_{\underset{\sim}{B}}$ is a geodesic. Differentiating again, we have "acceleration vectors" as follows:

$$\frac{d}{ds}E^*_{\underset{\sim}{B}}(t) = -d^2\cos(dt)\underset{\sim}{I} - d\sin(dt)\underset{\sim}{B} = -d^2(E_{\underset{\sim}{B}}(t)), \text{ if } \det \underset{\sim}{B} > 0.$$

$$\frac{d}{ds}E^*_{\underset{\sim}{B}}(t) = d^2\cosh(dt)\underset{\sim}{I} + d\sinh(dt)\underset{\sim}{B} = d^2(E_{\underset{\sim}{B}}(t)), \text{ if } \det \underset{\sim}{B} < 0.$$

But $\pm E_{\underset{\sim}{B}}(t)$ is always orthogonal to the tangent plane <u>at</u> $E_{\underset{\sim}{B}}(t)$, by Ex. VIII.1.4, hence in each case $E_{\underset{\sim}{B}}$ is a geodesic by Th^m 5.01.

<u>6.05 Other geodesics</u>

Our "positive definite" visual habits and Fig. 6.1 may not suggest it, but SL(2;R) is quite as symmetrical as a sphere, <u>in the metric tensor we are using</u>. (Indeed, its definition $\{\underset{\sim}{A} \mid \underset{\sim}{A}.\underset{\sim}{A} = 1\}$ is just like that of a sphere.) Multiplication $\underset{\sim}{M}_{\underset{\sim}{A}}$ by any unimodular operator $\underset{\sim}{A}$ is an orthogonal operator on $L(R^2;R^2)$ by Ex. IV.2.3, since for any $\underset{\sim}{B}$

$$(\underset{\sim}{M}_{\underset{\sim}{A}}(\underset{\sim}{B})).(\underset{\sim}{M}_{\underset{\sim}{A}}(\underset{\sim}{B})) = (\underset{\sim}{A}\underset{\sim}{B}).(\underset{\sim}{A}\underset{\sim}{B}) = \det(\underset{\sim}{A}\underset{\sim}{B}) = \det\underset{\sim}{A}\,\det\underset{\sim}{B} = \det\underset{\sim}{B} = \underset{\sim}{B}.\underset{\sim}{B}$$

so it maps SL(2;R) isometrically to itself, carrying geodesics to geodesics (and its inverse carries them back). Thus since it takes $\underset{\sim}{I}$ to $\underset{\sim}{A}$, we find all the geodesics through $\underset{\sim}{A}$ (Figs.6.3b, c) by applying $\underset{\sim}{M}_{\underset{\sim}{A}}$ to those through $\underset{\sim}{I}$ (Fig. 6.3a) which we have found already.

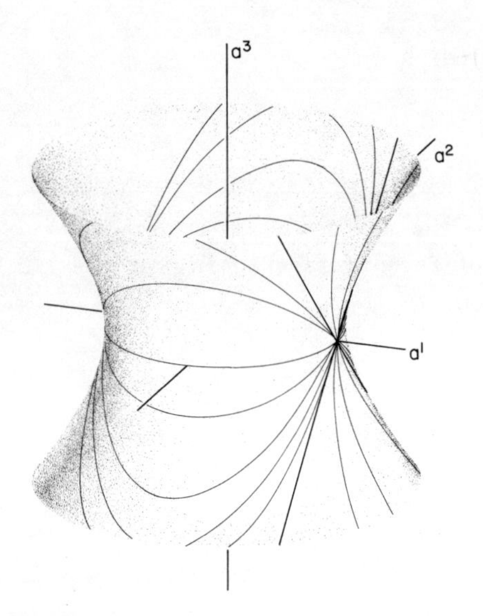

(a)

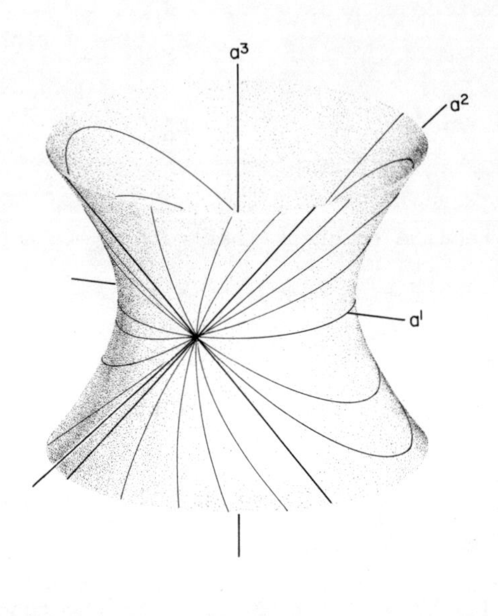

(b)

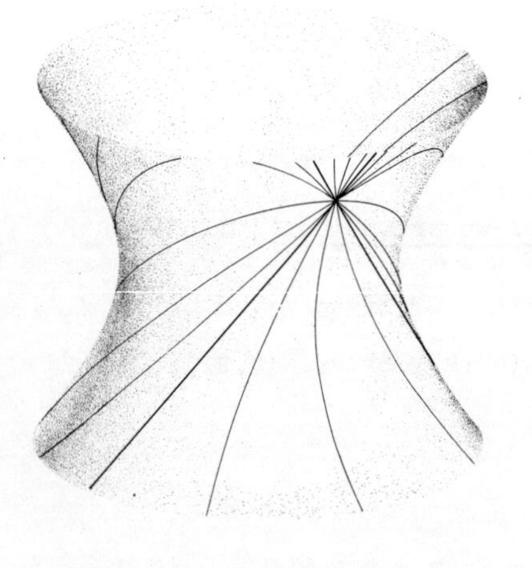

(c) Fig. 6.3

6.06 Gaps

The trace function on $L(R^2;R^2)$ is as "natural" as the determinant (and as this example illustrates, is closely related to it in general). Since it respects addition, and for $\underset{\sim}{B} \in T_{\underset{\sim}{I}}(SL(2;R))$ we have $tr(\bar{\underset{\sim}{B}}) = 0$, 6.02 gives

$$tr(Exp(\underset{\sim}{B})) = \begin{cases} 2\cos(d), & \underset{\sim}{B} \text{ timelike} \\ 2, & \underset{\sim}{B} \text{ null} \\ 2\cosh(d), & \underset{\sim}{B} \text{ spacelike.} \end{cases}$$

Thus for all $\underset{\sim}{B}$ we have $tr(Exp(\underset{\sim}{B})) \geq -2$, with equality only if $\sqrt{\det\underset{\sim}{B}}$ is real and an odd multiple of π, giving $\mathrm{Exp}(\underset{\sim}{B}) = -\underset{\sim}{I}$. (This is clearly analogous to the situation of Fig. 2.3; on the unit sphere likewise it is exactly the sets $\{\underset{\sim}{x} | \underset{\sim}{x}.\underset{\sim}{x} = (2k-1)^2\pi^2\}$, for each $k \in N$, that $\exp_p$ maps to the point opposite p. But in that case the sets so defined are spheres. Here they are not compact)

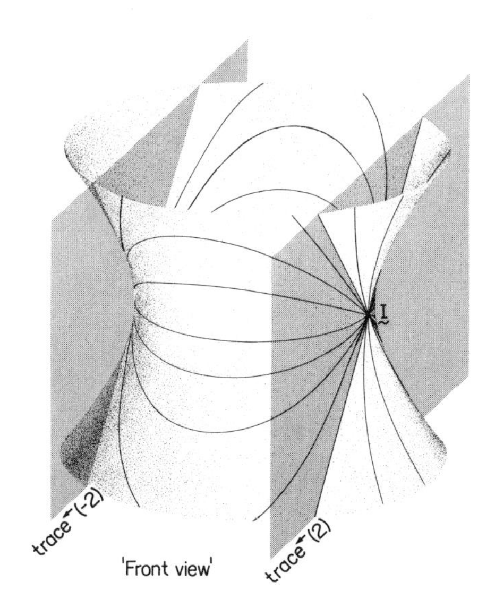

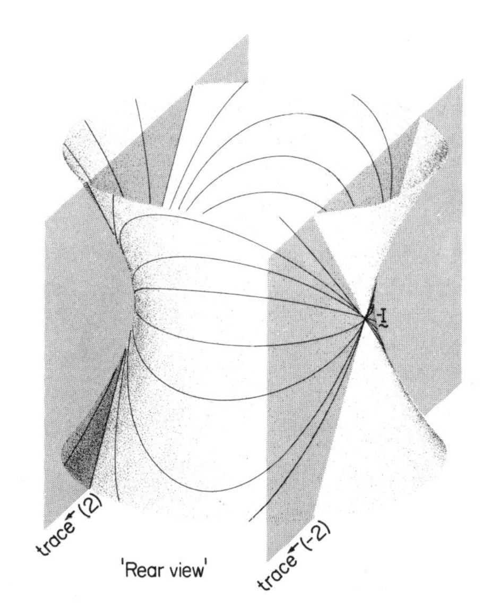

Fig. 6.4

So the trace slices SL(2;R) up (Fig. 6.4) into the Exp images of spacelike, null and timelike vectors. No point on a null or spacelike geodesic through $-\underset{\sim}{I}$ (except $-\underset{\sim}{I}$ itself) is reached by a geodesic from $\underset{\sim}{I}$, though SL(2;R) is clearly geodesically complete, cf. 2.03.

However, such a point $\underset{\sim}{A}$ can easily be reached from $\underset{\sim}{I}$ by a timelike curve. Indeed, it can be reached by a timelike

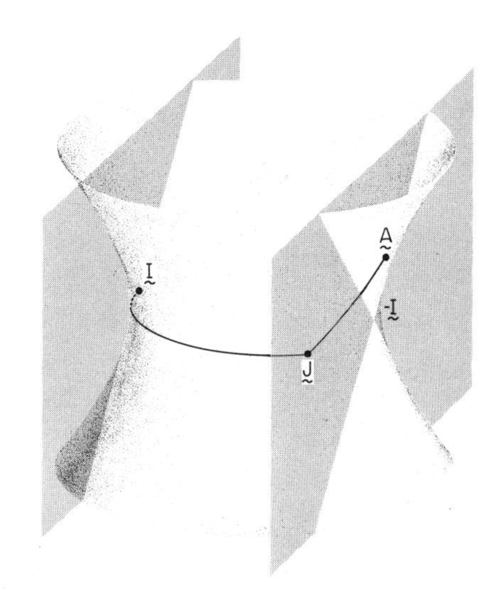

Fig. 6.5

"broken geodesic" with a single jump in direction, say at $\underset{\sim}{J}$. So if "physical effects" are considered to propagate along timelike geodesics, this means that events at $\underset{\sim}{I}$ can have effects at $\underset{\sim}{A}$ if some "interaction" happens at $\underset{\sim}{J}$.

6.07 The covering space

If we were seriously considering SL(2;R) as a model for physical spacetime, our previous sentence would lead at once to a genuine logical difficulty (unlike 4.05). If the effects of an event can propagate into that event's own past, we have the Time Travel Paradox: what is to stop a man murdering his mother at a time before his own conception? (This is stricter than the usual, Oedipal formulation. It's a wise child that knows its own father.) Some modern physical theories grapple with this problem, but most simply forbid it. The geometrical analogue, not paradoxical until physically interpreted in this particular way, is the existence of closed timelike curves like those in SL(2;R). (Other physical applications of pseudo-Riemannian geometry, such as to electrical circuit theory [Smale], do not associate paradoxes with such curves.) It happens that SL(2;R) has a close relative without this feature, whose construction we may sketch as follows.

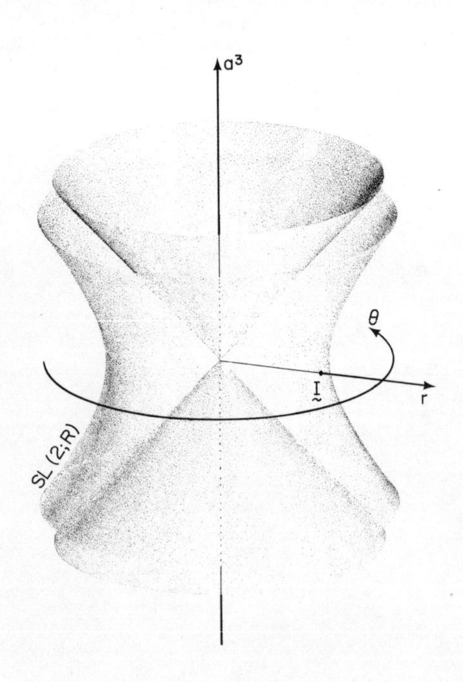

Fig. 6.6

Replace (Fig. 6.6) the rectangular coordinates (a^1,a^2,a^3,a^4) we have been using on $L(R^2;R^2)$ by "cylindrical coordinates" (r,θ,a^3,a^4) defined at all points except where $a^1=a^2=0$. This is not strictly a chart on

$$M = L(R^2;R^2)\backslash\{(0,0,a^3,a^4)\,|\,a^3,a^4 \in R\},$$

since θ takes values in the circle S^1 of possible angles, with 0 and 2π identified. It gives a map

$$M \to R \times S^1 \times R^2, \quad \text{not} \quad M \to R^4.$$

But since $SL(2;R)$ lies in M, this "pseudochart" lets us treat it as a submanifold of $R \times S^1 \times R^2$. Using the exponential map $\exp_1 : R \to S^1 : x \rightsquigarrow$ (angle of e^{ix}) of Fig. 2.2, we can define

$$P : R^4 \to R \times S^1 \times R^2 : (w,x,y,z) \rightsquigarrow (w,\exp_1(x),y,z)$$

which is obviously a local diffeomorphism (VII.2.02). This gives

$$\widetilde{SL}(2;R) = P^{\leftarrow}(SL(2;R))$$

as a sub-3-manifold of R^4, and

$$p : \widetilde{SL}(2;R) \to SL(2;R)$$

defined by restricting P is again a local diffeomorphism. We then "lift" the metric tensor field on $SL(2;R)$ to one on $\widetilde{SL}(2;R)$. If $\underset{\sim}{v},\underset{\sim}{w}$ are tangent vectors in $T_x(\widetilde{SL}(2;R))$ we define their dot product by $(\underset{\sim}{D}_x p(\underset{\sim}{v}))\cdot(\underset{\sim}{D}_x p(\underset{\sim}{w}))$. The result is non-degenerate, and has the same signature as that on $SL(2;R)$, since $\underset{\sim}{D}_x p$ is an isomorphism for each $x \in \widetilde{SL}(2;R)$.

The effect is to "unwind" $SL(2;R)$ as R "unwinds" the circle (Fig. 6.7a). The space $\widetilde{SL}(2;R)$ may be thought of as a "spiral copy" (Fig. 6.7b) of $SL(2;R)$, as long as it is understood that it does not "spiral in" or "spiral out". Each small piece has exactly the geometry of a corresponding piece of $SL(2;R)$.

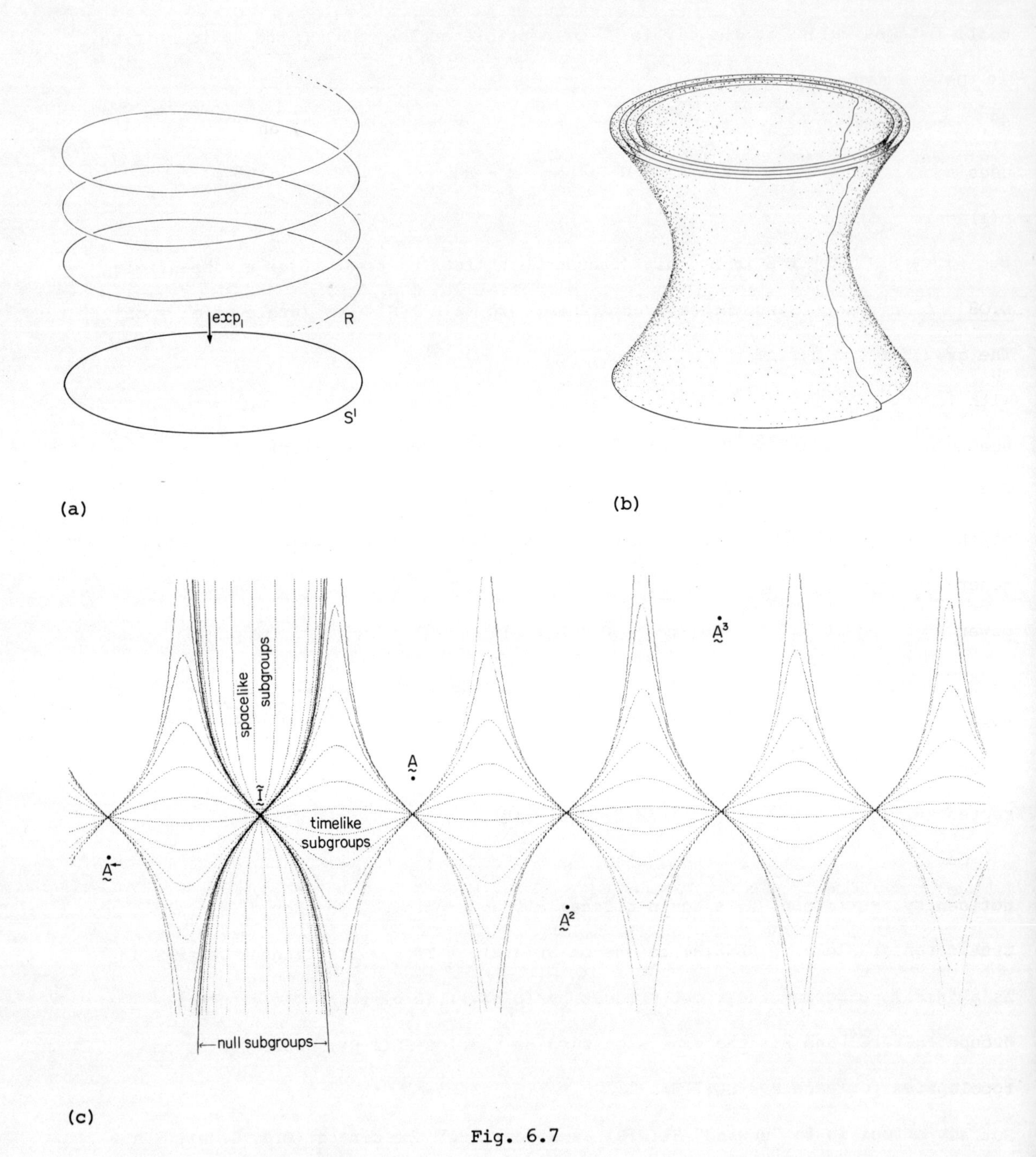

Fig. 6.7

This is an ad hoc construction of the universal cover of SL(2;R). A systematic construction, for the universal cover of a general space, a very useful tool, is given in most topology and Lie group texts.

Unlike SL(2;R), $\widetilde{SL}(2;R)$ has an atlas consisting of a single chart using all of R^3. The slice corresponding to Fig. 6.1 can be charted with R^2, and Fig. 6.7c uses such a chart to draw the geodesics through a particular point $\tilde{\underset{\sim}{I}}$ with $p(\tilde{\underset{\sim}{I}}) = \underset{\sim}{I}$.

6.08 Aside on Lie group theory

The group structure of SL(2;R) can also be "lifted", to make $\widetilde{SL}(2;R)$ a Lie group with identity $\tilde{\underset{\sim}{I}}$. It is then simple to prove that if $\underset{\sim}{A} \in \widetilde{SL}(2;R)$ does not lie in the image of $\widetilde{SL}(2;R)$'s exponential map (Fig. 6.7c), then nor does $\underset{\sim}{A}^k$ for any integer $k \neq 0$. This illustrates a geometric difference between <u>algebraic</u> groups, those defined by an algebraic matrix equation as SL(2;R) is by $\det\underset{\sim}{A} = 1$, and Lie groups in general. By a recent theorem [Markus], in an algebraic group every $\underset{\sim}{A}$ has some power $\underset{\sim}{A}^k$ in the image of Exp (a result Markus was led to by questions in differential equations). Thus unlike SL(2;R), $\widetilde{SL}(2;R)$ cannot turn up as an algebraic group - a result which happens to be weaker than the fact that $\widetilde{SL}(2;R)$ cannot turn up as a group of n×n matrices <u>at all</u>. It has no "faithful finite-dimensional representations", even non-algebraic ones.

Our definition of the pseudo-Riemannian structure on SL(2;R) and $\widetilde{SL}(2;R)$ does not generalise, since det is only quadratic on n×n matrices when n=2. But it is true for Lie groups in general that the Exp defined by power series can be realised as a "geodesic" exponential map with a suitable metric tensor field. For some groups this is Riemannian (for instance Spin(3), the unit quaternion group, is topologically the 3-sphere S^3 and its exponential map is the analogue of Fig. 2.3). But since Exp is always defined on the whole tangent space at $\underset{\sim}{I}$ (the <u>Lie algebra</u>) of the group, the group must become geodesically complete. In a connected, geodesically complete Riemannian manifold any two points are joined by a geodesic (see for instance [Spivak] for a proof), so no Riemannian structure fits this and similar examples.

6.09 "Relativistic SHM"

This remark is not physics (not that of our universe, anyway) and not mathematics ("observer" etc being physical notions). We make it, briefly and loosely, as exercise for the imagination.

The space SL(2;R) is wrong-dimensional to satisfy Einstein's equation, and the 4-dimensional analogue we could get by starting with

$$\{\underset{\sim}{x} \in R^5 \mid (x^1)^2 + (x^2)^2 - (x^3)^2 - (x^4)^2 - (x^5)^2 = 1\}$$

would need "negative energy density everywhere" (points the reader may elaborate after digesting Ch. XII). But <u>if</u> we interpret some particular geodesic c as "the history of an observer Q", and another as "the history of a particle P watched by Q", Fig. 6.7c shows that Q "sees P go to and fro, returning to him at times π apart as measured along c". The periodicity is independent of the velocity that Q imputes to P at their meetings.

This behaviour of particles is what a physicist at the centre of a linear, Newtonian inward field of force would see. But here, <u>every</u> world-line shows "Simple Harmonic Motion" relative to <u>every</u> other, with the same apparent period. Every physicist is equally "central".

6.10 Effects on R^2

Exercises 4 - 6 analyse the nature of $\{\mathrm{Exp}(t\underset{\sim}{B}) \mid t \in R\}$ as a family of operators on R^2. According as $\underset{\sim}{B}$ is timelike, null or spacelike the flows

$$R^2 \times R \to R^2 : (x,t) \rightsquigarrow (\mathrm{Exp}(t\underset{\sim}{B}))\underset{\sim}{x}$$

look in suitable coordinates (x,y) like Fig. 6.8a, b or c respectively.

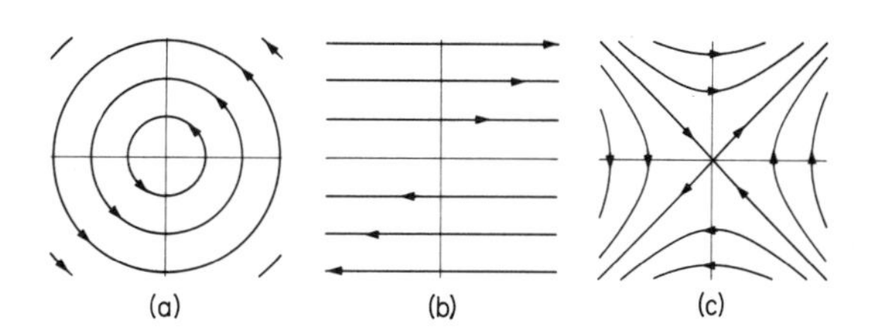

Fig. 6.8

Each is a family of "rotations" with respect to some "metric tensor" (degenerate in the case $\det\underset{\sim}{B} = 0$), which is unique up to a scalar factor. In the original coordinates on R^2, of course, they may look as in Fig. 6.9. It may be seen how, as $\det\underset{\sim}{B}$ tends to 0 from either side, the geometry of the flow tends to the degenerate case. The non-surjectivity of Exp now appears more geometrically natural. If $\mathrm{tr}\underset{\sim}{A} = -2$, $\underset{\sim}{A} = -\mathrm{Exp}(\underset{\sim}{B})$ for some null $\underset{\sim}{B}$ and cannot be reached from $\underset{\sim}{I}$ by such a family unless it is $-\underset{\sim}{I}$, as it switches the sides of the fixed line L in Fig. 6.9b and reverses it. If $\mathrm{tr}\underset{\sim}{A} < -2$, similar remarks apply, in terms of Fig. 6.9c.

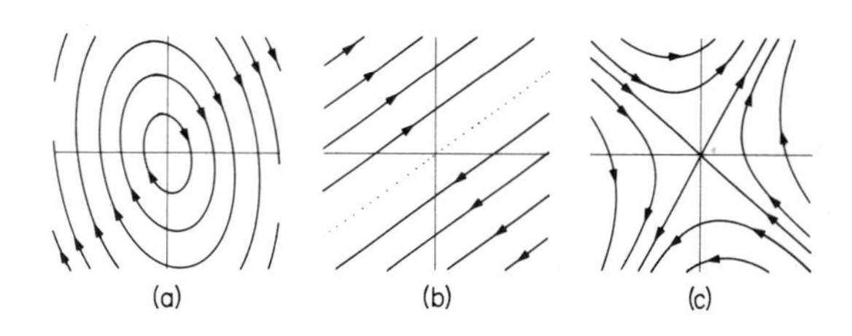

Fig. 6.9

6.11 Aside on crystal symmetries

A crystal has at most the symmetries of a lattice of dots like those in Fig. 6.10a, b, and their 3-dimensional analogues. (It may have fewer, as in c.) Choose one dot as origin, and a basis like the pairs of vectors

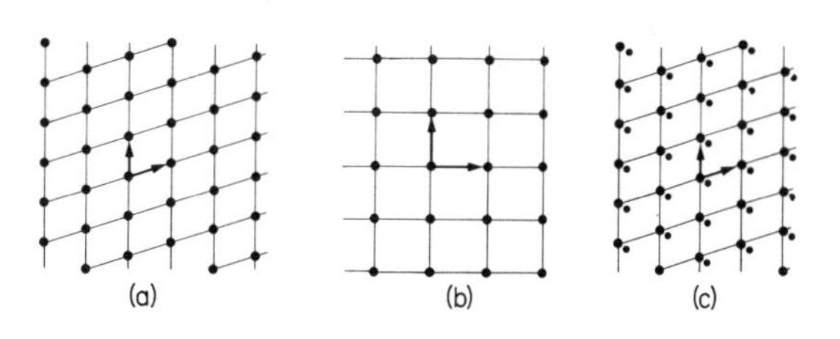

Fig. 6.10

shown in 6.10. Then any linear operator carrying dots to dots must have a matrix with only integer entries, since its columns give the (integer) coordinates of lattice points. Hence irrespective of basis, $\mathrm{tr}\underset{\sim}{A}$ is an integer.

A crystal symmetry A in two [three] dimensions must obviously preserve area [volume], at least up to sign. Suppose $\underset{\sim}{A} : R^2 \to R^2$ has $\det \underset{\sim}{A} = 1$, and is a rotation with respect to some inner product on R^2. Exercises 4 - 6 show that if $\underset{\sim}{A} \neq \pm\underset{\sim}{I}$, then $-2 < \mathrm{tr}\underset{\sim}{A} < 2$. Thus if $\mathrm{tr}\underset{\sim}{A}$ is an integer, it must be 0 or ± 1 (Fig. 6.6a). Hence (Ex. 7), by the unique appropriate metric it is a turn through $\pi/3$, $\pi/4$ or $2\pi/3$. Nothing like a $\pi/5$ turn is possible.

Hence, even in three dimensions, by Ex. 8 any crystal symmetry $\underset{\sim}{A}$ (even with $\det \underset{\sim}{A} < 0$) keeping some point fixed and preserving Euclidean lengths must have (at least) one of $\underset{\sim}{A}^2$, $\underset{\sim}{A}^3$, $\underset{\sim}{A}^4$, $\underset{\sim}{A}^6$ equal to $\underset{\sim}{I}$.

The operators in SL(2;R) with $|\mathrm{tr}\underset{\sim}{A}| > 2$ (Fig. 6.11b) have been called <u>relativistic crystal symmetries</u> [Ascher and Janner]. Crystallographic symmetries in Euclidean n-space, even including translations, can be systematically

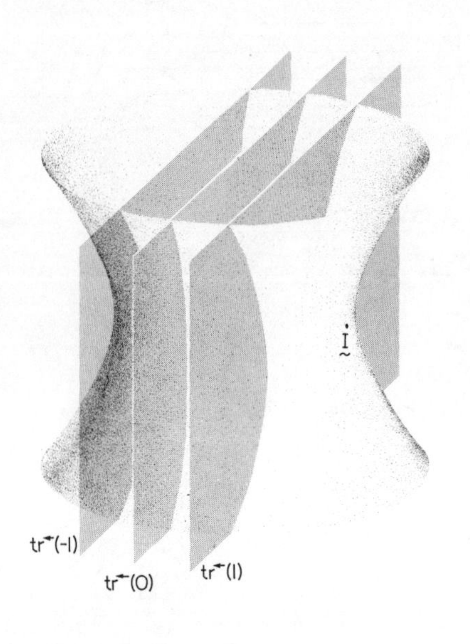

(a)

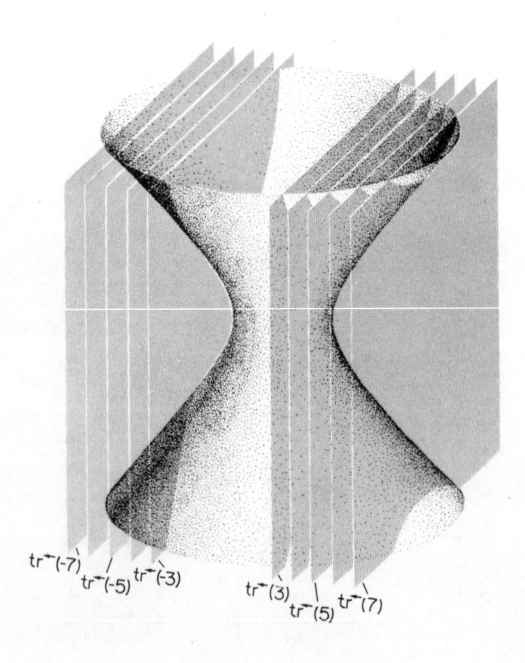

(b)

Fig. 6.11

enumerated [Schwarzenberger (2)]. But not even the crystallographic "point groups" (symmetries keeping a given point fixed) in 4 dimensions have been classified for Minkowski space.

Exercises IX.6

All operators in these exercises are on R^2, unless otherwise indicated.

1) Show by writing out the matrices that for all $\underset{\sim}{B} \in T_{\underset{\sim}{I}}(SL(2;R))$, we have $\det(\text{Exp}(\underset{\sim}{B})) = 1$. (Use 6.02 and the identities $\cos^2 x + \sin^2 x = 1 = \cosh^2 x - \sinh^2 x$.)

2) EITHER

Show from 6.02 and the standard identities

$$\sin(t+s) = \sin(t)\cos(s) + \cos(t)\sin(s)$$

$$\cos(t+s) = \cos(t)\cos(s) - \sin(t)\sin(s)$$

$$\sinh(t+s)=\sinh(t)\cosh(s)+\cosh(t)\sinh(s)$$

$$\cosh(t+s)=\cosh(t)\cosh(s)+\sinh(t)\sinh(s)$$

that $\text{Exp}(t\underset{\sim}{B})$ has the 1-parameter subgroup property

$$\bigstar \quad \text{Exp}((t+s)\underset{\sim}{B}) = (\text{Exp}(t\underset{\sim}{B}))\circ(\text{Exp}(s\underset{\sim}{B}))$$

OR

If you already know $\bigstar$ in greater generality from Lie group theory, derive the above formulae from it.

3) Show that every $\underset{\sim}{A} \in SL(2;R)$ with $\text{tr}\underset{\sim}{A} > -2$ is $\text{Exp}(\underset{\sim}{B})$ for some $\underset{\sim}{B}$.

4a) If $\text{tr}\underset{\sim}{B} = 0$, $\det\underset{\sim}{B} = 1$, then by 6.01 $\underset{\sim}{B}^2 = -\underset{\sim}{I}$. Deduce that if $\underset{\sim}{x}$ is any non-zero vector in R^2, $\underset{\sim}{x}$ and $\underset{\approx}{Bx}$ are linearly independent. (Hint: if $\underset{\sim}{v} = a\underset{\sim}{x} + b\underset{\approx}{Bx} = \underset{\sim}{0}$, then $\underset{\approx}{Bv} = \underset{\sim}{0}$ also.)

b) Using coordinates (x^1,x^2) referred to the basis $\{\underset{\sim}{x},\underset{\approx}{Bx}\}$, show that any symmetric bilinear form $\underset{\sim}{G}$ on R^2 such that

$$\underset{\sim}{G}(\underset{\sim}{B}\underset{\sim}{x},\underset{\sim}{B}\underset{\sim}{y}) = \underset{\sim}{G}(\underset{\sim}{x},\underset{\sim}{y}) \quad \forall\ \underset{\sim}{x},\underset{\sim}{y} \in R^2,$$

can be written (for some $g \in R$), as

$$\underset{\sim}{G}((x^1,x^2),(y^1,y^2)) = g(x^1y^1+x^2y^2).$$

c) Deduce that if $\underset{\sim}{A} = \mathrm{Exp}(\underset{\sim}{B})$ for some timelike $\underset{\sim}{B}$, and $\underset{\sim}{A} \neq \pm\underset{\sim}{I}$, the same holds for symmetric bilinear forms $\underset{\sim}{G}$ with

$$\underset{\sim}{G}(\underset{\sim}{A}\underset{\sim}{x},\underset{\sim}{A}\underset{\sim}{y}) = \underset{\sim}{G}(\underset{\sim}{x},\underset{\sim}{y}), \quad \forall\ \underset{\sim}{x},\underset{\sim}{y} \in R^2.$$

d) Show that if $\det \underset{\sim}{B} = d^2$, $d \in R$, then coordinates on R^2 referred to a basis $\underset{\sim}{x}$, $\frac{1}{d}\underset{\sim}{B}\underset{\sim}{x}$ give $\mathrm{Exp}(t\underset{\sim}{B})$ the matrix

$$\begin{bmatrix} \cos(td) & \sin(td) \\ -\sin(td) & \cos(td) \end{bmatrix}.$$

5) Suppose $|\mathrm{tr}\underset{\sim}{A}| > 2$, $\det\underset{\sim}{A} = 1$.

a) Show using the characteristic equation of $\underset{\sim}{A}$ (cf. I.3.13) that $\underset{\sim}{A}$ has two distinct real eigenvalues, and that choosing basis vectors $\underset{\sim}{b}_1$, $\underset{\sim}{b}_2$ belonging to them gives $\underset{\sim}{A}$ the matrix $\begin{bmatrix} \lambda & 0 \\ 0 & \frac{1}{\lambda} \end{bmatrix}$, some $\lambda \neq 0, \pm 1$.

b) Deduce that in $\underset{\sim}{b}_1,\underset{\sim}{b}_2$ coordinates, any non-zero symmetric bilinear form $\underset{\sim}{G}$ on R^2 such that

$$\underset{\sim}{G}(\underset{\sim}{A}\underset{\sim}{x},\underset{\sim}{A}\underset{\sim}{y}) = \underset{\sim}{G}(\underset{\sim}{x},\underset{\sim}{y}) \quad \forall\ \underset{\sim}{x},\underset{\sim}{y} \in R^2$$

has the formula

$$\underset{\sim}{G}((x^1,x^2),(y^1,y^2)) = g(x^1y^2 + x^2y^1), \text{ some } g \neq 0.$$

Show that $\underset{\sim}{G}$ is then a metric tensor on R^2 with signature 0, and that

$\underset{\sim}{a}_1 = \frac{1}{\sqrt{2g}}(\underset{\sim}{b}_1+\underset{\sim}{b}_2)$, $\underset{\sim}{a}_2 = \frac{1}{\sqrt{2g}}(\underset{\sim}{b}_1-\underset{\sim}{b}_2)$ is an orthonormal basis for it.

c) Show that in $\underset{\sim}{a}_1$, $\underset{\sim}{a}_2$ coordinates $\underset{\sim}{A}$ has the matrix

$$\frac{1}{2\lambda}\begin{bmatrix} 1+\lambda^2 & 1-\lambda^2 \\ 1-\lambda^2 & 1+\lambda^2 \end{bmatrix}, \text{ and that this equals } \pm\begin{bmatrix} \cosh t & \sinh t \\ \sinh t & \cosh t \end{bmatrix}$$

for some unique $t \in R$.

d) Deduce that $\underset{\sim}{A} = \pm\mathrm{Exp}(t\underset{\sim}{B})$, where in $\underset{\sim}{a}_1$, $\underset{\sim}{a}_2$ coordinates $[\underset{\sim}{B}] = \begin{bmatrix} 0 & 1 \\ 1 & 0 \end{bmatrix}$.

6a) Show, similarly to Ex. 5, that any $\underset{\sim}{A} \in SL(2;R)$ with $\mathrm{tr}\underset{\sim}{A} = \pm 2$ preserves exactly one (degenerate) symmetric bilinear form $\underset{\sim}{G}$ on R^2, up to a scalar factor. Prove that in suitable coordinates on R^2

$$\underset{\sim}{G}((x^1,x^2),(y^1,y^2)) = g(x^1y^1), \quad [\underset{\sim}{A}] = \pm\begin{bmatrix} 1 & t \\ 0 & 1 \end{bmatrix}$$

for some $g, t \in R$.

b) Deduce that $\underset{\sim}{A} = \pm\mathrm{Exp}(t\underset{\sim}{B})$, where $[\underset{\sim}{B}] = \begin{bmatrix} 0 & 1 \\ 0 & 0 \end{bmatrix}$ in the same coordinates.

7a) Deduce from 6.06 that if $\mathrm{tr}(\mathrm{Exp}(\underset{\sim}{B})) = 1, 0, \text{ or } -1$, then

$$\mathrm{Exp}(\underset{\sim}{B}) = \cos(\tfrac{\pi}{3})\underset{\sim}{I} \pm \sin(\tfrac{\pi}{3})\bar{\underset{\sim}{B}}, \quad \pm\bar{\underset{\sim}{B}}, \quad \text{or } \cos\left(\frac{2\pi}{3}\right)\underset{\sim}{I} \pm \sin\left(\frac{2\pi}{3}\right)\bar{\underset{\sim}{B}}$$

respectively.

b) Deduce using Ex. 3, Ex. 4 that if $\det\underset{\sim}{A} = 1$ and $\mathrm{tr}\underset{\sim}{A} = 1, 0, -1$, then with respect to some inner product (unique up to a scalar) $\underset{\sim}{A}$ is a turn through $\frac{\pi}{3}$, $\frac{\pi}{4}$ or $\frac{2\pi}{3}$ respectively.

8a) Show that an orthogonal operator $\underset{\approx}{A}$ on R^3 with its usual inner product has in some coordinates the matrix

$$\begin{bmatrix} \pm 1 & 0 & 0 \\ 0 & a & b \\ 0 & c & d \end{bmatrix}, \text{ where } \begin{bmatrix} a & b \\ c & d \end{bmatrix} \text{ is the matrix of a plane rotation.}$$

b) Deduce from Ex. 7 that if $\text{tr}\underset{\approx}{A}$ is an integer, $\underset{\approx}{I} = \underset{\approx}{A}^2, \underset{\approx}{A}^3, \underset{\approx}{A}^4$ or $\underset{\approx}{A}^6$.

For those with a bit of Lie group theory :-

9a) Find a metric tensor field on $GL^+(2;R) = \{\underset{\approx}{A} : R^2 \rightarrow R^2 \mid \det\underset{\approx}{A} > 0\}$ such that $\exp_{\underset{\approx}{I}}$ coincides with Exp. (Hint: show that

$$GL^+(2;R) \rightarrow SL(2;R) \times R : \underset{\approx}{A} \rightsquigarrow ((\det\underset{\approx}{A})^{-\frac{1}{2}}\underset{\approx}{A}, \log(\det\underset{\approx}{A})),$$

with the additive structure on R, is a Lie group isomorphism.)

b) Is only one signature possible for such a metric tensor, up to sign?

X Curvature

PANORAMIX: Alors, Obelix, l'Helvétie, c'est comment?
OBELIX: Plat.

Asterix Chez Les Helvètes.

1. FLAT SPACES

In treating the geometry of manifolds that were not simply nice flat affine spaces we have paid major attention to parallel transport along curves; the feature of general spaces that distinguishes them most dramatically is the disappearance of "absolute" parallelism. This prompts

1.01 Definition

A connection ∇ on a manifold M is locally flat (or "M with ∇" is), if any $p \in M$ has a neighbourhood U_p such that for $q \in U_p$, parallel transport gives the same result along any curve in U_p from p to q. If M has a metric tensor, M is locally flat if the corresponding Levi-Cività connection is.

M is globally flat if parallel transport between any $p,q \in M$ is the same for any curve in M from p to q. Fig. 1.1 illustrates a locally but not globally flat M. Parallel transport along any curve confined to U_1 or to U_2 gives the same result as along another confined to the same region. However, circumnavigating M gives a result different from the identity that is parallel transport along a curve that stays at one point. (Examples without "edges" are harder to draw, since none embeds in Euclidean 3-space with a locally flat metric.)

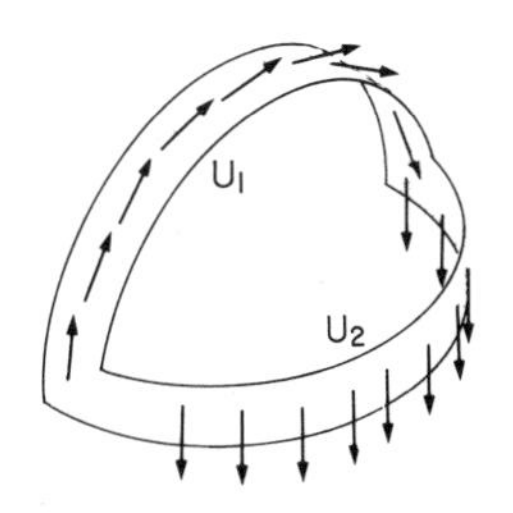

Fig. 1.1

If M is simply connected (that is any curve from p to q can be deformed, via curves from p to q, into any other) local flatness implies global. This is an example of the wide field of relations between the possible metrics/connections/curvatures on a manifold and its topological "shape". Practically all these relations involve algebraic topology. This handles "india-rubber geometry" rather as linear algebra handles another kind, but is a lot more complicated and a lot less complete. We shall illustrate the relationship further (1.04) by quoting again one of the few results of algebraic topology whose statement, at least, does not require machinery that would require another book to explain adequately.

1.02 Parallel fields and flatness

On a general M, no non-zero parallel vector fields (defined in VIII.4.01) exist; or there may be only a few. For example on the Klein bottle with an appropriate metric (not that induced by the position in R^3 shown, for which no non-zero field is parallel) the vector field shown in Fig. 1.2 is parallel. But for no metric does this manifold admit more than one parallel vector field (why?) up to scalar multiplication, and for most it admits none.

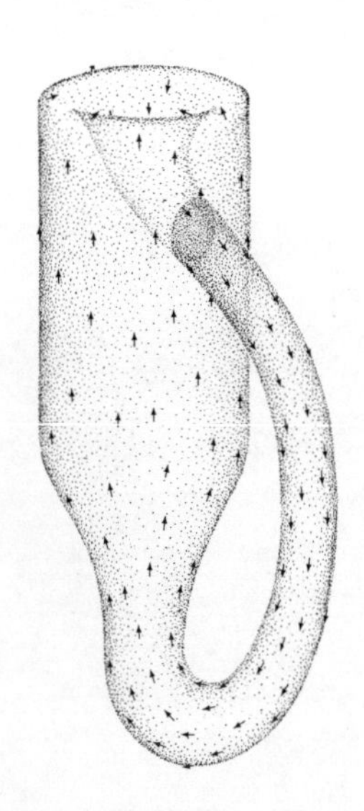

Fig. 1.2

In general:

1.03 Lemma

The set of all parallel vector fields on a manifold M with connection forms a vector space PM of dimension ≤ dim M, with equality if and only if M is globally flat. (Similarly for open subsets of M, such as the domain of a chart.)

Proof

By the linearity of covariant differentiation, sums and scalar multiples of parallel fields are again parallel. Since clearly any parallel vector field is specified by its value at any $p \in M$,

$$_p : PM \to T_pM : \underset{\sim}{v} \mapsto \underset{\sim}{v}_p = \underset{\sim}{v}(p)$$

is thus linear and injective, so that $\dim(PM) \leq \dim(T_pM) = \dim M$.

If equality holds, $_p$ is an isomorphism and hence any $\underset{\sim}{v}_p \in T_pM$ is part of a unique parallel vector field $\underset{\sim}{v}$. Thus $\underset{\sim}{v}$ restricted to any curve c from p to q is parallel, hence the result of parallel transport to q is $\underset{\sim}{v}(q)$, independently of c, so M is globally flat.

Conversely, if M is globally flat we can extend any $\underset{\sim}{v}_p \in T_pM$ to a parallel field $\underset{\sim}{v} : q \mapsto \underset{\sim}{v}_q$ by parallel transport along arbitrary curves, smooth by Ex. 2, so $_p$ is an isomorphism and equality holds. ■

1.04 Corollary

The two-dimensional sphere S^2 does not admit any metric which makes it globally flat.

Proof

By the Hairy Ball Theorem (cf. p.255) any vector field is zero somewhere on the sphere. Hence a parallel one is zero everywhere, so $\dim(PS^2) = 0 \neq 2 = \dim(S^2)$ for any metric. ■

Since S^2 is simply connected this means that it cannot be locally flat either; we shall not go into details, but the reader acquainted with the fundamental group can readily supply them. (The reader who is not can gain insight from considering how to construct an ad hoc proof from the material of this chapter.) The sphere can of course be flat somewhere - Fig. IX.1.1 shows two embeddings each with three flat regions - but not around every point.

We shall not be using this result, but it illustrates well the constraints the global topology of a manifold can put on its local structure. (A similar use of the Hairy Ball Theorem, incidentally, shows that S^2 admits no indefinite metric tensor field.) We encounter implications of local structure for global topology in XII.2.03. When a local structure is interpreted as the presence of hydrogen atoms or physicists, the algebraic topology needed to discuss such relationships exactly becomes another mathematical theory the cosmologist should be at home in.

The reader may have felt unhappy about Fig. 1.1 being described as flat, even locally. We are concerned, though, with intrinsic geometry; if we cut it we could spread it flat without wrinkles. Its geodesics would then appear as straight lines, the angle-sum of any triangle left intact by the cut would be 180°, and so forth: its local geometry is just that of the plane. This is in violent contrast to the sphere, no part of which can be matched to flat paper without distorting one or the other - as cartographers and anyone who has watched a shop-assistant gift-wrap a beachball have particularly good reason to know.

Our next results show that Definition 1.01 amounts to requiring M to be exactly like an affine space, as far as local internal measurements are concerned. From inside M, you can't say flatter than that.

1.05 Lemma

M with metric $\underset{\sim}{G}$ is locally flat if and only if around every $p \in M$ there is a chart $\phi : U \to X$ such that $\underset{\sim}{G}|U$ is given by ϕ as a constant metric tensor in the affine sense (VII.3.05).

Proof

If around every point there is such a chart, parallel transport within its domain corresponds to the usual parallelism in X, which is independent of curves (Ex. VIII.6.4), so M is locally flat.

If M is locally flat, let V be an open region around $p \in M$ in which parallel transport is independent of curves. Choose a basis $(u_1)_p, \ldots, (u_n)_p$ for T_pM. Since V with $G|V$ is globally flat we can extend the basis vectors to parallel vector fields $u_1, \ldots, u_n$ on V. By the symmetry of the Levi-Cività connection,

$$0 = T(u_i, u_j) = \nabla_{u_i} u_j - \nabla_{u_j} u_i - [u_i, u_j], \quad \forall i,j.$$

But since u_i, u_j are parallel, their covariant derivatives with respect to all vectors, and hence to each other, vanish.

Thus

$$0 = [u_i, u_j], \quad \forall i,j.$$

And so by VII.7.04 there is a chart $\phi : U \to R^n$, $U \subseteq V$, around p whose ∂_i are exactly the u_i. With respect to ϕ, then, for any $q \in U$ we have

$$g_{ij}(q) = G_q(\partial_i, \partial_j) = G_q(u_i(q), u_j(q))$$

$$= G_q(\tau(u_i(p)), \tau(u_j(p))) \quad \text{where } \tau \text{ is parallel transport along any curve from p to q,}$$

$$= G_p(u_i(p), u_j(p)) \quad \text{since } \nabla \text{ is compatible with } G$$

$$= g_{ij}(p).$$

Thus $\underset{\sim}{G}$ is given by ϕ as the metric tensor on R^n with constant coefficients g_{ij}, which is itself constant. ■

Notice that both defining characteristics of the Levi-Cività connection are involved in this proof.

1.06 Corollary

M is locally flat if and only if around every point there is a chart $\phi : U \to X$, such that $c : J \to M$ with $c(J) \subseteq U$ is a geodesic if and only if $\phi \circ c$ is affine.

Proof

If M is locally flat, the geodesics are thus by 1.05 and Ex. IX.4.2.

Conversely, if U is such a chart, without loss of generality suppose $X = R^n$, and use coordinates. At any $(x^1,\ldots,x^n)$ the vector ∂_i is represented by the affine curve $c : t \leadsto (x^1,\ldots,x^i+t,\ldots,x^n)$. By hypothesis c is geodesic. It follows that

$$\underset{\sim}{0} = \nabla_{c*}c* = \nabla_{\partial_i}\partial_i, \quad \text{so } \Gamma^k_{ii} = 0, \quad \forall i,k.$$

Similarly $t \leadsto (x^1,\ldots,x^i+t,\ldots,x^j+t,\ldots,x^n)$ represents $\partial_i + \partial_j$, whence

$$\begin{aligned}\underset{\sim}{0} &= \nabla_{(\partial_i+\partial_j)}(\partial_i+\partial_j)\\ &= \nabla_{\partial_i}\partial_i + 2\nabla_{\partial_i}\partial_j + \nabla_{\partial_j}\partial_j \quad \text{by linearity and symmetry}\\ &= 2\nabla_{\partial_i}\partial_j = 2\Gamma^k_{ij}\,\partial_k.\end{aligned}$$

So $\qquad 0 = \Gamma^k_{ij}, \quad \forall i,j,k$

everywhere (not just at one point only as in IX.2.06).

$$\partial_k(g_{ij}) = \Gamma_{ijk} + \Gamma_{jki} \qquad \text{(VIII.6.06)}$$

$$= 0 \quad \forall i,j,k$$

everywhere; the functions g_{ij} are therefore constant, and so therefore is the metric on R^n that they define.

Thus by 1.05 M is locally flat. ■

These results show at once that all the discussion in IX of energy and length in affine spaces applies locally in a locally flat manifold, confirming that local measurements within it give results exactly like affine geometry. (Globally the geometry may differ, however, even on a globally flat manifold: Ex. 4a,b.)

1.06 can be refined slightly for a spacetime, where not all kinds of geodesic can currently be related to measurements (nothing yet having been shown to go faster than light). The above proof nowhere required the ∂_i orthonormal, and we can always choose a basis $\underset{\sim}{b}_1,\ldots,\underset{\sim}{b}_n$ for the vector space of X such that all the $\underset{\sim}{b}_i$ and $\underset{\sim}{b}_i+\underset{\sim}{b}_j$ are timelike (as in Minkowski space: (1,0,0,0),(2,1,0,0),(2,0,1,0) and (2,0,0,1)). Hence the proof gives also

1.07 Corollary

A pseudo-Riemannian manifold is locally flat if and only if around every point there is a chart by which timelike geodesics correspond to timelike affine curves. ■

1.08 Curved round what?

At least he does not call it a "paradox": but the kind of Philosopher of IX.4.05 is often sure that space cannot be "curved", which he equates with "bent" (not with "not flat" in our sense of flatness), without a higher space providing directions to bend in. This can lead to remarkable ideas! The two chief points to hang on to, talking to him, are that flatness may well demand more dimensions than curvature when it comes to embeddings (Ex. 5) and that curvature in a geometry may arise physically in ways very different from bending (Ex. 6).

Exercises X.1

1) Is there a meaningful definition of a locally parallel vector field, as distinct from one parallel over its entire domain?

2) Show that the map $M \to TM : q \rightsquigarrow \underset{\sim}{v}_q$ defined in 1.03 on a globally flat manifold is C^∞. (Use Ex. VII.7.1c)

3a) Any connection on the circle is symmetric and locally flat.

b) The connection of Ex. VIII.4.4 and Ex. VIII.6.7 is not globally flat.

4a) The cylinder $\{(x,y,z) \in R^3 | x^2+y^2 = 1\}$, with the metric induced from the standard Riemannian one on R^3, is globally flat. (Either use Definition 1.01 and results about parallel transport, or the results of this section.)

b) Find a pair of geodesics in this manifold with infinitely many intersections and a pair that do not meet.

c) Show the open cone $\{(x,y,z) \in R^3 | x^2+y^2 = az^2, z < 0\}$, is a manifold, and flat locally but not globally in the induced metric from R^3. Show that any geodesic not pointing straight at (0,0,0) can be infinitely extended, and that whether it has self-intersections depends on whether $a \geq$ or $< 1/3$. (Hint: consider the cone cut and laid out flat.) Can a geodesic have infinitely many self-intersections? Are there any closed geodesics?

5a) Show by considering parallel transport as rolling without slipping, or otherwise, that the torus in R^3 defined by $(x^2+y^2+z^2+3)^2 = 16(x^2+y^2)$, with the metric induced from the standard one on R^3, is not flat.

b) The subset $\{\underset{\sim}{x} \in R^4 | (x^1)^2+(x^2)^2 = (x^3)^2+(x^4)^2 = 1\}$, is a manifold diffeomorphic to the torus above. The metric on it induced from the standard Riemannian one on R^4 is globally flat. (Label x by the two points (x^1,x^2) and (x^3,x^4) in the circle, to get coordinates.)

6a) Show that the metric of your answer to Ex. IX.4.6 is not globally flat, by applying 1.06. Is it locally flat?

b) Find an embedding of enough of the plane to contain Fig. IX.4.5 in R^3 to induce your metric on a vertical section through the camel and the palm tree. (Or at least find one that reproduces its qualitative features as regards geodesics: Fig. 1.3).

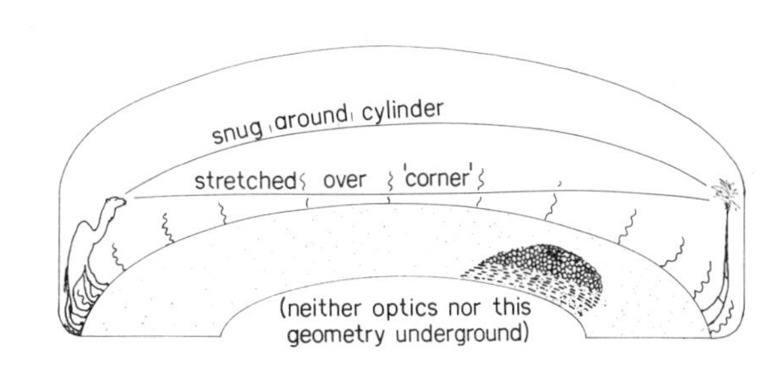

Fig. 1.3

Do you suppose that hot air "bends space" in this way, in some R^n or R^m unknown?

2. THE CURVATURE TENSOR

We now know the theory of locally flat spaces. But for the purposes of the "geometrising" of gravitation we mentioned in 4.03, this is like knowing the behaviour of light in a Newtonian vacuum: not helpful when there is matter present, and not leading to the theory of mirages and microscopes. If we are content to suppose spacetime flat, and write in "forces" that "act" in it, we can locally do optics and - very artificially - celestial mechanics as though in an affine space (though globally spacetime may be topologically more interesting than that). But to subsume gravitational forces, like optical effects, in the geometry, we must handle curvature.

One reason for interest in this approach is cosmological: one may regard the transformation of dynamics into geometry as "fiction" if prejudiced, but the result carries information not easily accessible in the flat approach. Just as topological type is limited by possession of a particular local structure (1.04 et seq.), so therefore does admission of it. If the dynamics can be described by giving spacetime a particular metric, whether or not it "really" has that metric, then spacetime is a manifold such as can have that metric: the topological implications follow.

So the cosmos is rather different from the classical solar system with its mathematically equivalent descriptions using either earth or sun as unmoving centre, and only a taste for simplicity dictating choice of the sun. In the small, "flat" and "unflat" physics may or may not be equivalent (depending on the theories) but in the large the unflat has implications that cannot be obtained directly from the flat, which is either local or assumes that spacetime is R^4. Let us look, then, at curved spaces.

2.01 Local curvature

If "curved" intrinsically in a manifold M is to mean "not flat" in the sense of "flat" explored in §1, a measure of curvature must be a measure of the breakdown of the defining property of flatness; parallel transport from p to q independent of the path between them.

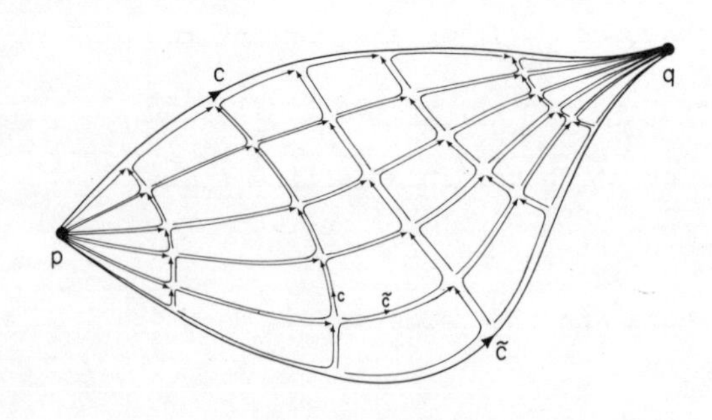

Fig. 2.1

The first thing to observe is that if we know how far this breaks down for small curves, we know how it breaks down for all curves: Fig. 2.1 illustrates this. Since parallel transport along a piece of curve, followed by parallel transport back along it, is the identity on tangent vectors at the starting point, it is easy to see that the difference between parallel transport along c and and $\tilde{c}$ is in a suitable sense the "sum" of the differences between curves like and . Curvature is thus a local property of M: if we know about the differences between curves up to an arbitrary small size, we know the difference between parallel transport along any two curves, between any p and q, that can be related by a picture like Fig. 2.1. (If M is simply connected, this means any two curves at all from p to q; otherwise, algebraic relations like those between local and global flatness are involved.)

Fig. 2.1 suggests immediately a natural approach to measurement of curvature. First, make all the curves involved lie smoothly together in a parametrised surface (IX.3.03)

$$S : [0,1] \times [a,b] \to M$$

with $S(t,a) = p$, $S(t,b) = q$, $\forall t \in [0,1]$ and $S(0,s) = c(s)$, $S(1,s) = \tilde{c}(s)$, $\forall s \in [a,b]$. and define the and curves as the composites of S with little rectangular paths in $[0,1] \times [a,b]$. Now the lack of route-independence of parallel transport, between p and q is obviously equally given by the difference of parallel transport along c and back along $\tilde{c}$, from the identity on T_pM. So we are led to say that the total curvature along any piece of surface is the difference from the identity of parallel transport once around its edge.

The difference between transport along c and $\tilde{c}$, then, is given by the total curvature of any surface between them, and this is obtained by adding the total curvature of all the small pieces ▱ of the surface, in the manner of Fig. 2.1. Now, in the manner of an electromagnetism course proving Stokes's Theorem, all we need do is substitute "infinitesimal" for "small" and "integrating" for "adding". Then the "infinitesimal curvature" at a point $x \in M$ becomes the assignment, to each infinitesimal piece P of surface with a corner at x, of the "infinitesimal rotation" away from the identity $T_xM \to T_xM$ that results from parallel transport around the edge of P. Integrating this over the whole surface gives the total curvature.

This approach works perfectly. However, it needs to be somewhat more precisely expressed. Apart from being careful about limits, and replacing infinitesimals by tangent space constructions (or setting up infinitesimals rigorously), one must be precise about the "adding" of total curvatures, and the integration of infinitesimal rotations, at different places, that we were so glib about just now. This can be done, and extremely elegantly: the main tool is the "moving frame" and the associated formulation of connections, due to Cartan. Unfortunately, this involves differential forms, and some theory of Lie groups (neither of which we have space to treat properly), for the geometry to be rigorous and visible; without them it means a relapse to coordinate manipulations and gropings in the dark. So we shall not be integrating curvature†, but we shall think of its meaning as above. This yields directly some of the properties it should have and guides our search for a rigorous definition in terms of the machinery we already have; once found this will let us prove these and other properties, and perform computations. We shall continue to carry the above geometry as a way to explain what is happening, even without having the formalism to show why it is what is happening, rigorously. The other choices would be to omit adequate geometrical explanation, or to cover so much purely mathematical material as never to approach physics: both defeating our objective of making geometry available in undergraduate mathematics for physics courses like that from which this book evolved. So we shall accompany our less geometric formal proofs with more geometrical handwaving, to be made more precise in a later volume, or by the reader's digging into wholly maths-oriented texts such as [Spivak] or [Kobayashi and Nomizu].

So what properties should $p \rightsquigarrow$ (curvature at p) have?

---

†Strictly one integrates the <u>curvature form</u>.

First, if at all $p \in M$ it vanishes, then M should be locally flat. Round any $p \in M$ we can find a simply connected neighbourhood U (say, the unit ball in some chart). For any two curves between $q, q' \in U$ we can find a surface between them, integrate curvature over it, and get the difference between parallel transport along them. But this integral of a vanishing quantity should be zero, so there is no difference: M is locally flat. (The proof using our rigorous definitions is Th^m. 2.05, but this is why it is true.)

Secondly, what <u>is</u> an "infinitesimal rotation"? By the above, it is an infinitesimal displacement from the identity on T_pM. We have observed, in discussing Th^m. VIII.4.06, that an infinitesimal displacement S from p in the old language means a tangent vector $\underset{\sim}{v}$ at p to M in the new. In this case $\underset{\sim}{v}$ is tangent, at the identity $\underset{\sim}{I}_p : T_pM \to T_pM$, to the vector space $L(T_pM;T_pM)$ and we can free it to be an <u>element</u> $\underset{\sim}{V}$ of $L(T_pM;T_pM)$. What kind of an element?

T_pM has metric tensor $\underset{\sim}{G}_p$, and since ∇ is compatible with $\underset{\sim}{G}$ parallel transport along any small finite curve from p to p must be a rotation. An "infinitesimal" rotation then will be tangent at $\underset{\sim}{I}_p$ to $O(T_pM)$, the collection of <u>orthogonal</u> operators on T_pM, which sits naturally as a manifold embedded in $L(T_pM;T_pM)$. (As the rotations of the plane, essentially the circle of angles to turn through, embed in 4-dimensional $L(R^2;R^2)$.) Not stopping to prove this submanifold property, we look at the properties that $\underset{\sim}{V}$ must have. Representing $\underset{\sim}{v}$, as in VIII §1, by a curve c with $c(0) = \underset{\sim}{I}_p$, to be sure that $\underset{\sim}{v}$ is tangent to $O(T_pM)$ we must have $c(t) \in O(T_pM)$, $\forall t$. ($c^*(0)$ is certainly tangent to $O(T_pM)$ if c is <u>in</u> $O(T_pM)$.) So for $\underset{\sim}{x}, \underset{\sim}{y} \in T_pM$, $(c(t)(\underset{\sim}{x})).(c(t)(\underset{\sim}{y}))$ is the same as $\underset{\sim}{x}.\underset{\sim}{y}$ for all t; $(c(\ )(\underset{\sim}{x})).(c(\ )(\underset{\sim}{y}))$ is constant. Thus for given $\underset{\sim}{x},\underset{\sim}{y}$

$$0 = \frac{d}{dt}(c(\ )(\underset{\sim}{x})).(c(\ )(\underset{\sim}{y}))\ (0)$$

$$= \left[\nabla_{c*} c(\ )(\underset{\sim}{x})(0)\right].c(0)(\underset{\sim}{y}) + c(0)(\underset{\sim}{x}).\left[\nabla_{c*} c(\ )(\underset{\sim}{y})(0)\right],$$

with the usual connection on $L(T_pM;T_pM)$.

Hence, $0 = \underset{\sim}{V}(\underset{\sim}{x}).\underset{\sim}{y} + \underset{\sim}{x}.\underset{\sim}{V}(\underset{\sim}{y})$ (why? think about transport in $L(T_pM;T_pM)$)

so $\underset{\sim}{V}(\underset{\sim}{x}).\underset{\sim}{y} = -\underset{\sim}{x}.\underset{\sim}{V}(\underset{\sim}{y})$ always.

Equivalently (p.438),

$$\underset{\sim}{V}(\underset{\sim}{x}).\underset{\sim}{x} = 0 \quad \forall \underset{\sim}{x} \in T_pM.$$

Such a $\underset{\sim}{V}$ is called skew-self-adjoint, since $\underset{\sim}{V}$ is exactly the negative of its adjoint. Any "infinitesimal rotation" at p can thus be given by a skew-self-adjoint operator on T_pM (and in fact any such operator can be realised this way). Analogously, the traceless operators studied in IX. §6 are "infinitesimal unimodular operators".

The coordinate form is immediate from IV.3.14; with respect to an orthonormal basis when $\underset{\sim}{G}_p$ is an inner product, skew-self-adjointness is equivalent to the condition $[\underset{\sim}{V}]^i_j = -[\underset{\sim}{V}]^j_i$ and is called skew-symmetry, a usage (like "symmetry") which it is wiser to avoid in the indefinite case.

The reader should convince himself that in a rotation in ordinary 2 or 3-space, every vector's tip is at any moment moving at right angles to the vector (this is the condition $\underset{\sim}{V}(\underset{\sim}{x}).\underset{\sim}{x} = 0$),and interpret similarly the equivalent condition $\underset{\sim}{V}(\underset{\sim}{x}).\underset{\sim}{y} = -\underset{\sim}{x}.\underset{\sim}{V}(\underset{\sim}{y})$. Skew-self-adjointness just extends these intuitive facts about vectorial rates of rotation to general dimensions and metric tensors.

What, finally, should an "infinitesimal piece of area" be?

Before leaving "small" for "infinitesimal" we were talking about the images of little rectangles. As "small" gets smaller, these

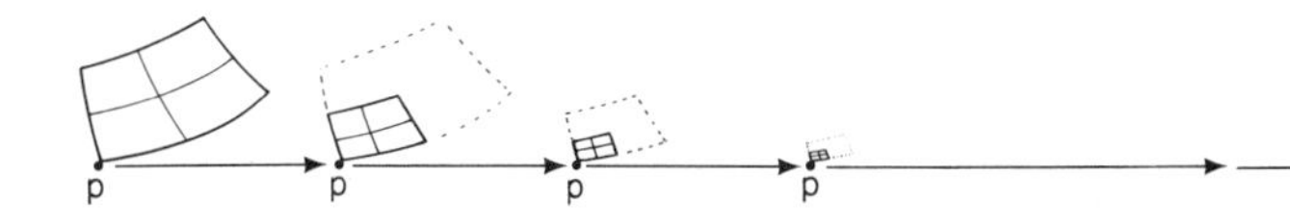

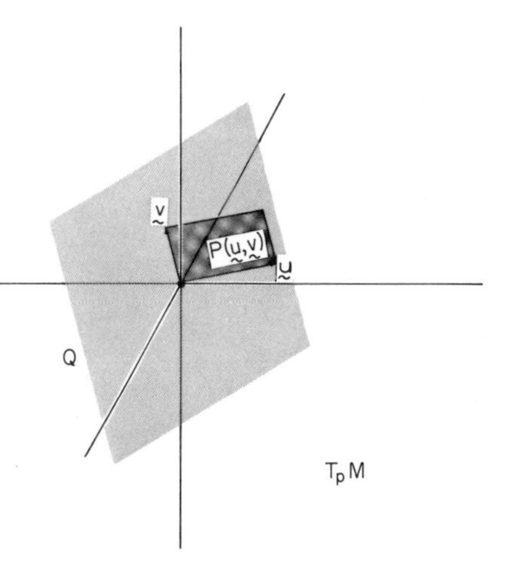

Fig. 2.2

images look more and more like parallelograms, with straighter and straighter sides. "In the limit" then, the sides meeting at p turn into infinitesimal displacements - tangent vectors - at p ($\underset{\sim}{u}, \underset{\sim}{v}$ say), and we have an actual parallelogram $P(\underset{\sim}{u}, \underset{\sim}{v})$ defined by them in their common plane Q (which is the "linear approximation" at p to the surface itself). So curvature at p should give us for each way Q that the surface can pass through P, an "infinitesimal rotation per area" as we shall be "integrating over an area". (Clearly it must depend on the attitude of Q: if S^1 is the equator in S^2, and $S^2 \times R \subseteq R^4$ has the induced metric, the surface $S^1 \times R$ is just a cylinder, and parallel transport of vectors in $T(S^1 \times R)$ is clearly independent of route (prove it!) while $S^2 \times \{0\}$ which meets it in any $p \in S^1 \times \{0\}$ carries the "nonflatness" of $S^2 \times R$.) This "attitude-dependent, per-area-of-parallelogram" behaviour means we want on each Q a skew-symmetric bilinear form (cf. Ex. V.1.11) with values in the space of "infinitesimal rotations". Skew-symmetry is very natural here: switch $\underset{\sim}{u}$ and $\underset{\sim}{v}$ and we reverse the way we go round the parallelogram $P(\underset{\sim}{u}, \underset{\sim}{v})$, which should obviously give minus the rotation.

It turns out that the rotation per area, though it depends on Q, does so linearly, so that the curvature is described by a skew-symmetric bilinear map assigning to any $(\underset{\sim}{u}, \underset{\sim}{v}) \in T_pM \times T_pM$ the "infinitesimal rotation" that results from parallel transport around the "infinitesimal parallelogram" $P(\underset{\sim}{u},\underset{\sim}{v})$.

To sum up, we are looking for a skew-symmetric bilinear map from $T_pM \times T_pM$ to the space of skew-self-adjoint operators $T_pM \to T_pM$. Equivalently, we want a linear map

$$\underset{\sim}{R}_p : T_pM \otimes T_pM \to L(T_pM;T_pM)$$

which is the same by V.1.08 as a map

$$T_pM \otimes T_pM \to (T_pM)^* \otimes T_pM$$

which corresponds to an element of

$$(T_pM \otimes T_pM)^* \otimes ((T_pM)^* \otimes T_pM) = (T^1_3M)_p,$$

namely a $\binom{1}{3}$-tensor at p.

We shall use $\underset{\sim}{R}$ to indicate both the collection of maps $\underset{\sim}{R}_p$ as above, and the $\binom{1}{3}$-tensor field to which it naturally corresponds. When the distinction matters the context will make clear which aspect we are using.

We evidently want $\underset{\sim}{R}_p$ to depend smoothly on p (we always want things to depend smoothly on p), and the skew-self-adjointness and skew-symmetry conditions give

$$((\underset{\sim}{R}_p(\underset{\sim}{u},\underset{\sim}{v}))(\underset{\sim}{x})).\underset{\sim}{y} = \underset{\sim}{x}.((\underset{\sim}{R}_p(\underset{\sim}{u},\underset{\sim}{v}))(\underset{\sim}{y})) \text{ as real numbers}$$

$$\underset{\sim}{R}_p(\underset{\sim}{u},\underset{\sim}{v}) = -\underset{\sim}{R}_p(\underset{\sim}{v},\underset{\sim}{u}) \text{ as operators.}$$

These identities are satisfied by the tensor we are about to set up, along with others that are hard to motivate geometrically without a deep analysis of the "torsion-zero" condition on the connection which yields them.

2.02 Small pieces

Let us look for what $\underset{\sim}{R}_p(\underset{\sim}{u}_p,\underset{\sim}{v}_p)$ "should" be for given $\underset{\sim}{u}_p,\underset{\sim}{v}_p \in T_pM$, by parallel transport of a typical $\underset{\sim}{t}_p \in T_pM$ around a small bent parallelogram that in the limit of smallness tends to the unbent one in T_pM defined by $\underset{\sim}{u}_p$ and $\underset{\sim}{v}_p$. Extending $\underset{\sim}{u}_p$, $\underset{\sim}{v}_p$ to vector fields $\underset{\sim}{u}$, $\underset{\sim}{v}$ with corresponding local flows ϕ, ψ, we transport $\underset{\sim}{t}_p$ along the solution curve of $\underset{\sim}{u}$ through p to $\phi_t(p)$ and then along a curve of $\underset{\sim}{v}$ to $\psi_s(\phi_t(p))$. Similarly, we can transport it via $\psi_s(p)$ to $\phi_t(\psi_s(p))$. If $\underset{\sim}{u}$ and $\underset{\sim}{v}$ commute we then have two vectors tangent to M at the same point $\phi_t(\psi_s(p)) = \psi_s(\phi_t(p)) = q$, say, and we can subtract one from the other to find how they differ. (If they don't commute we have to transport $\underset{\sim}{t}$ across a gap: hence in the limit we can expect a correction term involving $[\underset{\sim}{u},\underset{\sim}{v}]$, which is the limit per unit area of the gap (cf. VII.7.03). For the moment assume they commute.)

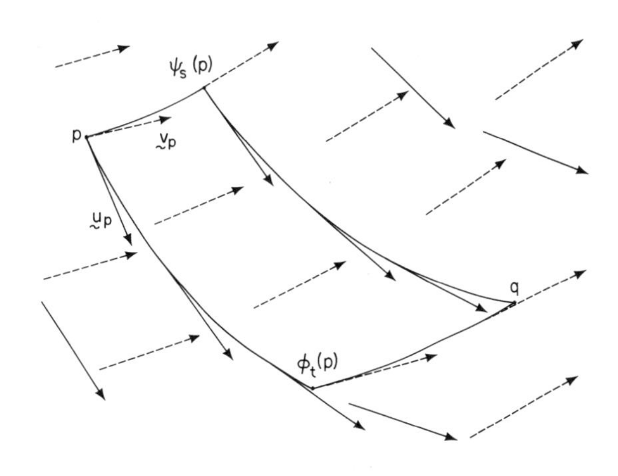

Fig. 2.3

Labelling the four parallel transport maps involved:

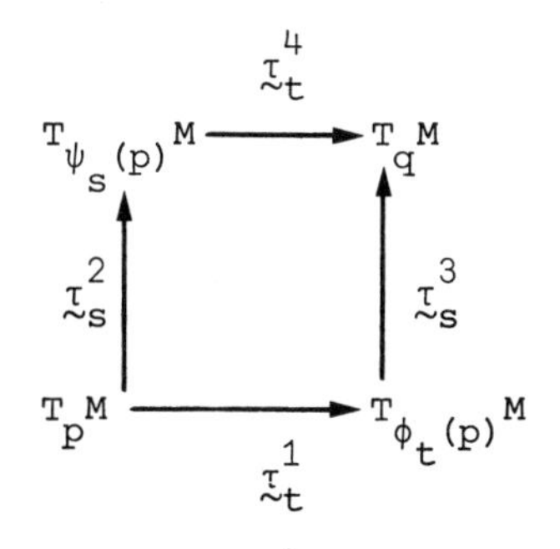

we get a "difference" vector $(\underset{\sim}{\tau}{}_s^3\underset{\sim}{\tau}{}_t^1(\underset{\sim}{t}_p) - \underset{\sim}{\tau}{}_t^4\underset{\sim}{\tau}{}_s^2(\underset{\sim}{t}_p))$ in T_qM.

More conveniently, let us extend $\underset{\sim}{t}_p$ also to a field $\underset{\sim}{t}$, and look at the vector

$$(\underset{\sim}{\tau}{}_s^3\underset{\sim}{\tau}{}_t^1)^{\leftarrow}\ \underset{\sim}{t}_q - (\underset{\sim}{\tau}{}_t^4\underset{\sim}{\tau}{}_s^2)^{\leftarrow}\ \underset{\sim}{t}_q$$

which equally measures the curvature of our little piece of area and stays in T_pM as we vary s and t. We expect curvature to be "per area", and the closer our bent piece of area approximates a parallelogram as it shrinks the more nearly its area goes as ts. This suggests that we look for $\underset{\sim}{R}_p(\underset{\sim}{u},\underset{\sim}{v})\underset{\sim}{t}$ in the limit as

$$\lim_{(s,t)\to 0} \frac{(\underset{\sim}{\tau}{}_s^3\underset{\sim}{\tau}{}_t^1)^{\leftarrow}\ \underset{\sim}{t}_q - (\underset{\sim}{\tau}{}_t^4\underset{\sim}{\tau}{}_s^2)^{\leftarrow}\ \underset{\sim}{t}_q}{ts}.$$

This can be expressed, if it exists, in terms of limits we have taken before. First, rewrite it as

$$\lim_{(s,t)\to 0} \frac{1}{ts}\Big[\underset{\sim}{\tau}{}_t^{1\leftarrow}\ \underset{\sim}{\tau}{}_s^{3\leftarrow}\ \underset{\sim}{t}_q - \underset{\sim}{\tau}{}_t^{1\leftarrow}\ \underset{\sim}{t}_{\phi_t(p)} - (\underset{\sim}{\tau}{}_s^{2\leftarrow}\ \underset{\sim}{t}_{\psi_s(p)} - \underset{\sim}{t}_p)$$

$$-(\underset{\sim}{\tau}{}_s^{2\leftarrow}\ \underset{\sim}{\tau}{}_t^{4\leftarrow}\ \underset{\sim}{t}_q - \underset{\sim}{\tau}{}_s^{2\leftarrow}\ \underset{\sim}{t}_{\psi_s(p)} - (\underset{\sim}{\tau}{}_t^{1\leftarrow}\ \underset{\sim}{t}_{\phi_t(p)} - \underset{\sim}{t}_p))\Big].$$

Since subtraction is continuous, this gives (using Ex. 1)

$$\lim_{t\to 0}\left\{\lim_{s\to 0}\left(\frac{\underset{\sim}{\tau}{}_t^{1\leftarrow}\ \underset{\sim}{\tau}{}_s^{3\leftarrow}\ \underset{\sim}{t}_q - \underset{\sim}{\tau}{}_t^{1\leftarrow}\ \underset{\sim}{t}_{\phi_t(p)}}{ts}\right) - \lim_{s\to 0}\left(\frac{\underset{\sim}{\tau}{}_s^{2\leftarrow}\ \underset{\sim}{t}_{\psi_s(p)} - \underset{\sim}{t}_p}{ts}\right)\right\}$$

$$-\lim_{s\to 0}\left\{\lim_{t\to 0}\left(\frac{\underset{\sim}{\tau}{}_s^{2\leftarrow}\ \underset{\sim}{\tau}{}_t^{4\leftarrow}\ \underset{\sim}{t}_q - \underset{\sim}{\tau}{}_s^{2\leftarrow}\ \underset{\sim}{t}_{\psi_s(p)}}{ts}\right) - \lim_{t\to 0}\left(\frac{\underset{\sim}{\tau}{}_t^{1\leftarrow}\ \underset{\sim}{t}_{\phi_t(p)} - \underset{\sim}{t}_p}{ts}\right)\right\}.$$

By continuity of $\underset{\sim}{\tau}{}_t^{1\leftarrow}$, $\underset{\sim}{\tau}{}_s^{2\leftarrow}$ and division by non-zero s and t, this is

$$\lim_{t\to 0}\frac{1}{t}\left\{\underset{\sim}{\tau}_t^{1\leftarrow}\lim_{s\to 0}\left(\frac{\underset{\sim}{\tau}_s^{3\leftarrow}\underset{\sim}{t}_q-\underset{\sim}{t}_{\phi_t(p)}}{s}\right)-\lim_{s\to 0}\left(\frac{\underset{\sim}{\tau}_s^{2\leftarrow}\underset{\sim}{t}_{\psi_s(p)}-\underset{\sim}{t}_p}{s}\right)\right\}$$

$$-\lim_{s\to 0}\frac{1}{s}\left\{\underset{\sim}{\tau}_s^{2\leftarrow}\lim_{t\to 0}\left(\frac{\underset{\sim}{\tau}_t^{4\leftarrow}\underset{\sim}{t}_q-\underset{\sim}{t}_{\psi_s(p)}}{t}\right)-\lim_{t\to 0}\left(\frac{\underset{\sim}{\tau}_t^{1\leftarrow}\underset{\sim}{t}_{\phi_t(p)}-\underset{\sim}{t}_p}{t}\right)\right\}$$

which by VIII.4.06 is exactly

$$\lim_{t\to 0}\frac{1}{t}\left(\underset{\sim}{\tau}_t^{1\leftarrow}(\nabla_{\underset{\sim}{v}_{\phi_t(p)}}\underset{\sim}{t})-\nabla_{\underset{\sim}{v}_p}\underset{\sim}{t}\right)-\lim_{s\to 0}\frac{1}{s}\left(\underset{\sim}{\tau}_s^{2\leftarrow}(\nabla_{\underset{\sim}{u}_{\psi_s(p)}}\underset{\sim}{t})-\nabla_{\underset{\sim}{u}_p}\underset{\sim}{t}\right)$$

alias, by VIII.4.06 again,

$$\nabla_{\underset{\sim}{u}_p}(\nabla_{\underset{\sim}{v}}\underset{\sim}{t})-\nabla_{\underset{\sim}{v}_p}(\nabla_{\underset{\sim}{u}}\underset{\sim}{t}).$$

So we have a formula for $\underset{\sim}{R}_p(\underset{\sim}{u}_p,\underset{\sim}{v}_p)\underset{\sim}{t}_p$ (if we can prove it independent of the extensions $\underset{\sim}{u}$, $\underset{\sim}{v}$, $\underset{\sim}{t}$), when $\underset{\sim}{u}$ and $\underset{\sim}{v}$ commute: $\underset{\sim}{R}_p(\underset{\sim}{u}_p,\underset{\sim}{v}_p)$ is exactly the extent to which $\nabla_{\underset{\sim}{u}}$ and $\nabla_{\underset{\sim}{v}}$ fail to do the same. Rather than compute directly the correction factor when $\underset{\sim}{u}$ and $\underset{\sim}{v}$ do not commute (which would involve the proof of the assertion left unproved in VII.7.03) we treat the above as motivation for our _definition_ of curvature with a correction factor that is clearly reasonable, and that by Ex. 2e is the only choice compatible with the desired independence of extensions.

2.03 Definition

The _curvature_ of a connection ∇ on M is the map

$$\underset{\sim}{R} : T^1M \times T^1M \to L(T^1M;T^1M)$$

defined by

$$\underset{\sim}{R}(\underset{\sim}{u},\underset{\sim}{v})\underset{\sim}{t} = \nabla_{\underset{\sim}{u}}(\nabla_{\underset{\sim}{v}}\underset{\sim}{t}) - \nabla_{\underset{\sim}{v}}(\nabla_{\underset{\sim}{u}}\underset{\sim}{t}) - \nabla_{[\underset{\sim}{u},\underset{\sim}{v}]}\underset{\sim}{t}.$$

It is immediate that $\underset{\sim}{R}(\underset{\sim}{u},\underset{\sim}{v})$ is indeed linear, and so does lie in the real vector space $L(T^1M;T^1M)$. Furthermore (Ex. 2) the vector $(\underset{\sim}{R}(\underset{\sim}{u},\underset{\sim}{v})\underset{\sim}{t})_p$ depends only on $\underset{\sim}{u}_p$, $\underset{\sim}{v}_p$ and $\underset{\sim}{t}_p$, so we may write it as $\underset{\sim}{R}_p(\underset{\sim}{u}_p,\underset{\sim}{v}_p)\underset{\sim}{t}_p$ to indicate independence of the extensions. (We cannot expect to <u>define</u> $\underset{\sim}{R}_p$ without taking extensions beyond p and T_pM, as it is not T_pM that is curved. It is remarkable that though all three terms in the definition depend on the choice of extensions, $\underset{\sim}{R}_p$ does not.) As above we may consider $\underset{\sim}{R}$ as a $\binom{1}{3}$-tensor field on M, the <u>curvature tensor field</u> of ∇. If ∇ is the Levi-Cività connection, $\underset{\sim}{R}$ is the <u>Riemann tensor</u> on M (not "pseudo-Riemann", even if M is pseudo-Riemannian).

(WARNING: some writers call <u>minus</u> our $\underset{\sim}{R}$ the Riemann tensor. As with the Lorentz metric, it is not common in the journals to say which sign you are using.)

The first property we argued that $\underset{\sim}{R}$ should have is that its vanishing everywhere should imply flatness. This depends on the following lemma.

<u>2.04 Lemma</u>

Let $S : I \times J \to M$ be a parametrised surface, $\underset{\sim}{v}$ a vector field along S. Then in the notation of IX.3.03, if $S(t,s) = x$,

$$\frac{\partial}{\partial t}\Delta_s\underset{\sim}{v}(t,s) - \frac{\partial}{\partial s}\Delta_t\underset{\sim}{v}(t,s) = \underset{\sim}{R}_x(S_t^*(t,s),S_s^*(t,s))\underset{\sim}{v}(t,s).$$

<u>Proof</u>

If at $(t,s) \in I \times J$ either S_t^* or S_s^* vanishes, both sides are zero. If not, $\underset{\sim}{D}_{(t,s)}S$ is injective and we can use Ex. VII.2.7a to find a chart $U \to R^n$ around $S(t,s)$ in which S_t^*, S_s^* are the restrictions to S of ∂_1, ∂_2, which commute. Extend any $\underset{\sim}{w}$ along S to $\tilde{\underset{\sim}{w}}$ on U with $\tilde{w}^i(x^1,\ldots,x^n) = v^i(x^1,x^2)$, $i = 1,\ldots,n$. The result follows since the definitions then imply $\Delta_t\underset{\sim}{w} = (\nabla_{\partial_1}\tilde{\underset{\sim}{w}})\circ S$, $\Delta_s\underset{\sim}{w} = (\nabla_{\partial_2}\tilde{\underset{\sim}{w}})\circ S$;

applying this with $\underset{\sim}{w} = \underset{\sim}{v}$, $\Delta_t \underset{\sim}{v}$ and $\Delta_s \underset{\sim}{v}$ in turn makes the left hand side equal the definition of the right, applied to $\tilde{S}^*_t$, $\tilde{S}^*_s$, $\tilde{\underset{\sim}{v}}$.

■

2.05 Theorem

A symmetric connection ∇ on a manifold M is locally flat if and only if its curvature tensor field $\underset{\sim}{R}$ is identically zero.

Proof

If M has a chart around p such that parallel transport is given by the usual affine one in R^n, we have commuting basis vector fields ∂_i. Using the expansion in 2.02 of the definition of $\underset{\sim}{R}_p(\underset{\sim}{u},\underset{\sim}{v})$ when $\underset{\sim}{u},\underset{\sim}{v}$ commute, it is immediate that the operators $\underset{\sim}{R}_p(\partial_i,\partial_j)$ on T_pM are zero. Hence $\underset{\sim}{R}_p(\underset{\sim}{u},\underset{\sim}{v}) = \underset{\sim}{0}$ in general, by linearity.

Suppose that in the domain of the chart $\phi : U \to R^n$ around p the curvature tensor vanishes. Then without loss of generality suppose $\phi(p) = (0,\ldots,0)$ and $\phi(U) = \{(x^1,\ldots,x^n) \in R^n \,\big|\, |x^i| < 1, \forall_i\}$, an open box around the origin.

If $\underset{\sim}{v}_p$ is any vector at p, define $\underset{\sim}{v}$ on U as follows. At points "on" the x^1-axis (via ϕ), define it by parallel transport along that axis.

For points in a line $L = \{(x^1,t,0,\ldots,0) \,\big|\, |t| < 1\}$, define it by parallel transport along L from $(x^1,0,\ldots,0)$ where it was defined in the first step.

Inductively, for points of the form $(x^1,\ldots,x^i,0,\ldots,0)$ define it by

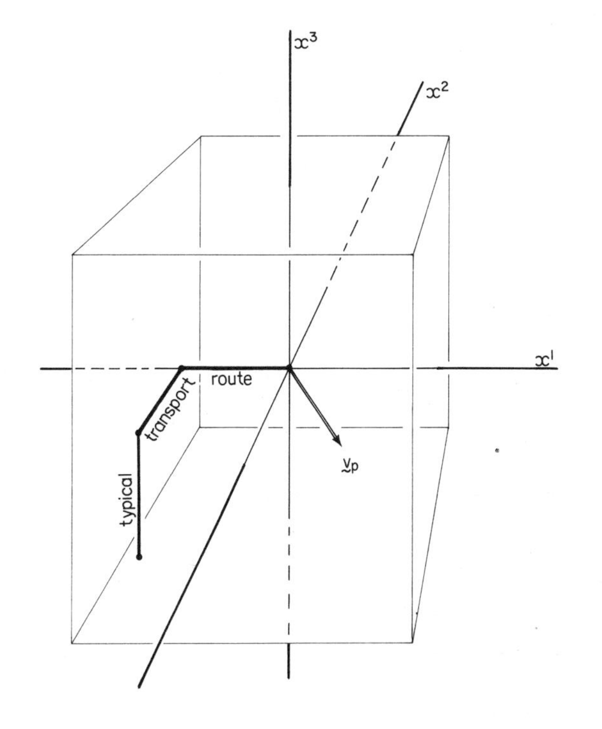

Fig. 2.4

parallel transport along $\{(x^1,\ldots,x^{i-1},t,0,\ldots,0) \,\big|\, |t| < 1\}$ from $(x^1,\ldots,x^{i-1},0,0)$, until $i=n$ (Fig. 2.4). Evidently $\underset{\sim}{v}$ is a smooth vector field with value $\underset{\sim}{v}_p$ at p; we claim it is parallel.

Now at points $(x^1,0,\ldots,0)$ we have $\nabla_{\partial_i}\underset{\sim}{v} = \underset{\sim}{0}$ for all i, by construction. At a point $(x^1,x^2,0,\ldots,0)$ we have only $\nabla_{\partial_1}\underset{\sim}{v}$ to check. Defining $S(t,s) = (x^1+t,x^2+s,0,\ldots,0)$ we have

$$\begin{aligned}
\underset{\sim}{0} &= \underset{\sim}{R}(S_t^*,S_s^*)\underset{\sim}{v} \quad \text{since } \underset{\sim}{R} = \underset{\sim}{0} \text{ by assumption} \\
&= \Delta_t(\Delta_s\underset{\sim}{v}) - \Delta_s(\Delta_t\underset{\sim}{v}) \quad \text{by 2.04} \\
&= -\Delta_s(\Delta_t\underset{\sim}{v}) \text{ since } \Delta_s\underset{\sim}{v} \text{ is } \underset{\sim}{0} \text{ by construction of } \underset{\sim}{v}.
\end{aligned}$$

Thus $\Delta_t\underset{\sim}{v}$ is parallel along $c : s \rightsquigarrow (x^1,x^2+s,0,\ldots,0)$; but at $c(-x^1) = (x^1,0,\ldots,0)$ it is zero by construction, so it is zero also at $(x^1,x^2,0,\ldots,0)$. But $\Delta_t\underset{\sim}{v}(t,s)$ is exactly $\nabla_{(\partial_1)_q}\underset{\sim}{v}$, where $q = S(t,s)$, so we have shown that $\nabla_{\partial_1}\underset{\sim}{v}$ is $\underset{\sim}{0}$.

Onward by induction: we parallel transport $\nabla_{\partial_1}\underset{\sim}{v},\ldots,\nabla_{\partial_i}\underset{\sim}{v}$ from $(x^1,\ldots,x^i,x^{i+1},0,\ldots,0)$ to $(x^1,\ldots,x^i,0,\ldots,0)$ where the first $(i-1)$ are zero by induction and the i-th by construction. All of $\nabla_{\partial_{i+1}}\underset{\sim}{v},\ldots,\nabla_{\partial_n}\underset{\sim}{v}$ vanish at $(x^1,\ldots,x^{i+1},0,\ldots,0)$ by construction.

Thus $\nabla_{\partial_i}\underset{\sim}{v} = \underset{\sim}{0}$ everywhere in U for all i, so by linearity at each point $q \in U$ we have $\nabla_{\underset{\sim}{u}}\underset{\sim}{v} = \underset{\sim}{0}$ for all $\underset{\sim}{u} \in T_qM$, so $\underset{\sim}{v}$ is parallel along all curves in U. Thus parallel transport of $\underset{\sim}{v}_p$ from p to $q \in U$ is independent of the curve in U.

But $\underset{\sim}{v}_p$ was arbitrary, hence U is globally flat. Finally, p was arbitrary so M is locally flat.

■

2.06 Corollary

A pseudo-Riemannian manifold has curvature identically zero if and only if around every point we can so choose coordinates that the g_{ij} become constant functions $\pm\delta_{ij}$.

Proof

Apply 1.05 and choose orthonormal coordinates on the affine space. ■

We next check the skew-symmetry and skew-self-adjointness we obtained loosely in 2.01 for $\underset{\sim}{R}$, along with two other properties that come from the symmetry of ∇.

2.07 Lemma

The Riemann tensor satisfies, for any $\underset{\sim}{u}$, $\underset{\sim}{v}$, $\underset{\sim}{w}$, $\underset{\sim}{x}$ on M,

A) $\underset{\sim}{R}(\underset{\sim}{u},\underset{\sim}{v}) = -\underset{\sim}{R}(\underset{\sim}{v},\underset{\sim}{u})$

B) $\underset{\sim}{R}(\underset{\sim}{u},\underset{\sim}{v})\underset{\sim}{w}\cdot\underset{\sim}{x} = -\underset{\sim}{R}(\underset{\sim}{u},\underset{\sim}{v})\underset{\sim}{x}\cdot\underset{\sim}{w}$

C) $\underset{\sim}{R}(\underset{\sim}{u},\underset{\sim}{v})\underset{\sim}{w} + \underset{\sim}{R}(\underset{\sim}{v},\underset{\sim}{w})\underset{\sim}{u} + \underset{\sim}{R}(\underset{\sim}{w},\underset{\sim}{u})\underset{\sim}{v} = \underset{\sim}{0}$

D) $\underset{\sim}{R}(\underset{\sim}{u},\underset{\sim}{v})\underset{\sim}{w}\cdot\underset{\sim}{x} = \underset{\sim}{R}(\underset{\sim}{w},\underset{\sim}{x})\underset{\sim}{u}\cdot\underset{\sim}{v}$

(A holds for the curvature tensor of any connection, C for that of any symmetric connection, B for that of any connection compatible with the metric. D requires that $\underset{\sim}{R}$ be the Riemann tensor.)

Proof

A) is an immediate consequence of the definition, since $\nabla_{\underset{\sim}{w}}$ depends linearly on $\underset{\sim}{w}$ and $[\underset{\sim}{u},\underset{\sim}{v}] = -[\underset{\sim}{v},\underset{\sim}{u}]$.

C) is almost as immediate. It suffices to prove it at any $p \in M$ for $\underset{\sim}{u}_p$, $\underset{\sim}{v}_p$, $\underset{\sim}{w}_p$, which we extend arbitrarily to commuting vector fields $\underset{\sim}{u}$, $\underset{\sim}{v}$, $\underset{\sim}{w}$ in a neighbourhood of p. (Alternatively, work with completely arbitrary $\underset{\sim}{u}$, $\underset{\sim}{v}$, $\underset{\sim}{w}$ and use the Jacobi identity (Ex. VII.7.6).) Thus, with the Lie bracket terms vanishing,

$$\begin{aligned} &\underset{\sim}{R}(\underset{\sim}{u},\underset{\sim}{v})\underset{\sim}{w} + \underset{\sim}{R}(\underset{\sim}{v},\underset{\sim}{w})\underset{\sim}{u} + \underset{\sim}{R}(\underset{\sim}{w},\underset{\sim}{u})\underset{\sim}{v} \\ &\quad = (\nabla_{\underset{\sim}{u}}\nabla_{\underset{\sim}{v}}\underset{\sim}{w}-\nabla_{\underset{\sim}{v}}\nabla_{\underset{\sim}{u}}\underset{\sim}{w})+(\nabla_{\underset{\sim}{v}}\nabla_{\underset{\sim}{w}}\underset{\sim}{u}-\nabla_{\underset{\sim}{w}}\nabla_{\underset{\sim}{v}}\underset{\sim}{u})+(\nabla_{\underset{\sim}{w}}\nabla_{\underset{\sim}{u}}\underset{\sim}{v}-\nabla_{\underset{\sim}{u}}\nabla_{\underset{\sim}{w}}\underset{\sim}{v}) \\ &\quad = (\nabla_{\underset{\sim}{u}}\nabla_{\underset{\sim}{v}}\underset{\sim}{w}-\nabla_{\underset{\sim}{v}}\nabla_{\underset{\sim}{w}}\underset{\sim}{u})+(\nabla_{\underset{\sim}{v}}\nabla_{\underset{\sim}{w}}\underset{\sim}{u}-\nabla_{\underset{\sim}{w}}\nabla_{\underset{\sim}{u}}\underset{\sim}{v})+(\nabla_{\underset{\sim}{w}}\nabla_{\underset{\sim}{u}}\underset{\sim}{v}-\nabla_{\underset{\sim}{u}}\nabla_{\underset{\sim}{v}}\underset{\sim}{w}) \qquad \text{by VIII.5.02} \\ &\quad = \underset{\sim}{0}. \end{aligned}$$

This result is called Bianchi's first identity.

B) we deduce as follows. Any linear operator $\underset{\sim}{A}$ is skew-self-adjoint if and only if $\underset{\sim}{A}\underset{\sim}{w}.\underset{\sim}{w} = \underset{\sim}{0}$ for all $\underset{\sim}{w}$, since this implies

$$\underset{\sim}{0} = \underset{\sim}{A}(\underset{\sim}{x}+\underset{\sim}{y}).(\underset{\sim}{x}+\underset{\sim}{y}) = \underset{\sim}{A}\underset{\sim}{x}.\underset{\sim}{x} + \underset{\sim}{A}\underset{\sim}{x}.\underset{\sim}{y} + \underset{\sim}{A}\underset{\sim}{y}.\underset{\sim}{x} + \underset{\sim}{A}\underset{\sim}{y}.\underset{\sim}{y} = \underset{\sim}{A}\underset{\sim}{x}.\underset{\sim}{y} + \underset{\sim}{A}\underset{\sim}{y}.\underset{\sim}{x}$$

and the converse is trivial. We can again suppose $\underset{\sim}{u}$, $\underset{\sim}{v}$ commuting, so it suffices to prove that

$$\underset{\sim}{R}(\underset{\sim}{u},\underset{\sim}{v})\underset{\sim}{w}.\underset{\sim}{w} = \underset{\sim}{0}$$

which is equivalent to

$$(\nabla_{\underset{\sim}{u}}\nabla_{\underset{\sim}{v}}\underset{\sim}{w}).\underset{\sim}{w} = (\nabla_{\underset{\sim}{v}}\nabla_{\underset{\sim}{u}}\underset{\sim}{w}).\underset{\sim}{w}$$

for commuting $\underset{\sim}{u}$, $\underset{\sim}{v}$, by applying definition 2.03. Now by VIII.6.03

$$\underset{\sim}{v}(\underset{\sim}{w}.\underset{\sim}{w}) = (\nabla_{\underset{\sim}{v}}\underset{\sim}{w}).\underset{\sim}{w} + \underset{\sim}{w}.\nabla_{\underset{\sim}{v}}\underset{\sim}{w} = 2(\nabla_{\underset{\sim}{v}}\underset{\sim}{w}).\underset{\sim}{w}$$

and $\underset{\sim}{u}(\underset{\sim}{v}(\underset{\sim}{w}.\underset{\sim}{w})) = \underset{\sim}{u}(2(\nabla_{\underset{\sim}{v}}\underset{\sim}{w}).\underset{\sim}{w}) = 2(\nabla_{\underset{\sim}{u}}\nabla_{\underset{\sim}{v}}\underset{\sim}{w}).\underset{\sim}{w} + 2\nabla_{\underset{\sim}{v}}\underset{\sim}{w}.\nabla_{\underset{\sim}{u}}\underset{\sim}{w}$, so $(\nabla_{\underset{\sim}{u}}\nabla_{\underset{\sim}{v}}\underset{\sim}{w}).\underset{\sim}{w} = \frac{1}{2}\underset{\sim}{u}(\underset{\sim}{v}(\underset{\sim}{w}.\underset{\sim}{w})) - \nabla_{\underset{\sim}{v}}\underset{\sim}{w}.\nabla_{\underset{\sim}{u}}\underset{\sim}{w}$.

Now $(\nabla_{\underset{\sim}{v}}\nabla_{\underset{\sim}{u}}\underset{\sim}{w}).\underset{\sim}{w} = \frac{1}{2}\underset{\sim}{v}(\underset{\sim}{u}(\underset{\sim}{w}.\underset{\sim}{w})) - \nabla_{\underset{\sim}{u}}\underset{\sim}{w}.\nabla_{\underset{\sim}{v}}\underset{\sim}{w}$ by similar reasoning

$= \frac{1}{2}\underset{\sim}{v}(\underset{\sim}{u}(\underset{\sim}{w}.\underset{\sim}{w})) - \nabla_{\underset{\sim}{v}}\underset{\sim}{w}.\nabla_{\underset{\sim}{u}}\underset{\sim}{w}$ ($\underset{\sim}{G}$ symmetric)

$= \frac{1}{2}\underset{\sim}{u}(\underset{\sim}{v}(\underset{\sim}{w}.\underset{\sim}{w})) - \nabla_{\underset{\sim}{v}}\underset{\sim}{w}.\nabla_{\underset{\sim}{u}}\underset{\sim}{w}$ since $\underset{\sim}{v}(\underset{\sim}{u}(f)) = \underset{\sim}{u}(\underset{\sim}{v}(f))$ $\forall f$,

by the assumption that $[\underset{\sim}{u},\underset{\sim}{v}] = \underset{\sim}{0}$

$= (\nabla_{\underset{\sim}{u}}\nabla_{\underset{\sim}{v}}\underset{\sim}{w}).\underset{\sim}{w}$ as required.

(But <u>why</u> it is true is that $\underset{\sim}{R}(\underset{\sim}{u},\underset{\sim}{v})$ is an "infinitesimal rotation", discussed in 2.01.)

D), a fact as simple as it is surprising (since the need to think about rotations with four directions to consider, and zero torsion to bear in mind, makes it less than easy to see geometrically why something so neat should hold), is an algebraic consequence of A, B and C. It is much more memorable than its proof, which is summarised as Ex. 3. ■

2.08 Components

As usual, just insert the ∂_i. Thus the components of $\underset{\sim}{R}$ are defined by

$$\underset{\sim}{R}(\partial_k,\partial_\ell)(\partial_j) = R^i_{jk\ell}\partial_i ,$$

so that $R^i_{jk\ell}$ is the i-th component of the image of ∂_j under $\underset{\sim}{R}(\partial_k,\partial_\ell)$.
(This order for the indices is a little odd in relation to the left-hand side, but is a lot older than the point of view which produced the latter and is utterly standard.) As important as the $R^i_{jk\ell}$ are the dot products in B and D above: in components we get

$$* \qquad (\underset{\sim}{R}(\partial_k,\partial_\ell)\partial_j).\partial_i = (R^h_{jk\ell}\partial_h).\partial_i = R^h_{jk\ell}(\partial_h.\partial_i) = g_{hi}R^h_{jk\ell} = R_{ijk\ell}$$

in the usual "lowered index" notation for an application of $\underset{\sim}{G}_\downarrow$ to one part of a tensor.

Though geometrically [illegible] primitive than the $R^i_{jk\ell}$, which are the components of what we geometrically decided curvature ought to be, the $R_{ijk\ell}$ carry the same information (since $R^i_{jk\ell} = g^{ih}R_{hjk\ell}$) and are manipulatively more convenient. The identities above become

A) $R_{ijk\ell} = -R_{ij\ell k}$

B) $R_{ijk\ell} = -R_{jik\ell}$

C) $R_{ijk\ell} + R_{ik\ell j} + R_{i\ell jk} = 0$ (Bianchi's first identity)

D) $R_{ijk\ell} = R_{k\ell ij}$ (using * for both sides and applying A and B).

(The $R^i_{jk\ell}$ express A and C equally well, but we need the ∂_i orthonormal for skew-symmetry to be just $R^i_{jk\ell} = -R^j_{ik\ell}$, and if $\underset{\sim}{R} \neq \underset{\sim}{0}$ we can't get them orthonormal everywhere, by 2.06. The + and - signs involved for general skew-self-adjointness, depending on whether pairs of vectors have the same type, complicate things further.)

Finding the components of $\underset{\sim}{R}$ in terms of those of ∇ we proceed as follows.

$$\begin{aligned}
R^i_{jk\ell}\partial_i &= \underset{\sim}{R}(\partial_k,\partial_\ell)\partial_j \\
&= \nabla_{\partial_k}\nabla_{\partial_\ell}\partial_j - \nabla_{\partial_\ell}\nabla_{\partial_k}\partial_j \quad \text{since } \partial_k, \partial_\ell \text{ commute} \\
&= \nabla_{\partial_k}(\Gamma^h_{\ell j}\partial_h) - \nabla_{\partial_\ell}(\Gamma^h_{kj}\partial_h) \quad \text{by definition of the } \Gamma^i_{jk} \\
&= \left(\partial_k(\Gamma^h_{\ell j})\partial_h - \Gamma^h_{\ell j}\nabla_{\partial_k}(\partial_h)\right) - \left(\partial_\ell(\Gamma^h_{kj})\partial_h - \Gamma^h_{kj}\nabla_{\partial_\ell}(\partial_h)\right), \quad \text{by VIII.3.01.Civ)}
\end{aligned}$$

$$= (\partial_k\Gamma^i_{\ell j}-\Gamma^h_{\ell j}\Gamma^i_{kh}-\partial_\ell\Gamma^i_{kj}+\Gamma^h_{kj}\Gamma^i_{\ell h})\partial_i,$$ changing dummy index on the 1st and 3rd terms.

So

$$R^i_{jk\ell} = \partial_k\Gamma^i_{\ell j} - \partial_\ell\Gamma^i_{kj} + \Gamma^h_{kj}\Gamma^i_{\ell h} - \Gamma^h_{\ell j}\Gamma^i_{kh},$$ and

$$R_{ijk\ell} = g_{ih}R^h_{jk\ell}$$

$$= g_{ih}(\partial_k\Gamma^i_{\ell j} - \partial_\ell\Gamma^i_{kj}) + g_{ih}(\Gamma^m_{kj}\Gamma^h_{\ell m} - \Gamma^m_{\ell j}\Gamma^h_{km})$$

$$= g_{ih}(\partial_k\Gamma^i_{\ell j} - \partial_\ell\Gamma^i_{kj}) + (\Gamma^m_{kj}\Gamma_{\ell mi} - \Gamma^m_{\ell j}\Gamma_{kmi})$$

in terms of the components of $\underset{\sim}{G}$ and ∇. Substitution from VIII.6.06 gives formulae in terms of the g_{ij} ,and their first and second partial derivatives, alone; but if the reader likeshairy formulae for their own sake that much, by now he has forsaken this book for another.

2.09 Independent components

The space of $\binom{1}{3}$-tensors on an n-dimensional vector space has dimension n^4. Thus in coordinates $\underset{\sim}{R}$ will have on a surface, 3-manifold or spacetime respectively, 16, 81 or 256 component functions. The relevant number is in fact somewhat smaller. For instance since $R_{ijk\ell} = -R_{jik\ell}$ we must always have $R_{iik\ell} = 0$. In fact (Ex. 4) the space of $\binom{1}{3}$-tensors at each $p \in M$ with the symmetries required of a curvature tensor has dimension $\frac{1}{12}n^2(n^2-1)$. There is no very natural choice of $\frac{1}{12}n^2(n^2-1)$ basis vectors in terms of the ∂_i, however, for general n.

2.10 Bianchi's second identity

One of the striking points in 2.01 was that the difference between parallel transport along two curves from p to q is given by the integral (suitably defined) of $\underset{\sim}{R}$ over <u>any</u> surface between them, independently of the choice of surface. This fact absolutely requires the "curvature form" viewpoint for its proof and clear exposition. But it should recall a familiar fact to students of electromagnetism: the integral of the "curl" of a given vector field over a surface in R^3 depends only on the boundary of the surface, by Stokes's Theorem. This holds moreover locally for the integral over the surface of any field $\underset{\sim}{v}$ whose divergence is zero, since this implies that $\underset{\sim}{v}$ is the curl of some field, at least locally. (And therefore, in topologically simple R^3, globally - an implication that does <u>not</u> hold in general, contrary to too many books.) So we have a condition on the derivatives of $\underset{\sim}{v}$, zero divergence, integrating to an "independence of surface" result. Something very similar applies here: $\underset{\sim}{R}$ is in a very precise sense a generalised "curl" of ∇, and the independence of its integral of the surface integrated over, for a fixed boundary, has as a local equivalent (analogous to "zero divergence") the condition

$$(\nabla_{\underset{\sim}{u}}\underset{\sim}{R})(\underset{\sim}{v},\underset{\sim}{w}) + (\nabla_{\underset{\sim}{v}}\underset{\sim}{R})(\underset{\sim}{w},\underset{\sim}{u}) + (\nabla_{\underset{\sim}{w}}\underset{\sim}{R})(\underset{\sim}{u},\underset{\sim}{v}) = \underset{\sim}{0}$$

for $\underset{\sim}{u}, \underset{\sim}{v}, \underset{\sim}{w} \in T^1M$. ($\nabla_{\underset{\sim}{u}}R$ etc are of course $\binom{1}{3}$-tensor fields like $\underset{\sim}{R}$, and like $\underset{\sim}{R}$ can be treated as maps $T^1M \times T^1M \to L(T^1M;T^1M)$.) In components, this becomes evidently

$$R^h_{ijk;\ell} + R^h_{ik\ell;j} + R^h_{i\ell j;k} = 0$$

which is the classical form of <u>Bianchi's second identity</u> (sometimes known simply as <u>Bianchi's identity</u>). The proof is straightforward, and left to the reader (Ex. 5). This equation is not obviously related to integrating curvature over surfaces, but in the "curvature form" context appropriate to such integration it does become so.

The reader is probably familiar with a number of conservation laws, and should note that they all have their differential and integral forms. For example if $\underset{\sim}{v}$ is a field of force it is locally equivalent to say "curl $\underset{\sim}{v} = \underset{\sim}{0}$" or that "work done in going along a path depends only on its boundary" (that is, on its end points), and if it is a static magnetic field the fact that div $\underset{\sim}{v} = \underset{\sim}{0}$ is equivalent to the fact that the magnetic flux through a surface depends only on the boundary ("between two surfaces with the same boundary, no lines of force get lost"). The similarity of all such laws, both in their differential and integral forms, cannot however be displayed without a coherent language for integration; it is part of the more perfect approach we mentioned in 2.01.

2.11 Definition

If the curvature tensor on M is constant in the sense of VIII.7.10, then we say M has constant curvature (cf. Ex. 6).

Exercises X.2

1) Suppose that $f(s,t)$ is defined for $(s,t) \in R^2$, $s,t > 0$, and there exists $x = \lim_{(s,t)\to(0,0)} f(s,t)$, in the sense of Ex. VII.1.1a (here R^2 has its usual topology, and f takes values in any Hausdorff space X).

Show that both of

$$\lim_{s\to 0}\left(\lim_{t\to 0} f(s,t)\right), \quad \lim_{t\to 0}\left(\lim_{s\to 0} f(s,t)\right)$$

must exist and be equal to x if the limits

$$\lim_{t\to 0} f(s,t), \quad \lim_{s\to 0} f(s,t)$$

exist whenever we fix s,t respectively near enough to 0.

2a) Show from the definition that

$$\underset{\sim}{R}(\underset{\sim}{u},\underset{\sim}{v})(\underset{\sim}{t}+\underset{\sim}{t}') = \underset{\sim}{R}(\underset{\sim}{u},\underset{\sim}{v})\underset{\sim}{t} + \underset{\sim}{R}(\underset{\sim}{u},\underset{\sim}{v})\underset{\sim}{t}'$$

$$\underset{\sim}{R}(\underset{\sim}{u},\underset{\sim}{v})(f\underset{\sim}{t}) = f\underset{\sim}{R}(\underset{\sim}{u},\underset{\sim}{v})\underset{\sim}{t}$$

for any vector fields $\underset{\sim}{u}$, $\underset{\sim}{v}$, $\underset{\sim}{t}$, $\underset{\sim}{t}'$ on M, $f : M \to R$. Deduce by expressing $\underset{\sim}{t}$ as $t^i\partial_i$, where $t^i : M \to R$, that $(\underset{\sim}{R}(\underset{\sim}{u},\underset{\sim}{v})\underset{\sim}{t})$ depends for fixed $\underset{\sim}{u}$, $\underset{\sim}{v}$ only on $\underset{\sim}{t}_p$, not $\underset{\sim}{t}$. Define $(\underset{\sim}{R}(\underset{\sim}{u},\underset{\sim}{v}))_p$, and show it to be linear.

b) Show that $(\underset{\sim}{u},\underset{\sim}{v}) \rightsquigarrow (\underset{\sim}{R}(\underset{\sim}{u},\underset{\sim}{v}))_p$ is bilinear, and depends only on $\underset{\sim}{u}_p$ and $\underset{\sim}{v}_p$. (If $\underset{\sim}{u}'$, $\underset{\sim}{v}'$ have $\underset{\sim}{u}'_p = \underset{\sim}{u}_p$, $\underset{\sim}{v}'_p = \underset{\sim}{v}_p$, $(\underset{\sim}{u}-\underset{\sim}{u}')_p = \underset{\sim}{0} = (\underset{\sim}{v}-\underset{\sim}{v}')_p$. Consider $\underset{\sim}{R}(\underset{\sim}{w},\underset{\sim}{x})$ where $\underset{\sim}{w}_p = \underset{\sim}{0} = \underset{\sim}{x}_p$, and use bilinearity.)

c) Deduce that $(\underset{\sim}{R}(f\underset{\sim}{u},g\underset{\sim}{v})h\underset{\sim}{t})_p = f(p)g(p)h(p)(\underset{\sim}{R}(\underset{\sim}{u},\underset{\sim}{v}))_p\underset{\sim}{t}_p$. Define $\underset{\sim}{R}_p$.

d) Show that any map

$$\mathbb{R} : T^1M \times T^1M \to L(T^1M;T^1M),$$

which preserves addition and satisfies

$$\bigstar \qquad \mathbb{R}(f\underset{\sim}{u},g\underset{\sim}{v})h\underset{\sim}{t} = fgh\,\mathbb{R}(\underset{\sim}{u},\underset{\sim}{v})\underset{\sim}{t}$$

for any f, g, h : M → R, can be specified by a $\binom{1}{3}$-tensor field.

e) Show that if $\mathbb{R}$ as in (d) satisfies

$$\mathbb{R}(\underset{\sim}{u},\underset{\sim}{v})\underset{\sim}{t} = \nabla_{\underset{\sim}{u}}\nabla_{\underset{\sim}{v}}\underset{\sim}{t} - \nabla_{\underset{\sim}{v}}\nabla_{\underset{\sim}{u}}\underset{\sim}{t}$$

when $\underset{\sim}{u}$ and $\underset{\sim}{v}$ commute, it satisfies

$$\mathbb{R}(\underset{\sim}{u},\underset{\sim}{v})\underset{\sim}{t} = \nabla_{\underset{\sim}{u}}\nabla_{\underset{\sim}{v}}\underset{\sim}{t} - \nabla_{\underset{\sim}{v}}\nabla_{\underset{\sim}{u}}\underset{\sim}{t} - \nabla_{[\underset{\sim}{u},\underset{\sim}{v}]}\underset{\sim}{t}$$

in general. (Express $\underset{\sim}{u}$, $\underset{\sim}{v}$ as sums of functions multiplying the ∂_i, which commute, and use ★.)

f) Discuss the relationship between checks like (d) and VIII.5.04, that something defined using values <u>around</u> p actually only uses values <u>at</u> p (and so defines a tensor), with the checks common in physics texts that a set of functions defined in terms of a coordinate system transforms correctly (and so defines a tensor).

3) Deduce from the first Bianchi identity that for any $\underset{\sim}{u}, \underset{\sim}{v}, \underset{\sim}{w}, \underset{\sim}{x} \in T^1M$

$$\text{E1} \quad \underset{\sim}{R}(\underset{\sim}{u},\underset{\sim}{v})\underset{\sim}{w}\cdot\underset{\sim}{x} + \underset{\sim}{R}(\underset{\sim}{v},\underset{\sim}{w})\underset{\sim}{u}\cdot\underset{\sim}{x} + \underset{\sim}{R}(\underset{\sim}{w},\underset{\sim}{u})\underset{\sim}{v}\cdot\underset{\sim}{x} = 0$$

Write down the corresponding E2, E3, E4 with first terms $\underset{\sim}{R}(\underset{\sim}{x},\underset{\sim}{u})\underset{\sim}{v}\cdot\underset{\sim}{w}$, $\underset{\sim}{R}(\underset{\sim}{w},\underset{\sim}{x})\underset{\sim}{u}\cdot\underset{\sim}{v}$, $\underset{\sim}{R}(\underset{\sim}{v},\underset{\sim}{w})\underset{\sim}{x}\cdot\underset{\sim}{u}$. Subtract (E3+E4) from (E1+E2) and use 2.07A,B to deduce that

$$\underset{\sim}{R}(\underset{\sim}{x},\underset{\sim}{u})\underset{\sim}{v}\cdot\underset{\sim}{w} = \underset{\sim}{R}(\underset{\sim}{v},\underset{\sim}{w})\underset{\sim}{x}\cdot\underset{\sim}{u} \quad \text{for any } \underset{\sim}{u}, \underset{\sim}{v}, \underset{\sim}{w}, \underset{\sim}{x}$$

which is just 2.07D rewritten.

4) Express 2.07A,B,C as the condition that $\underset{\sim}{R}_p$ be in the kernel of a suitable linear map $\underset{\sim}{S} : (T^1_3M)_p \to R^m$, where $m = \frac{n^2}{12}(11n^2+1)$. (Hint: one component of $\underset{\sim}{S}(\underset{\sim}{R}_p)$ might be $R^1_{212} + R^1_{221}$). Show that $\underset{\sim}{S}$ is surjective, and deduce from I.2.10 that the space of tensors satisfying 2.07A,B,C,D has dimension $\frac{n^2}{12}(n^2-1)$.

5a) If $\underset{\sim}{A} \in T^1_kM$ and $\underset{\sim}{u}, \underset{\sim}{v}_1, \ldots, \underset{\sim}{v}_n \in T^1M$, show that

$$\nabla_{\underset{\sim}{u}}\underset{\sim}{A}(\underset{\sim}{v}_1,\ldots,\underset{\sim}{v}_n) = \nabla_{\underset{\sim}{u}}(\underset{\sim}{A}(\underset{\sim}{v}_1,\ldots,\underset{\sim}{v}_n)) - \sum_{i=1}^{k} A(\underset{\sim}{v}_1,\ldots,\nabla_{\underset{\sim}{u}}\underset{\sim}{v}_i,\ldots,\underset{\sim}{v}_n)$$

from VIII.7.03.

b) Use (a) and 2.07 to prove the second Bianchi identity in the form without coordinates, or use 2.08 and IX.7.08 to prove it in components. Show that the two forms are equivalent.

6) If M has constant curvature $\underset{\sim}{R} \neq \underset{\sim}{0}$, can some chart make its components $R^i_{jk\ell}$ constant?

3. CURVED SURFACES

3.01 Gaussian curvature

When n=2 the formula of 2.09 for the dimension of the space of possible curvature tensors reduces to $(\frac{1}{12})4(4-1) = 1$, so on a surface $\underset{\sim}{R}$ is essentially just a number proportional to area at each point. This is in keeping with our discussion in 2.01; an "infinitesimal rotation" of a tangent plane is exactly a scalar "rate of turn" since rotations of a plane are so much simpler than those of higher spaces. This "rate of turn per area" or "density of rate of turn" is correspondingly much simpler than curvature in higher dimensions and its study is older, as witness its name of Gaussian curvature. (Riemann laid the foundations of tensor geometry a generation after Gauss's work on surfaces, and some of the ideas we have presented date from as recently as the 1950's.) Integrating this quantity is likewise much simpler. Even here we shall not cover the technicalities† in this volume but it

---

†Not hard to find, since so many books start with the geometry of surfaces, usually embedded in R^3. The best modern one is [do Carmo]. But in our view it is easier to motivate $\underset{\sim}{R}$ directly, and see why it reduces to a scalar on surfaces, than motivate Gaussian curvature separately by geometric ideas special to it and then let it explode into a fourth-degree tensor when n goes to 3. Moreover, many results such as Schur's Theorem (§5) are true only when n > 2, so surfaces do not illustrate them.

is worth mentioning Gauss's result on such integration.

The effect of parallel transport around a closed curve is a rotation of the tangent space through an angle equal to the integral of the curvature over any surface whose boundary it is, allowing for orientation. (It may be that there is no such surface, as for the curve c of Fig. 3.1. The study of when this happens is the beginning of homology theory, part of algebraic topology.) In particular,if the curve is a "geodesic triangle" (defined as "three points joined by geodesics" which reduces to meaning an ordinary triangle if the space is affine),

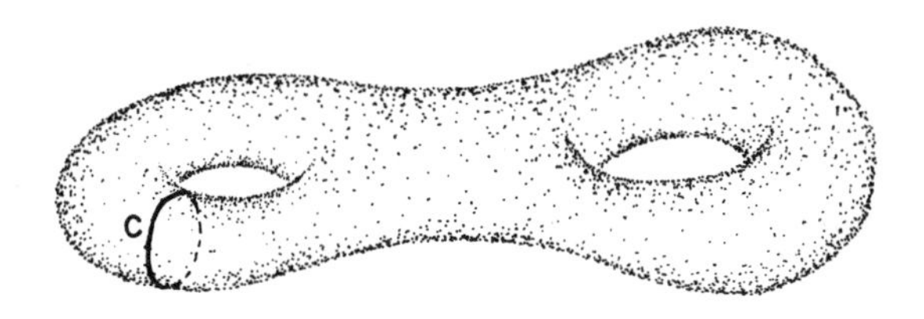

Fig. 3.1

we know how a tangent vector to a side is transported along the side and hence how <u>any</u> vector is, thanks to n=2. With its three jumps at corners (treatment of "piecewise smooth" curves is one of the technicalities we are skipping) it is thus clear that the tangent vector to the boundary turns through (Fig. 3.2) an angle of

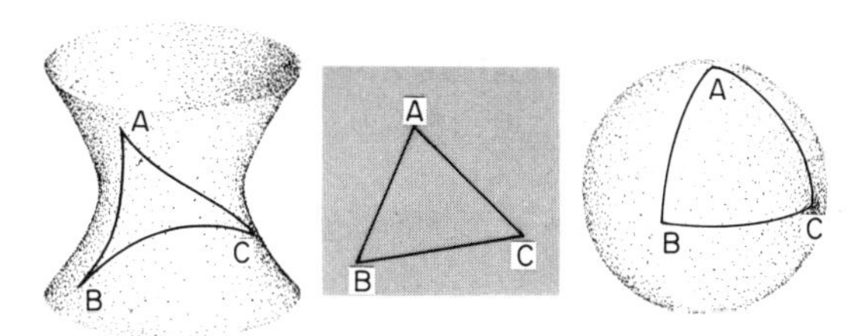

Fig. 3.2

$$\left((\pi-A) + (\pi-B) + (\pi-C) + \text{integral of curvature over inside of triangle}\right)$$

when we go round once, ending where it began. Since a turn of 2π is no turn, this gives Gauss's result

$$A + B + C = \pi + \text{integral}$$

which reduces to the Euclidean result (and is equivalent to the parallel postulate) for the plane with its usual, curvature-zero, metric, and generalises it to curved surfaces.

Notice that on a saddle or spindle shape the angle sum is less than π, corresponding to negative Gaussian curvature, and on a convex one such as a sphere it is greater. (Thus, on the Earth there is a very right-angled triangle with corners at the N. Pole, 0°N, 0°W and 0°N, 90°W, whose angle sum is $\frac{3}{2}\pi$; for negative curvature, test with string some geodesics on the middle of an adult human female chest.)

3.02 Deflection

Similarly, curvature can allow a "geodesic diangle" - two points joined by two distinct geodesics as in Ex. IX.4.6, Ex. 1.6 - and the integral of the curvature over the surface between them gives the amount one is "deflected" relative to the other. For example, suppose on R^2 we have a Riemannian metric with non-zero curvature (positive) only near the origin; such a metric is induced, for instance, by the embedding in Euclidean 3-space shown in Fig. 3.3. Then the two geodesics shown meet twice, despite each lying entirely in a flat part of the space, because of the curvature between them. Thus curvature in a small region can affect the global geometry of the whole space. It is the four-dimensional, pseudo-Riemannian analogue of this effect that is being analysed in the

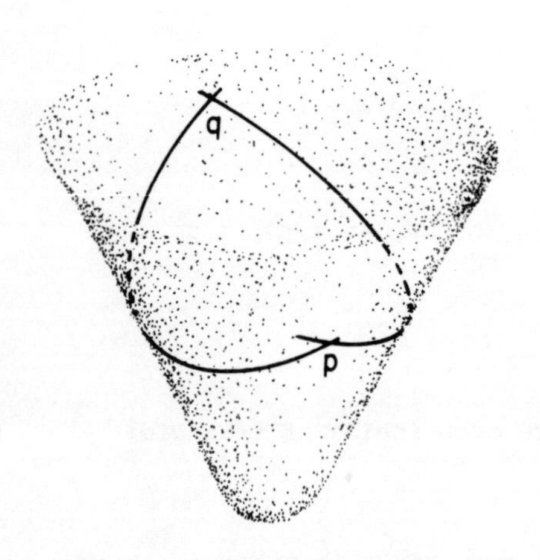

Fig. 3.3

extraordinarily careful measurements of stellar positions made at every solar eclipse, since general relativity links the presence of matter to curvature (Ch. XII). Since the amount of curvature imposed by even a solar mass is not great, two geodesics that meet twice are not much "deflected", hence are never very far apart, hence for there to be much curvature "between them" they must pass near the central region, which is why light following them is lost in sunlight except at eclipse. Indeed, since Earth is too close to the sun for null geodesics from the same star passing on either side to meet here, at least one must be blocked. So the deflection cannot be measured directly as a difference, but only as a change of apparent position; a very delicate job for such a small change.

This foretaste of general relativistic effects via analogy with ordinary surfaces is offered in hopes that the reader may find it helpful, but the reader should treat it with caution. Firstly, because of the differences between Riemannian and pseudo-Riemannian geometry (cf. the examples in IX §5, §6), though with due care the above discussion of triangles can be made precise almost as easily in the pseudo-Riemannian case. There one defines the "angle turned through" with the aid of cosh rather than cos (cf. IX. §6). Secondly, and more importantly, consider the differences between surfaces and higher manifolds. We cannot for instance have a <u>small</u> difference between the directions of two null geodesics through a point in two dimensions as we can in three. Moreover Fig. 3.3 shows an effect simpler than gravitation can be, since in general the presence of specified curvature within a bounded region is incompatible with having zero curvature outside it. This is exactly because of the independence of spanning surface we remarked on in 2.10. In two dimensions this independence is fairly trivial. (Even in a - physically uninteresting - <u>compact</u> 2-manifold we can find at most two spanning surfaces, and in a non-compact case at most one.) But in R^3, for instance, if non-zero curvature is confined to the unit ball $B = \{(x,y,z) \mid x^2+y^2+z^2 < 1\}$ any two curves from p to q that lie entirely outside B have a spanning surface also entirely outside B. On this the curvature to be integrated is identically zero (hence parallel transport along the two curves gives the same results.) Thus the integral must vanish over <u>any</u> surface between them, even one that <u>does</u> go through B, which severely limits the curvature we can have on B. A straightforward 3-dimensional analogue of Fig. 3.3 is thus impossible. We could allow curvature in a region $C = \{(x,y,z) | x^2+y^2 < 1\}$ with a metric on each z=constant plane like that induced on one plane in Fig. 3.3 (though to realise it by an embedding we would need four flat dimensions) and get "deflected" geodesics as in Fig. 3.4 (Ex. 2). The independence of surface of the integral of curvature between f and g then says that whether we go through C high or low we get the same answer. This is reasonable, smacking of a conservation law for whatever is causing the curvature in C, if we think of z as time. But in a four-dimensional spacetime, any analogue that confines the curvature to regions that are non-compact only in time keeps it dodgeable by spanning surfaces.

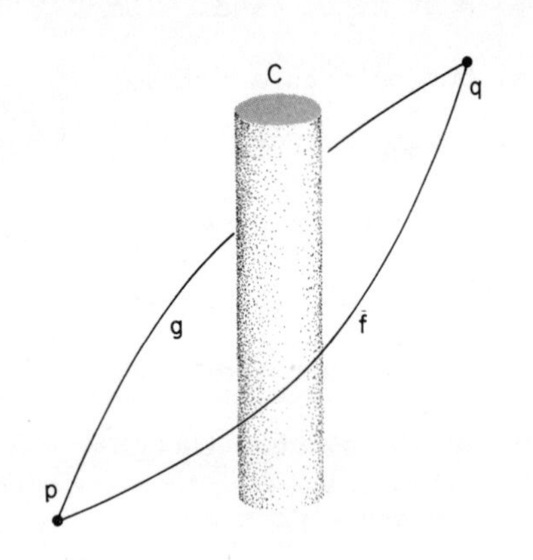

Fig. 3.4

We cannot then expect to describe gravity simply by making curvature a function of matter, vanishing in empty space. In the dimensions we live in, this would mean that geodesics outside a ball of matter would meet twice, or not, independently of whether the matter was there. Hardly a model for gravitation, which can make satellites going East meet satellites going West in a regular fashion which is clearly related to the presence of the Earth.

Exercises X.3

1) Use Gauss's theorem about geodesic triangles and prove or assume that any Riemannian surface can be broken up into such triangles (Fig. 3.5) to prove the Gauss-Bonnet Theorem: the integral of curvature over a compact surface S is equal to 2π times the Euler characteristic (v-e+f) of S, where v is the number of vertices, e the number of edges, and f the number of triangular faces. (Euler first showed this number to be 2 for any such subdivision of S^2: it is independent of the triangulation for any surface.)

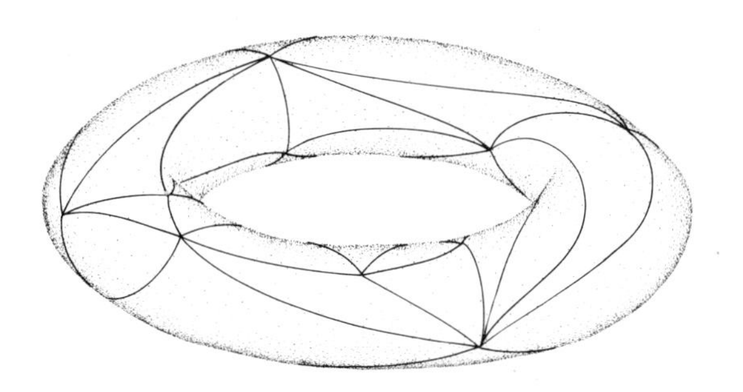

Fig. 3.5

This incidentally gives another proof of 1.04: since the curvature must integrate to 4π it cannot be everywhere zero, so S^2 cannot be locally flat. Of all compact surfaces, in fact, only the torus and Klein bottle have Euler characteristic zero, so only these are possibilities for locally flat metrics; both in fact can possess them.

The Gauss-Bonnet Theorem is the earliest of all results relating curvature to an algebraic-topological quantity.

2a) Let (r,θ,z) be cylindrical coordinates on R^3. Show that the Euclidean metric on R^3 is given by the line element

$$(ds)^2 = (dz)^2 + r^2(d\theta)^2 + (dr)^2$$

where $r \neq 0$. (cf. Ex. IX.5.1)

b) Let $f : R \to R$ be smooth and satisfy $f(0) = 1$, $f(x) = .99$ if $x^2 \geq 1$.

(Hint : try $f(x) = \frac{1}{100}(.99 + e^{\left(\frac{x^2}{x^2-1}\right)})$, for $x^2 < 1$.)

Define a metric $\underset{\sim}{\tilde{G}}$ on R^3 by the line element formula

$$(ds)^2 = (dz)^2 - r^2 f(r)(d\theta)^2 - (dr)^2 \quad \text{for } r \neq 0.$$

Show that it extends smoothly to points where $r = 0$ (though this particular coordinate representation does not). Give its form in (x,y,z) coordinates.

c) Compute the Levi-Cività Γ^i_{jk} for $\underset{\sim}{\tilde{G}}$, and the components of the Riemann tensor, and give the geodesic equation in terms of r, θ, and z only, at points where $r > 1$.

d) Give explicitly a typical geodesic $c : R \to R^3$ with respect to $\underset{\sim}{\tilde{G}}$ that does not pass through $c = \{(r,\theta,z) | r \leq 1\}$.
(Hint: reduce the problem to that of geodesics in the plane $z = 0$, using Ex. IX.1.4. Prove that the geometry of this plane is that of the surface of Fig. 3.3 in Euclidean 3-space, up to the sign of the metric, where the conical part is generated by lines at $\cos^{\leftarrow}(.99)$ to the z-axis. Find a map from $U \subseteq R^2$ to this cone, $(r,\theta) \rightsquigarrow (r, \frac{100}{99}\theta, ?) \in R^3$, by which straight lines correspond to geodesics.)

e) There are two distinct timelike geodesics with this metric, from $(50,0,0)$ to $(60,\pi,200)$, that do not pass through C.

f) There are no geodesics in this manifold with domain R and image contained in $\{(r,\theta,z) | 1 < r < k\}$, for any $k \in R$.

g) Show that while passing geodesics are "deflected by C", a curve of the form $t \rightsquigarrow (r,\theta,t)$ for r,θ fixed is a geodesic. (This geometry does not make things "fall".)

3) Gauss was so pleased to discover that his curvature depends only on a surface's metric tensor - not on its embedding in R^3 - he called the result his

remarkable theorem, or Theorema Egregium. The name has stuck. Why is VIII.6.07 a generalisation of the Theorema Egregium?

4. GEODESIC DEVIATION

Suppose in a spacetime we have F, a parametrised surface (Def^n. IX.3.03) with each s-constant curve F_s a timelike geodesic parametrised by arc length, and each F_t a null geodesic. Then suppose an observer Q following $F_{s'}$ of Fig. 4.1 is watching a particle P following F_s. He watches via light rays. That is, information about P at point A_0 in spacetime reaches Q at A_1, by parallel transport along F_t. Information about P at A_0' comes similarly to Q at A_1' along $F_{t'}$. Suppose Q's interest is in P's velocity F_s^* (That is, P's spacetime velocity, which can be turned into a "space per time" velocity by a choice of chart - alias frame of reference, sometimes.) This by definition of "geodesic" is parallel transported along F_s. Then we see that the change he records is exactly the difference between parallel transport $F_s(t)$ to A_1' via A_1 or A_0', since Q "remembers" the previous value by parallel transport along his own world line $F_{s'}$. So the total acceleration Q perceives in P between A_0 and A_0' is given by the "total curvature", in the sense of 2.01, of the piece of the surface bounded by the four bits of geodesic.

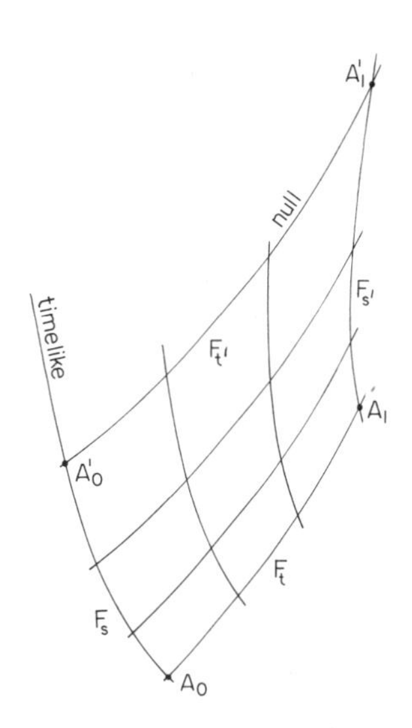

Fig. 4.1

Going to the limit, then, the relative acceleration of two "infinitely close" geodesics is given directly by the curvature tensor. (In practice this means "sufficiently close"; consider the definition of limits, and the bars to absolute accuracy of measurement.) In particular $\underset{\sim}{R}$ tells us whether geodesics are locally inclined to approach or separate; it also describes how they move around each other.

With a number of "sufficiently close" geodesics passing near a point, conversely, $\underset{\sim}{R}$ itself can be determined to any given level of accuracy.

Physically this means that for a bunch of freely-falling particles with negligible gravitational effect on each other, the relative motions (produced in Newtonian terms by "tidal forces") which spread out, flatten, rotate etc. the bunch are in this description determined by the local value of $\underset{\sim}{R}$, and themselves suffice - given enough particles - to determine $\underset{\sim}{R}$ in their vicinity. The curvature tensor is thus a physical "quantity" which can be directly determined from measurements in the neighbourhood of a point, not just a construct from the metric tensor $\underset{\sim}{G}$. (It would be hard to measure $\underset{\sim}{G}$ accurately enough locally to get useful experimental values for the second derivatives of it involved in $\underset{\sim}{R}$.)

Computing the "relative motion" of two actual geodesics, for which we could take the limit as they approach, involves integral techniques. We therefore defer the precise treatment of geodesic deviation.

5. SECTIONAL CURVATURE

Computation with the whole curvature tensor is somewhat unwieldy in components, particularly as 2.09 cannot in general be utilised in any convenient way to reduce the number of component functions to $\frac{n^2}{12}(n^2-1)$. That the $2n^2(n-1)$ functions $R_{iik\ell}$ and R_{ijkk} are necessarily zero allows us to restrict attention to the other $n^2(n-1)^2$ of the $R_{ijk\ell}$, but on a 4-D spacetime this still leaves 144 functions. For the rest of this chapter, then, we consider ways of "cutting down" the Riemann tensor to get various kinds of significant information.

5.01 Definition

One of these ways is, since curvature is so much simpler for surfaces, to consider the surfaces S in M passing through the point $p \in M$ of interest: what curvature does that of M impose on them? To avoid curvature that is not imposed, we want a curve in S through p to be as straight as M permits - a geodesic. Thus we consider the images under the exponential map $\exp_p$ of non-degenerate planes (2-dimensional vector subspaces) of T_pM, and examine their curvature with the induced metric. This turns out to be, at p, a restriction of the curvature tensor on M. More precisely, $\underset{\sim}{u}_p, \underset{\sim}{v}_p, \underset{\sim}{w}_p, \underset{\sim}{x}_p$ vectors in such a plane $P \subseteq T_pM$; $\underset{\sim}{R}_M$ is the Riemann tensor of the metric $\underset{\sim}{G}$ on M, and $\underset{\sim}{R}_p$ is the Riemann tensor on P of the metric (non-degenerate in a neighbourhood of $\underset{\sim}{0} \in P$) given by

$$* \qquad \underset{\sim}{t}.\underset{\sim}{s} = \underset{\sim}{G}(\underset{\sim}{D}_x \exp_p(\underset{\sim}{t}), \underset{\sim}{D}_x \exp_p(\underset{\sim}{s})) \quad \text{for } x \in P,\ \underset{\sim}{t}, \underset{\sim}{s} \in T_xP.$$

Then we have

$$** \qquad (\underset{\sim}{R}_p(\underset{\sim}{0})(\underset{\sim}{u}_p, \underset{\sim}{v}_p))\underset{\sim}{w}_p.\underset{\sim}{x}_p = (\underset{\sim}{R}_M(p)(\underset{\sim}{u}_p, \underset{\sim}{v}_p))\underset{\sim}{w}_p.\underset{\sim}{x}_p$$

considering the vectors involved as on the left tangent vectors to P at $\underset{\sim}{0}$, on the right to M at p, in the natural way. (Ex. 1)

Now if $\underset{\sim}{u}_p, \underset{\sim}{v}_p$ are not linearly independent, both sides vanish, by skew-symmetry. If they are, we know all such $\underset{\sim}{R}_p(\underset{\sim}{u}_p, \underset{\sim}{v}_p)\underset{\sim}{w}_p.\underset{\sim}{x}_p$ if we know $\underset{\sim}{R}_p(\underset{\sim}{u}_p, \underset{\sim}{v}_p)\underset{\sim}{u}_p.\underset{\sim}{v}_p$, by skew-self-adjointness, since $\underset{\sim}{u}_p, \underset{\sim}{v}_p$ form a basis for $T_{\underset{\sim}{0}}P$. (Just write $\underset{\sim}{w}_p = w^1\underset{\sim}{u}_p + w^2\underset{\sim}{v}_p$, $\underset{\sim}{x}_p = x^1\underset{\sim}{u}_p + x^2\underset{\sim}{v}_p$ and expand.) So only this matters for the curvature of $\exp_p(P)$.

In 2.01 we used "proportional to area of a parallelogram" as a way of seeing "bilinear and skew-symmetric in the vectors defining the parallelogram" geometrically. This is legitimate even without a specific measure of area in mind, since we can only change our measure by scalar multiplication, which leaves such proportionality intact. But in a plane P with a metric tensor we do have a natural measure of area. Namely, choose any orthonormal basis $\underset{\sim}{b}_1$, $\underset{\sim}{b}_2$ for P and set the area of the parallelogram defined by $\underset{\sim}{u}$, $\underset{\sim}{v}$ equal to the determinant of the map $P \to P$ defined by $\alpha\underset{\sim}{b}_1 + \beta\underset{\sim}{b}_2 \mapsto \alpha\underset{\sim}{u} + \beta\underset{\sim}{v}$. (cf. I.3.05 et seq. By Ex. IV.2.2b it is immediate that this does not depend on the choice of orthonormal $\underset{\sim}{b}_1$, $\underset{\sim}{b}_2$ except up to sign, which will disappear in the squaring that follows.) Call this area $\|\underset{\sim}{u},\underset{\sim}{v}\|$. It then follows from skew-symmetry and skew-self-adjointness that the number

$$k(P) = \frac{\underset{\sim}{R}(\underset{\sim}{u},\underset{\sim}{v})\underset{\sim}{u}.\underset{\sim}{v}}{\|\underset{\sim}{u},\underset{\sim}{v}\|^2}$$

depends only on the plane P defined by $\underset{\sim}{u}$ and $\underset{\sim}{v}$ in T_pM, if they are linearly independent (Ex. 2). (If $\underset{\sim}{u} = \lambda\underset{\sim}{v}$, neither P nor this number is defined). We call k(P) the sectional curvature of M at p for the section $P \subseteq T_pM$.

Now suppose we know k(P) for any P. This determines $\underset{\sim}{R}(\underset{\sim}{u},\underset{\sim}{v})\underset{\sim}{u}.\underset{\sim}{v}$ for any pair $\underset{\sim}{u}$, $\underset{\sim}{v}$, by

$$\underset{\sim}{R}(\underset{\sim}{u},\underset{\sim}{v})\underset{\sim}{u}.\underset{\sim}{v} = \|\underset{\sim}{u},\underset{\sim}{v}\|^2 \; k(P(\underset{\sim}{u},\underset{\sim}{v}))$$

where $P(\underset{\sim}{u},\underset{\sim}{v})$ is the plane of $\underset{\sim}{u}$ and $\underset{\sim}{v}$. Then this determines $\underset{\sim}{R}(\underset{\sim}{u},\underset{\sim}{v})\underset{\sim}{w}.\underset{\sim}{x}$ for any $\underset{\sim}{w},\underset{\sim}{x} \in T_pM$ (not just in $P(\underset{\sim}{u},\underset{\sim}{v})$ as above), as a consequence of the symmetry 2.07D of $\underset{\sim}{R}(\underset{\sim}{u},\underset{\sim}{v})\underset{\sim}{w}.\underset{\sim}{x}$ in the pairs $(\underset{\sim}{u},\underset{\sim}{v})$ and $(\underset{\sim}{w},\underset{\sim}{x})$. This fact is very similar to the polarization identity (Ex. IV.1.7d: if a bilinear form is symmetric we can express its value on any pair $(\underset{\sim}{x},\underset{\sim}{y})$ in terms of its values on pairs of the form $(\underset{\sim}{z},\underset{\sim}{z})$) but it is somewhat more awkward to construct a formula as we are dealing with "bivectors" rather than vectors. Thus the fact that the $\underset{\sim}{R}(\underset{\sim}{u},\underset{\sim}{v})\underset{\sim}{u}.\underset{\sim}{v}$ determine the $\underset{\sim}{R}(\underset{\sim}{u},\underset{\sim}{v})\underset{\sim}{w}.\underset{\sim}{x}$ (and hence the function k on the space of planes determines $\underset{\sim}{R}$) is most quickly established less directly (Ex. 3).

(The analogy with the polarization identity helps to avoid a curiously common error. A symmetric bilinear form $\underset{\sim}{A}$ on X has its values $\underset{\sim}{A}(\underset{\sim}{x},\underset{\sim}{y})$ fixed if we know <u>all</u> $\underset{\sim}{A}(\underset{\sim}{z},\underset{\sim}{z})$, $\underset{\sim}{z} \in X$. A somewhat smaller set of $\underset{\sim}{z}$ will suffice, but it is <u>not</u> enough to know only $\underset{\sim}{A}(\underset{\sim}{b}_i,\underset{\sim}{b}_i)$, $i = 1,\ldots,n$ for a given basis $\underset{\sim}{b}_1,\ldots,\underset{\sim}{b}_n$ (Ex. 4). In exactly the same way, knowing all the $\underset{\sim}{R}(\underset{\sim}{u},\underset{\sim}{v})\underset{\sim}{u}.\underset{\sim}{v}$ suffices to know $\underset{\sim}{R}$ completely, but knowing all the $\underset{\sim}{R}(\partial_i,\partial_j)\partial_i.\partial_j = -R_{ijij}$ does not suffice to determine $R_{ijk\ell}$ in general. This should be clear dimensionally. The $\frac{n}{2}(n-1)$ numbers $R_{ijij} = R_{jiji}$ can hardly suffice if $n > 2$ to fix a point in the $\frac{n^2}{12}(n^2-1)$-dimensional space of possible curvature tensors. But it is nice to have a simpler analogue of the way they fail.)

Thus the $\binom{1}{3}$-tensor field $\underset{\sim}{R}$ on M can be expressed in terms of a $\binom{0}{0}$-tensor field (real-valued function) on the $\frac{n}{2}(n+1)$-dimensional manifold of planes in the tangent spaces of M. It is often easier to deal with functions than with tensors, and there is a surprising result that concerns k directly:

<u>5.02 Schur's Theorem</u>

If $\dim M \geq 3$ and we have a function $\kappa : M \to R$ such that for any non-degenerate plane $P \subseteq T_pM$ we have $k(P) = \pm\kappa(p)$ according as $\underset{\sim}{G}$ induces a definite or indefinite metric on P, then κ is constant. (Thus, if sectional curvature is at every point independent of all but the signature of the section it is independent also of the point.)

<u>Proof</u>

Ex. 5. ■

<u>5.03 Corollary</u>

M satisfying the above conditions is of constant curvature (Def^n. 2.11). ■

It is a little surprising that 5.02 holds even for M pseudo-Riemannian. The definition of k(P) required P non-degenerate (so that we could find for P an orthonormal basis), and in a pseudo-Riemannian manifold of dimension > 2 every tangent space contains degenerate planes. On these we have no more a natural notion of "area" than we have of "length" on a null line, so that while we can still reduce $\underset{\sim}{R}$ to its values of the form $\underset{\sim}{R}(\underset{\sim}{u},\underset{\sim}{v})\underset{\sim}{u}\cdot\underset{\sim}{v}$ we cannot always further reduce these to a sectional curvature. (Note that $P(\underset{\sim}{u},\underset{\sim}{v})$ may be degenerate independently of whether either or both of $\underset{\sim}{u},\underset{\sim}{v}$ are null.) But it turns out that there are not, in a topological sense, "enough" planes where k is undefined, to block the theorem.

Exercises X.5

1) Prove the equation ** of 5.01. (You need to relate the connection on M for $\underset{\sim}{G}$ to that on P for the metric given by *. Notice that in general, M-parallel transport of tangent vectors to a surface S in M need not agree with S-parallel transport, along a curve in S - think of $M = R^3$, $S = S^2$ - thus there is something to prove. This is one place where a component argument, with the right chart, has its advantages.)

2) If $\underset{\sim}{w} = w^1\underset{\sim}{u} + w^2\underset{\sim}{v}$ and $\underset{\sim}{x} = x^1\underset{\sim}{u} + x^2\underset{\sim}{v}$ are linearly independent, then

$$\frac{\underset{\sim}{R}(\underset{\sim}{w},\underset{\sim}{x})\underset{\sim}{w}\cdot\underset{\sim}{x}}{\|\underset{\sim}{w},\underset{\sim}{x}\|^2} = \frac{\underset{\sim}{R}(\underset{\sim}{u},\underset{\sim}{v})\underset{\sim}{u}\cdot\underset{\sim}{v}}{\|\underset{\sim}{u},\underset{\sim}{v}\|^2} .$$

3a) Any $\binom{1}{3}$-tensor $\underset{\sim}{S}$ on a metric vector space X which satisfies A, B, C (and hence also D) of 2.07, together with

$$\underset{\sim}{S}(\underset{\sim}{u},\underset{\sim}{v})\underset{\sim}{u}\cdot\underset{\sim}{v} = 0, \quad \forall \underset{\sim}{u},\underset{\sim}{v} \in X,$$

is identically zero.

b) Deduce that if two $\binom{1}{3}$-tensors $\underset{\sim}{Q}$, $\underset{\sim}{R}$ on X satisfying A, B, C of 2.07 have

$$\underset{\sim}{Q}(\underset{\sim}{u},\underset{\sim}{v})\underset{\sim}{u}.\underset{\sim}{v} = \underset{\sim}{R}(\underset{\sim}{u},\underset{\sim}{v})\underset{\sim}{u}.\underset{\sim}{v} \quad \forall\ \underset{\sim}{u},\underset{\sim}{v} \in X$$

then $\underset{\sim}{Q} = \underset{\sim}{R}$.

c) Deduce that $\underset{\sim}{Q} = \underset{\sim}{R}$ even if "$\forall\ \underset{\sim}{u},\underset{\sim}{v} \in X$" is replaced by "$\forall\ \underset{\sim}{u},\underset{\sim}{v} \in X$ linearly independent and such that the plane containing $\underset{\sim}{u},\underset{\sim}{v}$ is non-degenerate". (Hint: take a sequence of non-degenerate planes tending to a typical degenerate.)

4) Show that on R^2 with the usual basis $\underset{\sim}{e}_1$, $\underset{\sim}{e}_2$, if we define $\underset{\sim}{G}((x^1,x^2),(y^1,y^2)) = x^1y^1+x^2y^2$, $\underset{\sim}{H}((x^1,x^2),(y^1y^2)) = x^1(y^1+\frac{1}{2}y^2)+x^2(\frac{1}{2}y^1+y^2)$, then both $\underset{\sim}{G}$ and $\underset{\sim}{H}$ are symmetric bilinear forms (in fact inner products) with $\|\underset{\sim}{e}_1\|_{\underset{\sim}{G}} = \|\underset{\sim}{e}_1\|_{\underset{\sim}{H}} = \|\underset{\sim}{e}_2\|_{\underset{\sim}{G}} = \|\underset{\sim}{e}_2\|_{\underset{\sim}{H}} = 1$, but $\underset{\sim}{G} \neq \underset{\sim}{H}$. Draw the sets $\{\underset{\sim}{v} \mid \|\underset{\sim}{v}\|_{\underset{\sim}{G}} = 1\}$, $\{\underset{\sim}{v} \mid \|\underset{\sim}{v}\|_{\underset{\sim}{H}} = 1\}$.

5a) Prove that $\|\underset{\sim}{u},\underset{\sim}{v}\|^2 = (\underset{\sim}{u}.\underset{\sim}{u})(\underset{\sim}{v}.\underset{\sim}{v}) - (\underset{\sim}{u}.\underset{\sim}{v})^2 = (g_{ac}g_{bd}-g_{ad}g_{bc})u^av^bu^cv^d$, if $\underset{\sim}{G}$ restricts to a definite metric on the plane of $\underset{\sim}{u}$ and $\underset{\sim}{v}$, minus this if indefinite.

b) Use Ex. 3c to show that if for all non-degenerate planes $P \subseteq T_pM$, $k(P) = \kappa(p)$ for P entirely spacelike, and $k(P) = -\kappa(p)$ otherwise, then

$$R(\underset{\sim}{u},\underset{\sim}{v})\underset{\sim}{w}.\underset{\sim}{x} = \kappa(p)[(\underset{\sim}{u}.\underset{\sim}{w})(\underset{\sim}{v}.\underset{\sim}{x})-(\underset{\sim}{u}.\underset{\sim}{x})(\underset{\sim}{v}.\underset{\sim}{w})],$$

or in coordinates

$$R_{ijk\ell} = \kappa(p)(g_{i\ell}g_{jk}-g_{ik}g_{j\ell}),\quad R^i_{jk\ell} = \kappa(p)(\delta^i_\ell g_{jk}-\delta^i_k g_{j\ell}).$$

c) Use (b), Ricci's Lemma (VIII.7.06), and the second Bianchi identity to show that if $\dim M \geq 3$, $\partial_h\kappa = 0$ for all h. Deduce that κ is constant.

d) Why is the assumption of the theorem trivially true for a surface? Give a counter-example in this dimension.

6) Show that the converse of 5.03 is false by proving that $S^2 \times R$ with its usual Riemannian metric has constant curvature, but $k(P)$ is not independent of $P \subseteq T_pM$.

6. RICCI AND EINSTEIN TENSORS

In this section we "cut down" the Riemann tensor to the part of the curvature that we shall associate with the presence of matter at a point, in Chapter XII.

6.01 Definition

The Ricci transformation at p with respect to a tangent vector $\underset{\sim}{v} \in T_pM$ is the map

$$\underset{\sim}{R}_{\underset{\sim}{v}} : T_pM \to T_pM : \underset{\sim}{u} \rightsquigarrow \underset{\sim}{R}(\underset{\sim}{u},\underset{\sim}{v})\underset{\sim}{v} \quad .$$

Evidently this is linear for each $\underset{\sim}{v}$, and if we know $\underset{\sim}{R}$ for all $\underset{\sim}{v}$ we know in particular

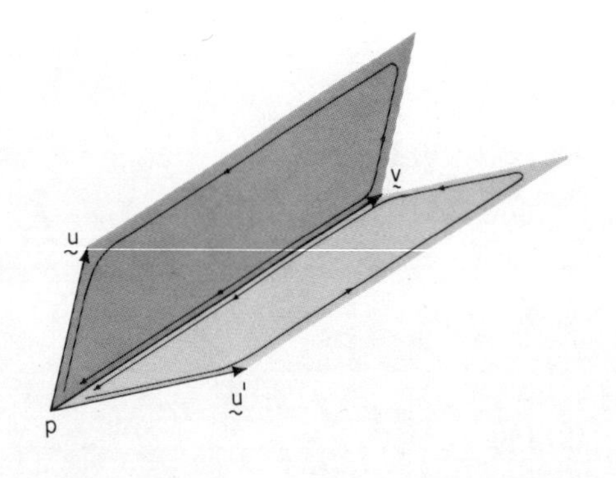

Fig. 6.1

$$-\underset{\sim}{R}_{\underset{\sim}{u}}(\underset{\sim}{v})\cdot\underset{\sim}{v}=\underset{\sim}{R}(\underset{\sim}{u},\underset{\sim}{v})\underset{\sim}{u}\cdot\underset{\sim}{v}=-\underset{\sim}{R}(\underset{\sim}{u},\underset{\sim}{v})\underset{\sim}{v}\cdot\underset{\sim}{u}=-\underset{\sim}{R}_{\underset{\sim}{v}}(\underset{\sim}{u})\cdot\underset{\sim}{u}$$

for all $\underset{\sim}{u},\underset{\sim}{v}$, hence by Ex. 5.3 the Ricci transformations suffice to determine $\underset{\sim}{R}$.

Geometrically, "$\underset{\sim}{R}_{\underset{\sim}{v}}$ takes each $\underset{\sim}{u}$ to the difference produced in $\underset{\sim}{v}$ by parallel transport around the infinitesimal parallelogram fixed by $\underset{\sim}{u}$ and $\underset{\sim}{v}$ (Fig. 6.1)" - a statement the reader is left to make precise by an appropriate reexpression in terms of limits.

Thus $\underset{\sim}{R}_{\underset{\sim}{v}}$ gives us a measure of how curved M is in each of the planes containing $\underset{\sim}{v}$, and hence it is a kind of "curvature along $\underset{\sim}{v}$". But it is still a bit elaborate for many purposes, so we sacrifice some information by reducing it to

6.02 Definition

The Ricci curvature $R_{\underset{\sim}{v}}$ of M along a tangent vector $\underset{\sim}{v}$ is the trace $\mathrm{tr}\underset{\sim}{R}_{\underset{\sim}{v}}$ of the corresponding Ricci transformation.

Now, the trace of a linear operator is the sum of the diagonal elements in its matrix, with respect to any basis whatever (cf. I.3.14). So let us choose a basis to make the geometrical interpretation of this example as simple as possible.

If $\underset{\sim}{v}$ is non-null, we have $\underset{\sim}{v} = \lambda\underset{\sim}{w}$ for some unit vector $\underset{\sim}{w}$ and $\underset{\sim}{R}_{\underset{\sim}{v}} = \lambda^2\underset{\sim}{R}_{\underset{\sim}{w}}$. Extend $\underset{\sim}{w}$ to an orthonormal basis $\underset{\sim}{w} = \underset{\sim}{b}_1, \underset{\sim}{b}_2, \ldots, \underset{\sim}{b}_n$. In the notation of §5, we then have $\| \underset{\sim}{b}_i, \underset{\sim}{w} \|^2 = 1$, for all $i \neq 1$. Now in an orthonormal basis the i-th component x^i of any vector $\underset{\sim}{x}$ is $\pm\underset{\sim}{x}.\underset{\sim}{b}_i$. ($\pm$ depending on whether $\underset{\sim}{b}_i$ is timelike or spacelike. If $\underset{\sim}{b}_i.\underset{\sim}{b}_i = -1$, that is minus the i-th component of $\underset{\sim}{b}_i$; other vectors go likewise.)

Suppose $\underset{\sim}{v}$ is timelike and $\underset{\sim}{G}$ has signature 2-dimM (the simplest and most interesting non-Riemannian case). Then all of $\underset{\sim}{b}_2, \ldots, \underset{\sim}{b}_n$ are spacelike, so $x^i = -\underset{\sim}{x}.\underset{\sim}{b}_i$, $i = 2,\ldots,n$. In particular the i-th diagonal entry, $i \neq 1$, in the matrix of $\underset{\sim}{R}_{\underset{\sim}{v}}$ for this basis is the i-th component of $\underset{\sim}{R}_{\underset{\sim}{v}}(\underset{\sim}{b}_i)$, which is thus

$$-\underset{\sim}{R}(\underset{\sim}{b}_i,\underset{\sim}{v})\underset{\sim}{v}.\underset{\sim}{b}_i = -\lambda^2 \frac{\underset{\sim}{R}(\underset{\sim}{b}_i,\underset{\sim}{w})\underset{\sim}{w}.\underset{\sim}{b}_i}{\| \underset{\sim}{b}_i,\underset{\sim}{w} \|^2}, \quad \text{since } \| \underset{\sim}{b}_i,\underset{\sim}{w} \|^2 = 1$$

$$= \lambda^2 \frac{\underset{\sim}{R}(\underset{\sim}{b}_i,\underset{\sim}{w})\underset{\sim}{b}_i.\underset{\sim}{w}}{\| \underset{\sim}{b}_i,\underset{\sim}{w} \|^2}.$$

That is λ^2 times the sectional curvature for a section in the plane of $\underset{\sim}{b}_i$ and $\underset{\sim}{v}$. The first diagonal entry is

$$\underset{\sim}{R}(\underset{\sim}{b}_1,\underset{\sim}{v})\underset{\sim}{v}.\underset{\sim}{b}_1 = \underset{\sim}{R}(\underset{\sim}{w}.\underset{\sim}{v})\underset{\sim}{v}.\underset{\sim}{w} = \lambda\underset{\sim}{R}(\underset{\sim}{v}.\underset{\sim}{v})\underset{\sim}{v}.\underset{\sim}{w} = \underset{\sim}{0}.\underset{\sim}{w} = 0.$$

Thus the trace $R_{\underset{\sim}{v}}$ is λ^2 times the sum of the sectional curvatures of M in any orthogonal set of (n-1) planes containing $\underset{\sim}{v}$ (the λ^2 just allowing for the size of $\underset{\sim}{v}$). A similar discussion shows that if M is Riemannian we get $-\lambda^2$ times the sum of these curvatures.

The Ricci curvature $R_{\underset{\sim}{v}}$ thus gives in either case (dividing appropriately by $\pm\lambda^2(n-1)$ a kind of "arithmetic mean curvedness" of M in the direction $\underset{\sim}{v}$ at p. If $\underset{\sim}{G}$ induces an indefinite metric on $v^{\perp}$ the mean is oddly weighted by + and - signs, and if $\underset{\sim}{v}$ is null we have no "normalising factor" λ^2, but $R_{\underset{\sim}{v}}$ is still the same thing: a convenient scalar measure of curvedness along $\underset{\sim}{v}$. (If we had defined sectional curvatures using $(\underset{\sim}{u}.\underset{\sim}{u})(\underset{\sim}{v}.\underset{\sim}{v})-(\underset{\sim}{u}.\underset{\sim}{v})^2$ instead of $\|\underset{\sim}{u},\underset{\sim}{v}\|^2$ - cf. Ex. 5.5a - we would get $-\lambda^2\times$(their sum) for all non-null $\underset{\sim}{v}$.)

In a similar way we can further "average" the curvature at p by adding the $R_{\underset{\sim}{v}_i}$ for any choice of orthonormal basis $\underset{\sim}{v}_1,\ldots,\underset{\sim}{v}_n$ for T_pM, but to see why the result does not depend on the choice we backtrack slightly, generalising the definitions above for algebraic convenience.

6.03 Definition

The Ricci transformation at p with respect to a pair of vectors $\underset{\sim}{u},\underset{\sim}{v} \in T_pM$ is the linear map

$$\underset{\sim}{R}_{\underset{\sim}{u},\underset{\sim}{v}} : T_pM \to T_pM : \underset{\sim}{w} \rightsquigarrow \underset{\sim}{R}(\underset{\sim}{w},\underset{\sim}{u})\underset{\sim}{v} \ .$$

(Thus the $\underset{\sim}{R}_{\underset{\sim}{v}}$ of 6.01 becomes short for $\underset{\sim}{R}_{\underset{\sim}{v},\underset{\sim}{v}}$.)

The Ricci tensor of M at p is the bilinear map

$$\underset{\sim}{\tilde{R}}_p : T_pM \times T_pM \to \mathbb{R} : (\underset{\sim}{u},\underset{\sim}{v}) \rightsquigarrow \mathrm{tr}\, \underset{\sim}{R}_{\underset{\sim}{u},\underset{\sim}{v}}$$

so that the Ricci curvature $R_{\underset{\sim}{v}}$ is exactly $\underset{\sim}{\tilde{R}}_p(\underset{\sim}{v},\underset{\sim}{v})$. Clearly $\underset{\sim}{\tilde{R}}$ is a $\binom{0}{2}$-tensor, and we have

6.04 Lemma

The Ricci tensor is symmetric; $\tilde{R}(u,v) = \tilde{R}(v,u)$.

Proof

Choose an orthonormal basis $b_1,\ldots,b_n$ for T_pM. Then the i-th diagonal entry in the matrix of $R_{u,v}$ becomes, using $\sigma = b_i \cdot b_i$ as a sign factor,

$$\begin{aligned}\left[R_{u,v}\right]_i^i &= \sigma R_{u,v}(b_i)\cdot b_i = \sigma R(b_i,u)v\cdot b_i && \text{by definition}\\ &= \sigma R(v,b_i)b_i\cdot u && \text{by 2.07D}\\ &= \sigma R(b_i,v)u\cdot b_i && \text{by 2.07A,B}\\ &= \sigma R_{v,u}(b_i)\cdot b_i\ .\end{aligned}$$

Thus the individual diagonal entries in the matrix are symmetric functions of u and v, hence so is their sum, $\tilde{R}$. ■

6.05 Corollary

The Ricci tensor is determined by the Ricci curvatures, and vice versa.

Proof

By the polarisation identity, since $\tilde{R}$ is symmetric,

$$R(u,v) = \tfrac{1}{4}R_{u+v} - \tfrac{1}{4}R_{u-v}.$$

The converse is trivial. ■

Algebraically and manipulatively the Ricci tensor is more convenient than the Ricci curvature, though geometrically the latter is a simpler idea. (This is just like the relation between an inner product $\underset{\sim}{G}$ and the length $\sqrt{\underset{\sim}{G}(\underset{\sim}{v},\underset{\sim}{v})}$ that it defines for single vectors, except that there is no need to look at the square root of $\tilde{\underset{\sim}{R}}(\underset{\sim}{v},\underset{\sim}{v})$.)

6.06 Definition

The scalar curvature $R(p)$ of M at p is the contraction twice over of the $\binom{2}{2}$-tensor $\underset{\sim}{G}^* \otimes \tilde{\underset{\sim}{R}}$ (cf. IV.1.12). (Since both $\underset{\sim}{G}^*$ and $\tilde{\underset{\sim}{R}}$ are symmetric, the choice of which covariant factors to contract with which contravariant ones makes no difference.)

6.07 Components

It is immediate that if $\underset{\sim}{u} = u^i\partial_i$, $\underset{\sim}{v} = v^i\partial_i$, $\underset{\sim}{w} = w^i\partial_i$,

$$\underset{\sim}{R}_{\underset{\sim}{u},\underset{\sim}{v}}(\underset{\sim}{w}) = \underset{\sim}{R}(\underset{\sim}{w},\underset{\sim}{u})\underset{\sim}{v} = \underset{\sim}{R}(w^k\partial_k, u^\ell\partial_\ell)(v^j\partial_j) = u^\ell v^j w^k R^i_{jk\ell}\partial_i$$

(cf. 2.08), so the components of the Ricci transformation $\underset{\sim}{R}_{\underset{\sim}{u},\underset{\sim}{v}}$ are $u^\ell v^j R^i_{jk\ell}$, those of $\underset{\sim}{R}_{\underset{\sim}{v}}$ are $v^\ell v^j R^i_{jk\ell}$. It follows that

$$\tilde{\underset{\sim}{R}}(\underset{\sim}{u},\underset{\sim}{v}) = \mathrm{tr}[u^\ell v^j R^i_{jk\ell}] = u^\ell v^j R^i_{ji\ell}$$

and the components of the Ricci tensor are the sums $R^i_{ji\ell}$ (= $R^i_{\ell ij}$, by symmetry). These are normally denoted by $R_{j\ell}$ (or R_{ij}, $R_{\alpha\beta}$ etc.) the fact of having only two indices sufficing to distinguish them from the components of the Riemann tensor.

Another useful expression of them is

$$R_{ij} = R^h_{ihj} = \delta^\ell_h R^h_{i\ell j} = g^{k\ell} g_{kh} R^h_{i\ell j} = g^{k\ell} R_{ki\ell j}.$$

6.08 Warning

Some authors define $R_{ij} = R^h_{ijh}$. Since R^k_{ijh} is skew-symmetric in j and h, this gives minus the Ricci tensor we have defined. If spacetime is given a metric of signature +2 - spacelike vectors having $\underset{\sim}{v}.\underset{\sim}{v}$ positive - then the same $\underset{\sim}{R}$ results (VIII.6.08). The sign of $\underset{\sim}{\tilde{R}}$ depends on which contraction is used, and that of $\underset{\sim}{G}^* \otimes \underset{\sim}{\tilde{R}}$ and hence of its contraction R depends on both these choices. If both choices are opposite to ours, the result for R is the same; if only one, the result is minus our scalar curvature.

6.09 Many contractions

The trace function is equivalent to contraction (V.1.12) so both $\underset{\sim}{\tilde{R}}$ and R are contractions of $\underset{\sim}{R}$. It follows easily from the symmetries of R that any other contraction is zero or expressible in terms of the Ricci tensor, so no other contraction can catch any of the information lost in taking this one. Clearly in general information is lost (we consider what in §7) as we are going from the $\frac{n^2}{12}(n^2-1)$-dimensional space of possible curvature tensors at a point to the $\frac{n}{2}(n+1)$-dimensional space of symmetric bilinear forms.

The scalar curvature at p is a sum of n^2 terms, $g^{ij}R_{ij}$. If we choose an orthonormal basis for T_pM (or make $\partial_1(p),\ldots,\partial_n(p)$ one by taking normal coordinates around p) the g^{ij} becomes δ^{ij} at p (only!) and $R(p) = \sum_{i=1}^{n} R_{ii}$. So R(p) is the sum of Ricci curvatures with respect to n orthonormal vectors, referred to at the end of 6.02; independence of choice follows, since Def^n. 6.06 makes no use of bases to define R(p).

6.10 Ricci directions

If M is Riemannian, by IV.4.09 we can always find an orthonormal basis for T_pM (and hence normal coordinates by IV.2.06, around p) making $R_{ij}(p) = 0$ for $i \neq j$. If M is pseudo-Riemannian we may be able to do this, but not necessarily (IV.4.11).

If we can, and the principal directions of $\tilde{\underset{\sim}{R}}_p$ are uniquely defined, they are called the Ricci directions of M at p and the corresponding R_{ii} are the principal curvatures.

If $\underset{\sim}{G}$ has signature 2-dimM, then $\underset{\sim}{v}^{\perp}$ for a timelike vector $\underset{\sim}{v}$ (the set of "entirely spacelike" or "infinite velocity" vectors, according to an observer who defines "zero velocity" by $\underset{\sim}{v}$, cf. XI. §2) inherits a negative definite metric. Since $\tilde{\underset{\sim}{R}}$ restricts to a symmetric form on $\underset{\sim}{v}^{\perp}$ we can apply IV.4.09 to get always a set of (n-1) spacelike Ricci directions relative to v.

If $\tilde{\underset{\sim}{R}}_p$ is isotropic (IV.4.10) at all $p \in M$, M is called an Einstein manifold (we explain why in XII.2.02). Just as "independence of plane" for sectional curvature implies independence of position also (Schur's Theorem), isotropy of $\tilde{\underset{\sim}{R}}_p$, which makes Ricci curvature independent of direction, implies that it is similarly independent of position. Before we prove this (6.14) we need to discuss an important consequence of the second Bianchi identity.

The tensor involved is $\tilde{\underset{\sim}{R}}$ "with one index raised", which we shall denote by $\bar{\underset{\sim}{R}}$. ($\bar{\underset{\sim}{R}}(p)$ is essentially the map $T_pM \to T_pM : x \rightsquigarrow \underset{\sim}{G}_{\uparrow}(\underset{\sim}{y} \rightsquigarrow \tilde{\underset{\sim}{R}}(\underset{\sim}{x},\underset{\sim}{y}))$.) If $\underset{\sim}{x}$ points in one of the Ricci directions, its image is just $\underset{\sim}{x}$ times the corresponding principal curvature: consider the proof of IV.4.09.) We shall call $\bar{\underset{\sim}{R}}$ the bivariant Ricci tensor when we want a special name, and denote its components in the usual way by $R^i_j = g^{ik}R_{kj}$. Notice that its sign involves that of $\underset{\sim}{G}$ (cf. 6.08), that R(p) is just $\mathrm{tr}(\bar{\underset{\sim}{R}}(p))$, and that $\bar{\underset{\sim}{R}}$ is self-adjoint since $\underset{\sim}{R}$ is symmetric.

Now the Bianchi identity implies the following, considering the $\binom{1}{1}$-tensor $\nabla_{\underset{\sim}{u}}\bar{\underset{\sim}{R}}$ as a map $T_pM \to T_pM$ and the $\binom{1}{2}$-tensor $\nabla\bar{\underset{\sim}{R}}$ as the map $T_pM \to L(T_pM;T_pM) : \underset{\sim}{v} \rightsquigarrow (\underset{\sim}{u} \rightsquigarrow (\nabla_{\underset{\sim}{u}}\bar{\underset{\sim}{R}})\underset{\sim}{v})$ (cf. VIII.7.02, 7.05):

6.11 Lemma

For any $\underset{\sim}{v} \in T_pM$, $\underset{\sim}{v}(R) = 2\mathrm{tr}((\nabla\bar{\underset{\sim}{R}})\underset{\sim}{v})$.

In coordinates, since $\underset{\sim}{v}(R) = dR(\underset{\sim}{v}) = v^j\partial_jR$, this means (VIII.7.08) that

$$\partial_j R = 2R^{\ell}_{j;\ell}.$$

Proof

Summing n 2nd Bianchi identities,

$$R^h_{ihj;\ell} + R^h_{ij\ell;h} + R^h_{i\ell h;j} = 0.$$

By skew-self-adjointness and ∇'s commuting with contractions, this gives

$$R_{ij;\ell} + R^h_{ij\ell;h} - R_{i\ell;j} = 0,$$

and since ∇ commutes with raising and lowering indices,

$$* \qquad (g^{ik}R_{ij})_{;\ell} + (g^{ik}R^h_{ij\ell})_{;h} - (g^{ik}R_{i\ell})_{;j} = 0.$$

So summing over k and ℓ,

$$R^{\ell}_{j;\ell} + (g^{i\ell}R^h_{ij\ell})_{;h} - (R)_{;j} = 0.$$

But for any function f we have $(f)_{;j} = f_{,j} = \partial_j f$, by VIII.7.08, and

$$g^{i\ell}R^h_{ij\ell} = g^{i\ell}g^{kh}R_{kij\ell} = g^{kh}(g^{i\ell}R_{ik\ell j})$$

$$= g^{kh}R_{kj} \quad \text{by the last identity of 6.07}$$

$$= R^h_j.$$

So, changing one dummy index h to ℓ,

$$\partial_j R = 2R^{\ell}_{j;\ell}$$

as required. ■

This result can equally be formulated as

$$(R^{\ell}_{j} - \tfrac{1}{2}\delta^{\ell}_{j}R)_{;\ell} = 0$$

applying VIII.7.09, which leads us to make

6.12 Definition

For any $\binom{1}{n}$-tensor field $\underset{\sim}{J}$, the $\binom{0}{n}$-tensor field

$$(\underset{\sim}{v}_1,\ldots,\underset{\sim}{v}_n) \rightsquigarrow \mathrm{tr}[\underset{\sim}{u} \rightsquigarrow (\nabla_{\underset{\sim}{u}}\underset{\sim}{J})(\underset{\sim}{v}_1,\ldots,\underset{\sim}{v}_n)]$$

with components $J^{\ell}_{i_1\ldots i_n;\ell}$ is called its divergence (we see why in XI.3.07) and denoted by div $\underset{\sim}{J}$. (In the case of $\underset{\sim}{J}$ a contravariant vector field on R^3 with a constant metric this reduces to the familiar "$\underset{\sim}{\nabla}.\underset{\sim}{J}$").

6.11 thus deduces from the Bianchi identity, which as we mentioned resembles a conservation law (2.10), that the divergence of the Ricci tensor is half the gradient of the scalar curvature, and equivalently that the divergence of the Einstein $\binom{1}{1}$-tensor field

$$\underset{\sim}{E} = \bar{\underset{\sim}{R}} - \tfrac{1}{2}R\underset{\sim}{I} \qquad (E^i_j = R^i_j - \tfrac{1}{2}R\delta^i_j \text{ in coordinates})$$

is identically zero. This result is the conservation equation of general relativity. Geometrically it is a necessary condition for $\underset{\sim}{R}$ to be the Riemann tensor of some metric (as div $\underset{\sim}{v} = 0$ is necessary for a vector field $\underset{\sim}{v}$ to be a "curl" in R^3); physical meaning it will acquire (XI.3.07) when we reach the physical theory that uses it.

6.13 Lemma

If dim M = n > 2, $\bar{\underset{\sim}{R}} = \underset{\sim}{E} - \frac{1}{n-2}(\text{tr } \underset{\sim}{E})\underset{\sim}{I}$. (And hence, $\underset{\sim}{E} = \bar{\underset{\sim}{R}} - \frac{1}{2}(\text{tr } \bar{\underset{\sim}{R}})\underset{\sim}{I}$, so that $\bar{\underset{\sim}{R}}$ and $\underset{\sim}{E}$ determine each other and thus carry the same information.) In particular if n=4, as it will often in the two remaining chapters, $\bar{\underset{\sim}{R}} = \underset{\sim}{E} - \frac{1}{2}(\text{tr } \underset{\sim}{E})\underset{\sim}{I}$.

Proof

tr $\underset{\sim}{I}$ = n, as it is the sum of the n 1's on the diagonal of the identity n × n matrix.

So $\text{tr } \underset{\sim}{E} = \text{tr}(\bar{\underset{\sim}{R}} - \frac{1}{2}R\underset{\sim}{I}) = R - \frac{1}{2}R \text{ tr } \underset{\sim}{I} = -\frac{n-2}{2} R$.

Hence $\underset{\sim}{E} = \bar{\underset{\sim}{R}} + \frac{1}{n-2}(\text{tr } \underset{\sim}{E})\underset{\sim}{I}$, and therefore,

$$\bar{\underset{\sim}{R}} = \underset{\sim}{E} - \frac{1}{n-2}(\text{tr } \underset{\sim}{E})\underset{\sim}{I}.$$

6.14 Lemma

An Einstein n-manifold, for n > 2, has constant scalar curvature.

Proof

We have $\tilde{\underset{\sim}{R}}_p = \lambda(p)\underset{\sim}{G}_p$ $\forall p \in M$, for some $\lambda : M \to R$. Hence

$$\underset{\sim}{G}^* \otimes \tilde{\underset{\sim}{R}} = \lambda \underset{\sim}{G}^* \otimes \underset{\sim}{G}.$$

Contracting, $R = n\lambda$.

Hence $\tilde{\underset{\sim}{R}} = \frac{1}{n} R\underset{\sim}{G}$,

so $\bar{\underset{\sim}{R}} = \frac{1}{n} R\underset{\sim}{I}$, "raising one index".

Therefore

$$dR = 2 \operatorname{div} \bar{\underset{\sim}{R}} = \frac{2}{n} \operatorname{div} (R\underset{\sim}{I}).$$

But

$$\begin{aligned} \operatorname{div} (R\underset{\sim}{I}) &= (R\delta^i_j)_{;i} dx^j \quad \text{in coordinates} \\ &= R_{;i} \delta^i_j \, dx^j \quad \text{by VIII.7.09} \\ &= (\partial_i R) dx^i \\ &= dR \end{aligned}$$

hence

$$dR = \frac{2}{n} dR \, .$$

Thus

$$dR = \underset{\sim}{0} \quad \text{if } n \neq 2, \text{ hence R is constant.}$$

■

Note the similarity to Schur's Theorem, not only in the result but in the use made of the Bianchi identity.

6.15 Corollary

An Einstein n-manifold, for n > 2, has constant Ricci curvature.

Proof

$\tilde{\underset{\sim}{R}} = \frac{1}{n} R\underset{\sim}{G}$ is a constant multiple of $\underset{\sim}{G}$: apply Def^n. VIII.7.10.

Exercises X.6

1) Show that the coordinate expression for the divergence of a $\binom{1}{n}$-tensor $t^k_{j_1 \dots j_n}$, using VIII.7.08, is

$$t^k_{j_1 \dots j_m ; k} = t^k_{j_1 \dots j_m , k} + t^s_{j_1 \dots j_m} \Gamma^k_{ks} - \sum_{i=1}^{m} t^k_{j_1 \dots j_{i-1} s j_{i+1} \dots j_m} \Gamma^s_{k j_i} .$$

In particular, for a $\binom{1}{1}$ and $\binom{1}{0}$ field respectively,

$$t^k_{j;k} = t^k_{j,k} + t^s_j\Gamma^k_{ks} - t^k_s\Gamma^s_{kj}\ , \qquad t^k_{;k} = t^k_{,k} + t^s\Gamma^k_{ks} = \partial_k t^k + t^s\Gamma^k_{ks}.$$

2) Show that on a 2-manifold the Einstein tensor vanishes identically.

7. THE WEYL TENSOR

If the dimension n of M is 1, the curvature tensor necessarily vanishes, since $\underset{\sim}{R}(\underset{\sim}{u},\underset{\sim}{v})$ is skew-symmetric in $\underset{\sim}{u}$ and $\underset{\sim}{v}$. If n=2, $\frac{n^2}{12}(n^2-1) = 1$; since contraction down to R is thus a linear, non-zero map between 1-dimensional spaces it is an isomorphism, and $\underset{\sim}{R}$ reduces to a scalar as in §3. In three dimensions,

$$\frac{n^2}{12}(n^2-1) = 6 = \frac{n}{2}(n+1)$$

so that $\underset{\sim}{R}$ and $\underset{\sim}{\tilde{R}}$ live in spaces of the same dimension. It is not hard to show directly that contraction from $\underset{\sim}{R}$ to $\underset{\sim}{\tilde{R}}$ is surjective, hence for n=3 an isomorphism. Thus on 3-manifolds the Riemann tensor is determined by the Ricci tensor (cf. Ex. 3).

On a 4-manifold, however,

$$\frac{n^2}{12}(n^2-1) = 20, \quad \frac{n}{2}(n+1) = 10$$

so the contraction has a non-zero kernel (and so on for n > 4, since n^4 increases much faster than n^2). What is lost?

Consider what information is lost in general by contraction, or equivalently by taking a trace. Any linear operator on an n-dimensional space X can be expressed as

$$\underset{\sim}{A} = \underset{\sim}{S} + \underset{\sim}{T}, \text{ where } \underset{\sim}{S} = \underset{\sim}{A} - \frac{1}{n}(\text{tr } \underset{\sim}{A})\underset{\sim}{I}, \quad \underset{\sim}{T} = \frac{1}{n}(\text{tr } \underset{\sim}{A})\underset{\sim}{I}.$$

Then $\text{tr } \underset{\sim}{T} = \frac{1}{n}(\text{tr } \underset{\sim}{A}) \text{ tr } \underset{\sim}{I} = \frac{1}{n}(\text{tr } \underset{\sim}{A})n = \text{tr } \underset{\sim}{A}$,

$$\text{tr } \underset{\sim}{S} = 0$$

so that we have expressed $\underset{\sim}{A}$, in a natural way as a sum of "traceless" and "traceable" parts: essentially decomposing $L(X;X)$ as the direct sum $(\ker(\text{tr})) \oplus \{x\underset{\sim}{I} \mid x \in R\}$, (cf. Ex. VII.3.1).

We can decompose the Ricci transformations in this way, to get $\underset{\sim}{R}_{\underset{\sim}{u},\underset{\sim}{v}} = \underset{\sim}{S}_{\underset{\sim}{u},\underset{\sim}{v}} + \underset{\sim}{T}_{\underset{\sim}{u},\underset{\sim}{v}}$ and have the trace $\underset{\sim}{\tilde{R}}(\underset{\sim}{u},\underset{\sim}{v})$ carried by $\underset{\sim}{T}_{\underset{\sim}{u},\underset{\sim}{v}}$ while $\underset{\sim}{S}_{\underset{\sim}{u},\underset{\sim}{v}}$ represents the information lost by taking the trace since $\text{tr } \underset{\sim}{S}_{\underset{\sim}{u},\underset{\sim}{v}} = 0$. This does not quite give a satisfactory decomposition of $\underset{\sim}{R}$, since the two reconstructed 4-tensors $\underset{\sim}{S}$ and $\underset{\sim}{T}$ do not have the symmetries 2.07. However, by <u>imposing</u> the symmetries on them (in the way that for a bilinear $\underset{\sim}{A} : X \times X \to R$, for instance, $\underset{\sim}{B}(\underset{\sim}{u},\underset{\sim}{v}) = \underset{\sim}{A}(\underset{\sim}{u},\underset{\sim}{v}) + \underset{\sim}{A}(\underset{\sim}{v},\underset{\sim}{u})$ is its <u>symmetrised</u> and $\underset{\sim}{F}(\underset{\sim}{u},\underset{\sim}{v}) = \underset{\sim}{A}(\underset{\sim}{u},\underset{\sim}{v}) - \underset{\sim}{A}(\underset{\sim}{v},\underset{\sim}{u})$ its <u>skew-symmetrised</u> form), and seeking a similarly symmetrised term to allow for the remaining contraction down to scalar curvature, we are led to

7.01 Definition

The <u>Weyl tensor</u> on an n-manifold ($n > 2$) with metric tensor field $\underset{\sim}{G}$ is the $\binom{1}{3}$-tensor field defined by

$$\underset{\sim}{C}(\underset{\sim}{u},\underset{\sim}{v})\underset{\sim}{w} = \underset{\sim}{R}(\underset{\sim}{u},\underset{\sim}{v})\underset{\sim}{w} - \frac{1}{n-2}\left(\underset{\sim}{\tilde{R}}(\underset{\sim}{v},\underset{\sim}{w})\underset{\sim}{u} - \underset{\sim}{\tilde{R}}(\underset{\sim}{u},\underset{\sim}{w})\underset{\sim}{v} - (\underset{\sim}{u}\cdot\underset{\sim}{w})\underset{\sim}{r}(\underset{\sim}{v}) + (\underset{\sim}{v}\cdot\underset{\sim}{w})\underset{\sim}{r}(\underset{\sim}{u})\right)$$

$$+ \frac{R}{(n-1)(n-2)}\left((\underset{\sim}{v}\cdot\underset{\sim}{w})\underset{\sim}{u} - (\underset{\sim}{u}\cdot\underset{\sim}{w})\underset{\sim}{v}\right)$$

where $\underset{\sim}{r}(\underset{\sim}{v}) = \underset{\sim}{G}_{\uparrow}(\underset{\sim}{x} \rightsquigarrow \underset{\sim}{\tilde{R}}(\underset{\sim}{u},\underset{\sim}{x}))$, or equivalently by

$$\underset{\sim}{C}(\underset{\sim}{u},\underset{\sim}{v})\underset{\sim}{w}.\underset{\sim}{x} = \underset{\sim}{R}(\underset{\sim}{u},\underset{\sim}{v})\underset{\sim}{w}.\underset{\sim}{x} - \frac{1}{n-2}\left((\underset{\sim}{u}.\underset{\sim}{x})\tilde{\underset{\sim}{R}}(\underset{\sim}{v},\underset{\sim}{w})-(\underset{\sim}{v}.\underset{\sim}{x})\tilde{\underset{\sim}{R}}(\underset{\sim}{u},\underset{\sim}{w})-(\underset{\sim}{u},\underset{\sim}{w})\tilde{\underset{\sim}{R}}(\underset{\sim}{v},\underset{\sim}{x})+(\underset{\sim}{v}.\underset{\sim}{w})\tilde{\underset{\sim}{R}}(\underset{\sim}{u},\underset{\sim}{x})\right)$$

$$+ \frac{R}{(n-1)(n-2)}\left[(\underset{\sim}{u}.\underset{\sim}{x})(\underset{\sim}{v}.\underset{\sim}{w})-(\underset{\sim}{v}.\underset{\sim}{x})(\underset{\sim}{u}.\underset{\sim}{w})\right].$$

In coordinates, then

$$C_{ijk\ell} = R_{ijk\ell} - \frac{1}{n-2}\left(g_{ik}R_{j\ell}-g_{i\ell}R_{jk}-g_{jk}R_{a\ell}+g_{j\ell}R_{ik}\right)$$

$$+ \frac{R}{(n-1)(n-2)}\left(g_{ik}g_{j\ell}-g_{i\ell}g_{jk}\right).$$

$\underset{\sim}{C}$ is the analogue of $\underset{\sim}{S}$ in the simpler example above, since $C^i_{jik} = 0$ (Ex. 1c); it is the "traceless" or "contractionless" part of $\underset{\sim}{R}$. $\underset{\sim}{R}$ is determined by $\underset{\sim}{C}$ and its "traceable part" $\tilde{\underset{\sim}{R}}$, using $R = \text{tr}\,\tilde{\underset{\sim}{R}}$ and any of the above three equations, thus $\underset{\sim}{C}$ contains exactly the information lost in contracting $\underset{\sim}{R}$ to $\tilde{\underset{\sim}{R}}$. If the Ricci tensor (or equivalently by 6.13 the Einstein tensor) vanishes, it is immediate that $\underset{\sim}{R} = \underset{\sim}{C}$. Physically, this permits the Weyl tensor to emerge in Chapter XII as the "vacuum curvature".

Exercises X.7

1a) Show that the three equations given above to define the Weyl tensor are equivalent, and that the map $\underset{\sim}{R} \rightsquigarrow \underset{\sim}{C}$ is linear.

b) Show that $\underset{\sim}{C}$ has the symmetries 2.07.

c) Show that $C^i_{jki} = 0$, and deduce that all contractions of $\underset{\sim}{C}$ vanish.

2a) The space $L(X;X) \cong X^* \otimes X$ inherits the metric $\underset{\sim}{G}^* \otimes \underset{\sim}{G}$ from a metric $\underset{\sim}{G}$ on X (cf. IV.1.12, V.1.08); is $\underset{\sim}{A} \rightsquigarrow (\underset{\sim}{A} - \frac{1}{n}(\text{tr}\,\underset{\sim}{A})\underset{\sim}{I})$ orthogonal projection onto ker(tr), with respect to this metric?

b) Is $\underset{\sim}{R}_p \rightsquigarrow \underset{\sim}{C}_p$ orthogonal projection onto ker $(\underset{\sim}{R} \rightsquigarrow \tilde{\underset{\sim}{R}})$, with respect to the analogous metric?

3a) Show that on a 3-manifold the Weyl tensor vanishes identically.

b) Deduce an expression for the Riemann tensor in terms of the Ricci tensor, on a 3-manifold.

XI Special Relativity

"Regard motion as though it were stationary, and what becomes
of motion?
Treat the stationary as though it moved, and that disposes of
the stationary.
Both these having been disposed of, what becomes of the One?"

Seng-ts'an

In this chapter and the next we examine the specific models of physical phenomena that grew from the considerations discussed in Ch. O. §3.

1. ORIENTING SPACETIMES

We start by introducing some ideas for general spacetimes, needed in this chapter and the next. Basically, our models for spacetime are Lorentz 4-manifolds (cf. VII.3.04). Now we add some definitions that will allow formal models for the motion of physical "particles".

1.01 Definition

Let M be a connected Lorentz manifold.

Choose one timelike vector $\underset{\sim}{v}$ (VII.3.04) at some $p \in M$ as forward in time. Then a timelike or non-zero null vector $\underset{\sim}{w}$ at $q \in M$ is also forward if there is a continuous curve $\underset{\sim}{c} : [a,b] \to TM$ from $\underset{\sim}{v}$ to $\underset{\sim}{w}$ such that for no $s \in [a,b]$ is $\underset{\sim}{c}(s)$ a spacelike or zero vector. (Notice that $\underset{\sim}{c}$ is not a curve in M. Its projection $\Pi \circ \underset{\sim}{c}$ by the bundle map (VII.3.03), which is in M, is not required to be a like curve.)

For any non-spacelike $\underset{\sim}{w} \neq 0$, such a curve will exist from $\underset{\sim}{v}$ to either $\underset{\sim}{w}$ or $-\underset{\sim}{w}$ (Ex. 1a). If for no $\underset{\sim}{w}$ does both happen, M is time-orientable. The choice of a $\underset{\sim}{v}$ is then a time-orientation of M, and M with such a choice made is time-oriented. In a time-oriented manifold, if a non-spacelike $\underset{\sim}{w} \neq \underset{\sim}{0}$ is not forward it is backward.

A curve or path c in a time-oriented manifold M is forward [respectively, backward] if its tangent vector $c^*(t)$ (VII.5.02) is forward [respectively, backward] for all t. We shall usually parametrise timelike forward curves by proper time (IX.4.05) denoted by σ.

M is causal if there is no forward curve $c : [a,b] \to M$ with $c(a) = c(b)$. (Ex. 1c,d,e; cf. also IX.6.07.) Physically, if M is not causal you can go forward in time to meet yourself starting out (or something can). Normal ideas of causality break down and physics becomes very complicated. For example, in an initial-value problem the data cannot be given arbitrarily, since they must form part of their own solutions for time ahead and past. A compact Lorentz manifold cannot be causal (why not?) so spacetime must be noncompact, be noncausal or somewhere break down in its Lorentz structure.

Minkowski space is time-orientable and causal (Ex. 2), hence so is the 4-dimensional affine space X (without intrinsic coordinates) with a constant Lorentz metric, which special relativity takes as a model for physical spacetime. (X will have this standard meaning throughout the chapter.)

1.02 Definition

A spacelike section of a time-oriented spacetime M is a smoothly embedded 3-manifold $S \subseteq M$ such that

(i) the induced metric on S is everywhere negative definite

(ii) for every $p \in M$ there is a timelike curve through p that meets S, and either all such curves are forward from x to S, all are backward, or $p \in S$.

A forward curve $c :]a,b[\to M$ is called a <u>history</u>. A history c with some $c(t) \in S$ is <u>at rest relative</u> to S, or at $_S$<u>rest</u>, if $c^*(t).\underset{\sim}{v} = 0$ for all $\underset{\sim}{v}$ tangent to S.

The term "history" is suggested by the idea of c as a potential "trajectory through spacetime of a particle (or physicist)". Some books implicitly use the term "world line" for the same concept (cf. IX.1.02). In fact (XII. §4) the notion of "particle" is problematical in classical relativistic physics, which strictly should work only with fields. The approximate concept of a "particle" as an entity at a point simply provides useful motivation in some discussions. The "forward curve" in contrast is a <u>mathematically</u> precise concept that we can discuss rigorously in what follows, even if it lacks a strictly precise physical interpretation. (As indeed does even the definition of "derivative", for analogous reasons.)

<u>Exercises XI.1</u>

1a) In a general connected Lorentz manifold M with timelike $\underset{\sim}{v} \in T_pM$, and timelike or null $\underset{\sim}{w} \in T_qM$, choose a path $\alpha : [a,b] \to M$ from p to q and show by working in successive charts along its image that there is a path $\tilde{\alpha}$ in TM such that $\tilde{\alpha}(t)$ is always timelike and at $\alpha(t)$. Then show that there is such an affine path γ in T_qM from $\tilde{\alpha}(b)$ either to $\underset{\sim}{w}$ or to $-\underset{\sim}{w}$, and combine $\tilde{\alpha}$ and γ to get c as in Definition 1.01. (Hint for finding γ: show that there is such a γ from $\tilde{\alpha}(b)$ to $\underset{\sim}{w}$ if and only if $\tilde{\alpha}(b).\underset{\sim}{w}$ is positive.)

b) Show that M is time-orientable if and only if there is no curve $\underset{\sim}{c}$ in TM from some arbitrarily chosen non-spacelike $\underset{\sim}{v}$ to $-\underset{\sim}{v}$ with $\underset{\sim}{c}(t)$ never spacelike or zero.

c) Show that $S^1 \times R^3$ (where S^1 is the circle), with the metric given by $(ds)^2 = (d\theta)^2 - (dx)^2 - (dy)^2 - (dz)^2$ in the obvious coordinates, is time-orientable but not causal. Is it flat?

d) Show that whether a time-orientable manifold is causal does not depend on the orientation.

e) Find a locally flat spacetime which is not time-orientable (think of the Möbius strip or the Klein bottle $\times R^2$). Can a non-time-orientable spacetime be "causal", in the sense of having no timelike closed curves?

2. Prove that Minkowski space M^4 is time-orientable and causal. (Divide the non-spacelike vectors in its vector space L^4 into forward and backward, and carry the distinction to each T_xM^4 by $\overleftarrow{\underset{\sim}{d}}_x$.)

2. MOTION IN FLAT SPACETIME

2.01 Definition

An inertial frame F for the affine space X, with constant Lorentz metric, that we consider throughout this chapter is a choice ${}_F\underset{\sim}{e}_0$ of a unit (hence non-null) forward vector in the vector space T of X. We shall also denote by ${}_F\underset{\sim}{e}_0$ the vector $\overleftarrow{\underset{\sim}{d}}_x({}_F\underset{\sim}{e}_0) \in T_xX$, for any $x \in X$. (The reader may add precision if he wishes by writing ${}_F\underset{\sim}{e}_{0x}$, ${}_F\underset{\sim}{e}_{0y}$ etc, but this multiplies suffixes beyond comfort - particularly when x is replaced by c(t).) A particle whose motion is described by a forward curve, or more precisely a history c in X is at rest relative to F, or at ${}_F$rest, at c(t) if $c^*(t) = \lambda({}_F\underset{\sim}{e}_0) \in T_{c(t)}X$ for some $\lambda \in R$. An affine history c at ${}_F$rest for some (and hence all) c(t) will also be called an inertial observer, to whom the frame F is appropriate.

This amounts to the choice of a "rest velocity", relative to which others are to be measured. To measure them, we define the time-component $t_F(\underset{\sim}{w})$ relative to F of a vector $\underset{\sim}{w}$ in T or any T_xX to be $\underset{\sim}{w}\cdot{}_F\underset{\sim}{e}_0 \in R$, the space-component to be $\underset{\sim}{s}_F(\underset{\sim}{w}) = \underset{\sim}{w} - t_F(\underset{\sim}{w})\,{}_F\underset{\sim}{e}_0 \in ({}_F\underset{\sim}{e}_0)^\perp$ in T or T_xX. We call vectors $\underset{\sim}{w}$ with $t_F(\underset{\sim}{w}) = 0$ entirely spacelike relative to F, or entirely ${}_F$spacelike.

The time difference relative to F between $x, y \in X$ is (repeating the label t_F for a function of the two arguments x and y) $t_F(x,y) = t_F(\underset{\sim}{d}(x,y))$, the time-component of their vector separation in X. If $t_F(x,y) = 0$, x and y are $_F$simultaneous. Their space-separation relative to F is $\underset{\sim}{d}_F(x,y) = \underset{\sim}{s}_F(\underset{\sim}{d}(x,y)) \in T$, and their $_F$distance is $d_F(x,y) = \| \underset{\sim}{d}_F(x,y) \|$. (The quantity $t_F(x,y)$ is of course only a matter of direct physical measurement when $\underset{\sim}{d}_F(x,y) = \underset{\sim}{0}$, since only then can there be an observer at rest relative to F (that is a physicist whose motion is approximated by a history at $_F$rest) who measures the time between x and y along his world line. For events off his world line he has to infer a time label by allowing for the time he takes to learn of them. However, $t_F(x,y)$ does give exactly the difference in the time labels he uses, and $\underset{\sim}{d}_X(x,y)$ the separation of his space labels.)

The velocity relative to F or $_F$velocity of a forward curve $c : [a,b] \to X$ at $c(t)$ is the vector, entirely spacelike (relative to F),

$$c_F^*(t) = \frac{\underset{\sim}{s}_F(c^*(t))}{t_F(c^*(t))} \in T_{c(t)}X.$$

Evidently this is equal to

$$\overleftarrow{\underset{\sim}{d}}_{c(t)} \left(\left(\frac{d}{ds} t_F(c(a),c(\ ))(t) \right)^{-1} (\underset{\sim}{d}_F(c(a),c(\ ))^*(t) \right).$$

Here the scalar differential, equal to $t_F(c^*(t))$, allows for variety in parametrisation (c may not be going ahead in time at unit speed according to F) and the vector (by abuse of language considered free in T rather than bound at $\underset{\sim}{d}_F(c(a),c(t))$ to simplify the expression) is the derivative of "spatial position" according to F.

The $_F$speed of c at c(t) is the real number

$$v_F(t) = \sqrt{|c_F^*(t) . c_F^*(t)|}, \quad (|\ | \text{ needed since } c_F^*(t) \text{ spacelike}).$$

Evidently $v_F(t) = 0$ if and only if c is at ${}_F$rest at c(t).

2.02 Time dilation

By IX.4.02, if $\underset{\sim}{d}_F(c(a),c(b)) = \underset{\sim}{0}$ but c is not always at ${}_F$rest, then time measured along c is less than $t_F(c(a),c(b))$. This is physically interpreted as the (experimentally verified) statement that "time passes more slowly" or is "dilated" for a clock (atomic, or heartbeat, or...) whose motion is described within experimental error by the history c. We now have the notation to derive some classical formulae for the results obtained geometrically in IX.4. Assume that $c^*(t)$ never vanishes and reparametrise as follows:

Define $f : [a,b] \to [0, t_F(c(a),c(b))] : t \rightsquigarrow t_F(c(a),c(t))$. Then

$$\frac{df}{ds} = t_F(c^*)$$

which never vanishes, since ${}_F\underset{\sim}{e}_0$ is orthogonal only to spacelike vectors. Thus f has a smooth inverse g and we can define $\tilde{b} = t_F(c(a),c(b))$ and

$$\tilde{c} = c \circ g : [0,\tilde{b}] \to X$$

has $\tilde{c}(t)$ as "that position in X, on the curve c, which according to an observer at ${}_F$rest is t later than c(a)". Then if (cf. VII.5.02)

$$\sigma : [0,\tilde{b}] \to R$$

is arc length (proper time) along c we have (cf. IX.4.05)

$$\frac{d\sigma}{ds}(t) = \sqrt{\tilde{c}^*(t).\tilde{c}^*(t)}$$

$$= \sqrt{[(t_F(\tilde{c}^*(t)))^2 + (\underset{\sim}{s}_F(\tilde{c}^*(t))).(\underset{\sim}{s}_F(\tilde{c}^*(t)))]} \qquad \text{(why?)}$$

$$= \sqrt{[(t_F(\tilde{c}^*(t)))^2 + (t_F(\tilde{c}^*(t))\tilde{c}^*_F(t)).(t_F(\tilde{c}^*(t))\tilde{c}^*_F(f))]}$$

by definition of $\tilde{c}^*_F$ (2.01)

$$= (t_F(\tilde{c}^*(t)))\sqrt{(1 + \tilde{c}^*_F(t).\tilde{c}^*_F(t))}$$

$$= \sqrt{1-v_F^2} \quad \text{since } t_F(\tilde{c}^*(t)) = 1 \text{ by construction,}$$

$$\text{and } -v_F^2 = \tilde{c}^*_F.\tilde{c}^*_F.$$

This is the classic formula for "relativistic time dilation". The formula applies even when c does not "$_F$return" by having $\underset{\sim}{d}_F(c(a),c(b)) = \underset{\sim}{0}$ and in particular, when c is at rest in some other inertial frame F'. But if $w_{F'}$ is the speed relative to F' of a history h at $_F$rest, the same formula applies; relative to F', the time for h is dilated. This symmetry between two inertial frames, and consequently between two affine histories (two inertial observers), helps the feeling of "paradox" that persists in some quarters. It was tempting to transfer the symmetry to the case of one history <u>not</u> affine. But the formula applies only when F is constant. If interpreted as the frame used by an observer P following c, who sets $_F\underset{\sim}{e}_0 = c^*$, this requires that $\underset{\sim}{d}_{c(t)}c^*(t)$ be constant. Otherwise $_F\underset{\sim}{e}_0$ changes, and the affine subspace S_t of points in a purely spacelike relation to c(t) ("$_F$simultaneous with c(t)") turns, so that P's time labels

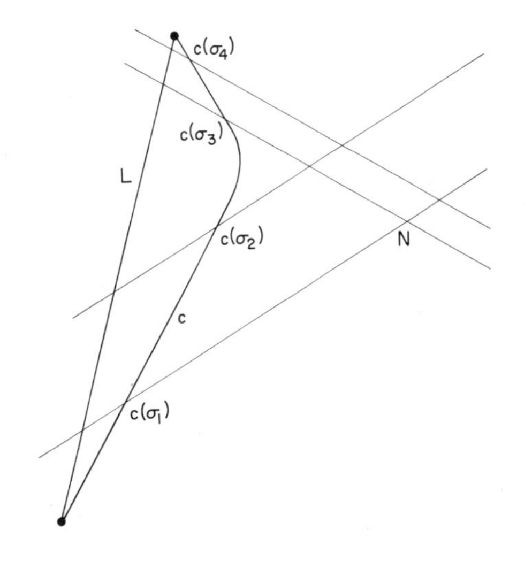

Fig. 2.1

become more complicated. A sharp acceleration turns S_t rapidly. In Fig. 2.1, L is the path of an inertial observer Q between the same points in X. While P is turning, his labelling system considers proper time along L to change <u>faster</u> than P's own time labels for the points.

Computing a correct formula, that would reduce to $\sqrt{1-v_F^2}$ for P inertial, requires bringing in the rate of change of c*. The result is messy, and physically meaningless anyway: $\frac{d\sigma'}{d\sigma}$ would mean "how fast his proper time σ', is changing relative to mine σ, <u>right now</u>", namely at $L \cap S_\sigma$ using the frame of reference ${}_F\underset{\sim}{e}_0 = c^*$. But as we can't properly compare watches till we reach the same point, and can't <u>detect</u> what is happening "right now" except where we are, this is essentially a fiction. Only the integrals of the proper times signify for comparison. (Note how P's labelling system breaks down at points like N.)

It is thus safer to treat comparisons of proper time along curves in the manner of IX.4: use $\sqrt{1-v_F^2}$ only with a <u>fixed</u> inertial frame F - and then cautiously.

We come now to an even earlier formula, predating Einstein:

<u>2.03 Lemma</u> (Lorentz-Fitzgerald contraction)

Let $x, y \in X$ have spacelike separation $\underset{\sim}{d}(x,y) \in T$, with ${}_F$distance ℓ where F is a frame of reference in which $\underset{\sim}{d}(x,y)$ is entirely spacelike (x,y are ${}_F$simultaneous). Then in the frame appropriate to an inertial observer Q whose ${}_F$speed is v_F, their distance is

$$\ell' = \frac{\ell}{\sqrt{1-v_F^2}} .$$

In general, if Q's ${}_F$velocity $\underset{\sim}{v}_F$ is linearly independent of $\underset{\sim}{d}(x,y)$,

$$\ell' = \frac{\ell}{\sqrt{1-w_F^2}}$$

where $w_F = |\underset{\sim}{v}_F \cdot \underset{\sim}{e}|$, $\underset{\sim}{e}$ being the unit vector in the $\underset{\sim}{d}(x,y)$ direction.

Proof

Ex. 1. ■

(2.03 is a simpler statement than we could make about measurements of the length of a rod - the usual description - because that would require going into what point in the history of one end we compare with a given point for the other to get "length". Not even the ends of a stick can be simultaneous absolutely.)

The time and space measurement alterations of 2.02 and 2.03 were at the core of the original formulation of special relativity, from the postulate that no experiment whatever can prove one observer "at rest" rather than another, in particular that all observers must obtain the same value for the velocity of light. Minkowski was the first to replace time and space with these "corrections" by a single flat geometric spacetime, now called Minkowski space, which different observers resolve differently into separate "space" and "time".

2.04 Four-velocity

We shall interpret the mathematical models of this chapter and the next in the usual physical way. We suppose that a physicist whose motion we describe by a history c perceives a motion into the future at unit speed (one second per second), using at each point c(t) the frame appropriate to the affine history tangent there to c. So the perceived "velocity" through spacetime is the unit tangent vector in the direction of $c_*(t)$. If c is parametrised by proper time σ, (arc length along timelike curves, IX.4.05), the perceived vector at $c(\sigma)$ is thus exactly $c_*(\sigma)$. For the conveniences this offers, we shall always assume parametrisation by arc length (rather than, for instance, the F-dependent parametrisation $\tilde{c}$ given in 2.02, "parametrisation by F-time") unless otherwise stated.

The vector $c^*(\sigma)$ is often called the <u>4-velocity</u> of c at $c(\sigma)$. For it was discovered as a set of four components (Ex. 2),

$$\left(\frac{dx^0}{d\sigma}, \frac{dx^1}{d\sigma}, \frac{dx^2}{d\sigma}, \frac{dx^3}{d\sigma}\right) = \left(\frac{1}{\sqrt{1-v_F^2}}, \frac{v^1}{\sqrt{1-v_F^2}}, \frac{v^2}{\sqrt{1-v_F^2}}, \frac{v^3}{\sqrt{1-v_F^2}}\right),$$

(where v^1, v^2, v^3 are the components of the space-velocity c_F^*), which "transform as a vector". That is, the rule on the right produces four functions for each choice of affine chart $X \to R^4$, with the results for different bases appropriately related (cf. VII.4.04). (When one considers how many rules producing four functions for each chart do <u>not</u> have this property, it always seems rather wonderful when it appears. Unless one defines the vector first, then derives the components, so that it is true automatically.)

2.05 Momentum

Notation: we shall henceforth avoid the use of p to denote a point.

One usually first meets momentum as the contravariant vector $\underset{\sim}{p} = m\underset{\sim}{v}$, "mass times velocity". A little surprisingly, it is fundamentally a <u>covariant</u> vector as it arises physically.

There are a number of reasons for this. For example, in Newtonian mechanics a typical force acting on a particle is the gradient of a potential, Φ say. But $d\Phi$ is a one-form. So to have

"rate of change of momentum = force"

one must either have $\underset{\sim}{p}$ covariant or $d\Phi$ contravariant -

either $\quad$ "$\frac{d}{dt}(m\underset{\sim}{v}) = \underset{\sim}{G}_{\uparrow}(d\Phi)$" $\quad$ or $\quad$ "$\frac{d}{dt}(\underset{\sim}{G}_{\downarrow}(m\underset{\sim}{v})) = d\Phi$".

The second approach has several advantages. For instance, a simpler right hand side to integrate when we want the total change in the quantity differentiated on the left (whatever the exact meanings of the differentiation and integration). Moreover, it turns out that a Hamiltonian is best defined as a function $H : T^*M \to R$, where M is the space of possible configurations for some system. Then $\underset{\sim}{f} \in T^*_qM$ rather than $\underset{\sim}{G}_{\uparrow}(\underset{\sim}{f})$ is used as a description of the momentum and H suffices to define a flow ϕ on T^*M entirely without further reference to $\underset{\sim}{G}$. ($\underset{\sim}{G}$ is of course usually involved in defining H.) If the system has configuration $q \in M$ and momentum $\underset{\sim}{p} \in T^*_qM$ at time 0, it will have momentum $\phi(\underset{\sim}{p},t) \in T^*M$ at time t and position $\Pi^*(\phi(\underset{\sim}{p},t))$ in M (cf. VII.3.03). Thus ϕ fully describes the motions of the system. If we worked in TM, $\underset{\sim}{G}$ would be spuriously present in our computations, obscuring the essential geometry of what is happening. (The geometric treatment of classical mechanics needs differential forms, which we have had to defer to a later Volume. The reader is referred to [Abraham and Marsden], [Souriau] or - most easily read - [Maclane (1)] for good geometric accounts.)

This irrelevance of $\underset{\sim}{G}$ once H is defined is not hard to prove in classical mechanics, but it seems as strange there as does the fact that variational principles work (cf. IX. §4, initial remarks). The explanation is again that classical theory is a limiting case of that more deeply comprehensible system, quantum mechanics.

Let us therefore give a sketchy account of quantum momentum, hoping to make the reader more receptive to covariant momentum vectors in what follows. The quantum description of something's motion is a wave. A complex wave function ψ determines the probability of finding the something near a given point. The simplest non-trivial solution of all to the simplest wave equation is a scalar plane wave filling flat space,

$$\psi(x,t) = \cos(f(x)-\omega t) + i \sin(f(x)-\omega t)$$

in "space S and time T separate" language, where $\omega \in R$ and $f : S \to R$ is affine. (In "spacetime X" language it is even simpler, just $\psi(x) = \cos(g(x)) + i\sin(g(x))$ where $g : X \to R$ is affine. But we shall stay non-relativistic for the present.)

Now, this describes the wave, and hence the motion. But unless we know the particular metric, we do not know what to mean by the direction of the motion described. We know how the phase planes "$f(x) - t = \text{constant}$" change with t. But we can say how they move only if we assume that they do not "slip sideways" (Fig. 2.2): that they move in a direction orthogonal to themselves. Thus if we define $\underset{\sim}{f}$ on free vectors as the linear part of f, the velocity is $\underset{\sim}{v} = \underset{\sim}{G}_{\uparrow}(\underset{\sim}{f})$ for $\underset{\sim}{G}$ our choice of metric. But, up to a constant m say, the covariant momentum $\underset{\sim}{G}_{\downarrow}(m\underset{\sim}{v}) = m\underset{\sim}{f}$ is already contained in the geometry, independently of $\underset{\sim}{G}$.

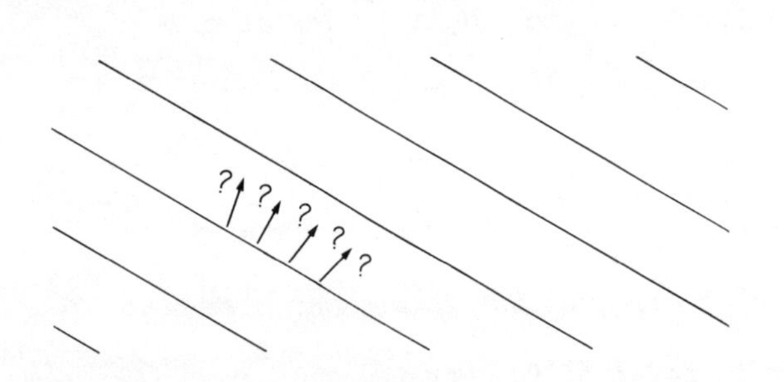

Fig. 2.2

It is appropriate in the plane wave solution to have $\underset{\sim}{v}$ a free vector, since this solution corresponds to knowing momentum exactly and correspondingly (satisfying the uncertainty principle) position not at all. A more complicated, but more useful solution can govern the motion by a "wave packet" localised at some time in a region U of S, like a short train of waves going down a canal. This can still be approximated by a plane wave at each point, if the differential calculus is to be trusted, but the results at different places will be different (think of ripples on a pond). So our momentum $\underset{\sim}{f}$ becomes a distinct linear functional for each $x \in S$, hence a cotangent vector field (zero outside U). What we measure as momentum will be $\underset{\sim}{f}_x$ at the point x where we happen to find the particle (roughly speaking) and this becomes less accurately predictable as position becomes more so, as the smaller a U is allowed for $\underset{\sim}{f}$ to be non-zero the more it must vary.

The quantum-mechanical Hamiltonian governs the evolution of ψ by the Schrödinger equation. Thereby it controls the changes in the region $U \subseteq S$, in which the particle is localised, and the covariant vector field $\underset{\sim}{f}$, with $\underset{\sim}{f}|S\backslash U$ zero. In the classical limit (as Planck's constant is taken to 0) U shrinks to a point position $q \in S$ and $\underset{\sim}{f}$ yields a classical momentum $\underset{\sim}{p} \in T^*_q S$. Correspondingly the classical Hamiltonian controls directly the evolution of q and $\underset{\sim}{p}$, not q and $\underset{\sim}{v}$ or $m\underset{\sim}{v}$.

Non-relativistic momenta, then, are covariant vectors. Seeking to make the closest possible analogy:

2.06 Definition (temporary)

The 4-momentum at $c(\sigma)$ of a history c parametrised by proper time σ is the covariant vector $\underset{\sim}{p}(\sigma) = m\underset{\sim}{G}_{\downarrow}(c^*(\sigma)) \in T^*_{c(\sigma)}X$. Here m is a positive real number associated at $c(\sigma)$ with the history, called its mass or rest mass. (The physical definition of mass for a particle P depends on the interactions of P with other things. We will redefine it mathematically in 2.09, for a closer relation with the way it is used in physics.)

In coordinates, $\underset{\sim}{p}$ has components $mg_{ij}\dfrac{dc^i}{d\sigma}$, where $c(\sigma) = (c^0(\sigma), c^1(\sigma), c^2(\sigma), c^3(\sigma))$ by the normal formula for $\underset{\sim}{G}_{\downarrow}$ (IV.3.02). Choose an inertial frame F, an orthonormal basis ${}_F\underset{\sim}{e}_0$, $\underset{\sim}{e}_1$, $\underset{\sim}{e}_2$, $\underset{\sim}{e}_3$ for T, and a chart $\phi : X \to R^4$ giving $\underset{\sim}{d}_x(\partial_i) = \underset{\sim}{e}_i$, for all $x \in X$. This means that $\underset{\sim}{p}$ is represented by

$$(p_0, p_1, p_2, p_3) = \left(\frac{m}{\sqrt{1-v_F^2}}, \frac{-mv^1}{\sqrt{1-v_F^2}}, \frac{-mv^2}{\sqrt{1-v_F^2}}, \frac{-mv^3}{\sqrt{1-v_F^2}}\right)$$

in the notation of 2.01, 2.04, since $g_{00} = 1$, $g_{11} = g_{22} = g_{33} = -1$. Now

$$p_0 = m(1-v_F^2)^{-\frac{1}{2}} = m\left(1 + \frac{1}{2}v_F^2 + \frac{3}{8}v_F^4 + \frac{5}{16}v_F^6 + \frac{9}{32}v_F^8 + \ldots\right).$$

If v_F is small compared to the speed 1 ascribed in any frame to a null vector, this is well approximated by

$$p_0 = m + \tfrac{1}{2} m v_F^2.$$

The second term is just the classical one for <u>kinetic energy</u> of a particle. The higher order terms only become significant at "relativistic speeds" as measured by F (those for which the "correction factor" $\sqrt{1-v_F^2}$ becomes important) so they may be thought of as a correction for higher speeds to the kinetic energy.

Thus we may call (p_0-m) the ${}_F$<u>relativistic kinetic energy</u> of the history at $c(\sigma)$. This indeed agrees with the energy needed to accelerate a particle described by c from ${}_F$rest to $c^*(\sigma)$, as measured in F. But energy is not really a relativistic notion, depending as it does on the frame. (Even in Newtonian mechanics the energy $\tfrac{1}{2}mv^2$ is not absolutely defined, as there has been no meaning attached to "absolute zero velocity" since physics abandoned the viewpoint of Aristotle.)

The restriction of $\underset{\sim}{p} : T_{c(\sigma)}X \to R$ to entirely ${}_F$spacelike vectors is

$$\underset{\sim}{u} \rightsquigarrow \frac{m}{\sqrt{1-v_F^2}} \underset{\sim}{u}.c^*_F(\sigma),$$

$$\text{or in components } u^i\partial_i \rightsquigarrow \frac{m\, u^i p_i}{\sqrt{1-v_F^2}} \quad \text{(summing over } i = 1, 2, 3).$$

We call $p_0 = m/\sqrt{1-v_F^2}$ the ${}_F$<u>relativistic mass</u> $m(v_F)$. Using it the map is

$$\underset{\sim}{u} \rightsquigarrow m(v_F)\underset{\sim}{u}.c^*_F(\sigma) \quad \text{or} \quad u^i\partial_i \rightsquigarrow m(v_F)u^i p_i$$

which is just the classical, space and time separate, momentum $m\underset{\sim}{G}_{\downarrow}(\underset{\sim}{v})$ except for the "corrected" mass (and our choice of sign for the metric). Notice that this device translates the "kinetic energy" part of the p_0 into "mass", not just the "rest mass" part: p_0 can be regarded as entirely "mass", and justifies this in terms of F by giving the "resistance to acceleration". In 2.07 we translate the "rest mass" part into "energy".

Classical momentum, like classical energy, depends on a choice of rest velocity and so is not absolutely defined. Special relativity fuses these two observer-dependent quantities into the geometrically defined 4-momentum, which requires no arbitrary choices for its definition. In this way the theory is much more "absolute" than Newtonian mechanics, most of whose propositions are relative to a choice of inertial frame.

2.07 Collisions, mergers, splits.

We have not yet modelled any forces as influences on our histories. Until we do so, we shall assume that the histories are the geodesics of flat spacetime - straight lines, affinely parametrised by arc length - in an obvious analogy to Newton's First Law. What happens when two or more collide?

Without considering the short range forces involved, it is natural to require the sum of the 4-momenta afterwards to equal the sum of the 4-momenta before. This corresponds to the Newtonian conservations of energy and momentum separately, and implies them as low speed approximations in a particular inertial frame. But it does not reduce to them, even in the low speed approximation - it is more general, and simpler.

Newtonian mechanics requires momentum to be conserved in all collisions. But for conservation of energy it requires either that the collision be "perfectly elastic" or that the description of the particles include details of the ways their internal structure can absorb the energy that does not reappear as gross movement. These ways always involve heat, and hence raise questions of statistical mechanics and thermodynamics. Thus Newtonian conservation of energy is simple enough for school texts when describing idealised billiard balls, but it becomes very complicated to say anything interesting about the collision of two balls of wet putty.

Relativistic mechanics, by contrast, requires conservation of the whole 4-momentum for all collisions - whether the histories bounce "elastically" (Ex. 3) or soggily, or stick together, the sum of the 4-momenta must remain unaltered.

Of course, if a vector is unaltered its individual components in any given basis cannot change either. Thus for a given frame F the sum of the $p^1 = mv^1/\sqrt{1-v_F^2}$ for the various histories must be conserved, and similarly for the p^2 and p^3; for v_F small this approximates the conservation of Newtonian momentum. We may therefore call the purely spacelike vector $\underset{\sim}{p}_F = p_1 dx^1 + p_2 dx^2 + p_3 dx^3$ the $_F$<u>momentum</u>. But conservation of p_0 implies something new. Consider for instance appropriate histories for two blobs of putty which travel towards each other with great speed, meet and merge into one blob at $_F$rest. Let them have $_F$velocities $\underset{\sim}{v}_1$, $\underset{\sim}{v}_2$, $_F$relativistic kinetic energies E_1, E_2, and rest masses m_1, m_2 before the collision. Then

$$\text{Total } p_0 \text{ before merger} = (m_1+E_1) + (m_2+E_2).$$

$$\text{Total } p_0 \text{ after merger} = m \quad , \text{ the rest mass of the new blob.}$$

$$\therefore \; m = (m_1+m_2) + (E_1+E_2)$$

so the total rest mass present has increased by E_1+E_2, "energy has become mass". Rest mass is not conserved, though $\underset{\sim}{p}$ and p_0 are.

Conversely, run the same collision backward: we end up with less rest mass, more kinetic energy than we started with. Making a structure of static matter above a city fly apart into two (or many more) pieces at high speeds turns a few micrograms of its mass into a disastrous quantity of energy. (Many of the "pieces", in practice, may have <u>zero</u> rest mass: cf. 2.10).

We shall therefore refer to m equally as rest <u>mass</u> or as rest <u>energy</u> and p_0 as

$_F$energy; mass and energy, as concepts, merge. Notice that both are anyway relative to a choice of inertial frame - changing frame changes them and their sum, along with momentum. Mass-energy and momentum are not equivalent in the arithmetic way that mass and energy are, however, but related in the same more subtle way as time and distance measurements are to each other.

2.08 Unnatural units

In this subsection (but nowhere else in Chapters I - XII), c refers to the number "speed of light". Numerically it is close to 3×10^8 metres per second.

The discussion in Ch. O. §3, without the special choice of units, would lead to a Lorentz metric given by

$$\underset{\sim}{G}'(\underset{\sim}{x},\underset{\sim}{y}) = c^2x^0y^0 - x^1y^1 - x^2y^2 - x^3y^3$$

and to equivalent but more complicated formulae. The "one second per second in time, no change in space" perception of own movement by a physicist, whose motion we describe by an affine history, gives a 4-velocity (1,0,0,0) in an "appropriate frame" (cf. 2.01). The metric $\underset{\sim}{G}'$ assigns this a length c, not 1, so proper time σ differs by the factor c from "arc length", as parameters for curves. The time-dilation formula (2.02) becomes in these new units:

$$\frac{d\sigma}{ds}(t) = \sqrt{1-v_F^2/c^2}.$$

The Lorentz-Fitzgerald contraction (2.03) becomes

$$\ell' = \ell/\sqrt{1-v_F^2/c^2}.$$

The components of 4-velocity (2.04) become

$$\left(\frac{1}{\sqrt{1-v_F^2/c^2}}, \frac{v^1}{\sqrt{1-v_F^2/c^2}}, \frac{v^2}{\sqrt{1-v_F^2/c^2}}, \frac{v^3}{\sqrt{1-v_F^2/c^2}} \right),$$

and correspondingly for 4-momentum $m(\underset{\sim}{G}'_{\downarrow}(\underset{\sim}{v}))$

$$\left(\frac{mc^2}{\sqrt{1-v_F^2/c^2}}, \frac{-mv^1}{\sqrt{1-v_F^2/c^2}}, \frac{-mv^2}{\sqrt{1-v_F^2/c^2}}, \frac{-mv^3}{\sqrt{1-v_F^2/c^2}} \right),$$

by IV.3.02, since $g_{00} = c^2$ and otherwise $g_{ii} = -1$. Hence, expanding as in 2.06 we get

$$p_0 = mc^2 + \tfrac{1}{2} mv_F^2 + \tfrac{3}{8}\, m \frac{v_F^4}{c^2} + \ldots$$

and the famous equation

$$E = mc^2$$

for the rest energy. But we shall not use this "unnatural units" factor again. A clear discussion of conversions among commonly encountered systems of units will be found, among many other things, in [Synge].

2.09 Definition 2.06 reconsidered

In Newtonian terms, mass is the most fundamental quantity in sight. It is the "quantity of matter" in the particle, body, system ... under consideration, conserved by the dynamics and by any change of inertial frame. So we were led, by the Newtonian idea of particles as bits of mass flying about, into defining $\underset{\sim}{p}$ in terms of m. But since m is _not_ conserved in collisions, this is backwards. In physics, what is more conserved is more fundamental. Thus, 4-momentum is more fundamental than energy is more fundamental than heat is more fundamental than temperature. So we shall from now on think of particles as bits of 4-momentum flying about.

More precisely, since when a particle can interact with a field its 4-momentum can change continuously, we think of the particle as its history c and a non-zero cotangent vector field $\underset{\sim}{p}$ along c, such that $\underset{\sim}{G}_{\uparrow}(\underset{\sim}{p}(\sigma))$ is always a scalar multiple of $c_*(\sigma)$. (Since a classical particle, unlike a field or a probability wave, does have a well defined "direction" for its motion, 4-momentum should be "in that direction" according to $\underset{\sim}{G}$.) We can then <u>define</u> rest mass m by

$$m^2 = \underset{\sim}{p}\cdot\underset{\sim}{p}$$

(Ex. 4a) and let the question of whether it is conserved depend on the detailed physics of the particle and the ambient field. For a re-entering space module it is not, for an electron it is conserved as long as the electron is. To call either - or anything - a "particle" is an approximation, if "particle" is not given a new meaning. Relativistic quantum field theory may provide such a meaning.

2.10 Zero rest mass

In 2.06 we were restricted to a <u>timelike</u> history, because only on this had we a <u>unit</u> forward tangent vector as 4-velocity. So the history of a "light particle" or <u>photon</u> following a <u>null</u> curve c, however parametrized, will have

$$m^2 = \underset{\sim}{p}\cdot\underset{\sim}{p} = \underset{\sim}{G}_{\downarrow}(c_*(t))\cdot\underset{\sim}{G}_{\downarrow}(c_*(t)) = c_*(t)\cdot c_*(t) = 0,$$

<u>zero</u> rest mass. We know photons have momentum - light pushes things - but there is no frame-independent way of assigning them an m that makes the old $m\underset{\sim}{v}$ definition work.

Suppose a zero rest mass particle "slows down" onto a timelike curve from its initial null world line, to go at less than the limiting speed, while maintaining the zero rest mass character that is part of its identity. (If a gamma ray "slows" into an electron-positron pair then rest mass appears, but we consider that the gamma ray disappears.) It then satisfies

$$\underset{\sim}{p} = O\underset{\sim}{G}_{\downarrow}(c^*(t)) = \underset{\sim}{0}$$

(Ex. 4b). Having neither energy nor momentum, it no longer exists. So a zero rest mass particle can _only_ travel at the limiting speed.

(Photons, incidentally, are _not_ slowed down in their character as electromagnetic or probability waves by air, glass or whatever non-vacuum transparency they meet: only their _group velocity_ is. See [Feynman] for an excellent account of this distinction.)

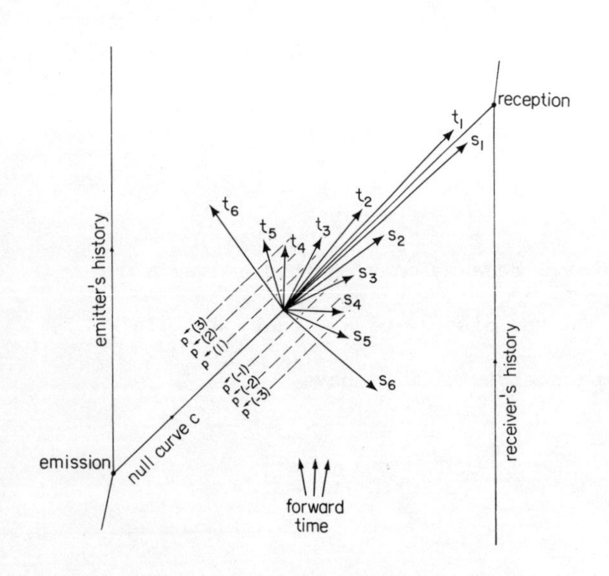

Fig. 2.3

Consider now a zero rest mass history c, with the wave aspect of the particle whose motion we wish to describe approximated around c(t) by a plane wave (cf. 2.05). Its 4-momentum $\underset{\sim}{p}$ at c(t) is the linear functional with the contours shown (Fig. 2.3) Notice that this pattern does correspond to something moving in the direction c*(t) : an observer passing through c(t) with timelike 4-velocity $\underset{\sim}{t}_i$ (choice of 6 shown) will perceive wavefronts moving to the right at limiting speed. As for any functional, its timelike component p_0 is given by its value on the timelike basis vector, here $\underset{\sim}{t}_i$, and its spacelike components p_1, p_2, p_3 by its values on the chosen spacelike vectors $(\underset{\sim}{s}_i)_1$, $(\underset{\sim}{s}_i)_2$, $(\underset{\sim}{s}_i)_3$ orthonormal to $\underset{\sim}{t}_i$. (One $\underset{\sim}{s}_i$ represents these three for each i in Fig. 2.3, for dimensional reasons.) So the observed energy E is just p_0, and the size p of the observed "ordinary momentum" has $p^2 = p_1{}^2 + p_2{}^2 + p_3{}^2$. Thus $E^2 = p^2$ necessarily, since $\underset{\sim}{p}$ is a null covector.

The energy E can be seen to be the number of wavefronts cut by a unit timelike vector (this is frequency). Likewise, p is the number of wavefronts cut by a unit purely spacelike (to the observer) vector in the direction of travel (this is wave number, 1/wavelength, up to sign).

E and p do correspond to the energy and momentum, as measured in the inertial frame F with ${}_F\underset{\sim}{e}_0 = \underset{\sim}{t}_i$, transferred from emitter to receiver. As can be seen, an observer fleeing the emitter will report lower energy and longer wavelength, an observer advancing on it will report the reverse. (These effects are the well known red and violet shifts respectively, so called because these colours are the low and high energy ends of the visible spectrum. Together they are called the Doppler effect.)

Exercises XI.2

1) Prove Lemma 2.03. Does this result apply to an accelerating observer, measuring distances by his "instantaneous inertial frame" F given by ${}_F\underset{\sim}{e}_0 = c^*(\sigma)$, or does it require corrections for this case like the time dilation formula (2.02)? What lengths would he assign to an inertially moving stick?

2) If F is an inertial frame, show that for an orthonormal basis $\underset{\sim}{e}_0$, $\underset{\sim}{e}_1$, $\underset{\sim}{e}_2$, $\underset{\sim}{e}_3$ with $\underset{\sim}{e}_0 = {}_F\underset{\sim}{e}_0$ and an affine chart $X \to R^4$ giving $\underset{\sim}{d}_x(\partial_i) = \underset{\sim}{e}_i$, $c^*(\sigma)$ has the component form given in 2.04.

3) What is the definition of a "perfectly elastic" Newtonian collision? Define

a perfectly elastic relativistic collision without referring to an inertial frame.

4a) Show that p defined as $m\underset{\sim}{G}_{\downarrow}(c^*(\sigma))$ satisfies $p.p = m^2$, if $c^*(\sigma)$ is timelike.

b) Show that m defined as $\sqrt{p.p}$ satisfies $p = m\underset{\sim}{G}_{\downarrow}(c^*(\sigma))$ if p is timelike and $c^*(\sigma)$ is a scalar multiple of $\underset{\sim}{G}_{\uparrow}(p)$.

5) Suppose two zero rest mass particles travelling the same null curve have energies E_1, E_2 at some point, as measured in a frame F. Show that their energies E_1', E_2' relative to another frame F' give

$$\frac{E_1'}{E_1} = \frac{E_2'}{E_2} .$$

(Red shifts are ratios, independent of energy, frequency, wavelength.)

3. FIELDS

Matter does not really come in particles of zero size. In Newtonian mechanics the centre of mass of a rigid body moves under gravity like a particle with the mass of the whole body, which makes the "particle" idealisation a handy one. Relativistically it is less reasonable: there is no way to define a rigid body, and the most interesting features of a zero rest mass entity are that its energy "is" frequency and its momentum "is" wave number, which cannot even be spoken of while it is regarded as entirely point-concentrated. We shall not refine into rigor the wave packet view used above by getting deeper into quantum mechanics, however. Rather, let us look at matter spread out smoothly, moving through spacetime, as classical hydrodynamics for instance considers it spread smoothly (and infinitely divisibly) through space.

3.01 Describing a flux of matter

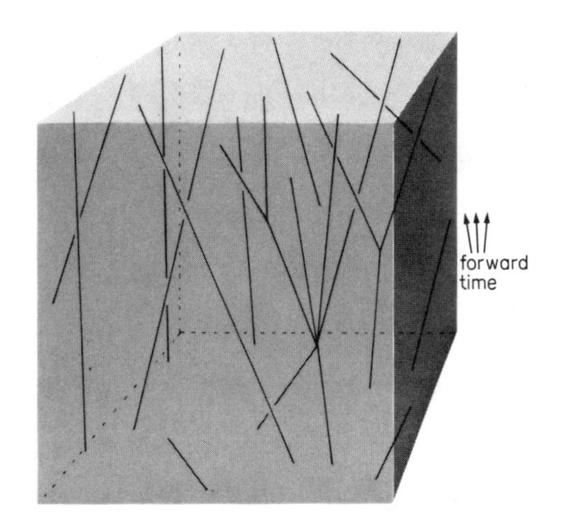

Fig. 3.1

First, let us mentally approximate a smooth flux of matter by a crowd of particles, colliding, merging and splitting (Fig. 3.1), following timelike or null curves. (This is a crutch towards a more precise concept.) Now, 4-momentum is carried along each path: thus we have a flux of 4-momentum forward in time along the network that the paths make. To know how the 4-momentum is distributed across a particular spacelike section is to know "where the matter is" in it. (Though not where each bit of matter in some earlier spacelike section "now is" - particles lose their individuality in an inelastic relativistic collision, just as in even a non-relativistic quantum one like two electrons bouncing off each other.) As in classical fluid mechanics, then, we wish in the smooth version to describe a flux. A flux of the vector quantity 4-momentum, not the scalar quantity mass; but let us consider for a moment the simpler problem of modelling a classical fluid.

While a description by the velocity vector field is informative (VII.6.01) it is not quite complete for practical purposes unless the fluid is incompressible. If the density ρ of the fluid in Fig. 3.2 is greater at the middle of the pipe than its value,

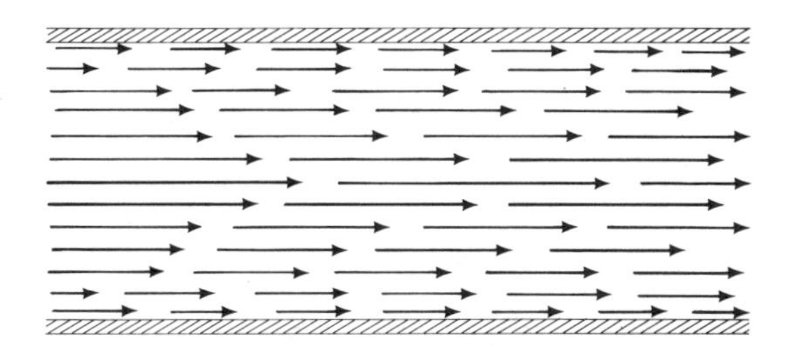

Fig. 3.2

ρ_B say, at the boundary, we get a larger total flux than if it is equal to or less than ρ_B. So the field we want should describe how <u>much</u> fluid is passing a given point, not just how <u>fast</u> it is going. If $\rho(x)$ is density at x, then $\rho\underset{\sim}{v}$ is a natural candidate field. But how, geometrically, do we use it in questions of "total flux"? Once again we have to get the variance right.

To find the flux through a given surface, we must integrate the flux per unit area through the surface. As in X.2.01, this means that we must assign at each point a quantity per unit area to each plane P in T_xX, which in this case will mean the "flux density at x through a surface passing through x with the attitude P". For a flux of mass this will be a scalar for each P, summed up by a skew-symmetric bilinear form $T_xX \times T_xX \to R$, just as curvature is skew-symmetric and bilinear $T_xM \times T_xM \to$ {"infinitesimal rotations"}. For the flux of a covariant vector quantity it will be $T_xM \times T_xM \to T_x^*M$.

But this is per unit <u>area</u>. Fine for 3-dimensional flows, where we can find the net flow into or out of a region by integrating over its 2-dimensional boundary but in four dimensions we shall have to integrate over 3-dimensional hypersurface boundaries. Which would make the flow of a vector quantity a 4-tensor. Fortunately there is a less cumbersome description of it. Rather than fix a plane in 3-space by giving two vectors that span it, or a hyperplane in 4-space by giving three, we can in both cases give it as the kernel of a cotangent vector, on 3-space or 4-space respectively. But if $P = \ker \underset{\sim}{f}$, then also $P = \ker(\lambda\underset{\sim}{f})$ for any $0 \neq \lambda \in R$. How do we choose one particular covector to label P?

In the Riemannian situation we can choose by requiring $\underset{\sim}{f}.\underset{\sim}{f} = 1$, and $\underset{\sim}{f}(c^*(t)) > 0$ when c with $c(t) = x$ is a curve crossing P in what has been chosen as the "positive" direction for the surface: just choose $\underset{\sim}{f} = \underset{\sim}{G}_\downarrow(\underset{\sim}{n})$, where $\underset{\sim}{n}$ is the "positive unit normal" to P, in the language of electromagnetism texts. But if $\underset{\sim}{G}$ is indefinite, P may be degenerate. In this case all the vectors normal to P are <u>in</u> P, not crossing it (Fig. IV.1.5d); equivalently $\underset{\sim}{f}.\underset{\sim}{f} = 0$ for any f with ker $\underset{\sim}{f}$ = P, so there is no unit covector to label P. (And a smooth closed hypersurface such as bounds a compact region in flat spacetime must have some degenerate tangent hyperplanes, by simple topological arguments: Fig. VII.3.7). So we must be a little more subtle.

Instead of simply representing P, let $\underset{\sim}{f}$ represent an actual piece A of area with attitude P: say the parallelogram fixed by two vectors $\underset{\sim}{u},\underset{\sim}{v}$ (Fig. 3.3). (In spacetime, a piece A of volume in the hyperplane P, fixed by three vectors $\underset{\sim}{u},\underset{\sim}{v},\underset{\sim}{x}$.) Then we can define $\underset{\sim}{f}(\underset{\sim}{w})$ as the volume (hypervolume) of the parallelepiped Q fixed by $\underset{\sim}{u},\underset{\sim}{v},\underset{\sim}{w}$ (hyperparallelepiped fixed by $\underset{\sim}{u},\underset{\sim}{v},\underset{\sim}{x},\underset{\sim}{w}$), which is well defined as $\underset{\sim}{G}$ is non-degenerate on the whole space, even if P is degenerate. The technicalities (in particular those that fix the sign of $\underset{\sim}{f}$) are gathered in Ex. 1, but geometric understanding is more important in what follows.

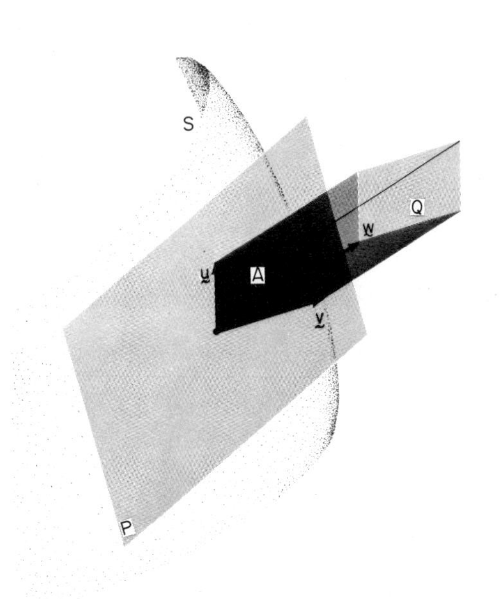

Fig. 3.3

The covector $\underset{\sim}{f}$ labels A in a thoroughly "per area" way: if A is doubled (say by doubling $\underset{\sim}{u}$) so is $\underset{\sim}{f}$, but another A in P with the same area (by any skew-symmetric bilinear measure) as A will give the same $\underset{\sim}{f}(\underset{\sim}{w})$, for any $\underset{\sim}{w}$, as A. If $\underset{\sim}{G}$ is Riemannian, we have a well defined idea of "unit area" on P (say if $\underset{\sim}{u},\underset{\sim}{v}$ are orthonormal, the area of A is 1: cf. X.5.01) and for A a unit area, $\underset{\sim}{f}(\underset{\sim}{n}) = 1$ (Ex. 1e) so in this case $\underset{\sim}{f}$ is $\underset{\sim}{G}_{\downarrow}(\underset{\sim}{n})$ as before - our new method agrees with the old, where that one works.

So, we label each "area (volume) in a plane (hyperplane) in T_xX, with a choice of which way through is positive" by a cotangent vector $\underset{\sim}{f} \in T^*_xX$. A description of the flux of any $\binom{k}{h}$-tensor quantity should say how much of this quantity goes positively through such an "infinitesimal area (volume)" at each x. That is, at each point $x \in X$ we should have a linear (why?) map taking $\underset{\sim}{f}$ to a $\binom{k}{h}$-tensor in $(T^k_h X)_x$. That is a map $T^*_x X \to (T^k_h X)_x$, equivalently an element of $L(T^*_x X; (T^k_h X)_x) \cong (T^{k+1}_h X)_x$. For a scalar flux such as mass or heat, this means a vector in $T_x X$, since $k = h = 0$; actually a contravariant vector at x, but used as a linear functional

$$T^*_x \to R: \left(\text{area (volume) labelled by } \underset{\sim}{f} \in T^*_x X\right) \leadsto \left(\text{flux through}\right).$$

In the case of classical fluid mechanics the contravariant field appropriate is exactly $\rho\underset{\sim}{v}$, where ρ is density and $\underset{\sim}{v}$ is velocity, as we saw at first. But in general such a decomposition is not possible.

3.02 The flux of 4-momentum.

We see then that a flux of the $\binom{0}{1}$-tensor quantity 4-momentum should be a field of maps $\underset{\sim}{T}_x : T^*_x X \to T^*_x X$, which are thus actually operators on $T^*_x X$: otherwise described via the natural isomorphism $L(V;V) \cong V^* \otimes V$ as a $\binom{1}{1}$-tensor field $\underset{\sim}{T}$ on X. (Not to be confused with $\underset{\sim}{T}$ the torsion $\binom{1}{2}$-tensor VIII.5.05, which is always zero in relativity theory as we use the Levi-Cività connection.) This is variously called the <u>matter tensor</u> for the flux of matter described, the <u>energy-momentum tensor</u> for it, or the <u>stress</u> or <u>stress-energy tensor</u> for reasons apparent in 3.04.

3.03 Components

Choose an inertial frame (cf. 2.01) F for X and a chart $X \to R^4$ with the ∂_i orthonormal everywhere, $\partial_0 = {}_F\underset{\sim}{e}_0$. In the manner of 3.01, the covector $dx^0 = \underset{\sim}{G}_\downarrow({}_F\underset{\sim}{e}_0)$, with kernel ${}_F\underset{\sim}{e}_0$, represents unit volume in the 3-space of entirely ${}_F$spacelike vectors at a point x. So

$$\underset{\sim}{T}(dx^0) = T^0_0 dx^0 + T^0_1 dx^1 + T^0_2 dx^2 + T^0_3 dx^3$$

is the amount of p of 4-momentum "passing through the $_F$present" of x, per unit volume. Since "passing through the present" just means "now", $\underset{\sim}{T}(dx^0)$ can be understood as the $_F$<u>density</u> of 4-momentum at x. (Notice that this <u>must</u> depend on F. Density as "amount per unit volume" depends on "unit volume" which depends on "unit spacelike length" which depends on the observer.) Separating this into $_F$timelike and $_F$spacelike parts as in 2.07, we see that T^0_0 then represents $_F$<u>energy density</u> and $(T^0_1 dx^1 + T^0_2 dx^2 + T^0_3 dx^3)$ represents $_F$<u>momentum density</u>, relative to F. If the matter whose motion is being described is a fluid or solid moving at less than the speed of light, the mass-energy density T^0_0 can be seen (relative to F) as a particular combination of rest mass density and density of energy stored in the elastic forces in the material.

Similarly for i = 1, 2 or 3, $\underset{\sim}{T}(dx^i) = T^i_j dx^j$ represents 4-momentum going "sideways"; the flux per unit area through the hypersurface orthogonal to ∂_i, in a sense determined by ∂_i. (Note the difference between "sense" and "direction". For instance, "We can finally see the Midnight Sun, having crossed the Arctic Circle, Northwards" refers to the <u>sense</u>. We might have been travelling in the <u>direction</u> East-North-East, not due North, as we crossed.) To an observer at $_F$rest, then, $\underset{\sim}{T}(dx^i)$ is seen as the flux of 4-momentum per unit time, per unit area - time and space being illusorily unglued by his intrinsic, local chart - across a surface orthogonal to ∂_i and at rest. Taking this apart as we did for $\underset{\sim}{T}(dx^0)$, T^i_0 is the i-component of the flux of $_F$energy he sees, which is thus described by the vector (as appropriate for the flux of a scalar) $T^1_0\partial_1 + T^2_0\partial_2 + T^3_0\partial_3$. The covector $T^i_1 dx^1 + T^i_2 dx^2 + T^i_3 dx^3$ is the i-component of $_F$momentum flux.

<u>3.04 Self-adjointness of $\underset{\sim}{T}$</u>

Just as $_F$momentum is $\underset{\sim}{G}_\downarrow$($_F$mass times $_F$velocity), for the above idealisation of particles going at less than lightspeed, one expects $_F$<u>momentum density</u> for continuous matter to be $\underset{\sim}{G}_{\downarrow}$ of

$$({}_F\text{mass-energy times }{}_F\text{velocity) per unit volume}$$

which can be rearranged as

$$({}_F\text{energy per unit volume) times }{}_F\text{velocity;}$$

namely, $_F$energy flux. In the component forms above for $_F$momentum density and $_F$energy flux, this gives the equation

$$* \qquad (T^0_1 dx^1 + T^0_2 dx^2 + T^0_3 dx^3) \;=\; \underset{\sim}{G}_{\downarrow}(T^1_0\partial_1 + T^2_0\partial_2 + T^3_0\partial_3).$$

Since $\underset{\sim}{G}_{\downarrow}(\partial_0) = dx^0$ by construction and $\underset{\sim}{G}_{\downarrow}$ is linear, this implies that

$$\begin{aligned}
(T^0_0 dx^0 + T^0_1 dx^1 + T^0_2 dx^2 + T^0_3 dx^3) &= \underset{\sim}{G}_{\downarrow}(T^0_0\partial_0 + T^1_0\partial_1 + T^2_0\partial_2 + T^3_0\partial_3),\\
\underset{\sim}{T}(dx^0) &= \underset{\sim}{G}_{\downarrow}(\underset{\sim}{T}^*(\partial_0)) \quad \text{by III.1.06}\\
&= \underset{\sim}{G}_{\downarrow}\underset{\sim}{T}^*\underset{\sim}{G}_{\uparrow}(dx^0)\\
&= \underset{\sim}{T}^T(dx^0) \quad (\text{cf. IV.2.08;}\\
&\qquad \text{also, } \underset{\sim}{T} \text{ is } T^*_xX \to T^*_xX, \text{ not } T_xX \to T_xX)
\end{aligned}$$

So for the unit timelike covariant vector dx^0, $\underset{\sim}{T}^T$ has the same effect as $\underset{\sim}{T}$. But <u>any</u> unit forward timelike covariant vector could have been dx^0: this is just a matter of labelling, not physics. So for any unit forward timelike covariant vector, $\underset{\sim}{f}$,

$$\underset{\sim}{T}(\underset{\sim}{f}) = \underset{\sim}{T}^T(\underset{\sim}{f}).$$

But since we can find a basis $\underset{\sim}{f}^0, \underset{\sim}{f}^1, \underset{\sim}{f}^2, \underset{\sim}{f}^3$ for T^*_xX consisting only of such vectors, this means that for any $\underset{\sim}{g} \in T^*_xX$ whatever,

$$\underset{\sim}{T}(\underset{\sim}{g}) = \underset{\sim}{T}(g_i\underset{\sim}{f}^i) = g_i(\underset{\sim}{T}\underset{\sim}{f}^i) = g_i(\underset{\sim}{T}^T\underset{\sim}{f}^i) = \underset{\sim}{T}^T\underset{\sim}{g},$$

$\underset{\sim}{T}$ is self-adjoint.

This illustrates beautifully the advantages of not restricting ourselves to the bases developed in VII §4, or always requiring orthogonality. Usually * is produced as above, and the equations

$$T^i_j = T^j_i \text{ for } 1 \le i, j \le 3$$

by an entirely separate physical argument. (See, for example, [Misner, Thorne and Wheeler].)

Unfortunately, <u>both</u> physical arguments are invalid. In fact, there is an extensive literature on <u>polar</u> materials (those with $\underset{\sim}{T}^T \neq \underset{\sim}{T}$), beginning in 1907. (See [Truesdell] for an account.) However it has been the concern chiefly of solid state physicists, in little contact with cosmologists who are overwhelmingly gaseous.

The catch in the argument for * was the purely $\underset{\sim}{G}_{\downarrow}$ (mass times velocity) view taken of momentum, for continuous matter as for particles. We have not discussed <u>angular</u> momentum, and cannot do so geometrically until we have the Lie group language deferred to a later volume. But if the particles in 3.1 have it, they carry <u>torque</u> as well as pressure, tension and shear effects. Of course the smaller a ball is, the faster it has to spin to have a given non-zero angular momentum; but the absurdity of the "limiting notion" of a point particle spinning infinitely fast just shows that angular momentum has to be more subtly conceived. The dipole (a point particle with an electric field but no net charge) is closely analogous. Ultimately the point mass gives far more trouble than any other attribute of a particle (cf. XII.4); as field quantities dipole moment and angular momentum, properly relativised, cause no more paradoxes than mass-energy density.

That being said, symmetric stress tensors are in the practical analysis of matter far more common. We shall return to the polar case in the next volume, when we examine the tensor geometry of material physics: but while in this one we have sophisticated language for discussing space and spacetime, for the matter that happens in it we have really only a few pictures. So for now we keep to a non-polar view of matter. In the next chapter - where we equate $\underset{\sim}{T}$ to a tensor self-adjoint for geometrical reasons - we use implicitly, without change of name, the symmetrised matter tensor $\frac{1}{2}(\underset{\sim}{T}+\underset{\sim}{T}^{T})$. This amounts to an extra physical hypothesis (usually made tacitly): that a density of angular momentum has no influence on the curvature of spacetime.

3.06 Signs

As we remarked above, T_1^1 represents force across a surface S orthogonal to ∂_1. Which sign corresponds to pressure in the ∂_i-direction, as opposed to tension?

Think of the momentum carried by the particle p in Fig. 3.4 leaving one side (pushing back the matter q it leaves) crossing S in the positive sense according to ∂_1, and arriving on the other side, pushing the matter r there forward. This evidently represents pressure, acting to separate the two sides. The particle's 4-velocity $\underset{\sim}{v}$ has $dx^1(\underset{\sim}{v}) > 0$, since this is what "the positive sense according to ∂_1" means. Now, $dx^1(\partial_1) = 1 > 0$ by the definition of the dual basis, VII. 4.01, 4.02, so $(m\underset{\sim}{G}_{\downarrow}(\underset{\sim}{v})).dx^1$ is positive too, since this is just $m dx^1(\underset{\sim}{v})$. So if the particle has ${}_F$energy E and ${}_F$momentum $\underset{\sim}{p}_F$, and thus 4-momentum $m\underset{\sim}{G}_{\downarrow}(\underset{\sim}{v}) = E dx^0 + \underset{\sim}{p}_F$, to transfer, we have

$$(E dx^0 + \underset{\sim}{p}_F).dx^1 > 0$$

and so $$\underset{\sim}{p}_F.dx^1 > 0 \quad \text{since } dx^0.dx^1 = 0.$$

So, since dx^1 is spacelike, the 1-component of $\underset{\sim}{p}_F$ is negative.

In the smooth version, T_1^1 is the 1-component of the momentum flowing across S, approximating our "network of particles" picture. So we see that T_1^1 is negative in the case of pressure, positive for tension.

Notice that if we used a metric of signature +2 then these meanings would be reversed, so the same material situation would be represented by minus the matter tensor appropriate with the Lorentz metric we use. But the fully covariant or fully contravariant forms, $T_{ij}dx^i \otimes dx^j$ or $T^{ij}\partial_i \otimes \partial_j$, would have the same sign in either version as the minus signs on $\underset{\sim}{G}$ and $\underset{\sim}{T}$ cancel.

3.06 Principal directions

If $\underset{\sim}{T}_x$ is self-adjoint and has a timelike eigenvector $\underset{\sim}{f} \in T^*_x X$, then by IV.4.13 $T^*_x X$ has an orthonormal basis of eigenvectors of $\underset{\sim}{T}_x$. Choose an inertial frame F by making ${}_F\underset{\sim}{e}_0$ the unit forward vector in the eigenspace $\{a\underset{\sim}{f} | a \in R\}$, and an affine chart making $\partial_0, \partial_1, \partial_2, \partial_3$, orthonormal eigenvectors of $\underset{\sim}{T}_x$ with $\partial_0 = {}_F\underset{\sim}{e}_0$. The parts of the matrix of $\underset{\sim}{T}_x$ representing ${}_F$momentum density and flux of ${}_F$energy vanish in these coordinates. F may then be considered as the "instantaneous rest frame" for the described matter at x, though nothing identifiable as <u>part</u> of the mass-energy may be at rest. (For example in a solid the "solid matter" may be moving one way, the "elastic energy" travelling along it the other way relative to F, resulting in a zero net flow.) In the spacelike part too, the off-diagonal elements vanish.

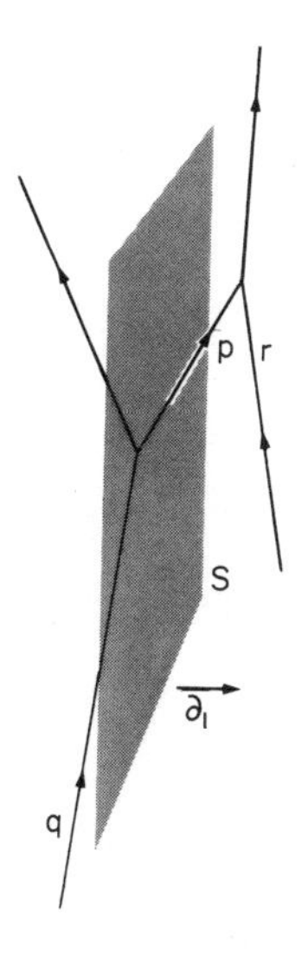

Fig. 3.4

This may be seen as the reduction of the stresses at x to pressure or tension in the three spacelike principal directions, with no shear stress in the planes orthogonal to these directions.

If $\underset{\sim}{T}_x$ describes a situation where the net flow of energy is at the speed of light (so the matter at x is present entirely as radiation, or equivalently as particles with zero rest mass, all going in precisely the same direction), it evidently has a null eigenvector. Clearly there is no frame for which the ${}_F$energy flow is zero, so it does not have a timelike one. Then IV.4.13 thus does not apply, and $\underset{\sim}{T}_x$ turns out to be in fact not diagonalisable. This highly special situation, however, is the only one among all those arising for fields so far observed in physics for which $\underset{\sim}{T}$ fails to have a full set of principal stresses.

3.07 Conservation

The conservation law for 4-momentum in collisions of particles (2.07) essentially treats a collision as happening in a small "black box": we do not know what happens in there, very often, but we insist that what goes in must equal what comes out. The same idea gives the conservation law for fields.

Consider a small parallel-sided box Q in spacetime: the sides are four pairs of parallelepipeds. We represent one pair by parallelograms in Fig. 3.5. Evidently the flow through P' does not equal that through P, in general: that would mean for instance that if P is in the "now" of x in the frame F, P' in the "now" of x', 4-momentum ${}_F$density does not change between x and x'. In fact $\underset{\sim}{T}$ would have to be a constant field, not just a conserved one. What we must require is that what gets lost (or appears) between P and P' must come out (go in) through the other six sides of the box. So if we take the four pairs P_μ, P'_μ, $\mu = 0,1,2,3$ of sides of Q we want the sum

$$\sum_{\mu=0}^{3} (\text{Flux through } P'_\mu - \text{Flux through } P_\mu)$$

to vanish. In the limit as Q approaches zero size, the 4-momentum flux through the various sides "becomes" (is increasingly well approximated by) the values of $\underset{\sim}{T}$ on the cotangent vectors $\underset{\sim}{f}^\mu$ labelling the "per area and attitude" of the sides P_μ, P'_μ. So for the limit of the above, a natural candidate is the limit of

$$\sum_{\mu=0}^{3} \left(\frac{\underset{\sim}{T}_{x'_\mu}(\underset{\sim}{f}^\mu) - \underset{\sim}{T}_x(\underset{\sim}{f}^\mu)}{\text{separation of } x \text{ and } x'_\mu} \right)$$

where x is the corner kept fixed as Q shrinks and the divisors stop the sum becoming trivially zero as the terms above approach each other.

To be a little more careful, notice that $\underset{\sim}{f}^\mu$ is strictly in $T^*_x X$, which is not the domain of $\underset{\sim}{T}_{x'_\mu}$, and that the image of $\underset{\sim}{T}_{x'_\mu}$ is not in the same vector space $T^*_x X$ as that of $\underset{\sim}{T}_x$, so they cannot strictly be subtracted. We must correct this by parallel transport. (In the particular case of flat spacetime we could go via the space of free vectors, since parallelism is independent of route, but let us be more general.) Then, if $_\mu\underset{\sim}{\tau}_h$ is parallel transport $T_x X \to T_{c_\mu(h)} X$ along the edge from x to x_μ parametrised by the curve c_μ with $c_\mu(0) = x$, the limit becomes

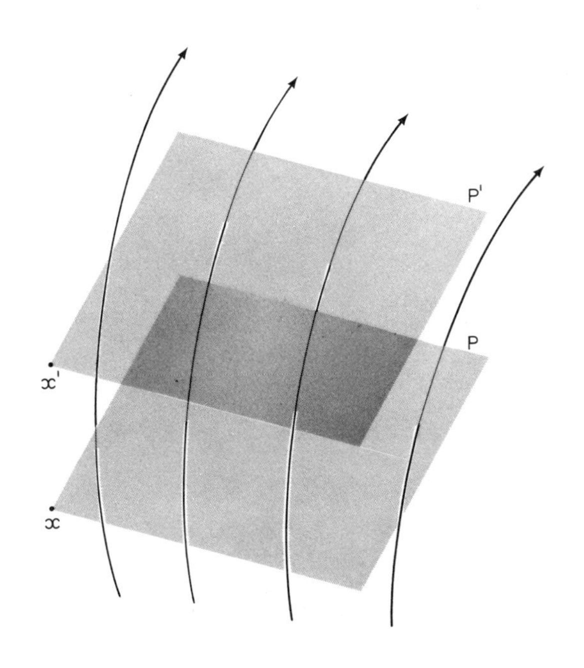

Fig. 3.5

$$\lim_{h\to 0} \sum_{\mu=0}^{3} \left(\frac{\left(\underset{\sim}{T}_{c_\mu(h)}(\underset{\sim}{f}^\mu \circ {_\mu\underset{\sim}{\tau}_h})\right) \circ {_\mu\underset{\sim}{\tau}_h} - \underset{\sim}{T}_x(\underset{\sim}{f}^\mu)}{h} \right)$$

$$= \sum_{\mu=0}^{3} \left(\lim_{h\to 0} \frac{\left[\underset{\sim}{T}_{c_\mu(h)} (\underset{\sim}{f}^\mu \circ {}_\mu\underset{\sim}{\tau}_h) \right] \circ {}_\mu\underset{\sim}{\tau}_h - \underset{\sim}{T}_x(\underset{\sim}{f}^\mu)}{h} \right)$$

$$= \sum_{\mu=0}^{3} (\nabla_{c^*_\mu(0)}\underset{\sim}{T})\underset{\sim}{f}^\mu ,$$

by reference to VIII.7.01, 7.02. Notice, though, that if we are to have a <u>box</u> as we go the limit, rather than some interlocking slices (Fig. 3.6) we must have each $\underset{\sim}{f}^\mu$ label the particular piece of area in P given by the three edge vectors $c^*_\nu(0)$, $\nu\neq\mu$. If we so parametrise the edges that the hypervolume of the box in T_xX fixed by the four $c^*_\mu(0)$ is unity, we have the $c^*_\mu(0)$ as a basis for T_xX with the corresponding $\underset{\sim}{f}^\mu$ simply the dual basis for T^*_xX (Ex. 2). Choosing a chart which gives these bases as ∂_μ, dx^μ at x,we have

Fig. 3.6

$$\nabla_{c^*_\mu(0)}\underset{\sim}{T} = \nabla_{\partial_\mu}(T^i_j\partial_i \otimes dx^j)$$

$$= T^i_{j;\mu}\ \partial_i \otimes dx^j \quad \text{by VIII.7.08,}$$

and so

$$(\nabla_{c^*_\mu(0)}\underset{\sim}{T})\underset{\sim}{f}^\nu = \text{Contraction of } (\underset{\sim}{T}^i_{j;\mu}\ \partial_i \otimes dx^j) \otimes dx^\nu \text{ over i and } \nu$$

$$= T^\nu_{j;\mu}dx^j .$$

Hence, provided the $c^*_\mu(0)$ fix a unit hypervolume,

$$\sum_{\mu=0}^{3} (\nabla_{c^*_\mu(0)} \underset{\sim}{T}) \underset{\sim}{f}^\mu = T^\mu_{j;\mu} dx^j ,$$

which by X.6.12 is called the <u>divergence</u> of the tensor field $\underset{\sim}{T}$ and is independent of the $c^*_\mu(0)$. We now have a geometric meaning for divergence in general: using the isomorphism

$$(T_x X)^1_h \cong L(T^*_x X; T^*_x X \otimes \ldots \otimes T^*_x X)$$

we think of a $\binom{1}{h}$-tensor field as describing the flux of a $\binom{0}{h}$-tensor quantity, following 3.01. Its divergence is thus "the amount of the $\binom{0}{h}$-tensor that is appearing or vanishing" per unit volume, at each point. The <u>field conservation law</u> for 4-momentum thus takes the form

$$\mathrm{div}\, \underset{\sim}{T} = \underset{\sim}{0}$$

or $$T^\ell_{i;\ell} = 0, \quad \text{in coordinates.}$$

The above reasoning, of course, did not establish this form for the law, but only motivated it. Of the quantities involved only the $\nabla_{\partial_\mu} \underset{\sim}{T}$ and $\mathrm{div}\, \underset{\sim}{T}$ have been strictly defined, in the absence of the theory of integration on manifolds that would let us talk rigorously about the flux through a hypersurface of a small but finite size. However, in flat spaces the integral of the divergence of a flux $\underset{\sim}{F}$ of any $\underset{\sim}{w}$, over a 4-dimensional region U, is exactly the total flux of $\underset{\sim}{w}$ into or out of U through its 3-dimensional boundary. So as we shrink U to a point, whether or not it is box-shaped, the limiting flux in or out per unit volume is indeed div $\underset{\sim}{F}$.

In a curved spacetime M we cannot simply add "tensors per unit volume, or hypervolume" at different points, so things are a little more complicated. But we can state the following "integral version" of the conservation law. Suppose the energy density $\underset{\sim}{T}_x(\underset{\sim}{f}).\underset{\sim}{f} \geq 0$ for all x and all choices of timelike "rest velocity" $\underset{\sim}{G}_{\uparrow}(\underset{\sim}{f})$ at x, and that the corresponding 4-momentum $_F$density is never a spacelike covector. Then, if $\underset{\sim}{T} = \underset{\sim}{0}$ everywhere on some spacelike hypersurface (cf. VII.2.03) S of M, we find that $\underset{\sim}{T} = \underset{\sim}{0}$ everywhere on M, assuming M itself has no singularities. So the absence of matter, at least, is conserved in a simple sense. From this we can deduce results on the way that determining non-zero $\underset{\sim}{T}$ on S determines $\underset{\sim}{T}$ on M, analogously to the way fixing Newtonian positions and momenta at time t_0 determines them for other t.

The conditions assumed, non-negative energy and no spacelike 4-momentum (such as would be possessed by particles going faster than light) are plainly necessary for this result. Without the first, we would find a solution where matter fields with stress tensors cancelling (some having negative energy) appear together forward of S and move off in different directions. Without the second, matter could "come in from infinity at infinite speed", following curves that stay forward of S, and decay into ordinary matter that then proceeds forward in time. So in neither case would the solution, even for vacuum initial conditions, be unique.

(It is in this sense, the uniqueness of solutions, that we referred to determining $\underset{\sim}{T}$ - the existence of solutions only holds locally. For example, in flat spacetime X start with a matter field in a compact region of some spacelike submanifold S with the matter having no internal forces, ever. That is with $T^i_j = 0$, $i,j > 0$, always, in some orthonormal coordinates. This is called for obvious reasons a "dust" stress tensor. Suppose all the 4-velocities of the dust particles point at the same element $x \in X$. Then the solution for $\underset{\sim}{T}$ grows without bound as we approach x, and so cannot exist at x, even without invoking gravitation and its effect on the metric - which produces singularities under far less artificial assumptions.)

Exercises XI.3

1a) Show that the set of bases of an n-dimensional vector space V fall into two classes, such that for two bases $\beta = \underset{\sim}{b}_1,\ldots,\underset{\sim}{b}_n$ and $\beta' = \underset{\sim}{b}'_1,\ldots,\underset{\sim}{b}'_n$ the linear operator $a^i\underset{\sim}{b}_i \rightsquigarrow a^i\underset{\sim}{b}'_i$ has positive determinant if β and β' belong to the same class, negative otherwise.

A choice of one particular class (say, for three-dimensional "physical" spaces by the right-hand rule) is called an orientation. With such a choice made, we say V is oriented, and a basis is called positively or negatively oriented according as it is in the chosen class or the other.

b) If V is an oriented n-dimensional metric vector space, use Ex. V.1.11 to show that there is a unique skew-symmetric n-linear form $\underset{\sim}{D}et$ on V (as distinct from the function det: $L(V;V) \rightsquigarrow R$) such that if $\underset{\sim}{b}_1,\ldots,\underset{\sim}{b}_n$ is an orthonormal basis, $\underset{\sim}{D}et(\underset{\sim}{b}_1,\ldots,\underset{\sim}{b}_n)$ is +1 or -1 according as $\underset{\sim}{b}_1,\ldots,\underset{\sim}{b}_n$ is positively or negatively oriented. What is the result of changing the order of the basis vectors?

For any ordered n-tuple $(\underset{\sim}{v}_1,\ldots,\underset{\sim}{v}_n)$ of vectors in V, we call $\underset{\sim}{D}et(\underset{\sim}{v}_1,\ldots,\underset{\sim}{v}_n)$ the volume, with respect to the particular orientation and metric, of the paralleliped fixed by $\underset{\sim}{v}_1,\ldots,\underset{\sim}{v}_n$.

c) If $|\underset{\sim}{D}et|$, the positive volume with respect to the metric, is defined by $|\underset{\sim}{D}et|(\underset{\sim}{v}_1,\ldots,\underset{\sim}{v}_n) = |\underset{\sim}{D}et(\underset{\sim}{v}_1,\ldots,\underset{\sim}{v}_n)|$, show that it is independent of the orientation used to define $\underset{\sim}{D}et$ but not multilinear.

d) If V is an n-dimensional metric vector space, $P \subseteq X$ a hyperplane, $\underset{\sim}{p}_1,\ldots,\underset{\sim}{p}_{n-1} \in P$ linearly independent, and $\underset{\sim}{v} \notin P$, show that there is exactly one $\underset{\sim}{f} \in V^*$ such that

(i) $\underset{\sim}{f}(\underset{\sim}{v}) > 0$

(ii) $|\underset{\sim}{f}(\underset{\sim}{w})| = |\underset{\sim}{D}et|(\underset{\sim}{p}_1,\ldots,\underset{\sim}{p}_{n-1},\underset{\sim}{w})$ for all $\underset{\sim}{w} \in V$.

Show that $\underset{\sim}{f}$ depends only on $\underset{\sim}{v}$ and the volume of the parallelepiped in P fixed by $\underset{\sim}{p}_1,\ldots,\underset{\sim}{p}_n$, whatever measure of volume is used in P.

e) Suppose $\underset{\sim}{G}$ is an inner product. Prove that there are exactly two unit vectors $\underset{\sim}{v}$ with $\underset{\sim}{v}^{\perp} = P$. If one is chosen as the "unit positive normal" and denoted by $\underset{\sim}{n}$, show that if $\underset{\sim}{p}_1,\ldots,\underset{\sim}{p}_{n-1}$ are orthonormal, the $\underset{\sim}{f}$ given by (d) with $\underset{\sim}{f}(\underset{\sim}{n}) > 0$, $|\underset{\sim}{f}(\underset{\sim}{w})| = |\underset{\sim}{\mathrm{Det}}|(\underset{\sim}{p}_1,\ldots,\underset{\sim}{p}_{n-1},\underset{\sim}{w})$ is exactly $\underset{\sim}{G}_{\downarrow}(\underset{\sim}{n})$, so $\underset{\sim}{f}(\underset{\sim}{w}) = \underset{\sim}{w}\cdot\underset{\sim}{n}$, and $\underset{\sim}{f}(\underset{\sim}{n}) = 1$.

2a) Show that if $\underset{\sim}{\mathrm{Det}}(\underset{\sim}{v}_1,\ldots,\underset{\sim}{v}_n) = 1$ and we use $\underset{\sim}{v}_1,\ldots,\underset{\sim}{v}_{i-1},\underset{\sim}{v}_{i+1},\ldots,\underset{\sim}{v}_n$ for $\underset{\sim}{p}_1,\ldots,\underset{\sim}{p}_{n-1}$ in Ex. 1d, and $\underset{\sim}{v}_i$ for $\underset{\sim}{v}$, then $\underset{\sim}{v}_1,\ldots,\underset{\sim}{v}_n$ are a basis for V with $\underset{\sim}{f}$ given exactly by the dual basis vector $\underset{\sim}{v}^i$.

b) Show that for any linearly independent $\underset{\sim}{v}_1,\ldots,\underset{\sim}{v}_n$ there exists a number $a > 0$ such that $|\underset{\sim}{\mathrm{Det}}|(a\underset{\sim}{v}_1,\underset{\sim}{v}_2,\ldots,\underset{\sim}{v}_n) = 1$. Deduce that we may parametrise the edges of the "box" in 3.07 from x to x'_μ, $\mu = 0,1,2,3$ so as to make

$$|\underset{\sim}{\mathrm{Det}}|(c_0^*(0),\ldots,c_3^*(0)) = 1.$$

3) Find the component forms of ${}_F$energy density etc (3.03) in unnatural units (2.08).

4. FORCES

So far we have used the word "force" only in the context of an inertial frame, to give Newtonian analogues for the "entirely ${}_F$spacelike to entirely ${}_F$spacelike" part of the stress tensor $\underset{\sim}{T}$. Is there an invariant, relativistic analogue for the force concept, as 4-velocity is analogous to Newtonian velocity?

For a matter field, the analogue is $\underset{\sim}{T}$ itself. Newtonian force means flow of Newtonian momentum (Ex. 1), and $\underset{\sim}{T}$ is exactly the flow of 4-momentum. (This point of view suggests thinking of the classical stress tensor as a single "force", which is entirely reasonable. Newtonian 3-space no more has innately given coordinates than does flat or bent spacetime, so the tensor is more fundamental than the set of components it has in some chart, which are interpreted as shear forces, etc. Thus even Newtonian "force" graduates from a $\binom{0}{1}$-tensor to a $\binom{1}{1}$-tensor.)

For a particle, the natural candidate for a relativistic or 4-<u>force</u> on it is "rate of change of its 4-momentum along its world line". (The Newtonian F = ma is really a definition of force more than a law of physics.) In a general spacetime, this rate of change should evidently be defined by covariant differentiation. Consider a history c with constant rest mass m, parametrised by proper time σ. Then

$$\underset{\sim}{p}(\sigma).\underset{\sim}{p}(\sigma) = m^2, \quad \text{constant}$$

$$\text{so} \quad \nabla_{c*}(\underset{\sim}{p}.\underset{\sim}{p})(\sigma) = 0, \quad \forall\sigma.$$

$$\text{Therefore} \quad \nabla_{c*}\underset{\sim}{p}.\underset{\sim}{p} + \underset{\sim}{p}.\nabla_{c*}\underset{\sim}{p} = 0, \quad \text{by VIII.7.03, 7.06 (just as for contravariant vectors).}$$

$$\text{Hence,} \quad \nabla_{c*}\underset{\sim}{p}.\underset{\sim}{p} = 0, \quad \text{by the symmetry of } \underset{\sim}{G}^*.$$

Thus for such a history the 4-force must be always orthogonal to its 4-momentum. Since no one non-zero cotangent vector at $x \in X$ can be orthogonal to $\underset{\sim}{G}_{\downarrow}(c^*(\sigma))$ for all curves c with $c(\sigma) = x$, the 4-force on the history must necessarily depend on its 4-velocity, or vanish.

This recalls the way that the Newtonian force on a charged particle in a magnetic field depends on its velocity; this indeed is the spacelike part of a relativistic example. The electromagnetic field is geometrically a 2-form, or skew-symmetric $\binom{0}{2}$-tensor field, on spacetime. (See, for example, [Misner, Thorne and Wheeler].) Contract this with the $\binom{1}{0}$-tensor $e(c^*(\sigma))$, where $e \in R$ is the charge on the particle, and the result is a $\binom{0}{1}$-tensor, the 4-force. This is the simplest possible relativistic "field of force". We must have a map taking 4-velocity (or 4-momentum) to 4-force, so we must have a tensor of total degree at least 2. (4-force might depend on other things, beside 4-velocity: we have shown only that it must vary with that at least, if we are to have particles with constant rest mass.) The Newtonian vector field of force, typified by the electric and gravitational fields, whose effect on a particle depends only on its position and perhaps a scalar such as charge, is impossible.

Electromagnetic forces thus "relativise" beautifully into special relativistic language (and indeed are simplified by it). In fact the group of Lorentz transformations was discovered before the notion of spacetime or the Lorentz metric, as exactly the transformations that left Maxwell's laws invariant. The behaviour of this field - in particular, of electromagnetic radiation, light especially - played a crucial role in the origin of special relativity.

What about the other great force field of classical physics: gravitation? It cannot be a $\binom{0}{1}$-tensor field as in Newton's theory, let alone take the ultra-convenient form $d\Phi$ for $\Phi : X \to R$, as long as we suppose m^2 fixed for each particle, for the reasons above. Letting rest mass vary leads to worse confusion. And it turns out that no effort to describe gravitation as a higher-order tensor field on flat spacetime has succeeded, either. All attempts have either broken down on inconsistencies, internal or with the facts, or made the flatness of the underlying space physically undetectable since no physical quantity is described as travelling by the parallel transport of the flat connection, which thus drops out of sight. Nor is a purely "force field" theory of gravitation greatly to be expected, as we see in the next section.

Exercises XI.4

1) In Newtonian terms, the force between two bits of matter is the flux of momentum between them : the net force on one bit is the net flux between it and all others.

a) Describe qualitatively the flux of momentum along the parts of the object in Fig. 4.1a, with no external forces and moving at constant velocity with no rotation.

b) Draw in a longitudinal cross-section the flux of each of the three components (in the obvious (x,y,z) coordinates) of the circular beam of Fig. 4.1b, at rest with equal and opposite axial forces pulling at the centres of its ends.

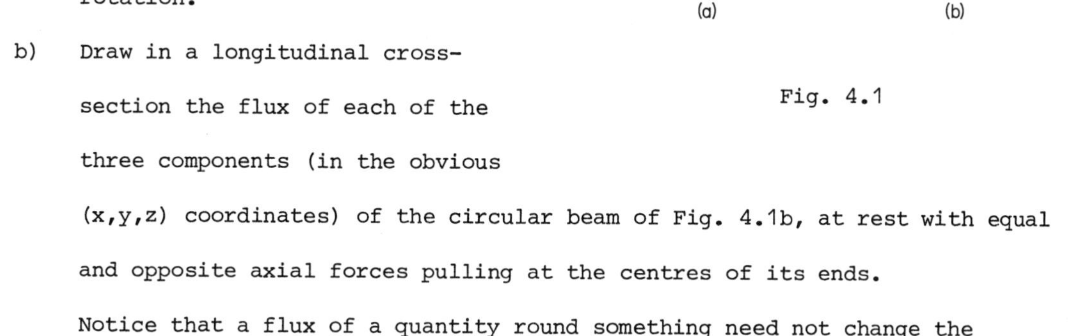

Fig. 4.1

Notice that a flux of a quantity round something need not change the quantity's density at any point (a), even if that density is zero (b).

c) Describe these two situations relativistically.

5. GRAVITATIONAL RED SHIFT AND CURVATURE

Suppose that spacetime is flat, and consider the gravitation due to an inhabitable ball B of matter at rest, everywhere, in some inertial frame F. Let L and U be experimenters at $_F$rest, L on the surface of B, U an $_F$distance directly above L. Suppose L has a perfectly efficient machine for turning rest mass into a tight beam of radiation: U has an equally efficient device for turning radiant energy into rest mass. L starts with a supply m of rest mass at $_F$rest, which he turns into radiation and beams up to U: she restores it to matter and drops it back to

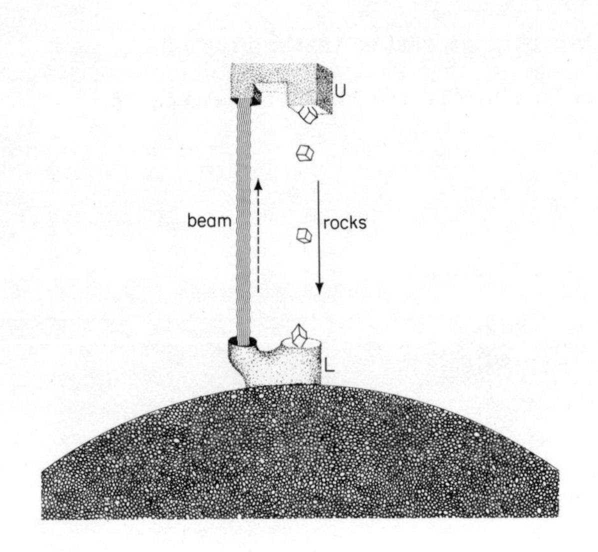

Fig. 5.1

him. If the amount of rest mass U reconstitutes is the same amount m that L started with, he gets back his original investment of mass/energy plus the kinetic energy gained by the mass in its fall. He can use this bonus to run his sewing-machine while beaming m rest mass back up to U for her to drop again, ... etc. L and U would have between them a perpetual motion machine. Even if their separate machines were not perfectly efficient (indeed, all such devices known are so far from it that L and U would have a net loss of useful energy; L would do much better to point his beam at a boiler) the arrangement would violate conservation of mass/energy, as measured in the frame F.

Rather than believe this, it is natural to suppose that not as much mass/energy reaches U as left L; the radiation reaches U with less energy than it had when it left L. This decrease is indeed experimentally observed, and in the right quantity, so conservation of mass/energy is inviolate.

Thus far, we could still think of gravity as a "force" against which the radiation does "work" in rising, and so loses energy. But for radiation, energy is proportional to frequency (2.09), so it must arrive with lower frequency than it left. This is the gravitational red shift, because it moves light towards the red end of the spectrum. It makes flat spacetime seem very unphysical, by the following argument, due to Schild.

Assume L and U in positions as before, following world lines w_L, w_U at ${}_F$rest one above the other (Fig. 5.2). Suppose L beams upward a continuous signal whose frequency he measures as ν.

L receives a continuous signal whose frequency she measures as ν'. Consider the track in spacetime of one wave crest emitted at e_1, received at r_1, and another crest going from e_2 to r_2, emitted N troughs later. (Or the analogue to crests and troughs for transverse radiation.)

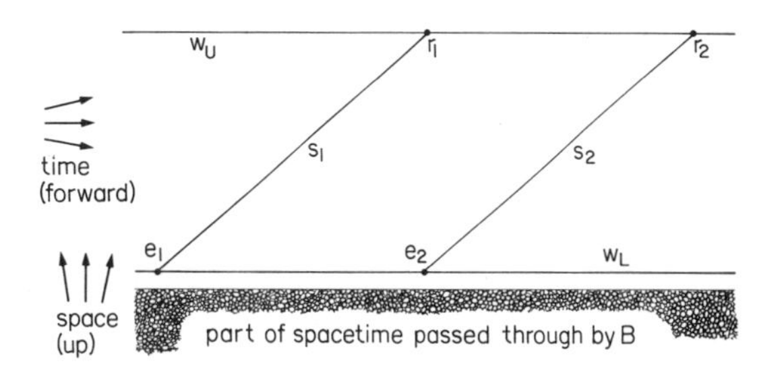

Fig. 5.2

If experiments are repeatable and spacetime has an affine structure, then the vectors $\underset{\sim}{d}(e_1,r_1)$ and $\underset{\sim}{d}(e_2,r_2)$ given by it must be equal. So the straight line S through e_2 and r_2 is parallel to S_1 through e_1 and r_1.

Now the world lines w_L and w_U are both at $_F$rest, and hence parallel, by assumption. So we have a parallelogram $e_1r_1r_2e_2$. But the length of the side e_1e_2 is measured by L as N/ν (N periods of radiation with frequency ν) and that of r_1r_2 by U as N/ν', which is greater, as $\nu' < \nu$ by the gravitational red shift. But a parallelogram with unequal opposite sides is as impossible in Minkowski space as in a Euclidean space (Ex. 1). Hence either there is curvature inside the quadrilateral fixed by the four world lines, or if the metric is flat at least one observer is measuring his proper time by something other than arc length. The metric given by measurements is essentially curved.

In the case of the metric we used for discussing mirages (Ex. IX 4.6, Ex. X.1.6) other experiments were possible, giving physical meaning to a metric for space without the curvature we found for $k^2\underset{\sim}{G}$. For spacetime no such experiments have been found: if there is an underlying flat metric it is lying very low.

Exercises XI.5

1) Let X be an affine space, and S_1, S_2, T_1, T_2 be one-dimensional affine subspaces of X such that S_1, S_2 are parallel translates of T_1, T_2 respectively and each $S_i \cap T_j$ consists of exactly one point $p_{ij} \in X$. Show that $\underset{\sim}{d}(p_{11},p_{12}) = \underset{\sim}{d}(p_{21},p_{22}), \underset{\sim}{d}(p_{11},p_{21}) = \underset{\sim}{d}(p_{12},p_{22})$ and deduce that for any constant metric tensor on X, opposite sides of a parallelogram have equal arc length (however parametrised).

XII General Relativity

NÂGASENA: Well, O king, will sticks and clods and cudgels and clubs find a resting-place in the air, in the same way as they do on the ground?

MILINDA: No, Sir.

NÂGASENA: But what is the reason why they come to rest on the earth, when they will not stand in the air?

MILINDA: There is no cause in the air for their stability, and without a cause they will not stand.

The Questions of King Milinda.

1. HOW GEOMETRY GOVERNS MATTER

Aristotle and Newton held that things fall because they are pulled to the earth; Nâgasana the sage and Einstein, that they fall because nothing stops them from falling. The difference is a profound one.

We saw at the end of last chapter the difficulty of describing gravity as a force in a flat spacetime. In this chapter we see that once we bring in curvature we do not need to call it a force at all. Newton's first law, stating that a particle moves on a straight line in space unless a force such as gravity acts on it, is replaced by the principle that it follows a geodesic in spacetime unless a 4-force acts on it, and gravity is not a 4-force. It is just the shape of space, which determines the geodesics.

1.01 The equivalence principle

Einstein's equivalence principle is often stated as: Experiments in a closed box cannot distinguish between the box being in a gravitational field, not changing with

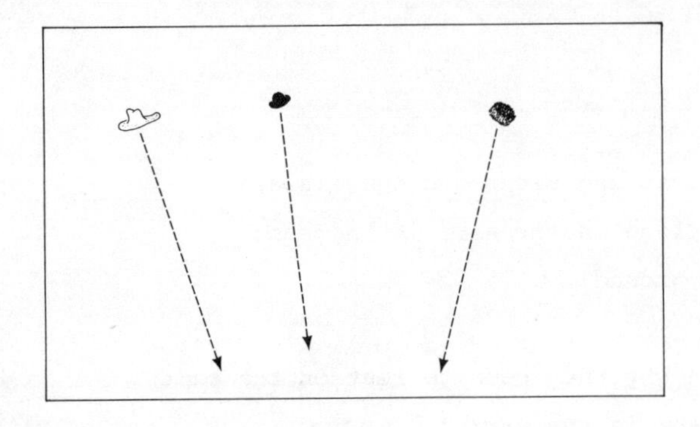

Fig. 1.1

time, and its being under uniform acceleration in a flat spacetime where no gravitational forces act.

Thus formulated it is obviously false: in Fig. 1.1 falling objects converge on each other, in a way barely affected by their own masses, inconsistently with acceleration of the box. We shall refine it in several stages to find a precise and tenable statement. First, evidently we should consider a sufficiently small box for such effects to be undetectable.

In the same way, if spacetime is curved in the metric given by distance and time measurements, as we have seen it is, its curvature will be there inside a small box to show that it is not in flat spacetime. The distortion involved in making flat maps of spheres is there even in a plane mapping of the surface and boundary of a duckpond. But the errors in measurement in mapping a pond are sufficient to conceal its non-planar character (indeed the wind- or duck-provoked variations in it swamp its sphericity) so we may for suitable local purposes treat it as flat. We do the same with a small piece of spacetime.

More technically, choose a chart $\phi : U \to R^n$ around any point x in spacetime M such that the resulting Γ^i_{jk} all vanish at x. (Use normal coordinates around x;

IX.2.05, 2.06.) Their vanishing for all $y \in U$ would mean U was flat, so we cannot arrange that in general. However, they are continuous, so we can choose U small enough to make all the Γ^i_{jk} smaller than any given $\varepsilon > 0$. If ε is given by the lower limit of our ability to detect curvature using, say, geodesic deviation, then the result is a region U that we may cautiously consider "flat for practical purposes". Cautiously: we could sew all of M up out of such "almost flat" pieces, and the result need not be flat. Consider a Buckminster Fuller geodesic dome; either the short segments must be curved, just a little, or the faces bent a little away from each other along the edges (otherwise the dome is a plane). Arguments from the equivalence principle must be strictly local to have meaning.

The chart ϕ with its "flat for local practical purposes" domain U is often called a local Lorentz reference frame. Since "Lorentz" is being used to mean flat it would perhaps be less misleading to call it approximately Lorentz. Reasoning aimed at exact results about events in a local Lorentz reference frame should never use the assumption that its domain is exactly flat.

Take then a box in U, sufficiently small that the measured accelerations due to gravity at different points are "parallel", with the ϕ-coordinates of its corners constant in time. ("Parallel" has a route-independent meaning in U up to our limits of measurement, by assumption.) Let V be the open set interior (at various times) to the box and $\psi = \phi|V$, to give us a chart $\psi : V \to R^4$ around x. Then the equivalence principle asserts that no experiment confined to V by an observer A whose world line is given by ψ as "rest in the box", with the above limit to its accuracy, can distinguish his situation as being

(1) That of an observer in a spacetime X he cannot in V distinguish from flat, with a velocity in X he cannot distinguish from constant, with a field present that acting alone would produce an acceleration of any matter in V parallel to its effect at any other point in V (though

it is not acting alone, at least on him; some other agency is balancing it to keep his velocity constant).

or (2) That of an accelerated observer in a flat spacetime (with some force such as a push from the floor of the box accelerating him), with no gravitational effects from outside U.

Correspondingly, it says that an observer B not experiencing any forces from the walls, floor, acceleration couch etc. of the box by standing, hanging or lying on them cannot distinguish by similar measurements whether his situation is

(1') That of a non-inertial observer in a flat spacetime where some field is acting that, in the absence of other forces, would give all matter near him an acceleration parallel to his own.

or (2') That of an inertial observer in a flat spacetime where no such field is acting.

(These forms of the principle are equivalent, since a permissible experiment for either A or B is to build a B or an A and have it report back.)

By our method of constructing ψ, (1') cannot be distinguished from

(1") That of an accelerated observer in a spacetime perhaps not flat, where some field is acting to accelerate him and, in the absence of other forces, matter near him, in a smoothly varying way.

so we reach the indistinguishability of (1") and (2').

The approximations involved vanish when we take the limit as the size of the box goes to zero, if all the "sufficiently small's " are defined carefully. The principle then says that an observer falling freely under gravity in any spacetime finds the same local physical laws - that means the same relationships between the values of sets of measurable quantities and their derivatives, at the points he actually passes through - as an inertial observer studying the behaviour of similar quantities in flat spacetime, in the absence of gravity. What happens must be independent of the chart, of course, so in the general spacetime we must use some connection to get well defined derivatives. The "natural" choice, in the strong sense outlined in VIII.6.09, is the Levi-Civitā connection for the metric given by measurement. In flat spacetime, of course, covariant differentiation is the same as "ordinary" differentiation. (Which for affine coordinates is given by just differentiating components, since the Γ^i_{jk} vanish.) Thus we have lost the various motivating travelling boxes, observers etc. and come to the invariant statement:

> The local (differential, as distinct from integral) forms of the laws of physics are identical in the "presence" or "absence" of gravity - no "gravitational force" need be allowed for - provided covariant differentiation is used.

This is the principle's precise form, and any imprecisions in our earlier formulations can be resolved by reference to it. It is a vital tool in finding general relativistic forms for physical laws already studied in flat spacetime, often used in component form (1.02).

The equivalence principle is not, of course, a necessary geometric fact like the conservation equation div $\underset{\sim}{E} = \underset{\sim}{0}$ (X.6.12), but a scientific hypothesis to be tested. It currently seems impossible to describe gravitation without curved spacetime, but the principle asserts that gravitation consists only in the relation between matter and the curvature of spacetime. It is quite legitimate to suppose that there is a tensor field involved as well, associated with all matter as the electromagnetic field is associated with charged matter. This would make gravitation more complicated, since we cannot get rid of the curvature aspect, but perhaps it just is that complicated. For instance, differentiating with anything but the Levi-Cività ∇ amounts to using ∇ plus a $\binom{1}{2}$-tensor field, by Ex. VIII.6.2.

See [Misner, Thorne and Wheeler] for a discussion of experimental tests of the equivalence principle. Here we shall assume it is true, since by the above it gives the simplest account of gravitation, and we investigate its consequences.

1.02 Components

In flat spacetime with any affine chart, the ∂_i are all parallel fields and the Γ^i_{jk} vanish everywhere. Hence the components $w^{i_1 \ldots i_k}_{j_1 \ldots j_n;\eta}$ of the covariant derivative of a $\binom{k}{h}$-tensor $\underset{\sim}{w}$ reduce to the derivatives $w^{i_1 \ldots i_k}_{j_1 \ldots j_n,\eta}$ of the components (cf. VIII.7.08), and differentiation can be treated in an entirely component-by-component fashion. For this reason the equivalence principle is sometimes stated as "semi-colons must reduce to commas in the case of flat spacetime".

(Note that with "curvilinear coordinates" in flat spacetime, the ∂_i are not parallel and the Γ^i_{jk} thus do not vanish (Ex. 1).)

1.03 Free fall

The simplest law of physics in flat spacetime is that a history c, on which no 4-force acts, has constant 4-velocity. This is actually a trivial consequence of the definition of 4-force which we made, being conditioned by Newton to seek a force as cause for any change in velocity. (Aristotle, by contrast, held that a force is needed to maintain velocity, which is closer to everyday experience. Only Newtonian relativity - the idea that any velocity can be chosen as "rest" - makes the newer idea intuitive.)

The local form of this law is (VIII.3.05),

$$\nabla_{c^*} c^* = 0.$$

So the equivalence principle generalises "a particle moving under no force in flat spacetime travels along an affine straight line" to "a history in a general spacetime, influenced only by gravitation, is a geodesic".

The movement of stars and planets has thus become less "forced" in our minds over the centuries. Mediaeval descriptions had the planets mounted on revolving crystal spheres, mounted on revolving spheres, mounted on ... etc., driven by some ultimate Primum Mobile (prime mover, somewhat identified with God) which supplied the Aristotelean force to keep them going. Newton had them falling freely around the sun, with no push on them; the only force active was universal gravitation. Finally, in general relativity even that constraining force disappears and the planets simply take their own course, not straying, at one with the geometrical Tao of spacetime.

From this point of view, then, an observer lying in a hammock is, exactly, an accelerated observer. The only force acting on him is the upward push of the hammock, just as in Newtonian mechanics without gravity a stone in a sling is accelerated inward away from its inertial movement, until the sling is released and the stone flies off tangentially on a straight path at constant speed. "Straight" now becomes "geodesic" and the absence of gravity is not required. Things fall, not for a cause, but since without a cause they will not stand.

Exercises XII.1

1) Does the $; \rightsquigarrow ,$ form 1.02 for the equivalence principle hold up if the chart used on flat spacetime is not affine? Compute the Γ^k_{ij} for the usual connection on R^2 in polar coordinates.

2. WHAT MATTER DOES TO GEOMETRY

2.01 Einstein's equation

We have decided that an apple falls because it is guided by the shape of spacetime. But why is this shape, around the earth, such that so many timelike geodesics meet the ground twice? (The past and future of a travelling golf ball both usually touch grass.) What form does the relation between the presence of matter and the curvature of spacetime take?

First, we observe that the nature of the matter seems unimportant. Whether the matter is charged or uncharged, matter or antimatter, solid or gaseous etc., has no influence on its gravitational effect as far as any experiments have indicated. In Newton's theory only mass was important; relativistically mass is inextricably mixed with energy and momentum. So the natural hypothesis is that however high or low the order of tensor needed to describe any "matter field", its interaction with the geometry of spacetime depends only on the concomitant flow of 4-momentum; that is its stress tensor.

Secondly, the curvature of spacetime is clearly non-zero even at points where no matter is present. One can see this by the Schild argument of XI.5 or, once gravitation is assumed to be entirely a curvature effect, by the "tidal stresses" associated with geodesic deviation in X.4 or by the more general considerations of X.3.02. Now the purely geometric argument of the latter (the others appeal to experimental evidence, available only for our spacetime) arises only when a spacetime of at least three spacelike dimensions is considered, as we saw. With only two, matter could affect the geometry outside the histories of solid bodies without imposing curvature there, as in Ex. X.3.2; we could describe a "gravitation" that involved curvature only at its source, the matter. (Though whether intelligent creatures in a three-dimensional spacetime could ever find such a theory valid is another question. Ex. X.3.2f suggests that a star could not have planets in stable orbits around it, and g casts doubt on whether stars would even form. In this case there could _be_ no "life as we know it" to consider the theory. Thus flatlanders would either be _very_ different from us or find that whatever they used for gravity could be detected by purely local effects -"local" being bigger than V of 1.01 - away from its source, like ours with red shifts and tidal effects. Compare the end of [Misner, Thorne and Wheeler], on "biological selection of physical constants".)

Now in a 3-manifold curvature is completely determined by the Ricci tensor, by X. §7. As we go to four dimensions, more possibilities unfold:

(1) The Weyl tensor, that gives the "non-Ricci" part of the curvature, no longer vanishes identically.

(2) The "spanning surface" argument shows that for matter to bend space around it, $\underset{\sim}{R}$ cannot vanish wherever matter isn't.

This suggests that we make the Ricci tensor the "locally determined part" of the curvature, whose value at x is determined by $\underset{\sim}{T}_x$, and leave the Weyl tensor as the "non-locally determined part", influenced here and now by the presence of the sun's matter at a point 93 million miles and eight minutes off in spacetime by our usual labels. (Only _suggests_ of course - we are motivating, not deriving, Einstein's equation. Similarly, Maxwell's equations come not from deduction but from Maxwell. Equations you can prove are either laws of geometry, not physics, or mere consequences of more fundamental principles. Such a derivation is possible for Einstein's equation. For instance, from very little more than physically reasonable symmetries plus the hypothesis that _only_ geometric effects appear in gravitation it is done in [Hojman, Kuchar and Teitelboim]. This work ought to be intelligible to a reader of this book who is familiar with Hamiltonian dynamics.)

The simplest idea, then would be to equate the Ricci tensor (adjusted to have the same variance) to the stress tensor. But by X.6.11 the divergence of $\bar{\underset{\sim}{R}}$ is dR, which can only be zero if R is constant. Since $\bar{\underset{\sim}{R}} = \underset{\sim}{T}$ would give R = 0 where there is no matter, this and the conservation law div $\underset{\sim}{T} = \underset{\sim}{0}$ (brought over from XI.3.07 by the equivalence principle) would imply tr $\underset{\sim}{T}$ equal to tr $\bar{\underset{\sim}{R}}$ equal to R equal to 0 everywhere. But $\underset{\sim}{T}$ can very easily, physically, have all its diagonal entries (energy density and three pressure terms) positive in some coordinates, which gives tr $\underset{\sim}{T} > 0$.

Thus it is not physically plausible simply to equate $\bar{\underset{\sim}{R}}$ to $\underset{\sim}{T}$. However, the Einstein tensor $\underset{\sim}{E}$ introduced in X.6.12 has identically vanishing divergence, always, and by X.6.13 determined the Ricci curvature. So the equation

$$(1) \qquad \underset{\sim}{E} = 8\pi\underset{\sim}{T}$$

(8π being purely a convenience to simplify units, like 4π in Maxwell's equations) both describes a local effect of matter on curvature, and implies the conservation law

$$\text{div } \underset{\sim}{T} = \underset{\sim}{0}.$$

(1) is the original <u>Einstein's equation.</u> It is the most general relation possible between $\underset{\sim}{T}$ and curvature that implies the conservation law ($\underset{\sim}{E}$ being essentially the only $\binom{1}{1}$-tensor with zero divergence constructible from $\underset{\sim}{R}$) except for the modification

$$(2) \qquad \underset{\sim}{E} = 8\pi\underset{\sim}{T} + \Lambda\underset{\sim}{I}$$

where $\Lambda \in \mathbb{R}$ and $\underset{\sim}{I}$ is the identity tensor field $T^*M \to T^*M$. Einstein transferred his affections for a while to this, as we see in 2.02, but we shall call only (1) by the name "Einstein's equation". Note that either version is often referred to in the plural, because it is represented by sixteen equations in coordinates.

2.02 Static solutions?

Let us look for a solution where spacetime M can be sliced into spacelike hypersurfaces S (cf. VII.2.03), each containing only dust at ${}_S$rest. That is, if we choose around x a chart with the timelike vector ∂_0 orthogonal to S we should have $\underset{\sim}{T}_x$ given as a matrix with the "energy density" T^0_0 as its only non-zero entry: no "energy flow" and no "internal forces". If we also require the ∂_i orthogonal at x, then $(T^0_0)_x$ subject to this "${}_S$rest" condition for the chart is well defined, giving a function $\mu : M \to R$ with $(T^0_0)_x = \mu(x)$. Einstein's equation gives, by X.6.13,

$$\bar{\underset{\sim}{R}} = \underset{\sim}{E} - \tfrac{1}{2}(\mathrm{tr}\,\underset{\sim}{E})\underset{\sim}{I} = 8\pi(\underset{\sim}{T} - \tfrac{1}{2}(\mathrm{tr}\,\underset{\sim}{T})\underset{\sim}{I})$$

so using the above chart at x we see that

$$R^0_0 = 8\pi(T^0_0 - \tfrac{1}{2}T^0_0) = 4\pi\mu, \quad \text{since tr } \underset{\sim}{T} = T^0_0$$

$$R^i_i = 8\pi(0 - \tfrac{1}{2}T^0_0) = -4\pi\mu, \quad i = 1,2,3 \quad \text{(no sum)}.$$

More realistically, let the sections contain gas at ${}_S$rest with pressure p, so that $T^0_0 = \mu$, $T^1_1 = T^2_2 = T^3_3 = -p$, off-diagonal terms vanishing (cf. XI.3.05). Then

$$R^0_0 = 8\pi(T^0_0 - \tfrac{1}{2}\mathrm{tr}\,\underset{\sim}{T}) = 8\pi(\mu - \tfrac{1}{2}(\mu-3p)) = 4\pi(\mu+3p)$$

$$R^i_i = 8\pi(T^i_i - \tfrac{1}{2}\mathrm{tr}\,\underset{\sim}{T}) = 8\pi(-p - \tfrac{1}{2}(\mu-3p)) = 4\pi(p-\mu)$$

in the same coordinates at $x \in M$.

Now, Einstein, convinced that the heavens endure from everlasting to everlasting and seeking to approximate the thin scattering of matter observed in the universe by such a dust or gas, wanted a static solution. That is, like the above dust or gas ones with the further condition that M can be expressed as the product manifold $S \times R$, for a particular model S of space, and that the maps

$$\bar{t} : S \times R \to S \times R : (x,r) \rightsquigarrow (x,t-r)$$

preserve all geodesics, etc. for all $t \in R$. This would allow spacelike hypersurfaces of the "constant" form $S \times \{t\} \subseteq M$, $t \in R$.

However, this implies that the Ricci curvature in the direction of the "static" timelike vectors is zero (Ex. 1), and hence that $R_0^0 = 0$ in the above coordinates at $x \in M$. So either no matter is present, in the "dust" solution, or the "gas" solution is under tension $\frac{1}{3}\mu$ (the second contrary to observation, the first contrary even to observers), or the solutions are not static. The only way this can be, if the matter is at rest in each S, is by differences between the sections S themselves. If the matter is evenly spread in S it has a finite size (cf. 2.03) and this size is increasing or decreasing with time - or perhaps just changing from increase to decrease. (Compare the way the North-South curvature of the Earth "forces" variation in the size of parallels of latitude.)

Einstein found this behaviour on the part of the solutions so reprehensible that he put a fudge factor in his equation to prevent it; he inserted the "cosmological constant" Λ to keep the cosmos constant (equation 2 of 2.01 above). If the density and pressure of the gas is nicely tailored to the cosmological constant of the host spacetime by arranging

$$4\pi(\mu+3p) = \Lambda \quad \text{everywhere, then}$$

$$
\begin{aligned}
R_0^0 &= T_0^0 - \tfrac{1}{2}(\mathrm{tr}\ \underset{\sim}{T}) \\
&= (8\pi E_0^0 + \Lambda) - \tfrac{1}{2}(8\pi\mathrm{tr}\ \underset{\sim}{E} + 4\Lambda) \quad \text{since tr } \underset{\sim}{I} = 4 \\
&= 4\pi(\mu+3p) - \Lambda \\
&= 0,
\end{aligned}
$$

so the heavens can endure. Of course, with the new equation, if $\Lambda \neq 0$ an <u>empty</u> universe must expand or contract. The universe requires just the right amount of matter to keep it steady.

Notice that $4\pi(\mu+3p)$ must be constant over M for this to work, because for Λ a function on M we have $\mathrm{div}(\Lambda\underset{\sim}{I}) = d\Lambda$ (Ex. 2), so the equation does not guarantee $\mathrm{div}\ \underset{\sim}{T} = \underset{\sim}{0}$ if Λ is not a constant. In fact energy density and pressure must be constant individually, for the following reason.

The "static" requirement above implies here that parallel transport by ∇^M of a vector tangent to $S \times \{t\}$ along a curve in $S \times \{t\}$ <u>keeps</u> it tangent to $S \times \{t\}$ (in contrast to Euclidean parallel transport of tangent vectors around an embedded sphere, for example). It follows immediately that the Riemann and Ricci tensors of $S \times \{t\}$ with the induced metric are exactly given by restricting the Riemann and Ricci tensors of M (false for the sphere in R^3). If the $_S$spacelike part of $\underset{\sim}{T}$ is isotropic, which we have assumed for both the "dust" and "gas" solutions (specifically, taking it as $\underset{\sim}{0}$ and $p\underset{\sim}{I}$ respectively) then so is the $_S$spacelike part of the Ricci tensor of M, using Einstein's equation either with or without the cosmological constant. Hence the <u>whole</u> of the Ricci tensor of $S \times \{t\}$ is isotropic. It is therefore constant, by X.6.15. (It was in this physical context, of point isotropy of matter at rest implying global homogeneity, that Einstein manifolds first came to attention; hence the name.) Its components, $R_i^i = (4\pi(p-\mu) - \Lambda)$ in coordinates as above, are thus constant too. Combining this with the constancy of $4\pi(\mu+3p)$, demanded above, μ and p must be constant over each $S \times \{t\}$. Constancy over M follows easily by the conservation law.

This result, that if the pressure and energy density of a gas are not everywhere the same the situation is not static, would seem intuitively obvious. But intuition unsupported, even of the great, can be wrong: Einstein's intuition of the stability of the universe made him avoid predicting the recession of the further galaxies, and the consequent, now famous, red shift that Hubble observed some ten years later. Subsequent measurements show that if Λ is non-zero then it is very small indeed: we shall take it as zero.

2.03 The shape of space

Looking at the matter of the Universe, we see it getting less dense - the further a galaxy is from us, the faster the distance between us and it is growing. So distances on a spacelike hypersurface S between galaxies at ${}_S$rest (to the approximation involved in treating matter as a thin gas) are increasing, rather like distances on an inflating balloon. Can we say that the spacelike sections as a whole really are becoming larger with time, that "the universe itself is expanding", rather than that matter is spreading out in infinite space?

On certain assumptions, yes. Namely, assume that there is a spacelike section S passing through us with the same sort of complete homogeneity, averaging on a large enough scale, as the sections of the static solutions in 2.02. (Similar relations between "isotropic" and "homogeneous" apply, but more complicated since spacetime is not "flat in the time direction".) This is sometimes called the <u>Copernican principle</u>, by analogy with the way Copernicus dislodged Earth from the centre of things, then the sun became an average star off-centre in the galaxy, and finally our galaxy was seen as just an average member of an average galactic cluster. The existence at all of a spacelike hypersurface S for which all matter is more or less at ${}_S$rest is quite a strong assumption, and the transition from "<u>we</u>

are nowhere special" to "there is nowhere special" is a little suspect, particularly as it leads eventually to the conclusion (cf. 2.04) that there are spacetime points in our past and our future that are quite drastically special. However, we can see quite far these days with various devices, and what we see is homogeneity (allowing for the way signal delay shows us earlier, denser spacelike sections) over a very substantial volume of space. On the available evidence then, the Copernican principle is a plausible assumption.

Now, using the values for R_0^0 and R_i^i of 2.02 of the "gas" solution in 2.02, we get sectional curvatures for the (i,j)-planes, i, j > 0, of $4\pi(p - \frac{\mu}{3})$ (Ex. 3). Now in these circumstances, where the matter is so thinly spread, μ is very much larger than p; recall the c^2 term (XI.2.08) involved in non-geometrised units. (One microgram of hydrogen in a cubic metre represents H-bomb amounts of energy but a pressure needing fine instruments even to detect.) So the spacelike sectional curvatures are everywhere the same negative number, on S. Since our metric is negative on spacelike directions, reversing the sign of scalar curvature, this means that the spacelike sectional curvatures of M at points in S correspond to positive scalar curvature for the positive definite situation of the surfaces in X. §3. M is bent in spacelike directions in the manner of a sphere, rather than a flat space or a saddle shape.

Now it is a fact that only certain Riemannian manifolds are candidates for S, given constant curvature of this kind, up to a scalar constant multiplying the metric. These are the sphere S^3 and various "smaller" spaces constructed from it by identifying points. For example real projective 3-space RP^3 (the 3-dimensional analogue to Ex. IX.1.3) can be constructed by identifying opposite points of S^3, or as the group of rotations of Euclidean 3-space. The latter construction gives an easy way to specify further identifications. For example, identify rotations A and B if $A = B\circ C$, where C is a symmetry rotation of the dodecadron. Or if $A = B\circ C$ where C is a rotation of $2\pi/n$ about a given axis, or ... . All such constructions give candidates for S, since they are <u>locally</u> just like S^3 - which is what we know about S - and <u>only</u> spaces obtained by such identifications of points in S^3 (not always going via RP^3) are candidates. Their classification (essentially that of finite groups acting suitably on S^3 - of which there are infinitely many) is outside our scope. However, in each case S has a meaningful finite "circumference" and "volume" deducible from the local value of the scalar curvature, and so is "finite but unbounded" in the classic phrase of Einstein. It has finite "size" but no boundary. (The timelike aspects of curvature then imply that this "size" cannot be constant from spacelike hypersurface to spacelike hypersurface unless Λ is just so, as we have seen.)

We cannot prove global results of this kind in this volume, so we only mention further the fact that the isotropy/homogeneity condition on S can be weakened, as one would hope: if it had to be <u>exactly</u> true it would have little physical relevance, as the matter we see is not exactly a uniform thin gas. If sectional curvatures on a Riemannian manifold S vary between positive k and $K > k$, we can change scale to make the upper bound 1, and set $\delta = k/K$. (Then $0 \le \delta \le$ curvatures ≤ 1 and S is called a δ-<u>pinched manifold</u>.) The question of how small δ can be and leave intact the topological conclusion that the space is S^3 (with perhaps some points identified) is a topic of active mathematical research. In the case dim S = 3 the latest, smallest value for which we have heard of a proof at the time of writing is $\frac{1}{4}$. Evidently δ must be strictly positive, as curvature even <u>strictly</u> greater than $\delta = 0$ does not imply compactness, let alone sphericity. (Consider the "bowls" - draw one -

$$\{(x,y,z) \in R^3 | x^2+y^2 = z^2-1, z > 0\},$$

$$\{(x^1,x^2,x^3,x^4) \in R^4 | (x^1)^2 + (x^2)^2 + (x^3)^2 = (x^4)^2 - 1, x^4 > 0\},$$

with the metrics induced from the Euclidean ones on R^3 and R^4)

Notice that if S is an RP^3 it will not embed in R^4, so its curvature must be "around" several dimensions if "around" any (cf. X.1.08). Furthermore, if the Copernican principle is true then it implies that the "measurement" metric tensor given by anything remotely like general relativity must have the properties that lead to S being S^3 or a closely related space, which does not <u>admit</u> a locally flat metric. So it is very unlikely that any theory of gravitation in a flat spacetime is compatible with the Copernican principle, even if the flatness is not supposed physically detectable.

The favourite topology for spacelike sections among cosmologists is that of S^3 (the simplest of the above spaces, and the "universal cover" of them all, as $\widetilde{SL}(2;R)$ is of $SL(2;R)$; cf.IX.6.07). Another common choice is to deny the Copernican principle and suppose that the universe consists of a finite amount of matter in the midst of infinite darkness: that there is not enough matter to "close up" space by the curvature it causes, and that on a large enough scale spacetime approximates Minkowski space arbitrarily well. More exactly, that the geometry of $M\backslash K$, where M is a spacetime and K is a region (including most of the matter) that has a compact intersection with any spacelike hypersurface, approximates the geometry of a piece of Minkowski space arbitrarily well for K large enough. Such a spacetime is called <u>asymptotically flat</u>, and is nice for coordinate calculations (needing only one chart), though it feels somehow rather lonely. Many results have been proved for asymptotically flat spacetimes, see [Hawking and Ellis].

2.04 The shape of spacetime

The Copernican principle, plus mild and reasonable physical conditions on the matter tensor, implies that there are regions in both our past and our future (joined to us by backward and by forward curves from here and now) where $\underset{\sim}{T}$ and $\underset{\sim}{R}$ grow without bound. A manifold cannot have infinite curvature and still be a manifold, so time must have a stop for some observers (such as those falling into black holes) or for all (final collapse of the universe). Likewise the past contains singularities; probably a Big Bang, (cf. IX.3.03).

For more precise statements of these facts the reader is referred to [Hawking and Ellis], which is devoted almost entirely to their discussion and proof.

Exercises XII.2

1a) Show that the symmetries on p.530 imply also that

$$t^{+} : S\times R \to S\times R : (x,r) \leadsto (x,r+t)$$

preserves geodesics etc. for any t. Use the symmetry $\overline{2t_0}$ and the flow $Q : S\times R\times R \to S\times R : (x,r,t) \leadsto (x,r+t)$ to establish the nature of parallel transport around $S\times\{t_0\}$ (p.531) and deduce that of the connection coefficients and Ricci curvatures.

2) Show that for M with any metric tensor and $f : M \to R$, we always have $\operatorname{div}(f\underset{\sim}{I}) = df$. (Two lines in coordinates.)

3) In 2.03 assume that the sectional curvature for any plane tangent to S is the same number, k say. Further assume that the sectional curvature for any plane in T_xM containing the "$_S$rest velocity" timelike vector orthogonal to S is k'. Hence show that $k = 4\pi(p - \frac{\mu}{3})$. (Remember the minus signs in $\underset{\sim}{G}$ and $\underset{\sim}{G}_{\uparrow}$.)

3. THE STARS IN THEIR COURSES

How is spacetime shaped in vacuum around a concentrated body of matter, such as the earth or the sun? Since the Einstein and Ricci tensors vanish there by Einstein's equation, this means: what must the Weyl tensor (which is then all of $\underset{\sim}{R}$) look like in such a region? The answer is obviously not unique - for instance the curvature around the earth is affected by the distant presence of the sun - unless we set boundary conditions giving the effect of other bodies outside our region of solution, and any "background field".

This non-local determination of the Weyl tensor by matter, the governing equation being div $\underset{\sim}{C} = \underset{\sim}{J}$ where $\underset{\sim}{J}$ is a function of $\underset{\sim}{T}$ (Ex. 1), is analogous to the non-local determination of the electromagnetic field by moving charges, governed by Maxwell's equations. However since the Weyl tensor is coupled to the shape of the underlying spacetime, gravitational effects "add" in a much more complicated way than electromagnetic ones. As a solvable first approximation, then, assume that the sun is alone in an asymptotically flat spacetime, with only "test particles of negligible mass" moving around to study the geometry outside it.

3.01 The Schwarzschild solution

We look for a static, spherically symmetric solution, around an unrotating spherical star of radius r_0 alone in space. That is, we take spherical coordinates (r,θ,ϕ) on R^3 (Fig. 3.1) and corresponding "hyper-cylindrical" coordinates (t,r,θ,ϕ) on R^4. Then seek an asymptotically flat Lorentz metric on R^4, dependent only on r. (Notice that these coordinates are not everywhere defined. Nor are they, strictly, given by a chart, since ϕ takes values in S^1. But locally they correspond to a chart.) We assume further that the spheres of the form $\{(t,r,\theta,\phi)\,|\,t=t_\alpha,\ r=r_\alpha\}$ have the usual metric for spheres of radius r_α, given by $ds^2 = -r_\alpha^2((d\theta)^2 + \sin^2\theta(d\phi)^2)$. (There is no loss of generality in this assumption, since given the spherical symmetry they must have constant curvature and hence a scalar multiple of the usual metric for spheres, as long as their circumferences always increase for increases in r_α: we could always reparametrise r. The negative sign reflects the fact that changes purely in θ or ϕ are in spacelike directions.) Symmetry means that there can be no off-diagonal spacelike/spacelike terms in the matrix for $\underset{\sim}{G}$ in these coordinates. The static

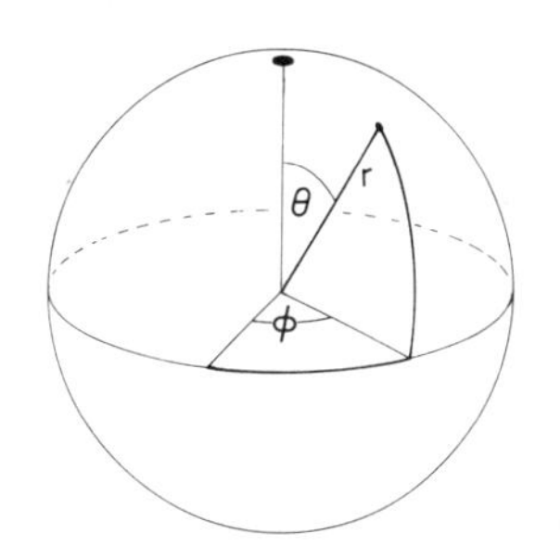

Fig. 3.1

condition tells us that for any t_0 we have a symmetry

$$(t,r,\theta,\psi) \rightsquigarrow (t_0-t,r,\theta,\psi)$$

which similarly implies that ∂_t is orthogonal to ∂_r, ∂_θ, ∂_ϕ (cf. VII §4 on "name" indices.) Thus there are no off-diagonal terms at all, and we are seeking a metric of the form

$$(ds)^2 = f(r)(dt)^2 - h(r)(dr) - r^2(d\theta)^2 - r^2\sin^2\theta(d\phi)^2 \quad \text{at } (t,r,\theta,\phi),$$

where f and h are functions $R \to R$.

We look first for a solution outside the region $\{(t,r,\theta,\phi)|r < r_0\}$ supposed to contain the matter, with f(r) and h(r) positive. The "asymptotically flat" requirement implies

$$\lim_{r\to\infty} f(r) = 1 = \lim_{r\to\infty} h(r),$$

since the usual Minkowski metric is in these coordinates

$$(ds)^2 = (dt)^2 - (dr)^2 - r^2(d\theta)^2 - r^2\sin^2\theta(d\phi)^2.$$

For technical convenience, we work with the natural logarithms of f and h, setting $f(r) = e^{\lambda(r)}$, $h(r) = e^{\xi(r)}$. Computation of the Ricci tensor (Ex. 2) gives as its non-identically-zero components

$$\text{R1)} \qquad R_{tt} = \left(\frac{1}{2}\frac{d^2\lambda}{dr^2} - \frac{1}{4}\frac{d\lambda}{dr}\frac{d}{dr}(\xi-\lambda) + \frac{1}{r}\frac{d\lambda}{dr}\right)\left(-e^{\lambda-\xi}\right)$$

R2) $$R_{rr} = \frac{1}{2}\frac{d^2\lambda}{dr^2} - \frac{1}{4}\frac{d\lambda}{dr}\frac{d(\xi-\lambda)}{dr} - \frac{1}{r}\frac{d\lambda}{dr}$$

R3) $$R_{\theta\theta} = (1 - \frac{r}{2}\frac{d(\xi-\lambda)}{dr} - e^{\xi})e^{-\xi}$$

R4) $$R_{\phi\phi} = (1 - \frac{r}{2}\frac{d(\xi-\lambda)}{dr} - e^{\xi})e^{-\xi}\sin^2\theta.$$

Setting these equal to zero, since the Ricci tensor must vanish by X.6.13 if the Einstein tensor vanishes, we get

$$\frac{d\lambda}{dr} = -\frac{d\xi}{dr}$$

by combining R1 and R2, hence $\lambda(r) = -\xi(r) + a$. (Since the domain of λ and ξ is connected, only one constant is needed: cf Ex. VII.5.01.) As $r \to \infty$, $f(r), h(r) \to 1$ by hypothesis, hence $\lambda(r)$, $\xi(r) \to 0$ so a can only be zero. Hence R3 gives

$$1 + r\frac{d\lambda}{dr} - e^{-\lambda} = 0.$$

since $\frac{df}{dr} = \frac{d}{dr}(e^{\lambda}) = e^{\lambda}\frac{d\lambda}{dr} = f\frac{d\lambda}{dr}$, this gives

$$1 + \frac{r}{f}\frac{df}{dr} - \frac{1}{f} = 0,$$

that is $$\frac{df}{dr} = \frac{1-f}{r},$$

whose general solution is

$$f(r) = (1 - \frac{k}{r}),$$

for some constant k. So the metric is given by the line element

$$(ds)^2 = (1 - \frac{k}{r})(dt)^2 - (1 - \frac{k}{r})^{-1}(dr)^2 - r^2(d\theta)^2 - r^2\sin^2\theta(d\phi)^2,$$

outside the star.

This is the Schwarzschild solution of Einstein's equation, satisfying it in vacuum. The constant k depends on the matter in the region $r < r_0$: if k is zero, the metric reduces to that of flat spacetime. Solving Einstein's equation for the inside of the star, for which we refer the reader to [Misner, Thorne and Wheeler], leads to the value $k = 2M$, where M is the "total mass-energy of the star" in appropriate units†. (A bad choice of units brings in a "universal gravitational constant" G, just as choosing units of length and time independently leads to a non-unity "universal limiting velocity" c in XI.2.08.) The concept of "total mass energy" is a bit more subtle than one might first guess - in fact only if the star is alone in asymptotically flat space, as here assumed, does it have an invariant meaning - but to examine it more closely would require the integral techniques we have agreed to defer. ("Total" implies that something is integrated over the star.) Therefore, we simply examine here the geometry of the metric

$$(ds)^2 = (1 - \frac{2M}{r})(dt)^2 - (1 - \frac{2M}{r})^{-1}(dr)^2 - r^2(d\theta)^2 - r^2\sin^2\theta(d\phi)^2$$

where M is some positive constant associated with the star. It turns out that this M coincides with the solar mass which, used in Newton's theory, gives the orbits best approximating the geodesics that we study below. In this sense, the "mass" can be found by analysis of orbits.

Notice that if $r_0 < 2M$ our region of interest includes points where f and h are negative and have no logarithms, so invalidating our method, but a direct check shows that we still have a solution. More critically, for $r = 2M$ the metric as given is undefined, and hence not a metric at all. The fault, however, is not in the metric but in the coordinates: it is strictly analogous to the way the spherical metric $((d\theta)^2 + \sin^2\theta(d\phi)^2)$ appears indefinite when θ is 0 or π. But we are here concerned with motions around an uncollapsed star with radius $r_0 > 2M$.

(Our own sun is an example for which the Schwarzschild radius 2M is about 2.95 kilometres.) We refer the reader especially to [Hawking and Ellis] for a careful treatment of that large subject, the fascinating geometry of black holes, as situations including criticalities like r = 2M in this solution are called.

3.02 Schwarzschild geodesics

The non-zero Christoffel symbols for the Levi-Cività connection of the Schwarzschild metric are, by Ex. 3,

$$\Gamma^t_{tr} = \Gamma^t_{rt} = \frac{M}{r(r-2M)}$$

$$\Gamma^r_{tt} = \frac{M}{r^2}\left(1 - \frac{2M}{r}\right)$$

$$\Gamma^r_{\theta\theta} = 2M - r$$

$$\Gamma^r_{\phi\phi} = \sin^2\theta(2M - r)$$

$$\Gamma^\theta_{r\theta} = \Gamma^\theta_{\theta r} = \Gamma^\phi_{r\phi} = \Gamma^\phi_{\phi r} = \frac{1}{r}$$

$$\Gamma^\theta_{\phi\phi} = \sin\theta\cos\theta$$

$$\Gamma^\phi_{\theta\phi} = \Gamma^\phi_{\phi\theta} = \cot\theta.$$

If a curve c is given by $c(\sigma) = (c^t(\sigma), c^r(\sigma), c^\theta(\sigma), c^\phi(\sigma))$ then the condition IX.1.04 that c be a geodesic thus becomes

---

†Eddington once caused consternation at a learned meeting by referring to the mass of the sun as about "1.45 kilometres" - perfectly valid in geometrised units, cf. Ex. 6. But its radius r_0 is much bigger than its mass.

(i) $$\frac{d^2c^t}{d\sigma^2} + \frac{2M}{c^r(c^r-2M)}\frac{dc^t}{d\sigma}\frac{dc^r}{d\sigma} = 0$$

(ii) $$\frac{d^2c^r}{d\sigma^2} + \frac{M(c^r-2M)}{(c^r)^3}\left(\frac{dc^t}{d\sigma}\right)^2 - \frac{M}{c^r(c^r-2M)}\left(\frac{dc^r}{d\sigma}\right)^2 - (c^r-2M)\left(\frac{dc^\theta}{d\sigma}\right)^2 - \sin^2 c^\theta (c^r-2M)\left(\frac{dc^\phi}{d\sigma}\right)^2 = 0$$

(iii) $$\frac{d^2c^\theta}{d\sigma^2} + \frac{2}{c^r}\frac{dc^r}{d\sigma}\frac{dc^\theta}{d\sigma} - \sin c^\theta \cos c^\theta \left(\frac{dc^\phi}{d\sigma}\right)^2 = 0$$

(iv) $$\frac{d^2c^\theta}{d\sigma^2} + \frac{2}{c^r}\frac{dc^r}{d\sigma}\frac{dc^\theta}{d\sigma} + \cot c^\theta \frac{dc^\theta}{d\sigma}\frac{dc^\phi}{d\sigma} = 0.$$

If for some $\sigma_0 \in R$ we have $c^\theta(\sigma_0) = \pi/2$, $\frac{dc^\theta}{d\sigma}(\sigma_0) = 0$, then for all σ, $c^\theta(\sigma) = \pi/2$ (why?) So we can without loss of generality suppose this, since we can always choose coordinates so as to <u>make</u> $c^\theta(\sigma_0) = \pi/2$ and $c_*(\sigma_0)$ orthogonal to ∂_θ. The equations reduce, for orbits thus "lying in the hyperplane $\theta = \pi/2$" (though strictly M has no hyperplanes, not being affine) to

G1) $$\frac{d^2c^t}{d\sigma^2} + \frac{2M}{c^r(c^r-2M)}\frac{dc^t}{d\sigma}\frac{dc^r}{d\sigma} = 0$$

G2) $$\frac{d^2c^r}{d\sigma^2} + \frac{M(c^r-2M)}{(c^r)^3}\left(\frac{dc^t}{d\sigma}\right)^2 - \frac{M}{c^r(c^r-2M)}\left(\frac{dc^r}{d\sigma}\right)^2 - (c^r-2M)\left(\frac{dc^\phi}{d\sigma}\right)^2 = 0$$

G3) $$\frac{d^2c^\phi}{d\sigma^2} + \frac{2}{c^r}\frac{dc^r}{d\sigma}\frac{dc^\phi}{d\sigma} = 0.$$

But $$\frac{d}{d\sigma}\left((c^r)^2\frac{dc^\phi}{d\sigma}\right) = (c^r)^2\left(\frac{d^2c^\phi}{d\sigma^2} + \frac{2}{c^r}\frac{dc^r}{d\sigma}\frac{dc^\phi}{d\sigma}\right),$$

hence G3 integrates by inspection to

G3') $(c^r)^2\frac{dc^\phi}{d\sigma}$ = constant = A, say. ("Conservation of angular momentum")

Similarly G1 integrates immediately to

G1') $$\left(1 - \frac{2M}{c^r}\right)\frac{dc^t}{d\sigma} = \text{constant} = B \quad \text{(cf Ex. 4.)}$$

The constants A and B are fixed if we know "initial conditions" $c(\sigma_0)$ and $c_*(\sigma_0)$ for some $\sigma_0 \in R$: evidently they may take any real values for an arbitrary geodesic, though B = 0, for instance, implies that the geodesic is spacelike.

3.03 Radial Motion

Equation G3' shows that if $\frac{dc^\phi}{d\sigma}(\sigma_0) = 0$ for some σ_0, c^ϕ is constant. Hence geodesics in the surface

$$S = \{(t,r,\theta,\phi) \mid \theta = \pi/2, \phi = \phi_0\}$$

with coordinates (t,r) and a metric given by

$$* \qquad (ds)^2 = \left(1 - \frac{2M}{r}\right)(dt)^2 - \left(1 - \frac{2M}{r}\right)^{-1}(dr)^2$$

coincide with geodesics in R^4 with the Schwarzschild metric.

We shall discuss this analytically in a moment. Notice first, however, that we have already encountered a 2-manifold with a metric of the form

$$(ds)^2 = f(r)(dx)^2 - h(r)(dr)^2$$

with $f(r) \to 0$ and $h(r) \to \infty$ as r descends to some $q \in R$: namely, the indefinite example of IX.5.03, with $x = \theta$, $q = 0$, and

$$f(r) = r^2, \quad h(r) = \frac{1}{r^2}\left(\frac{1}{r^2} + 1\right).$$

We have seen how the geometry of this "pushes geodesics inwards" so that they can rise from low r, reach a maximum and fall back. Only r and t are varying, so this models something "thrown straight up and falling straight back". Thus we have already a qualitative example of how spacetime curvature can guide timelike geodesics "downwards". That example is not asymptotically flat, however - indeed a radial curve, with t constant, has finite length - so this surface is rather different further out.

In R^3 with cylindrical coordinates (R,θ,z) (to reserve r for our radial coordinate in the Schwarzschild solution) and the same indefinite metric $\tilde{\underset{\sim}{G}}$ as in IX.5.02 we may take the surface

$$N = \{(R,\theta,z)\,|\,z = f(R),\ 0 < R < 1\}$$

(Fig. 3.2), where f is any indefinite integral of

$$]0,1[\ \to R : s \rightsquigarrow -\sqrt{1 + \frac{16M^2}{(1-s^2)^4}}\ .$$

Now consider a map defined by

$$\phi :]0,2[\times]2M,\infty[\ \to R^3\ :\ (t,r) \rightsquigarrow \left(\sqrt{1 - \frac{2M}{r}},\ t,\ f\left(\sqrt{(1 - \frac{2M}{r})}\right)\right),$$

still using cylindrical coordinates on R^3.

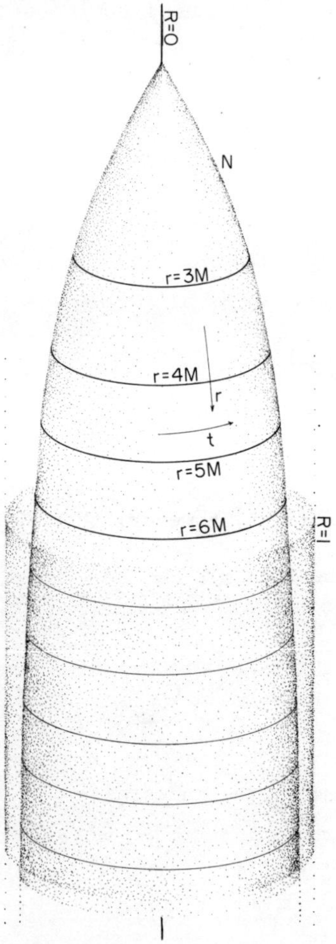

It is clear that the image U of ϕ in R^3 lies in N, that ϕ is injective and that $\overleftarrow{\phi}$ defines a chart on N with domain U and image in R^2. N with the coordinates (t,r) given by ϕ has the metric induced from $\tilde{\underset{\sim}{G}}$ given exactly by * above (Ex. 5). The same geometric reasoning as before then explains

Fig. 3.2

the "pull" on geodesics towards $r = 2M$. N is asymptotic to a cylinder's intrinsic flatness as $r \to \infty$.

Notice that we have been able to induce the Schwarzschild metric only by embedding just part of the (r,t)-plane in a flat R^3: if we tried to embed a longer t-interval than 2π, we would meet the same point in R^3 more than once. Fig. 3.2 is a realisation of the curvature of part of the surface, and has nothing to do with its cause, which is an embedding in R^4 with curvature intrinsically determined by Einstein's equation. (If a Philosopher of Science can be brought along as far as understanding Fig. 3.2, you will have cured him of "bent round what?" permanently. The embedding that gives this curvature locally will be evidently non-physical, even to him.)

For "radial motion" geodesics, G2 reduces to

$$\frac{d^2c^r}{d\sigma^2} + \frac{M(c^r-2M)}{(c^r)^3}\left(\frac{dc^t}{d\sigma}\right)^2 - \frac{M}{c^r(c^r-2M)}\left(\frac{dc^r}{d\sigma}\right)^2 = 0$$

For c timelike and σ proper time, we have

$$1 = c^*(\sigma).c^*(\sigma) = (1 - \frac{2M}{c^r})\left(\frac{dc^t}{d\sigma}\right)^2 - (1 - \frac{2M}{c^r})^{-1}\left(\frac{dc^r}{d\sigma}\right)^2$$

that is,
$$\left(\frac{dc^t}{d\sigma}\right)^2 = (1 - \frac{2M}{c^r})^{-1}\left[1 + (1 - \frac{2M}{c^r})^{-1}\left(\frac{dc^r}{d\sigma}\right)^2\right].$$

Combining these two equations,

$$\frac{d^2c^r}{d\sigma^2} + \frac{M}{(c^r)^2}\left[1 + (1 - \frac{2M}{c^r})^{-1}\left(\frac{dc^r}{d\sigma}\right)^2\right] - \frac{M}{c^r(c^r-2M)}\left(\frac{dc^r}{d\sigma}\right)^2 = 0,$$

which reduces to

$$\frac{d^2c^r}{d\sigma^2} + \frac{M}{(c^r)^2} = 0.$$

If at some σ_0 we have $\frac{dc^r}{d\sigma}(\sigma_0) = 0$, then $c^*(\sigma)$ is a scalar multiple of ∂_t, so that $c^*(\sigma).c^*(\sigma) = 1$ gives

$$\frac{dc^t}{d\sigma}(\sigma_0) = \left(1 - \frac{2M}{c^r(\sigma_0)}\right)^{-\frac{1}{2}},$$

$$\text{hence by G1'} \qquad \frac{dc^t}{d\sigma}(\sigma_1) = \frac{\left(1 - \frac{2M}{c_r(\sigma_0)}\right)^{\frac{1}{2}}}{\left(1 - \frac{2M}{c^r(\sigma_1)}\right)} \qquad \text{for all } \sigma_1.$$

Sometimes $c^r(\sigma)$ is a large enough multiple of M that we may approximate this by 1. (Here, at our distance from the sun, it is about one part in 600,000 less.) Then we may approximate c^t by σ, hence σ by t. The result (putting in the "gravitational constant" G for the sake of familiarity) is

$$\frac{d^2c^r}{dt^2} = -\frac{MG}{(c^r)^2}$$

as an approximation to the geodesic equation. This of course is exactly Newton's result for radial motion of a particle solely influenced by the gravitational effect of a fixed mass centred at r = 0.

3.04 Orbital motion

A timelike geodesic modelling the movements we see in the solar system has $dc^t/d\sigma$ very much larger than the other components of c^*. Thus at any point the geodesic is nearly tangent (Fig. 3.3) to a "radial motion" one with $dc^\theta/d\sigma = dc^\phi/d\sigma = 0$, and so by continuity has a nearly identical apparent "acceleration towards the sun". (Hence the applicability of Newtonian theory as an approximation in this case also.) Fig. 3.2 thus remains a better visualisation of why planets seem "pulled towards the sun" than any embedding in a flat space of the hypersurface $\theta = \pi/2$ that we might construct, since this would involve at least one more dimension than we can draw.

Consider a timelike geodesic c parametrised by proper time with $c^\theta = \frac{\pi}{2}$, $\frac{dc^\theta}{d\sigma} = 0$ identically, and $\frac{dc^\phi}{d\sigma} \neq 0$ for some and hence by G3' all σ. Using the "unit length" condition again, we have

$$1 = c^*(\sigma).c^*(\sigma) = (1 - \frac{2M}{c^r})\left(\frac{dc^t}{d\sigma}\right)^2 - (1 - \frac{2M}{c^r})^{-1}\left(\frac{dc^r}{d\sigma}\right)^2 - (c^r)^2\left(\frac{dc^\phi}{d\sigma}\right)^2$$

or,
$$\left(\frac{dc^t}{d\sigma}\right)^2 = \left(1 + (1 - \frac{2M}{c^r})^{-1}\left(\frac{dc^r}{d\sigma}\right)^2 + (c^r)^2\left(\frac{dc^\phi}{d\sigma}\right)^2\right)\left(1 - \frac{2M}{c^r}\right)^{-1}$$

Using this to eliminate c^t from G2,

$$\frac{d^2c^r}{d\sigma^2} + \frac{M}{(c^r)^2}\left(1+(1 - \frac{2M}{c^r})^{-1}\left(\frac{dc^r}{d\sigma}\right)^2 + (c^r)^2\left(\frac{dc^\phi}{d\sigma}\right)^2\right) - \frac{M}{c^r(c^r-2M)}\left(\frac{dc^r}{d\sigma}\right)^2 - (c^r-2M)\left(\frac{dc^\phi}{d\sigma}\right)^2 = 0$$

which reduces to

$$* \qquad \frac{d^2c^r}{d\sigma^2} - (c^r - 3M)\left(\frac{dc^\phi}{d\sigma}\right)^2 + \frac{M}{(c^r)^2} = 0.$$

Fig. 3.3

Even a spiral curve (a) at constant radius is nearly tangent to a purely radial motion (b) if $\frac{da^t}{d\sigma}$ is much greater than $\frac{da^\theta}{d\sigma}$.

This immediately gives "circular" orbits (helices in spacetime) of constant "radius" $c^r = a$: $\frac{d^2c^r}{d\sigma^2} = 0$ and G3' becomes

$$a^2 \frac{dc^\phi}{d\sigma} = A, \quad \text{so } \frac{dc^\phi}{d\sigma} = \frac{A}{a^2} = \frac{2\pi}{T}$$

where T is the period of the orbit, measured in proper time. Substituting in *, T must be precisely $2\pi a\sqrt{\frac{a}{M} - 3}$; cf. Ex. 6c. (If $A < 6M$, the orbit cannot be stable: the smallest perturbation inward will make it spiral down to $r = 2M$. See [Misner, Thorne and Wheeler].)

Substituting in * from G3' we have the precise form

$$\frac{d^2c^r}{d\sigma^2} = \frac{A}{(c^r)^4}(c^r - 3M) - \frac{M}{(c^r)^2}$$

for the "radial acceleration" of a timelike geodesic in Schwarzschild geometry, outside the Schwarzschild radius.

Let us now "reparametrise c by ϕ": since $\frac{dc^\phi}{d\sigma}$ never vanishes, c^ϕ has a smooth inverse at least locally by the Inverse Function theorem. Take $\psi :]0,2\pi[\to R$ such that $c^\phi(\psi(\phi)) = \phi$. ($c^\phi \circ \psi$ is often denoted simply by ϕ, as is c^ϕ. In this context the result is to denote three not-merely-formally-distinct objects, the real <u>number</u> (coordinate label) ϕ, and the two <u>maps</u> c^ϕ and $c^\phi \circ \psi$, by the same letter, in an attempt to "simplify".) We denote the corresponding reparametrisation $c^r \circ \psi$ of the r-coordinate function c^r by $\tilde{r}$.

Substituting in * the consequence (Ex. 7d)

$$\frac{d^2c^r}{d\sigma^2} = \left(\frac{dc^\phi}{d\sigma}\right)^2 \left(\frac{d^2\tilde{r}}{d\phi^2} - \frac{2}{\tilde{r}}\left(\frac{d\tilde{r}}{d\phi}\right)^2\right)$$

of G3, we get

$$\frac{d^2\tilde{r}}{d\phi^2} - \frac{2}{\tilde{r}}\left(\frac{d\tilde{r}}{d\phi}\right)^2 - \tilde{r} + 3M + \frac{M\tilde{r}^2}{A^2} = 0.$$

Setting $u(\phi) = 1/\tilde{r}(\phi)$, this is equivalent (Ex. 7) to

$$\frac{d^2u}{d\phi^2} + u - 3Mu^2 - \frac{M}{A^2} = 0.$$

By Ex. 8a this has an approximate solution in polar coordinates as a conic section with focus at $r = 0$, given by

$$\frac{1}{\tilde{r}(\phi)} = u(\phi) = \frac{1}{R}(1 + \varepsilon\cos(\phi-\phi_0)).$$

Here ε is the eccentricity (0, $0 < \varepsilon < 1$, 1, $\varepsilon > 1$ for circle, ellipse, parabola respectively), and ϕ_0 is the value of ϕ at closest approach to the origin (the <u>perihelion</u> - Greek for "near the sun").

An iterative procedure like the one used in the Appendix to converge on the solution of a differential equation yields the next approximation (Ex. 8b)

$$u(\phi) = \frac{1}{R}\left(1 + \frac{3M^2}{A^2}\right)(1 + \varepsilon\cos(\phi-p(\phi)), \quad \text{where } p(\phi) = \phi_0 + \frac{3M^2\phi}{A^2};$$

for $\varepsilon < 1$ this is "an ellipse whose perihelion is slowly rotating". This approximation is good enough for the purposes of solar system astronomy. It predicts precession rates of 43, 8 and 4 seconds of arc per century for geodesics modelling the orbits of Mercury, Venus and Earth respectively for instance, in good agreement with observation. For parabolic and hyperbolic orbits the difference from strict conic sections is too small to detect.

The analysis of null geodesics can be carried out on similar lines: light grazing the sun is "bent towards it" with an apparent deflection of 1.75 seconds of arc. Apart from its value as a test for general relativity, the effect of our local gravitation on light is thus not very significant or easy to detect; the geometry of optics in a vacuum only becomes dramatically different from the flat case in the vicinity of a black hole, or a body extending not far outside its Schwarzschild radius and so in danger of falling completely in, in gravitational collapse.

3.05 Spacelike geodesics

Consider the surface $S = \{(r,t,\phi,\theta) \mid t = t_0,\ r > 2M,\ \theta = \frac{\pi}{2}\}$. By G1' geodesics anywhere tangent to S remain in it, so G2 reduces, much as in 3.03, to

$$\frac{d^2c^r}{ds^2} - \frac{M}{c^r(c^r-2M)}\left(\frac{dc^r}{ds}\right)^2 - (c^r - 2M)\left(\frac{dc^\phi}{ds}\right)^2 = 0.$$

(We denote the arc length parameter along spacelike geodesics by s, since σ is proper <u>time</u>.) Clearly $\frac{d^2c^r}{ds^2}$ is always positive, unlike timelike radial motion where it is always negative or orbital motion where it is positive at perihelion, negative at aphelion (if any). So these geodesics have no aphelion (furthest point from the sun). As one might expect, something "infinitely fast" is above escape velocity. By Ex. 9 we can realise the negative definite metric

$$-(ds)^2 = (1 - \frac{2M}{r})^{-1}(dr)^2 + r^2(d\phi)^2$$

of S by embedding it in R^3 with minus the Euclidean metric and cylindrical coordinates (r,ϕ,z) as

$$N = \{(r,\phi,z) \mid z = -\sqrt{8M(r-2M)},\ r > 2M\}$$

(Fig. 3.4a) and using the r and ϕ of R^3 as coordinates on it. (For an uncollapsed star, with r_0 greater than the Schwarzschild radius, we have of course a positively curved "cap" in the region $r < r_0$; Fig. 3.4b).

The situation is clearly qualitatively similar to the Riemannian example in IX.5.03, for r greater than r_0 and 2M, and the shapes of these "instantaneous travel" orbits may be found experimentally by stretching strings on a model. Again, they are clearly "deflected towards the sun".

As no experimental test for the spacelike solutions of the Schwarzschild geodesics is in prospect, we leave the interested reader to investigate them. Are they sometimes (when?) approximately conics?

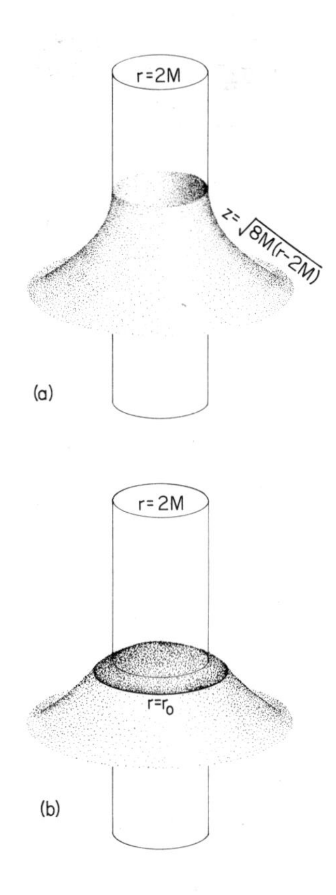

Fig. 3.4

Exercises XII.3

1a) If $\underset{\sim}{J}(\underset{\sim}{u},\underset{\sim}{v},\underset{\sim}{w}) = \underset{\sim}{v}(\bar{\underset{\sim}{R}}(\underset{\sim}{u},\underset{\sim}{w})) - \underset{\sim}{w}(\bar{\underset{\sim}{R}}(\underset{\sim}{u},\underset{\sim}{v})) + \frac{1}{6}(\underset{\sim}{w}(R)\underset{\sim}{u}.\underset{\sim}{v} - \underset{\sim}{v}(R)\underset{\sim}{u}.\underset{\sim}{w})$ (where, for example, $\underset{\sim}{w}(R)$ means $dR(\underset{\sim}{w})$; cf. VII.4.02), show that $\underset{\sim}{J}$ is a $\binom{0}{3}$-tensor field and the 2nd Bianchi identity is equivalent to div $\underset{\sim}{C} = \underset{\sim}{J}$.

b) Use Einstein's equation to give $\underset{\sim}{J}$ in terms of $\underset{\sim}{T}$.

2) Compute the connection coefficients, Riemann tensor and hence Ricci tensor of the metric

$$(ds) = e^{\lambda}(dt)^2 - e^{\xi}(dr)^2 - r^2(d\theta)^2 - r^2\sin^2\theta(d\phi)^2.$$

(You will need the fact that if $\tilde{g}(t,r,\theta,\phi) = g(r)$, then $\partial_r \tilde{g} = \frac{dg}{dr}$.)

3a) Inserting functions $\lambda(r) = \log(1 - \frac{2M}{r})$, $\xi(r) = -\lambda(r)$ in Ex. 2, or otherwise, find the connection coefficients of the Schwarzschild metric.

b) Substitute the result into the general geodesic equation to get (i) - (iv) of 3.02.

4) Use equation G1', and the Schild argument of XI.5 in reverse, to compute the ratio of the measured emission and reception frequencies for a photon going from $(t_1, r_1, \theta_1, \phi_1)$ to $(t_2, r_2, \theta_2, \phi_2)$ in the Schwarzschild solution, with $r_1, r_2 > r_0 > 2M$. (Note that for an observer in a "fixed position"

$$\frac{dc^r}{d\sigma} = \frac{dc^\theta}{d\sigma} = \frac{dc^\phi}{d\sigma} = 0, \text{ while } c^*(\sigma).c^*(\sigma) = 1$$

by definition of proper - measured - time.)

5) Compute the metric induced on the surface N of 3.03 by the metric $(ds)^2 = (dR)^2 + R^2(d\theta)^2 - (dz)^2$, using the chart given.

6a) In the case of radial motion, is there a difference between general relativistic and Newtonian values for escape speed at a point $x = (t,r,\theta,\phi)$? (The <u>escape speed</u> is the least speed, relative to $(\partial_t)_x$ as "rest", such that $\frac{dc^r}{d\sigma} > 0$ for all succeeding σ.)

b) Assume that the speed of light is 3.0×10^8 metres per second and also that one year is $10^7\pi$ seconds (both are true nearly to 3 significant figures). Suppose that a moon, earth and sun are in circular orbits of radii 3.9×10^8, 1.5×10^{11} and 3.0×10^{20} metres around an earth, sun and galaxy whose masses (in geometrised units) are 4.4×10^{-3}, 1.5×10^3 and 2.2×10^{14} metres respectively.

Show that their periods are approximately 1/13 year, 1 year and 10^8 years, respectively.

c) For non-radial motion, does escape speed depend on initial direction relative to $(\partial_t)_x$ as "rest"? (In Newtonian theory, it depends only on the orbit not hitting the sun. Is this still true? Compare circular orbits with radial motion.)

7a) Use the chain rule to state G3 in terms of the identity map $c^\phi \circ \psi$ and the reparametrisation $\tilde{r}$, and deduce the consequence used in 3.04.

b) Prove the equivalence of the differential equations stated in 3.04 for $\tilde{r}$ and $u = 1/\tilde{r}$.

8a) If u is a function $R \to R$, define a new function

$$D(u) = \frac{d^2u}{d\phi^2} + u - 3Mu^2 - \frac{M}{A^2}$$

and show that the function u given by the conic section equation satisfies $|D(u)| \leq k$ for some explicitly given real number k. When is it reasonable to treat k as zero (that is, use the conic as if it were an exact solution)?

b) Repeat (a) for the "precessing ellipse" equation.

c) What is the difference between "u is approximately a solution" as in (a) and (b), and "u approximates an exact solution $\tilde{u}$", in the sense that there is a small δ with $|\tilde{u}(\phi) - u(\phi)| < \delta$ for all ϕ? Are the two equivalent (i) for geodesics with domain R, (ii) for geodesics with domain a compact interval?

9) Compute the metric induced on the surface N of 3.05 by minus the Euclidean metric on R^3.

4. FAREWELL PARTICLE

We can now see how the result of 1.03 is a little fictitious.

Any physically significant particle must have 4-momentum. We cannot have, say, a particle with only charge: the charge can produce changes in the 4-momentum of a charged particle with rest mass, so if it persists in having zero 4-momentum it violates conservation. (If not we can consider it as coming into existence, like, say, an emitted photon). So either conservation is false, or a zero-4-momentum particle can interact with nothing we can interact with, and is thus physically meaningless as far as we are concerned.

Now a Newtonian particle of mass m, as the limit of little balls of radius r, density $3m/4\pi r^3$ as $r \to 0$, makes a moderate amount of sense. The strength of the gravitational field tends to infinity as we approach the particle, which is odd, and the energy to be obtained by letting two particles fall towards each other is infinite, which is odder. However, these oddities can be dodged. For relativistic particles there are more fundamental problems.

The particle has its own effect on spacetime. This means first that the metric of an asymptotically flat spacetime containing one sun and one particle is not exactly the Schwarzschild metric anywhere, any more than the Newtonian "central field of force" exactly allows for a space probe's effect on the sun. More importantly, the singularity of gravitation involves a singularity in spacetime itself: an infinity in the curvature is inconsistent with any pseudo-Riemannian structure. (The "singularity" at the Schwarzschild radius is an artefact of the chart, as remarked above, but the singularity at $r=0$ is not.) We can hardly say that the curve followed is a geodesic, if the structure by which we define "geodesic" breaks down at every point the particle visits.

We cannot avoid this problem, as we could the milder Newtonian analogue, by treating the particle as of arbitrarily small non-zero diameter and correspondingly high density. Once a body of matter, of any mass m, lies inside its Schwarzschild radius 2m it undergoes gravitational collapse (see [Hawking and Ellis] or [Misner, Thorne and Wheeler] for physics inside a black hole) and the singularity becomes physical, not a limiting fiction.

Nor can we say that the centre of mass of a larger body follows a geodesic, because "centre of mass" cannot be defined relativistically. Moreover there is no such thing as a rigid body, that is one such that a push at one side starts the whole body moving at once. (What does "at once" mean all over the body?) If it is "rigid in the frame of reference F", even in flat spacetime, this means that the $_F$speed of sound in it is infinite: and there are many frames in which the push travels through the body backwards in time. Thus while the assumption that planets are rigid spheres allows Newtonian mechanics to treat their orbits as those of points, there is no mathematically or physically practical way of ignoring their "internal vibrations" without ignoring relativity. (For a coherent treatment of the relativistic dynamics of classical matter without the simplifications/ approximations usual in cosmology texts, see [Dixon].)

The utility of geodesics, then, lies in the following rather complicated fact, which we state without proof. Suppose in some spacetime M we have a body of matter, or black hole, P, with mass and diameter (suitably approximately defined) small in comparison to its separation from the other parts of M with $T \neq 0$. Let U be a tubular region surrounding the track of P. Then we can approximate $M \backslash U$ by $M' \backslash U'$, where M' is a spacetime similar to M except that P is absent, and U' surrounds a geodesic. This, precisely formulated, is the more careful statement of the "particles follow geodesics" of 1.03, and the "planets follow geodesics" of §3. Strictly speaking, general relativity does not admit the point particles of classical mechanics.

Appendix: Existence and Smoothness of Flows

πάντα ρεi

Heraclitus of Ephesus.

1. COMPLETENESS

Axiom VI.4.01 was essentially one-dimensional. For general use we need more apparatus:

1.01 Definition

A Cauchy sequence in a metric space (X,d) (defined in VI.1.02) is a sequence $S : \mathbb{N} \to X : i \rightsquigarrow x_i$ in X such that for any $\varepsilon > 0$ there is an $M \in \mathbb{N}$ (generally larger for smaller ε) with the property that

$$m,n > M \;\Rightarrow\; d(x_m,x_n) < \varepsilon, \quad \text{(Ex. 1)}.$$

If some subsequence S' of a Cauchy sequence S converges, say to $x \in X$, so does S:

Any neighbourhood U of x contains an open ball $B(x,\varepsilon)$, by Definition VI.1.07 of the metric topology. Since S' converges and S is Cauchy, there are $L, M \in \mathbb{N}$ such that $d(x,x_i) < \frac{\varepsilon}{2}$ for $i > L$, x_i a point of S', and $d(x_i,x_n) < \frac{\varepsilon}{2}$ for $i, n > M$. So for $n > M$ the triangle inequality gives $d(x,x_n) \le d(x,x_i) + d(x_i,x_n) < \varepsilon$, for x_i any point of S' with $i > \max\{L,M\}$; hence $n > M \Rightarrow x_n \in U$. Since U was arbitrary, S thus converges to x.

Combining Ex. 1 and VI.4.04, a sequence in a compact interval $[a,b]$ with the usual metric converges if and only if it is Cauchy. But any Cauchy sequence in R lies in a compact interval ([min(K)-1, max(K)+1] where $M \in N$ has $m,n > M \Rightarrow |x_m - x_n| < \frac{1}{2}$, $K = \{x_1, x_2, \ldots, x_M\}$). So any Cauchy sequence in R converges. Conversely (Ex. 2) this fact implies the Intermediate Value Theorem. Thus VI.4.01 is a specialisation of

1.02 Definition

A metric space is <u>complete</u> if all Cauchy sequences in it converge. (cf. Ex. 3)

Exercises A1

1a) If $i \rightsquigarrow x_i$ in a metric space (X,d) converges to x, show that for any $\varepsilon > 0$ there is $M \in N$ such that $m,n > M \Rightarrow d(x_m, x) < \frac{\varepsilon}{2}$, $d(x, x_n) < \frac{\varepsilon}{2}$.

b) Deduce that a sequence in a metric space is necessarily Cauchy if it converges with respect to the metric topology.

2a) Suppose $f : R \to R$ is continuous, $f(R) = \{-1,+1\}$, and $f(a) = -1$, $f(b) = +1$. Construct a Cauchy sequence $S : i \rightsquigarrow x_i$ such that $i \rightsquigarrow f(x_i)$ does not converge. (Hint. Set $x_1 = a$, $x_2 = \frac{1}{2}(a+b)$, $x_3 = \frac{1}{4}(a+3b)$, ... until the first i with $f(x_i) = +1$; by continuity at b, this must happen for i finite. If $f(x_n) = -1$ get x_{n+1} by moving $\frac{|b-a|}{2^n}$ towards the most recent x_i with $f(x_i) = +1$, and vice versa. Prove S Cauchy and $f \circ S$ divergent.)

b) Deduce by VI.2.02 that S does not converge.

c) Deduce that if all Cauchy sequences in R <u>do</u> converge, the Intermediate Value Theorem is true of the real numbers.

3a) Deduce from Ex. VI.3.8c that if X is a finite-dimensional real vector space and $S : N \to X$ is Cauchy in one of the metrics of Ex. VI.3.5, for some basis, it is Cauchy in them all, for any basis. (N.B. There exist metrisations of the usual topology for which $\underset{\sim}{v}$, $2\underset{\sim}{v}$, $3\underset{\sim}{v}$, ... is Cauchy.)

b) Deduce that if R is complete (which we shall continue to assume), so is X in the metric given by any norm (cf. Ex. VI.3.8).

2. TWO FIXED POINT THEOREMS

Throughout this section f^n will mean $\overbrace{f\circ f\circ\ldots\circ f}^{n \text{ times}}$, not a component function, and $f^0(x)$ will mean x: similarly for F^n.

2.01 Definition

$p \in X$ is a fixed point of $f: X \to X$ if $f(p) = p$. For X a Hausdorff space, p is an attracting fixed point (Ex. 1) if for arbitrary $x \in X$, $\lim_{n\to\infty} f^n(x)$ exists and is p.

2.02 Definition

For (X,d) a metric space, $\lambda \in \,]0,1[$, a map $f:X \to X$ is a λ-contraction if $d(f(x),f(y)) \le \lambda d(x,y)$ for all $x,y \in X$.

2.03 Shrinking lemma

If (X,d) is complete and $f:X \to X$ is a λ-contraction, f has an attracting fixed point.

Proof

For any $x \in X$, $S_x : i \rightsquigarrow f^{i-1}(x) = x_i$ is Cauchy:

$$d(x_n,x_{n+1}) = d(f(x_{n-1}),f(x_n)) \le \lambda d(x_{n-1},x_n) = \ldots \le \lambda^{n-1} d(x_1,x_2) = \lambda^{n-1}k, \text{ say.}$$

If $m \ge n$, repeated use of the triangle inequality gives

$$d(x_n,x_m) \le d(x_n,x_{n+1}) + d(x_{n+1},x_{n+2}) + \ldots + d(x_{m-1},x_m)$$

$$\le (\lambda^{n-1} + \lambda^n + \ldots + \lambda^{m-2})\, k < \frac{\lambda^{n-1}k}{1-\lambda}, \text{ (Ex. 2).}$$

But $i \rightsquigarrow \lambda^{i-1}$ converges to 0 by Ex. VI.4.8, so for any $\varepsilon > 0$ there is $M \in N$ with

$$n > M \Rightarrow \lambda^{n-1} < \frac{(1-\lambda)\varepsilon}{k} \Rightarrow d(x_n,x_m) < \varepsilon \text{ for } m \geq n.$$

Similarly, $m > M \Rightarrow d(x_n,x_m) < \varepsilon$ for $n \geq m$. Combining these facts,

$$m, n > M \Rightarrow d(x_n,x_m) < \varepsilon.$$

Thus since X is complete, S_x converges to some $p \in X$. For $y \in X$, S_y similarly converges to $q \in X$. By Ex. 1b both p and q are fixed, thus $d(p,q) = d(f(p),f(q)) \leq \lambda d(p,q)$, so $d(p,q) = 0$ so $p = q$ by Axiom VI.1.02ii. Hence p is an attracting fixed point for f. ■

We need also a similar but more intricate result, first proved in [Hirsch and Pugh]:

2.04 Fibre Contraction Theorem

Let X be a Hausdorff space, (Y,d) a complete metric space, and $F : X\times Y \to X\times Y$ a fibre map over the projection $\pi_1 : X\times Y \to Y : (x,y) \rightsquigarrow x$. That means $\pi_1 F(x,y) = \pi_1 F(x,y')$ for all $x \in X$, $y,y' \in Y$ (Fig. 2.1). Equivalently, we can write F in the form:

$$F(x,y) = (f(x),f_x(y))$$

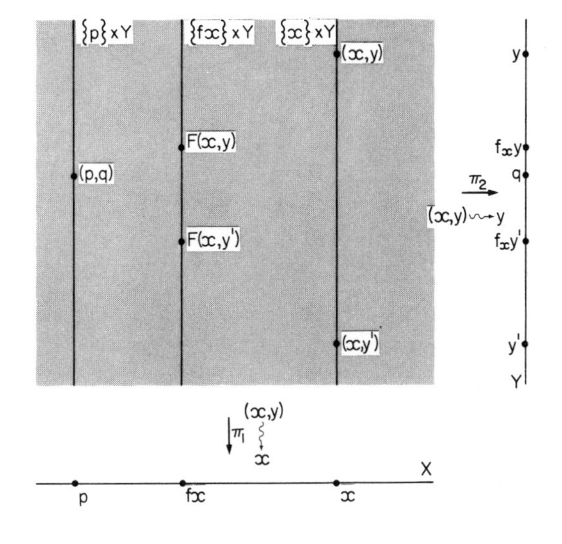

Fig. A2.1

where $f : X \to X$ and each $f_x : Y \to Y$.

Suppose that $\lambda \in]0,1[$ and

(a) For each $y \in Y$, the map $X \to Y : x \rightsquigarrow f_x(y)$ is continuous. (True if, for example, F is continuous.)

(b) f has an attracting fixed point $p \in X$. (True by 2.03 if X is a complete metric space, and f a λ-contraction.)

(c) Each f_x is a λ-contraction of (Y,d).

Then if $q \in Y$ is the attracting fixed point of f_p given by 2.03, the point $(p,q) \in X \times Y$ is an attracting fixed point of F.

Proof

Choose $x \in X$, set $x_i = f^{i-1}(x)$, $\delta_i = d(q, f_{x_i}(q))$. Then $\lim_{n\to\infty} \delta_i = 0$, for

$$\lim_{n\to\infty} f_{x_n}(q) = \lim_{n\to\infty} \pi_2(F(x_n,q)) = \pi_2(F(\lim_{n\to\infty}(x_n,q))) \text{ by VI.2.02 and (a)}$$

$$= \pi_2(F(p,q))$$

$$= f_p(q) = q.$$

For any $y \in Y$, $\pi_2(F^n(x,y)) = f_{x_n} \circ f_{x_{n-1}} \circ \ldots \circ f_{x_1}(y)$, so using the triangle inequality:

$$d(\pi_2(F^n(x,q)),q) \le d(f_{x_n} \circ \ldots \circ f_{x_1}(q),\ f_{x_n}(q)) + d(f_{x_n}(q),q)$$

$$\le \lambda d(f_{x_{n-1}} \circ \ldots \circ f_{x_1}(q),q) + \delta_n$$

$$\le \lambda[\lambda d(f_{x_{n-2}} \circ \ldots \circ f_{x_1}(q),q) + \delta_{n-1}] + \delta_n$$

$$\le \ldots \le \lambda^{n-1} d(f_{x_1}(q),q) + \lambda^{n-2}\delta_2 + \ldots + \lambda\delta_{n-1} + \delta_n$$

$$= \sum_{i=1}^{n} \lambda^{n-i}\delta_i = \sum\nolimits_n \text{ for short.}$$

But $\lim_{n\to\infty} \sum_n = 0$; setting $k = \frac{1}{2}n$, if n is even, $\frac{1}{2}(n-1)$ otherwise, and $M_k = \sup\{\delta_j \mid j \geq k\}$ (which exists for any k, since $\delta_j \to 0$) we have

$$\sum_n = \sum_{i=1}^{k} \lambda^{n-i}\delta_i + \sum_{i=k+1}^{n} \lambda^{n-i}\delta_i \leq (\lambda^{n-1}+\ldots+\lambda^{n-k})M_0 + (\lambda^{n-k-1}+\ldots+\lambda+1)M_k$$

$$\leq \frac{\lambda^{n-k}M_0}{1-\lambda} + \frac{M_k}{1-\lambda} \text{ by Ex. 2.}$$

As $n\to\infty$, so do k and $(n-k)$. $\lim_{(n-k)\to\infty} \lambda^{n-k} = 0$ by Ex. VI.4.8 and $\lim_{k\to\infty} M_k = 0$ by the convergence to zero of δ_j, so we have $\lim_{n\to\infty} \sum_n = 0$ also. Hence since

$$d(\pi_2(F^n(x,y)),q) \leq d(\pi_2(F^n(x,y)),\pi_2(F^n(x,q)))+d(\pi_2(F^n(x,q)),q) \leq \lambda^n d(y,q)+\sum_n ,$$

$\lim_{n\to\infty} d(\pi_2(F^n(x,y)),q) = 0$, so by Ex. VI.3.7b, $d(\lim_{n\to\infty} \pi_2(F^n(x,y)),q) = 0$. Thus $\pi_2(\lim_{n\to\infty} F^n(x,y)) = q$, and since $\pi_1(\lim_{n\to\infty} F^n(x,y)) = p$ we get $\lim_{n\to\infty} F^n(x,y) = (p,q)$.

Exercises A2

1a) Give examples of continuous maps $S^1 \to S^1$ with no fixed points, and with several .

b) If X is Hausdorff and $i \rightsquigarrow f^{i-1}(x)$ converges to p for some $x \in X$, show that $f(p) = p$, for f continuous.

c) If, further, all such sequences converge to p, show that p is the only fixed point of f (regardless of whether f is continuous).

2a) Show that for any $\lambda \in R$, $(1-\lambda)(1+\lambda+\ldots+\lambda^n) = (1-\lambda^{n+1})$.

b) Deduce that if $0<\lambda<1$, $m>n$, then $(\lambda^{n-1}+\lambda^n+\ldots+\lambda^{m-2}) < \dfrac{\lambda^{n-1}}{1-\lambda}$.

3. SEQUENCES OF FUNCTIONS

3.01 Definition

If X, Y are topological spaces and f_i are maps $X \to Y$, $i \in N$, a function $f : X \to Y$ is their (unique if Y is Hausdorff) <u>pointwise limit</u> if for every $x \in X$, $\lim_{n\to\infty} f_n(x)$ exists and equals $f(x)$.

Unfortunately, f may be less nice than the f_n. Thus, with $X = Y = R$, if $f_n(x) = (nx^2 + 1)^{-1}$ each f_n is C^∞ but their pointwise limit is the discontinuous function "$f(x) = 0$ if $x \neq 0$, $f(0) = 1$".

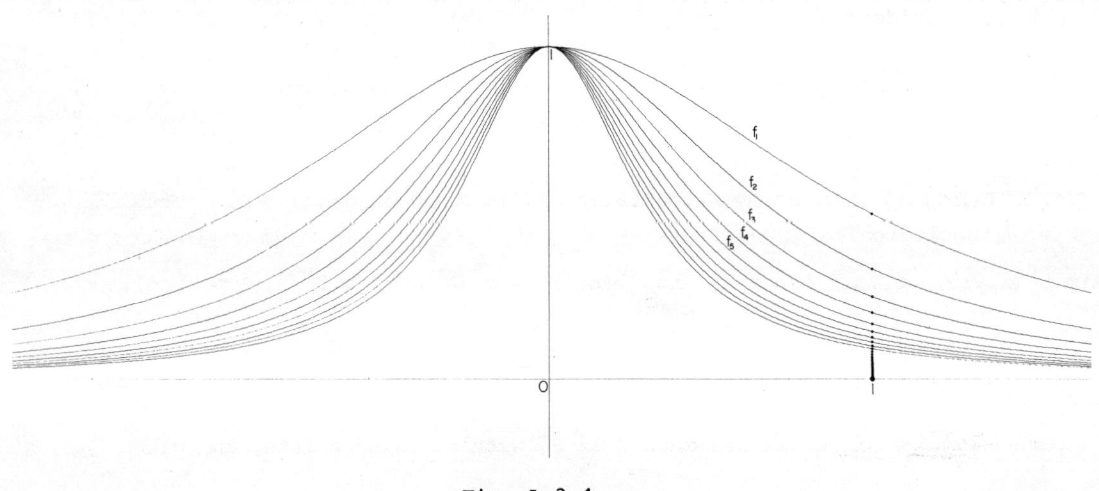

Fig. A.3.1

The trouble is that $i \rightsquigarrow f_i(x)$ converges more and more slowly as x approaches 0: for any n we can find $x \neq 0$ with $f_n(x)$ still arbitrarily close to 1. If Y is a metric space, we can define a stronger kind of convergence that behaves better.

3.02 Definition

If X is any set, (Y,d) a metric space, and $i \rightsquigarrow f_i$ a sequence of functions $X \to Y$, then $f : X \to Y$ is its <u>uniform limit</u>, and they <u>converge uniformly</u> to f, if for any $\varepsilon > 0$ there is an $M \in N$ such that

$$n > M,\ x \in X \Rightarrow d(f(x), f_n(x)) < \varepsilon. \quad \text{We write } f = \lim_{n\to\infty} f_n.$$

A function $g : X \to Y$ is <u>bounded</u> if for some (and hence - why? - any) $y \in Y$ the set $\{d(y,g(x)) | x \in X\} \subseteq R$ is bounded (VI.4.05). The <u>uniform metric</u> on the set of bounded functions $X \to Y$ is defined by

$$d_U(f,g) = \sup\{d(f(x),g(x)) | x \in X\}.$$

(This sup exists by the boundedness of f and g, the triangle inequality, and Ex. VI.4.6.) Evidently convergence of $i \rightsquigarrow f_i$ in the uniform metric is equivalent to uniform convergence (just combine the definitions).

<u>3.03 Lemma</u>

Let X be a topological space, Y a metric space. If $i \rightsquigarrow f_i$ converges uniformly to $f: X \to Y$ and the f_i are continuous, so is f.

<u>Proof</u>

For $x_0 \in X$, $\varepsilon > 0$, choose $M \in N$, such that $n > M, x \in X \Rightarrow d(f(x),f_n(x)) < \frac{\varepsilon}{3}$, a number $m > M$, and by continuity of f_m a neighbourhood U of x_0 such that

$$x \in U \Rightarrow d(f_m(x_0),f_m(x)) < \frac{\varepsilon}{3}. \quad \text{Then for } x \in U,$$

$$d(f(x_0),f(x)) \le d(f(x_0),f_m(x_0)) + d(f_m(x_0),f_m(x)) + d(f_m(x),f(x)) < \frac{\varepsilon}{3} + \frac{\varepsilon}{3} + \frac{\varepsilon}{3} = \varepsilon.$$

Hence f is continuous at x_0.

■

3.04 Lemma

Let X be a topological space, Y a complete metric space. Then the space F of bounded continuous maps $X \to Y$, with the uniform metric, is complete.

Proof

Let $i \rightsquigarrow f_i$ be a Cauchy sequence in F. Then clearly for any $x \in X$, $i \rightsquigarrow f_i(x)$ is a Cauchy sequence in Y, so we may define $f(x) = \lim_{n\to\infty} f_n(x)$. For $\varepsilon > 0$ we have M

$$\text{with} \quad m,n > M \;\Rightarrow\; d_U(f_m,f_n) < \frac{\varepsilon}{2} \;\Rightarrow\; d(f_m(x),f_n(x)) < \frac{\varepsilon}{2}, \quad \forall x \in X$$

$$\Rightarrow\; d(f(x),f_n(x)) \le \frac{\varepsilon}{2}, \quad \forall x \in X \text{ by Ex. 1a}$$

$$\Rightarrow\; d_U(f,f_n) \le \frac{\varepsilon}{2} \;\Rightarrow\; d_U(f,f_n) < \varepsilon.$$

Hence $i \rightsquigarrow f_i$ converges uniformly to f, which is continuous by 3.03, bounded by Ex. 1b, and hence in F.

3.05 Corollary

If $\bar{B}(y,\delta) = \{y' \mid d(y,y') \le \delta\} \subseteq Y$, and Y is complete, then so is the space F' of continuous functions $X \to \bar{B}(y,\delta)$, with the uniform metric.

Proof

Evidently $F' \subseteq F$, so a Cauchy sequence $i \rightsquigarrow f_i$ in F' has a uniform limit $f \in F$. Moreover $x \in X \;\Rightarrow\; f(x) = \lim_{n\to\infty} f_n(x) \in \bar{B}(y,\delta)$, since $\bar{B}(y,\delta)$ is closed. Hence $f \in F'$.

If X,Y have differential structures, so that the f_n can be differentiated, what can we say about the sequence $i \rightsquigarrow \underset{\sim}{D}f_i$? Even if $\lim_{n\to\infty} f_n$ and all the f_i are C^∞, we may not get $\lim_{n\to\infty} \underset{\sim}{D}f_n = \underset{\sim}{D}f$. If, for example $f_n : R \to R : x \rightsquigarrow \frac{1}{n}\sin(nx + n^2)$,

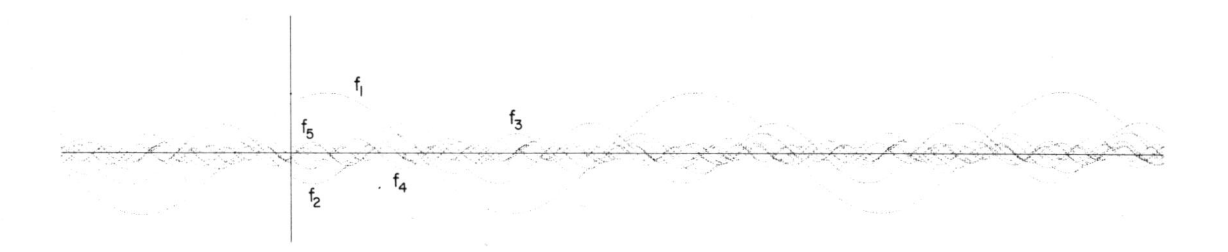

Fig. A.3.2

then $i \rightsquigarrow f_i$ converges uniformly to $0 : R \to R$ but for no x does $i \rightsquigarrow \frac{df_i}{dr}$ converge (why?). Likewise, the uniform limit of polynomials may be nowhere differentiable. However;

3.06 Lemma

Let X,Y be affine spaces with vector spaces S,T, $U \subseteq X$ be open, and $L(S;T)$ have a metric d given by a norm. If $i \rightsquigarrow (f_i : U \to Y)$ is a sequence of C^1 functions converging pointwise to f in the usual topology, and $i \rightsquigarrow (\tilde{D}f_i : U \to L(S;T))$ converges uniformly to $F : U \to L(S;T)$, then f is also C^1 and $D_x f = \overleftarrow{d}_{f(x)} \circ F \circ d_x$. (Recall $\tilde{D}$ notation, VII.1.02).

Proof

By Ex. 2 it suffices to choose affine charts $U \to R^m$, $Y \to R^n$ and correspondingly $L(S;T) \to R^{mn}$ (giving coordinate functions $f_n^j, f_n, \partial_i f_n^j, F_i^j : U \to R$ for $f_n, f, \tilde{D}f, F$) and the metric given in coordinates as

$$d([a_i^j],[b_i^j]) = \max\{|a_i^j - b_i^j| \mid i = 1,\ldots,m,\ j = 1,\ldots,n\}.$$

In this metric, uniform convergence of $i \rightsquigarrow \tilde{D}f_i$ to F means

* For any $\varepsilon>0$, $\exists$ M such that $n > M \Rightarrow |\partial_i f_n^j(x) - F_i^j(x)| < \varepsilon$, $\forall i, j, x$.

Since the f_n are C^1, the $\tilde{D}f_n$ are continuous, hence by 3.03 so are $\underset{\sim}{F}$ and the F_i^j. Hence (by Ex. VII.7c) existence of $\underset{\sim}{D}_x f$ follows if we prove $\partial_i f^j(x)$ exists and equals $F_i^j(x)$, each i,j : likewise continuity. Fix $x = (x^1,\ldots,x^n)$.

For $\varepsilon > 0$, apply * to $\frac{\varepsilon}{3}$ to get $M \in N$ such that

$$** \quad n > M \Rightarrow |\partial_i f_n^j(\tilde{x}) - F_i^j(\tilde{x})| < \frac{\varepsilon}{3}, \ \forall i,j,\tilde{x} \Rightarrow \left|\frac{d\tilde{f}_{ni}^j}{dt}(s) - \tilde{F}_i^j(s)\right| < \frac{\varepsilon}{3}, \ \forall i,j,s$$

wherever $\tilde{f}_{ni}^j(s) = f_n^j(x^1,\ldots,x^i+s,\ldots,x^m)$, $\tilde{F}_i^j(s) = F_i^j(x^1,\ldots,x^i+s,\ldots,x^m)$ are defined.

By continuity of F_i^j, there exists $\delta > 0$ such that

$$|s| < \delta \Rightarrow |\tilde{F}_i^j(s) - \tilde{F}_i^j(0)| < \frac{\varepsilon}{3}.$$

Combining this with ** by the triangle inequality, we get

$$*** \quad n > M, \ |s| < \delta \Rightarrow \left|\frac{d\tilde{f}_{ni}^j}{dt}(s) - \tilde{F}_i^j(0)\right| < \frac{2\varepsilon}{3}.$$

Now, suppose $r \in]-\delta,\delta[$, $n > M$, and $\left|\dfrac{\tilde{f}_{ni}^j(r)-\tilde{f}_{ni}^j(0)-r\tilde{F}_i^j(0)}{r}\right| \geq \frac{2\varepsilon}{3}$; applying the Mean Value Theorem to the function $t \rightsquigarrow [\tilde{f}_{ni}^j(t)-\tilde{f}_{ni}^j(0)-t\tilde{F}_i^j(0)]$, which is clearly C^1, this implies there exists s between 0 and r, hence with $|s| < \delta$, for which *** fails: contradiction. Therefore

$$|s| < \delta \Rightarrow \left|\frac{\tilde{f}_{ni}^j(s)-\tilde{f}_{ni}^j(0)}{s} - \tilde{F}_i^j(0)\right| < \frac{2\varepsilon}{3} \text{ for } n > M \Rightarrow \lim_{n\to\infty}\left|\frac{\tilde{f}_{ni}^j(s)-\tilde{f}_{ni}^j(0)}{s} - \tilde{F}_i^j(0)\right| \leq \frac{2\varepsilon}{3}.$$

Since $|\ |$ is continuous, this gives by VI.2.02.

$$|s| < \delta \;\Rightarrow\; \left| \frac{f^j(x^1,\ldots,x^i+s,\ldots,x^m)-f^j(x)}{s} - F^j_i(x) \right| \le \frac{2\varepsilon}{3} < \varepsilon .$$

Taking the limit as $\delta \to 0$, we see that $\partial_i f^j(x)$ exists and equals $F^j_i(x)$. ■

Exercises A3

1a) Deduce from Ex. VII.3.7 that if $d(x_m,x_n) < \varepsilon$ for all $m > M$, $d(\lim_{m\to\infty} x_m,x_n) \le \varepsilon$.

b) Shown by the triangle inequality that if $d_U(f_m,f) < \varepsilon$ and f_m is bounded, then so is f.

2a) Show that if $\| \ \|_S$, $\| \ \|_T$ are norms on finite-dimensional vector spaces S,T, then $\|\underset{\sim}{A}\| = \sup\{\|\underset{\sim}{A}\underset{\sim}{s}\|_T \mid \|\underset{\sim}{s}\|_S \le 1\}$ defines a norm on L(S;T).

b) Deduce from Ex. 1.3 that if $\underset{\sim}{D}f_i : U \to L(S;T)$ in 3.06 converge uniformly for one choice of norm, they converge for any. (This is false in infinite dimensions.)

4. INTEGRATING VECTOR QUANTITIES

One final addition to the technical background (§1-3 can be found in much greater detail in various Analysis texts) before we prove VII.6.04:

4.01 Definition

If X is a vector space and $\underset{\sim}{d} : X \times X \to X$ its natural affine structure (II.1.03), an indefinite integral for a curve $\underset{\sim}{c} : J \to X$ is a curve $\underset{\sim}{g} : J \to X$ such that $\underset{\sim}{d}_{\underset{\sim}{c}(t)}(\underset{\sim}{g}^*(t)) = \underset{\sim}{c}(t)$, $\forall t \in J$. The definite integral of $\underset{\sim}{c}$ from $a \in J$ to $b \in J$ is $\int_a^b \underset{\sim}{c}(s)ds = \underset{\sim}{g}(b) - \underset{\sim}{g}(a)$, where $\underset{\sim}{g}$ is an indefinite integral for $\underset{\sim}{c}$. If X is finite-dimensional and $\underset{\sim}{c}$ continuous, the existence of an indefinite integral follows directly from the $R \to R$ case, since for any $a \in J$ and basis $\underset{\sim}{b}_1,\ldots,\underset{\sim}{b}_n$ for X,

$$t \rightsquigarrow \left(\int_a^t c^i(s)\,ds\right)\underset{\sim}{b}_i \quad \text{is an indefinite integral for } \underset{\sim}{c}.$$

The uniqueness of the definite integral follows similarly.

5. THE MAIN PROOF

We prove a set of results that add up to VII.6.04. If you have difficulty following the argument, it may help first to read §6, where a similar but simpler use is made of the Fibre Contraction Theorem.

5.01 Theorem

Let Z be a finite-dimensional affine space with vector space T, $U \subseteq Z$ be open, and $\underset{\sim}{w} : U \to T$ be C^1. Then for any $z_0 \in U$ there exists $\varepsilon > 0$, a neighbourhood $N \subseteq U$ of z_0, and a continuous map $\phi_{\underset{\sim}{w}} : N \times]-\varepsilon,\varepsilon[\to U$ such that

(i) $\phi_{\underset{\sim}{w}}(z,0) = z, \quad \forall z \in N$

(ii) For any $z \in N$, $\phi_z :]-\varepsilon,\varepsilon[\to U : t \rightsquigarrow \phi_{\underset{\sim}{w}}(z,t)$ is differentiable and

$$\underset{\sim}{d}_{\phi_{\underset{\sim}{w}}(z,t)}(\phi_z^*(t)) = \underset{\sim}{w}(\phi_{\underset{\sim}{w}}(z,t)), \quad \forall t \in]-\varepsilon,\varepsilon[.$$

Proof

Pick coordinates, give T, $L(T;T)$ the corresponding square norms (as in 3.06) $\| \ \|_T$ and $\| \ \|_L$, and Z the metric $d(z,z') = \| \underset{\sim}{d}(z,z') \|_T$. Choose $b > 0$ such that $\bar{B}_b = \{z \mid d(z_0,z) \le b\} \subseteq N$. $\bar{B}_b$ is compact, so by $\underset{\sim}{w}$ C^1 and VI.4.11 there exist $m, \ell \in R$ such that

$$m = \sup\{\| \underset{\sim}{w}(z) \|_T \mid z \in \bar{B}_b\}, \quad \ell = \sup\{\| \underset{\sim\sim}{Dw} \|_L \mid z \in \bar{B}_b\}.$$

Choose ε, $\delta > 0$ such that $\varepsilon m+\delta<b, \ell\varepsilon<1$; set $\lambda=\ell\varepsilon$, $B_\delta = \{z|d(z_0,z)<\delta\}, W=B_\delta\times]-\varepsilon,\varepsilon[$. Let X be the space of continuous maps $W \to \bar{B}_b$, with the uniform metric d_U. By 3.05, X is complete. Define $f : X \to X$, for general $\phi \in X$, by

$$f(\phi)(z,t) = z + \int_0^t \underset{\sim}{w}(\phi(z,s))ds.$$

(Since $\phi(z,s) \in \bar{B}_b$ always, $\|\underset{\sim}{w}(\phi(z,s))\|\leq m$, so by Ex. 1b, $\|\int_0^t \underset{\sim}{w}(\phi(z,s))ds\|\leq m|t|<\varepsilon m$. So by the triangle inequality, $d(z\ ,f(\phi)(z,t))\leq d(z_0,z)+d(z,f(\phi)(z,t))<\varepsilon m+\delta<b$, hence $f(\phi)(z,t) \in \bar{B}_b$ $\forall(z,t) \in W$ and $f(\phi)$ is thus in X.)

Notice that any $\phi = f(\psi)$, $\psi \in X$, satisfies (i) automatically, and that $\underset{\sim}{d}_z\phi_z^*(0) = \underset{\sim}{v}(z) = \underset{\sim}{v}(\phi(z,0))$, already. If say $\phi_1(z,t) = z$, $\phi_2 = f(\phi_1)$, $\phi_3 = f(\phi_2)$,... one would hope that $i \rightsquigarrow \phi_i$ would converge to ϕ still satisfying (i), and (ii) for all $t \in]-\varepsilon,\varepsilon[$, not just 0. (Fig. A5.1 shows the images of ϕ_1 (namely just B_δ), ϕ_2, ϕ_3 and ϕ_4 with the images $a_1 = \{z\}$, a_2, a_3, a_4 of the corresponding

$$\phi_i\Big|\{z\}\times]-\varepsilon,\varepsilon[$$

which increasingly well approximates a solution curve through a typical $z \in B_\delta$.)

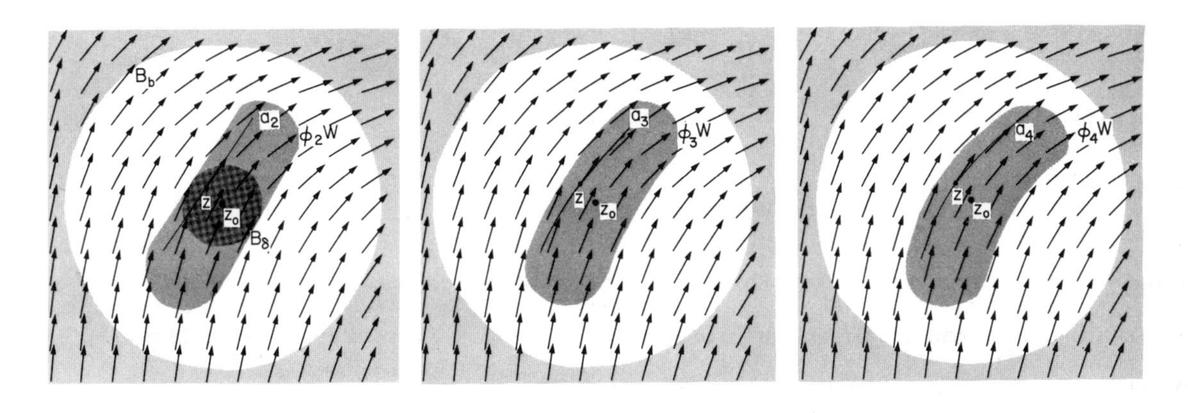

Fig. A5.1

This does happen. $\phi \in X$ is a fixed point of f if and only if (with $N = B_\delta$) it satisfies (i) and (ii), by an immediate check (for ii, just apply ∂_t to both sides of $\phi = f(\phi)$ written out fully) and f is a λ-contraction: for $\phi, \psi \in X$,

$$d_U(f(\phi),f(\psi)) = \sup\{d(f(\phi)(z,t),f(\psi)(z,t)) \mid (z,t) \in W\}$$

$$= \sup\left\{ \left\| \int_0^t \underset{\sim}{w}(\phi(z,s))ds - \int_0^t \underset{\sim}{w}(\psi(z,s))ds \right\|_T \middle| (z,t) \in W \right\}$$

$$= \sup\left\{ \left\| \int_0^t (\underset{\sim}{w}(\phi(z,s)) - \underset{\sim}{w}(\psi(z,s)))ds \right\|_T \middle| (z,t) \in W \right\}$$

$$\leq \sup\left\{ \left| \int_0^t \| \underset{\sim}{w}(\phi(z,s)) - \underset{\sim}{w}(\psi(z,s)) \|_T ds \right| \middle| (z,t) \in W \right\} \quad \text{by Ex. 1a}$$

$$\leq \sup\left\{ \ell \left| \int_0^t d(\phi(z,s),\psi(z,s))ds \right| \middle| (z,t) \in W \right\}$$ by the Mean Value Theorem on components of $\underset{\sim}{w}$, as in 3.06, and Ex. 1b,

$$\leq \ell\varepsilon \sup\left\{ d(\phi(z,s),\psi(z,s)) \mid (z,t) \in W \right\} = \lambda d_U(\phi,\psi).$$

Hence by the Shrinking Lemma f has an attracting fixed point $\phi_{\underset{\sim}{w}} \in X$.

5.02 Corollary

If $c :]-\varepsilon',\varepsilon'[\to U$ has (i)' $c(0) = z_0$,

(ii)' $\underset{\sim}{d}_{c(t)}(c^*(t)) = \underset{\sim}{w}(c(t)), \quad \forall t \in]-\varepsilon',\varepsilon'[$

then $c(t) = \phi_{\underset{\sim}{w}}(z,t)$ wherever both are defined.

Proof

The above proof holds equally with B_δ replaced by $\{z\}$, ε by $\min\{\varepsilon,\varepsilon'\}$. Again f is a λ-contraction, and $f(c) = c$ equivalent to (i)' and (ii)', so the result follows by the Shrinking Lemma and the uniqueness of attracting fixed points. ∎

5.03 Corollary

If $c : J \to U$ has $c(t_0) = z_0$, and $\underset{\sim}{d}_{c(t)}(c^*(t)) = \underset{\sim}{w}(c(t))$ when both sides are defined, then $c(t) = \phi_{\underset{\sim}{w}}(z_0, t-t_0)$ when both sides are defined.

Proof

$\tilde{c} : t \leadsto c(t+t_0)$ satisfies (i)', (ii)' of 5.02, hence $\tilde{c}(t) = \phi(z_0,t)$, so $c(t) = \tilde{c}(t-t_0) = \phi(z_0,t-t_0)$.

5.04 Corollary

If $\phi_{\underset{\sim}{w}} : N \times J \to U$, $\psi : N' \times J' \to U$ both satisfy (i) and (ii) of 5.01, then

$$\phi|(N \cap N') \times (J \cap J') = \psi|(N \cap N') \times (J \cap J').$$

Proof

For any $z \in N \cap N'$, ϕ_z and ψ_z both satisfy (i)', (ii)' of 5.02. So $\psi(z,t) = \psi_z(t) = \phi_{\underset{\sim}{w}}(z,t)$ when defined.

The $\phi_{\underset{\sim}{w}}$ existing by 5.01 is continuous by 3.03; more work is needed to show it differentiable. We know that the "vector partial derivative" $\underset{\sim}{\partial}_t\phi : W \to T : (z,t) \leadsto \overleftarrow{\underset{\sim}{d}}_{\phi(z,t)}\phi^*_z(t)$ exists and is continuous, being just $\underset{\sim}{w} \circ \phi_{\underset{\sim}{w}}|W$. It thus suffices by Ex. VII.7.1c to prove that if $\phi_t : B_\delta \to U : z \leadsto \phi_{\underset{\sim}{w}}(z,t)$ then $(z,t) \leadsto \underset{\sim}{D}_z\phi_t$ exists and is continuous on W. The following new proof of this is from [Sotomayor].

5.05 Theorem

The map $\phi_{\underset{\sim}{w}} : W \to U$ of 5.01 is C^1.

Proof

It suffices to show that

$$\partial_Z \phi_{\underset{\sim}{w}} : W \to L(T;T) : (z,t) \leadsto \underset{\sim}{d}_{\phi_{\underset{\sim}{w}}(z,t)} \circ D_z \phi_t \circ \overleftarrow{\underset{\sim}{d}}_z$$

exists and is continuous.

Let X, f be as in 5.01, L(T;T) have the norm $\| \ \|_L$, and Y be the space of continuous maps $W \to L(T;T)$ - i.e. candidates for $\underset{\sim}{\partial}_Z \phi_{\underset{\sim}{w}}$ - with the uniform metric. (Which we shall call d_U^Y, as we shall refer again to the metric d_U on X.)

(Y, d_U^Y) is complete by 3.04.

We want a fibre map $F : X \times Y \to X \times Y$ that will make any $F^n(\psi,\psi')$ converge to $(\phi, \underset{\sim}{\partial}_Z \phi)$. To get this we want - certainly for $\phi_{\underset{\sim}{w}}$ of 5.01, conveniently for general ϕ -

$$* \qquad F(\phi, \underset{\sim}{\partial}_Z \phi) = (f(\phi), \underset{\sim}{\partial}_Z (f(\phi))).$$

Evidently there are many such F's, since * involves only a small (in many senses) subset of X × Y. But the simplest is given just by differentiating f;

$$\begin{aligned}
\underset{\sim}{\partial}_Z (f(\phi))(z,t) &= \tilde{\underset{\sim}{D}}\left[I_Z + \int_0^t (\underset{\sim}{w} \circ \phi_s) ds \right](z) \\
&= \underset{\sim}{I} + \int_0^t \tilde{\underset{\sim}{D}}(\underset{\sim}{w} \circ \phi_s)(z) ds \quad \text{by Ex. IX.3.5 generalised} \\
&\qquad\qquad\qquad\qquad \text{(or for each component).} \\
&= \underset{\sim}{I} + \int_0^t (\tilde{\underset{\approx}{D}w}(\phi(z,s))) \circ \underset{\sim}{\partial}_Z \phi(z,s) ds.
\end{aligned}$$

Then if $f_\phi(\phi') = \underset{\sim}{I} + \int_0^t (\tilde{\underset{\approx}{D}w}(\phi(z,s))) \circ \phi'(z,s) ds$, and ϕ' is bounded by k, say,

$\| f_\phi(\phi')(z,t) \|_L \le \| \underset{\sim}{I} + |t| \ell k \| < 1 + \varepsilon \ell k$, so $f_\phi(\phi')$ is also bounded, clearly continuous, and thus in Y. So we can satisfy * by $F(\phi,\phi') = (f(\phi), f_\phi(\phi'))$. F so defined satisfies the conditions of the Fibre Contraction Theorem (2.04):

(a) $\bar{B}_b$ is compact; so by Ex. 2, $\tilde{\underset{\sim}{Dw}}|\bar{B}_b$ is uniformly continuous. So for ϕ' bounded by k, and any $\alpha > 0$ there exists $\mu > 0$ such that for $z, z' \in \bar{B}_b$

$$d(z,z') < \mu \Rightarrow \| \tilde{\underset{\sim}{Dw}}(z) - \tilde{\underset{\sim}{Dw}}(z') \|_L < \frac{\alpha}{\varepsilon k}, \quad \text{so}$$

$$d_U(\psi,\theta) < \mu \Rightarrow d_U^Y(f_\psi(\phi'), f_\theta(\phi'))$$

$$= \sup\left\{ \left\| \int_0^t [\tilde{\underset{\sim}{Dw}}(\psi(z,s)) - \tilde{\underset{\sim}{Dw}}(\theta(z,s))]\circ\phi'(z,s)ds \right\|_L \middle| (z,t) \in W \right\}$$

$$< \sup\left\{ \left| \frac{\alpha}{\varepsilon k} \int_0^t \| \phi'(z,s) \|_L ds \right| \middle| (z,t) \in W \right\} \leq \alpha.$$

So $\psi \rightsquigarrow f_\psi(\phi')$ is (uniformly) continuous, for each $\phi' \in Y$.

(b) is the proof of 5.01

(c) For $\phi', \psi' \in Y$, $\theta \in X$, we have

$$d_U^Y(f_\theta(\phi'), f_\theta(\psi')) = \sup\left\{ \left\| \int_0^t (\tilde{\underset{\sim}{Dw}}(\theta(z,s)))\circ(\phi'(z,s)-\psi'(z,s))ds \right\|_L \middle| (z,t) \in W \right\}$$

$$\leq \sup\left\{ \left| \ell \int_0^t \| \phi'(z,s)-\psi'(z,s) \|_L ds \right| \middle| (z,t) \in W \right\}$$

$$\leq \ell\varepsilon \sup\left\{ \| \phi'(z,s)-\psi'(z,s) \|_L \middle| (z,s) \in W \right\} = \lambda d_U^Y(\phi',\psi').$$

Hence F has an attracting fixed point $(\phi_{\underset{\sim}{w}}, \phi'_{\underset{\sim}{w}})$. If $\phi_1(z,t) = z$, $\phi_1'(z,t) = \underset{\sim}{I}$ we have $\partial_{\underset{\sim}{z}}\phi_1 = \phi_1'$: inductively defining $(\phi_n, \phi_n') = F(\phi_{n-1}, \phi_{n-1}')$ we have, by *, $\phi_n' = \partial_{\underset{\sim}{z}}\phi_n$ for all n. So applying 3.06, $\partial_{\underset{\sim}{z}}\phi$ exists and equals $\phi'_{\underset{\sim}{w}}$ which is continuous.

■

5.06 Theorem

If $\underset{\sim}{w}$ is C^k, so is $\phi_{\underset{\sim}{w}}$.

Proof

For $\theta \in X$, $1<n\leq k$, set $\underset{\sim}{\partial}_Z^n\phi(z,t) = \underset{\sim}{\tilde{D}}\phi_t(z) \in L^n(T;T) \tilde{=} L^{n-1}(T;L(T;T))$, where it exists. Using the square norm $\| \ \|_n$ on $L^n(T;T)$ corresponding to the coordinates in 5.01, let X be as above, $V^1 = Y$, and for $i > 1$ let V^i be the space of bounded continuous maps $W \to L^{i-1}(T;L(T;T))$ with the uniform metric given by $\| \ \|_i$. Set $X_n = X \times V^1 \times \ldots \times V^{n-1}$, $Y_n = V^n$, and define $F_n : X_n \times Y_n \to X_n \times Y_n$ inductively by

$$F_n\left[(\phi,\phi',\phi'',\ldots,\phi^{(n-1)}),\phi^{(n)}\right]$$

$$= \left[F_{n-1}((\phi,\ldots,\phi^{(n-2)}),\phi^{(n-1)}),f_{(\phi,\ldots,\phi^{(n-1)})}(\phi^{(n)})\right],$$

$$1 < n \leq k,$$

where $F_1 = F$ as defined in 5.01 and, for $\underset{\sim}{t}^1,\ldots,\underset{\sim}{t}^{n-1} \in T$,

$$\left[f_{(\phi,\ldots,\phi^{(n-1)})}(\phi^n)(z,t)\right](\underset{\sim}{t}^1,\ldots,\underset{\sim}{t}^{n-1})$$

$$= \sum_{i=1}^{n} \int_0^t \underset{\sim}{\tilde{D}}^{n-i+1}\underset{\sim}{w}(\phi(z,s))(\underset{\sim}{t}^1,\ldots,\underset{\sim}{t}^{n-1})\circ\phi^{(i)}(z,s)(\underset{\sim}{t}^{n-1+1},\ldots,\underset{\sim}{t}^n)ds \in L(T;T).$$

Then (Ex. 3) $\underset{\sim}{\partial}_Z^n(f(\phi)) = f_{(\phi,\underset{\sim}{\partial}_Z\phi,\ldots,\underset{\sim}{\partial}_Z^{n-1}\phi)}(\underset{\sim}{\partial}_Z^n\phi)$, $1<n\leq k$ *

and the proofs that each $f_{(\phi,\phi',\ldots,\phi^{(n-1)})}$ is a λ-contraction and each map

$(\phi,\phi',\ldots,\phi^{(n-1)}) \rightsquigarrow f_{(\phi,\phi',\ldots,\phi^{(n-1)})}(\phi^{(n)})$ continuous are just as in 5.05.

So the hypotheses of the Fibre Contraction Theorem are satisfied for F_2, appealing to 5.05 for (b), and hence inductively for F_n $1<n\leq k$. Using 3.06, we see that $\underset{\sim}{\partial}_Z^k \phi_{\underset{\sim}{w}}$ exists and is continuous.

For existence and continuity of $\underset{\sim}{D}^k\phi_{\underset{\sim}{w}}$ it remains to show that mixed partials up to order k exist and are continuous. But $\underset{\sim}{\partial}_Z \underset{\sim}{\partial}_t \phi_w = \underset{\sim}{\partial}_Z(\underset{\sim}{w}\circ\phi) = \tilde{\underset{\sim}{D}}(\underset{\sim}{w}\circ\phi_t)$ which exists and is continuous by hypothesis for $\underset{\sim}{w}$, 5.05 for ϕ_t, and the chain rule. Thus so does and is $\underset{\sim}{\partial}_t\underset{\sim}{\partial}_Z\phi_{\underset{\sim}{w}}$, by Ex. VII.7.1a. Hence we have $\underset{\sim}{\partial}_Z^2\phi_{\underset{\sim}{w}}$, $\underset{\sim}{\partial}_Z\underset{\sim}{\partial}_t\phi_{\underset{\sim}{w}}$, $\underset{\sim}{\partial}_t\underset{\sim}{\partial}_Z\phi_{\underset{\sim}{w}}$ and $\underset{\sim}{\partial}_t^2\phi_{\underset{\sim}{w}} = \underset{\sim}{\partial}_t(\underset{\sim}{w}\circ\phi)$ existing and continuous, and therefore also $\underset{\sim}{D}\phi_{\underset{\sim}{w}}$. Then similarly

$$\underset{\sim}{\partial}_Z\underset{\sim}{\partial}_t(\underset{\sim}{\partial}_Z\phi_{\underset{\sim}{w}}) = \underset{\sim}{\partial}_Z\underset{\sim}{\partial}_Z\underset{\sim}{\partial}_t\phi_{\underset{\sim}{w}} = \underset{\sim}{\partial}_Z\underset{\sim}{\partial}_Z(\underset{\sim}{w}\circ\phi)$$

which exists and is continuous since $\underset{\sim}{w}$ is C^k by hypothesis and we have just shown that $\phi_{\underset{\sim}{w}}$ is C^2. So $\underset{\sim}{\partial}_t\underset{\sim}{\partial}_Z(\underset{\sim}{\partial}_Z\phi_{\underset{\sim}{w}})$ also exists and is continuous, and so on. Inductively, $\phi_{\underset{\sim}{w}}$ is C^k. ■

5.07 Corollary

If $\underset{\sim}{w}$ is smooth, so is $\phi_{\underset{\sim}{w}}$. ■

5.08 Conclusion

Theorem VII.6.04 is true as stated. Around $x \in M$ modelled on the affine space Z, take a chart $\theta : V \to Z$. Let $U = \theta(V)$, $\underset{\sim}{w} : U \to T : z \rightsquigarrow \underset{\sim}{d}_z(\underset{\sim}{D}\theta)^{\leftarrow}\underset{\sim}{v}_z$; then if $\underset{\sim}{v}$ is C^k so is $\underset{\sim}{w}$. If $\phi_{\underset{\sim}{w}} : B_\delta \times]-\varepsilon,\varepsilon[\to \bar{B}_b$ is the map of the above results, and $N = \theta^{\leftarrow}(B_\delta)$, then

$$\phi_{\underset{\sim}{v}} : N \times]-\varepsilon,\varepsilon[\to M : (y,t) \rightsquigarrow \theta^{\leftarrow}\phi_{\underset{\sim}{w}}(\theta(y),t)$$

is well defined and a local flow for $\underset{\sim}{v}$ around x.

The uniqueness properties follow from 5.04, 5.03.

Exercises A5

1a) Show in the square norm, for any basis, on T, that for any $\underset{\sim}{c} : J \to T$ with O, $t \in J$ we have

$$I_t = \left\| \int_0^t \underset{\sim}{c}(s)\,ds \right\| \le \left| \int_0^t \| \underset{\sim}{c}(s) \|\, ds \right|.$$ (The $|\ |$ is needed in case $t < 0$ makes the integral negative.)

b) Deduce that if $\| \underset{\sim}{c}(s) \| \le m$ for all $s \in J$, $I_t \le m|t|$, hence if $J =]-\varepsilon,\varepsilon[$, $I_t \le m\varepsilon$.

2) If (X,d_X) and (Y,d_Y) are metric spaces, $K \subseteq X$ is compact, and $f : K \to Y$ is continuous, show that f is <u>uniformly</u> continuous. Namely, that for <u>any</u> $\varepsilon > 0$ there is a $\delta > 0$ such that $d_X(x,x') < \delta \Rightarrow d_Y(f(x),f(x')) < \varepsilon$ for any $x \in K$. (Hint: if not, show that for some $\varepsilon > 0$ there are sequences $i \mapsto p_i$, $i \mapsto q_i$ in K with $\lim_{n\to\infty} (d_X(p_n,q_n)) = 0$ but $d_Y(f(p_n),f(q_n)) > \varepsilon$ for all n, and obtain a contradiction via the convergent subsequence property.)

3a) Show that if $f : Z \to L(A;B)$, $g : Z \to L(B;C)$ are differentiable, where Z is an affine space, A,B,C vector spaces, then $h : z \mapsto g(z)\circ f(z)$ is also differentiable and $\tilde{\underset{\sim}{D}}h(z)(\underset{\sim}{t}) = \tilde{\underset{\sim}{D}}g(z)(\underset{\sim}{t})\circ f(z) + g(z)\circ\tilde{\underset{\sim}{D}}f(z)(\underset{\sim}{t})$.

b) Deduce * in the proof of 5.06.

6. INVERSE FUNCTION THEOREM

The Fibre Contraction Theorem also gives an exceptionally clean proof (again due to Sotomayor) of the Inverse Function Theorem (VII.1.04), as follows.

Suppose P,Q are affine spaces with vector spaces S,T, that $W \subseteq P$ is open and $h : W \to Q$ is C^1 and has $\underset{\sim}{D}_p h : T_pP \cong T_{h(p)}Q$ for some $p \in W$. Write $\underset{\sim}{q}(x) = \underset{\sim}{d}(f(x),A(x)) \in T$ for $x \in W$, where A is the affine map from P to Q with $A(p) = f(p) = q$, say, and linear part $\underset{\sim}{A} = \tilde{\underset{\sim}{D}}h(p)$. So we have $h(x) = A(x) + \underset{\sim}{q}(x)$, with $\underset{\sim}{D}_p\underset{\sim}{q} = \underset{\sim}{0}$. We look for a neighbourhood V of q in Q and $g : V \to U$ with

$g(h(x)) = x$ where defined, in a similar form $g(y) = \overleftarrow{A}(y) + \quad (y)$ for $y \in V$, with $\underset{\sim}{p} : V \to S$.

Give S an arbitrary basis $\underset{\sim}{s}_1, \ldots, \underset{\sim}{s}_n$, T the basis $\underset{\approx}{A}\underset{\sim}{s}_1, \ldots, \underset{\approx}{A}\underset{\sim}{s}_n$, and S,T, $L(S;T)$ and $L(T;S)$ the corresponding square norms $\| \; \|_S, \| \; \|_T, \| \; \|_L$ and $\| \; \|^L$. (Note that $\| \underset{\approx}{A}\underset{\sim}{s} \|_T = \| \underset{\sim}{s} \|_S \; \forall \underset{\sim}{s} \in S$, $\| \underset{\sim}{A} \|_L = \| \overleftarrow{\underset{\sim}{A}} \|^L = 1$.) By the assumed continuity of $\underset{\sim}{D}h$, there is an $\varepsilon > 0$ such that if $x \in B_\varepsilon = \{x | \; \| \underset{\sim}{d}(p,x) \|_S < \varepsilon\}$, then $\| \tilde{\underset{\approx}{D}}\underset{\sim}{g}(x) \|_L < \frac{1}{2}$. Set $V = A(B_{\varepsilon/2})$. The difference $\underset{\sim}{p}$ of g from $\overleftarrow{A}$ should be small near q, so we take as space X of candidates for $\underset{\sim}{p}$ the continuous maps $\underset{\sim}{\chi} : V \to S$ with $\| \underset{\sim}{\chi}(y) \|_S \leq \frac{\varepsilon}{2}$ for $y \in V$, with the uniform metric d_U. By 3.05 X is complete. Now if $g = \overleftarrow{\underset{\sim}{A}} + \underset{\sim}{\chi}$,

$$h(g(y)) = y \quad \Leftrightarrow \quad A(\overleftarrow{A}(y)+\underset{\sim}{\chi}(y))+\underset{\sim}{p}(\overleftarrow{A}(y)+\underset{\sim}{\chi}(y)) = y$$

$$\Leftrightarrow \quad \underset{\sim}{A}(\underset{\sim}{\chi}(y)) = -\underset{\sim}{p}(\overleftarrow{A}(y)+\underset{\sim}{\chi}(y))$$

Thus if we set $f(\underset{\sim}{\chi})(y) = -\overleftarrow{\underset{\sim}{A}}\,\underset{\sim}{p}(\overleftarrow{A}(y)+\underset{\sim}{\chi}(y))$, then $f(\underset{\sim}{\chi}) = \underset{\sim}{\chi}$ if and only if $h \circ g = I_V$. Evidently $f(\underset{\sim}{\chi})$ is continuous, so by Ex. 1 it is in X, and f is a map $X \to X$. Moreover f is a $\frac{1}{2}$-contraction:

$$d_U(\underset{\sim}{\chi},\underset{\sim}{\chi}') = \ell \quad \Rightarrow \quad \| \underset{\sim}{\chi}(y)-\underset{\sim}{\chi}'(y) \|_S \leq \ell, \; \forall y \in V$$

$$\Rightarrow \quad \| \underset{\sim}{d}(\overleftarrow{A}(y)+\underset{\sim}{\chi}(y), \overleftarrow{A}(y)+\underset{\sim}{\chi}'(y)) \|_S \leq \ell, \quad \forall y \in V$$

$$\Rightarrow \quad \| \underset{\sim}{p}(\overleftarrow{A}(y)+\underset{\sim}{\chi}(y))-\underset{\sim}{p}(\overleftarrow{A}(y)+\underset{\sim}{\chi}'(y)) \|_S \leq \frac{\ell}{2}, \quad \forall y \in V$$

by the Mean Value Theorem, since $\| \tilde{\underset{\approx}{D}}\underset{\sim}{p}(z) \| \leq \frac{1}{2}$ for $z \in B_\varepsilon$, and $(\overleftarrow{A}(y)+\underset{\sim}{\chi}(y))$, $(\overleftarrow{A}(y)+\underset{\sim}{\chi}'(y)) \in B_\varepsilon$. So

$$d_U(f(\underset{\sim}{\chi}),f(\underset{\sim}{\chi}')) = \sup\left\{ \| f(\underset{\sim}{\chi})(y)-f(\underset{\sim}{\chi}')(y) \|_S \,\middle|\, y \in V \right\}$$

$$= \sup\left\{ \| \overleftarrow{\underset{\sim}{A}}(\underset{\sim}{p}(\overleftarrow{A}(y)+\underset{\sim}{\chi}(y))-\underset{\sim}{p}(\overleftarrow{A}(y)+\underset{\sim}{\chi}'(y))) \|_S \,\middle|\, y \in V \right\} \leq \frac{\ell}{2} = \frac{1}{2}\, d_U(\underset{\sim}{\chi},\underset{\sim}{\chi}').$$

Thus by the Shrinking Lemma f has an attracting fixed point $\underset{\sim}{q}$, and $g = \overset{\leftarrow}{A}+\underset{\sim}{q}$ has $h \circ g = I_V$. If $\underset{\sim}{q}$ is differentiable so is g, with $\tilde{D}g(y) = \overset{\leftarrow}{\underset{\sim}{A}}+\tilde{\underset{\sim}{D}}\underset{\sim}{q}(y)$, so let Y be the space of candidates for $\underset{\sim}{D}\underset{\sim}{q}$ (namely, the bounded maps $V \to L(T;S)$) with the uniform norm. We want $F : X \times Y \to X \times Y$ with the property $F(\chi,\tilde{\underset{\sim}{D}}\chi) = \left(f(\chi),\tilde{\underset{\sim}{D}}(f(\chi))\right)$; since

$$\tilde{\underset{\sim}{D}}(f(\chi))(y) = \tilde{\underset{\sim}{D}}\left(\overset{\leftarrow}{A}\circ \underset{\sim}{p}\circ(\overset{\leftarrow}{A}+\chi)\right)(y) = -\overset{\leftarrow}{\underset{\sim}{A}}\circ[\tilde{\underset{\sim}{D}}\underset{\sim}{p}(\overset{\leftarrow}{\underset{\sim}{A}}(y)+\chi(y))]\circ(\underset{\sim}{A}+\tilde{\underset{\sim}{D}}\chi(y)),$$

we get it by setting

$$f_\chi(\chi') = -\overset{\leftarrow}{\underset{\sim}{A}}\circ[\tilde{\underset{\sim}{D}}\underset{\sim}{p}(\overset{\leftarrow}{\underset{\sim}{A}}(y)+\chi(y))]\circ(\overset{\leftarrow}{\underset{\sim}{A}}+\chi'(y)), \quad F(\chi,\chi') = (f(\chi),f_\chi(\chi')).$$

Now evidently (a) each $\chi \leadsto f_\chi(\chi')$, $\chi' \in Y$, is continuous. We have proved (b) that f has an attracting fixed point. Moreover (c) each f_χ is a ½-contraction:

$$\begin{aligned} d_U(f_\chi(\chi_1'),f_\chi(\chi_2')) &= \sup\left\{\| -\overset{\leftarrow}{\underset{\sim}{A}}\circ[\tilde{\underset{\sim}{D}}\underset{\sim}{p}(\overset{\leftarrow}{A}+\chi)]\circ(\chi_1'-\chi_2')(y)\|^L \,\middle|\, y \in V\right\} \\ &\le \sup\left\{\| \tilde{\underset{\sim}{D}}\underset{\sim}{p}(\overset{\leftarrow}{A}+\chi)(y)\|_L \|(\chi_1'-\chi_2')(y)\|^L \,\middle|\, y \in V\right\} \\ &< \tfrac{1}{2}\sup\left\{\| \chi_1'(y)-\chi_2'(y)\|^L \,\middle|\, y \in V\right\} = \tfrac{1}{2}d_U(\chi_1',\chi_2'). \end{aligned}$$

Thus applying the Fibre Contraction Theorem, if $\chi_1(y) = \underset{\sim}{0} \in S$, $\chi_1'(y) = \tilde{\underset{\sim}{D}}\chi_1'(y) = \underset{\sim}{0} \in L(T;S)$, $\forall y \in V$, and inductively $(\chi_n,\chi_n')=F(\chi_{n-1},\chi_{n-1}')=(\chi_n,\tilde{\underset{\sim}{D}}\chi_n)$, we have an attracting fixed point $(\underset{\sim}{q},\underset{\sim}{q}')$ with $i \leadsto \chi_i$ converging to $\underset{\sim}{q}$, $i \leadsto \tilde{\underset{\sim}{D}}\chi_i$ converging uniformly to $\underset{\sim}{q}'$. So $\underset{\sim}{q}'$ is continuous by 3.03, and $\underset{\sim}{D}\underset{\sim}{q}$ exists and is continuous by 3.06, and g is hence C^1.

We must show $g \circ h = I_U$, where $U = g(V)$, as well as $h \circ g = I_V$. $\tilde{\underset{\sim}{D}}g(q)$ is the isomorphism $\underset{\sim}{A}^{\leftarrow}$ (Ex. 2), so by what we have proved above there is a neighbourhood $U' \subseteq U$ of p and $h' : U' \to V$ such that $g \circ h = I_{U'}$. Then $h(x) = h(g(h'(x))) = (h \circ g)(h'(x)) = h'(x)$ for $x \in U'$, so $h' = h|U'$. Taking $V' = h(U')$, we have the required neighbourhood of $\underset{\sim}{q}$ and C^1 map $g' = g|V'$ with $g' \circ h = I_{U'}$, $h \circ g' = I_{V'}$.

Finally, we must show that if h is C^k, $k > 1$, so is g'. But if $k = 2$, $\underset{\sim}{D}h : TP \to TQ$ is C^1 and $\underset{\sim}{D}_{\underset{\sim}{0}_p}(\underset{\sim}{D}h) : T_{\underset{\sim}{0}_p}(TP) \to T_{\underset{\sim}{0}_q}(TQ)$ is an isomorphism (Ex. 3a), so $\underset{\sim}{D}h$ has a local C^1 inverse $G : W \to TP$, where $W \subseteq TQ$ is a neighbourhood of $\underset{\sim}{0}_q$. Since $\underset{\sim}{D}g'|W$ is also a local inverse for $\underset{\sim}{D}h$ (so $\underset{\sim}{D}h \circ \underset{\sim}{D}g' = \underset{\sim}{D}(h \circ g') = \underset{\sim}{D}(I) = I$), $\underset{\sim}{D}g|W = G$, so $\underset{\sim}{D}g'|W$ is C^1 and hence g' is C^2. Inductively, g' is C^k (Ex. 3c).

Exercises A6

1a) Show by the triangle inequality that $y \in V$, $\| \chi(y) \|_S \leq \frac{1}{2} \Rightarrow (A^{\leftarrow}(y) + \chi(y)) \in B_\varepsilon$.

b) Deduce by the Mean Value Theorem (compare proof of 3.06) that $\| \underset{\sim}{p}(A^{\leftarrow}(y) + \chi(y)) \|_T < \frac{\varepsilon}{2}$.

c) Deduce that $\| \underset{\sim}{A}^{\leftarrow} \underset{\sim}{p}(A^{\leftarrow}(y) + \chi(y)) \|_S \leq \frac{\varepsilon}{2}$.

2a) Show for any $(\chi, \chi') \in X \times Y$ that if $\chi(q) = \underset{\sim}{0}$, then $f(\chi)(q) = \underset{\sim}{0} \in S$, $f_\chi(\chi')(q) = \underset{\sim}{0} \in L(T;S)$.

b) Deduce that $\tilde{\underset{\sim}{D}}g(q) = \underset{\sim}{A}^{\leftarrow}$.

3a) Show that if $\underset{\sim}{D}_{\underset{\sim}{0}_p}(\underset{\sim}{D}h)$ exists, so does $\underset{\sim}{D}_{\underset{\sim}{v}}(\underset{\sim}{D}h)$ for every $\underset{\sim}{v} \in T_pP$.

b) Show that if h is C^2 and $\underset{\sim}{D}_{\underset{\sim}{0}_p} h$ is an isomorphism then so is $\underset{\sim}{D}_{\underset{\sim}{v}}(\underset{\sim}{D}h)$ an isomorphism for any $\underset{\sim}{v} \in T_pP$; deduce that so is $\underset{\sim}{D}_{\underset{\sim}{v}}(\underset{\sim}{D}h)$ for any $\underset{\sim}{v} \in T_xP$, $x \in U'$.

c) Deduce that g' is C^2 over its whole domain V', and, inductively, C^k.

Bibliography

R. Abraham and J.E. Marsden, Foundations of Mechanics, Benjamin 1967 (revised and updated edition promised shortly).

Manfredo Perdigão do Carmo, Differential Geometry of Curves and Surfaces, Prentice-Hall 1976.

P.A.M. Dirac, Quantum Mechanics, Oxford University Press 1958 (4th edition).

W.G. Dixon, Dynamics of extended bodies in general relativity

I. Momentum and angular momentum, A314 (1970) 499-527

II. Moments of the charge-current vector A319 (1970) 509-47

III. Equations of motion A277 (1974/75) 59-119

Philosophical Transactions of the Royal Society of London, Series A.

R.P. Feynman, R.B. Leighton, M. Sands, The Feynman Lectures on Physics (3 vols), Addison-Wesley 1964.

V. Guillemin and A. Pollack, Differential Topology, Prentice-Hall 1974.

S.W. Hawking and G.E.R. Ellis, The Large Scale Structure of Space-Time, Cambridge University Press 1973.

M.W. Hirsch and C.C. Pugh, Stable Manifolds for Hyperbolic Sets, Proc. Symposium in Pure Mathematics vol. XIV, AMS 1970.

W. Klingenberg, Existence of infinitely many closed geodesics, J.Diff.Geom 11(1976) 299-308.

L. Liusternik and L. Schnirelmann, Sur le problème de trois géodésiques fermées sur les surfaces de genre 0, Comptes Rendues Acad. Sci. Paris 189 (1929), 269-271.

S.A. Hojman, K. Kuchar, C. Teitelboim, Geometrodynamics Regained, Annals of Physics 96(1976) 88-135.

A. Janner and E. Ascher, Relativistic Crystallographic Point Groups in Two Dimensions, Physica, 45(1969) no. 1, 67-85.

N. Kagan, The Mathenauts, in Best of Science Fiction 10, ed. J. Merril, Mayflower 1967, 280-295.

S. Kobayashi and K. Nomizu, Foundations of Differential Geometry (2 vols), Interscience, NY 1969.

S. Maclane

1. Geometrical Mechanics, duplicated notes, Math.Dept., U.Chicago 1968

2. Hamiltonian Mechanics and Geometry, Amer.Math.Monthly 1970, 570-586.

L. Markus, Exponentials in Algebraic Matrix Groups, Advances in Math. 11, no. 3 (1973) 351-367.

R.M.F. Moss and G.T. Roberts, A Preliminary Course in Analysis, Chapman and Hall 1970.

I.R. Porteous, Topological Geometry, van Nostrand Reinhold 1969.

T. Poston and I.N. Stewart, Catastrophe Theory and its Applications, Pitman 1977.

R.L.E. Schwarzenberger

1. Elementary Differential Equations, Chapman & Hall 1969
2. Crystallography in Spaces of Arbitrary Dimension, Math.Proc.Cambr.Phil.Soc. 76(1974) 23-32.

G.E. Shilov, An introduction to the theory of linear spaces, Prentice-Hall 1961.

S. Smale, On the mathematical foundations of electrical circuit theory, J.Diff.Geom. 7(1972) 195-210.

J.-M. Souriau, Structure des Systèmes Dynamiques, Dunod, Paris 1970.

M. Spivak, A Comprehensive Introduction to Differential Geometry (5 vols), Publish or Perish Inc., 1975.

P. Stredder, Natural differential operators on Riemannian manifolds and representations of the orthogonal and special orthogonal groups, J.Diff.Geom. 10(1975) 647-660.

J.L. Synge, Relativity; The General Theory, N.Holland, Amsterdam 1960.

C. Truesdell and W. Noll, The Non-Linear Field Theories of Mechanics, vol.III/3, Handbuch der Physik, Springer-Verlag, Berlin 1965.

Index of Notations

Chapter V

Chapter VI

Chapter VII

Chapter VIII

Chapter IX

Chapter X

Chapter XI

Chapter XII

Index

C

"And further, by these, my son, be admonished: of making many books there is no end; and much study is a weariness of the flesh."

Ecclesiastes 12, 12.